FOOD ANALYSIS:
THEORY AND PRACTICE

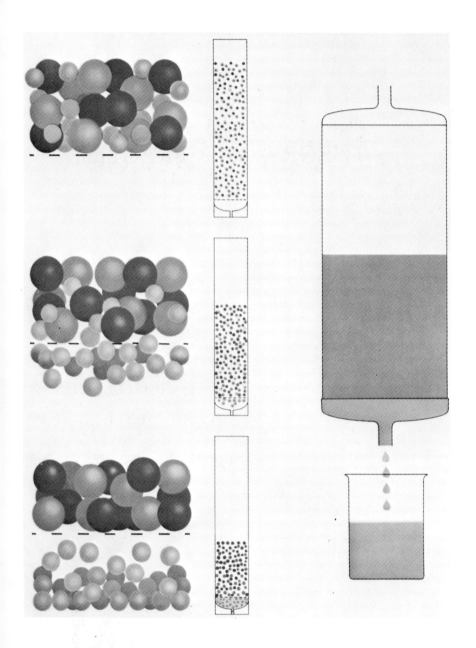

FOOD ANALYSIS: THEORY AND PRACTICE

Revised Edition

Yeshajahu Pomeranz

U.S. Grain Marketing Research Center

Clifton E. Meloan

Kansas State University

AVI PUBLISHING COMPANY, INC.
Westport, Connecticut

Library of Congress Cataloging in Publication Data

Pomeranz, Yeshajahu, 1922-
    Food analysis.

    Includes bibliographies and index.
    1. Food—Analysis.   I.   Meloan, Clifton E.,
joint author.   II.   Title.
TX541.P64 1978   664'.07   77-27867
ISBN 0-87055-238-4

Printed in the United States of America

Preface to
The Revised Edition

We are pleased with the reception of this book. Since it is in need of revisions here and there, we have taken this opportunity to update existing chapters and to add new sections on High Pressure Liquid Chromatography, Affinity Chromatography, Immobilized Enzymes, and Near Infrared Reflectance Spectroscopy.

As a result we feel that the book has been updated and will better serve the needs of the readers.

YESHAJAHU POMERANZ
Director, U.S. Grain Marketing Research Center,
U.S. Department of Agriculture and Professor,
Department of Grain Science and Industry,
Kansas State University, Manhattan

CLIFTON E. MELOAN
Professor of Analytical
Chemistry, Chemistry
Department, Kansas
State University,
Manhattan

March 1977

Preface To
The First Edition

This book has been designed for use as a text by undergraduate students majoring in food science and technology, and as a survey of modern analytical techniques and instruments for the worker and researcher in the field of food analysis.

In preparation of the manuscript, we followed the recommendations for subject coverage in a textbook of food analysis for students of food science and technology, as contained in the report of the Task Force on Food Analysis of the Institute of Food Technologists' Council Committee on Education.

Basic principles are stressed, rather than details of analytical methods. Although the emphasis has been on modern and sophisticated instruments and methods, we have described also the classical procedures that have been in use for many years. In chapters devoted to instrumentation, we have attempted to provide the background theory that is required for understanding the principles of each instrumental assay. Included are diagrams and descriptions of typical instruments, information on their application and precision, and sample problems with detailed solutions. Whenever applicable, we have compared the instrumental and assay procedures to evaluate their usefulness and limitations. The chapters on instrumentation end with problem questions (and answers) to further develop the subject matter covered in the text.

In the sections devoted to applications, qualitative and quantitative aspects of the basic instruments and procedures are discussed in terms of their numerous analytical and instrumental procedures, and discussions of their use in solving specific problems are followed by comprehensive and up-to-date bibliographic lists.

This book is not meant to replace standard methods of analysis; its main purpose is to explain the background and principles of those methods.   In writing this book, we were faced with several major questions.   The rapid advances in analytical instrumentation may make some of the sections out of date.   We hope by stressing fundamental principles, rather than details of methodology, that the book will be for some time a useful source of information both for the student and for the experienced researcher.   For understanding most of the material, the reader would be expected to have studied general, organic, analytical, and food chemistry, and to have an appreciation of biochemistry.   To allow for his limited training in physical chemistry, we have included the physicochemical principles, terminology, and detailed computations.

A major problem concerned the selection of methods and techniques that should be included.   While we realize that some of the instrumental techniques that require expensive equipment or more sophisticated biological methods may not be within the reach of many food chemists, we included them to stimulate the thinking of the searching student and the experienced researcher.   We hope that these new procedures will contribute to an appreciation of the scopes and potentialities of food analysis.   It might be worth pointing out that some of these techniques and procedures can yield excellent results with simplified adaptations—provided the principles are understood.   We hope most readers will agree with our selection of methods, techniques, and approaches, and would greatly appreciate comments from teachers, researchers, food analysts, and—most important—students in food science and technology.

We wish to thank the authors of articles and books, industrial companies, and publishers for permission to reproduce material.   Special appreciation is expressed to Dr. D. K. Tressler and the Avi Publishing Company for the many helpful suggestions in preparation of the book.

Y. POMERANZ
C. E. MELOAN

March 1971

Contents

Searching the Literature

INTRODUCTION

Keeping informed of current developments in his field is important to any professional worker. It is, however, particularly essential to scientists in a rapidly expanding area such as analytical chemistry. While no analyst can be familiar with all of the recent advances, he can learn how and where to look most effectively for the needed information. An attempt to survey the vast quantity of scientific literature may so frustrate the inexperienced worker that he decides to ignore it altogether. Alternatively, he may sacrifice bench work to do practically endless reading and searching the growing literature. Many books show how and where to find information; analyze the various sources; and describe the tools, elements, and theories of information storage and retrieval.

It is assumed that the analyst is "literature-conscious" and it is the purpose of this short chapter to outline the main sources of information and general approaches to a literature search.

For over 300 yr periodicals were the principal means for the exchange of scientific information. There are two basic types of scientific journals (1) primary sources which report original work and provide detailed experimental procedures, and data derived from them, and (2) secondary sources which process the original data into abstracts and reviews, and provide a condensed source of information.

Keys to the contents of periodicals are provided by abstracts and indexes. Abstracts which are short summaries of published articles keep the readers informed of current research; indexes, which may

1

list subjects, authors or formulas, assist in the search of published information. Most journals issue an index to each volume, generally annually, and some issue cumulative indexes covering several years. The most important compilations of abstracts and indexes cover a number of periodicals in a specific area, such as *Chemical Abstracts* (10,000 periodicals) in the area of chemistry.

To keep abreast of the current information in a familiar field, one generally reads a number of journals and then scans the abstracts for a wider coverage.

SEARCHING A NEW FIELD

General Background

In a new field, the first source is an advanced textbook. For general background information, several voluminous encyclopedia devoted to chemistry and technology can be consulted. These include the technical and detailed Kirk-Othmer's *Encyclopedia of Chemical Technology* (Interscience Publishers) and Ullmann's *Enzyklopedie der Technischen Chemie* (Urban and Schwarzenberg Publishing). *The Encyclopedia of Science and Technology* (McGraw-Hill Book Co.) is less detailed but covers a wider field.

An excellent source of information is the single-volume *Merck Index of Chemicals and Drugs*. This compilation revised every several years contains formulas, preparations, and properties of over 10,000 chemicals.

In the identification and determination of food composition, the analyst often has to determine the physical properties of the substance. The reference data can be obtained from various standard tables, handbooks, and some of the newer encyclopedias and dictionaries. Information on physical properties is often published by manufacturers of chemicals. In addition to industrial laboratories, government agencies (i.e. the U.S. National Bureau of Standards) publish authoritative compilations of physicochemical data.

The *Handbuch der Organischen Chemie*, initiated by F. K. Beilstein, is the largest and most comprehensive source of information in organic chemistry. The twenty-seven volumes of the main work published by the German Chemical Society included 200,000 entries covering the literature prior to 1910. Supplements are published periodically. The "Handbuch" deals with well-defined organic compounds and with natural materials of unknown structure. The coverage varies with the significance of the compound and the available information. Included are (whenever available) names, formulas, structures, history, occurrence, preparation, properties, technology, analysis, and reactions.   In recent years, several new references on organic chemistry have been published but the "Handbuch" continues to maintain its prominent position.

Reviews

The next stage is a survey or a critical review of the current knowledge in the particular field.   Monographs are basically comprehensive surveys of current knowledge on a specific subject.   One of the best known is the *Advances in Chemistry Series* published by the American Chemical Society.   Another series is published in *Annals of the New York Academy of Science*.   Both series are largely based on symposia organized by scientific societies.   There are several sources of review articles: annual reviews, special review periodicals, and special issues or selected parts of regular periodicals.   Since 1959, *Chemical Abstracts* of the American Chemical Society publishes annually a *Bibliography of Chemical Reviews*.   The number of such reviews is around 7,000 each year.

Since 1904, the Chemical Society (England) publishes authoritative summaries of the previous years' important papers in the form of *Annual Reports on the Progress of Chemistry*.   A parallel series is published by the Society of Chemical Industry in the form of *Reports on the Progress of Applied Chemistry*.   The reviews on foods are subdivided into the major areas and each section provides 100 to 200 references to original research papers, reviews, and proceedings of scientific conferences.   Each year, the April issue of *Analytical Chemistry* is devoted to review papers.   Various aspects of food analysis are covered every second year.   Additional background information is provided in *Analytical Chemistry* by reviews in other areas, such as biochemistry, clinical chemistry, water analysis, and various areas of inorganic, physical, and organic analyses.

One of the most useful sources of current information, for the reader with sufficient background knowledge, are articles in annual publications in the form of Advances in . . . , Annual Review of . . . , Progress in . . . , and Methods in . . . .   Those pertinent to food analyses are listed at the end of this chapter.

Theses

Theses and dissertations contain comprehensive reviews, generally in a limited area.   While most of the new information in theses is published in scientific journals, the literature reviews are shortened to reduce cost of publication.   Many countries publish lists of higher education degree theses.   Theses from U.S. universities are processed since 1938 for microfilming by University Microfilms of Ann Arbor, Michigan.   Abstracts of up to 600 words of such theses are published by *Dissertation Abstracts*.   Theses of a chemical nature are listed in Chemical Abstracts, and the entire theses can be purchased from University Microfilms.   An annual list of *U.S. Masters' Theses in the Pure and Applied Sciences* has been published since 1955–1956 by the Thermophysical Properties' Research Center, Purdue University, Lafayette, Indiana.

Symposia

Programs of conferences and symposia are published in several journals.   A comprehensive list is found in *Science;* the coverage of *Food Technology* is more selective and limited.   In the area of food science, they include proceedings of meetings sponsored by government departments (such as various agencies of the U.S. Department of Agriculture).   University Presses (i.e. MIT Press), professional groups, commercial publishing houses (i.e. Pergamon Press), and private companies (i.e. Campbell Soup Co.) frequently reproduce the lectures and stimulating discussions of meetings in which prominent scientists present invited review papers in the area of their competence.   Several renowned Symposia on Foods held at Oregon State University have been published as hard cover books by the Avi Publishing Co.

Abstracts of papers presented at scientific meetings are published for members in the form of books (i.e. by the American Chemical Society or Federation of Biological Sciences) or are included in periodicals (i.e. *Cereal Science Today* for AACC, and *J. American Oil Chemists' Society* for AOCS).   The abstracts are particularly useful in learning of the most recent developments.

Trade Publications

The purpose of most house organs and trade publications is to sell products.   Most such advertisements contain valuable information. The publications of many manufacturers are on a high scientific level.   Some of these publications are the best sources on properties and applications of specialized equipment and chemicals.   Several manufacturers issue periodically bibliographies and abstracts of technical-scientific articles in a specific area; some publish periodicals that reproduce pertinent articles from regular scientific journals; and some prepare detailed handbooks giving specifications, properties, and details of analytical procedures in selected areas (microbiology, enzymatic assay, electrophoresis, immunochemistry, automation in analytical chemistry). Industrial manuals are, of course, indispensable in installing, using, and servicing equipment. Several scientific journals prepare periodically lists of major commercial supply houses. A Comprehensive Guide to Scientific Instruments is published annually in *Science*.

Translations

Scientific journals continue to be the most important source of detailed information. Searching the literature becomes more difficult with the increased quantity of available literature and more publications in foreign languages. Many of the latter have a summary in English. In addition, active programs for translation from Slavic, Japanese, and Chinese literature are available.

Abstracts

Information on important publications is available in bibliographical lists or abstracting journals. *Chemical Abstracts* publishes over 200,000 abstracts a year, selected from about 10,000 journals in over 50 languages. The section on biochemistry covers various aspects of biology and chemistry of materials of plant, animal, and microbial origin. Included are informative abstracts of scientific original papers, patents, and some reviews; as well as lists of theses, monographs, books, reviews, and proceedings.

Several periodicals publish abstracts in a more restricted field. *Analytical Abstracts* and the abstract section of *Z. Analytische Chemie* are concerned primarily with analytical procedures. The abstract sections in *J. Science Food and Agriculture* and *Z. Lebensmitteluntersuchung Forschung* cover production, processing, storage, chemistry, and analyses of foods and agricultural products. The abstract sections in *J. American Oil Chemists' Society, J. Institute of Brewing, Wallerstein Laboratories Communications*, and many others cover in a broad sense the respective areas. Food patents are abstracted in *Food Technology*. Some research organizations prepare excellent abstracts for member companies (Corn Industries Research Foundation) or for the general public (Washington State University). Information on cereals is abstracted in Germany by *Documenta Cerealia* and published as a monthly addition to the weekly periodical *Die Muhle*.

Abstracts on current work (including theses) in East-European countries may be found as an appendix to the journal *Die Nahrung*. Abstracts in the *Zentr. Dok. Dienst Sozialist. Lander-Nahrung und*

Ernahrung—are an excellent source of information on scientific work in foods and nutrition.   Although some western periodicals are abstracted, the emphasis is on East-European publications.   Some of the latter journals are not abstracted by *Chemical Abstracts* and are generally unavailable to the English reader.

The journal *Food Science and Technology—Abstracts* has been published since 1969 by the International Food Information Service. The Institute of Food Technologists in the United States;   the Commonwealth Agricultural Bureau, headquartered in England; and the Institut fur Dokumentationswesen of the German Democratic Republic are the sponsors.

In addition to the comprehensive *Chemical Titles*, published by the American Chemical Society, several publications provide more limited bibliographical lists.   Selective lists are included in some periodicals such as *J. Chromatography* and *J. Lipid Research.*

Food analysts use primarily methods approved by various associations such as the Association of Official Analytical Chemists, American Association of Cereal Chemists, American Oil Chemists' Society, American Public Health Association, and American Society of Brewing Chemists (and their national or international counterparts).   Most of the methods recommended by the United States and other organizations have been developed after years of collaborative testing and are considered reliable and official.   To understand the background, limitations, and significance of analytical findings the analyst must acquaint himself with the current developments and periodically survey the pertinent sources.

A selected list of such sources is given at the end of this chapter.

BIBLIOGRAPHY

General Literature References

ANON.   1970.   The IFT World Directory and Guide. Institute of Food Technologists, Chicago, Ill.
BECKER, J., and HAYES, R. M.   1964.   Information Storage and Retrieval: Tools, Elements, Theories.   J. Wiley & Sons, New York.
BOTTLE, R. T. (Editor).   1962.   Use of the Chemical Literature.   Butterworths, London.
BURMAN, C. R.   1965.   How to Find Out in Chemistry.   Pergamon Press, New York.
CRANE, E. J., PATTERSON, A. M., and MARR, E. B.   1957.   A Guide to the Literature of Chemistry.   J. Wiley & Sons, New York.
DYSON, G. M.   1951.   Chemical Literature.   Longmans and Green, London.
GOULD, R. F. (Editor).   1961.   Searching the Chemical Literature.   Advan. Chem. Ser. *30.*   Am. Chem. Soc., Washington, D.C.
MELLON, M. G.   1965.   Chemical Publications: Their Nature and Use. McGraw-Hill Book Co., New York.

Handbooks, Dictionaries, General References

ABERHALDEN, E. 1920. Handbuch der Biologischen Arbeitsmethoden. Urban and Schwarzenberg, Berlin.

ABERHALDEN, E. (Editor). 1910–1933. Biochemisches Handlexikon. Springer-Verlag, Berlin.

ADAMS, R. (Editor). 1942. Organic Reactions. Chapman and Hall, London.

ANON. 1964. Heinz Nutritional Data. H. J. Heinz Co., Pittsburgh.

BORGSTROM, G. 1968. Principles of Food Science. The MacMillan Co., New York.

BURTON, B. T. (Editor). 1968. The Heinz Handbook of Nutrition. McGraw Hill Book Co., New York.

CAMERON, E. J., and ESTY, J. R. 1950. Canned Foods in Human Nutrition. National Canners Association, Washington, D.C.

DIEMAIR, W. (Editor). 1965. Handbook of Food Chemistry. Springer-Verlag, Berlin. (German)

FURIA, T. E. (Editor). 1968. Handbook of Food Additives. Chemical Rubber Co., Cleveland.

JACOBS, M. B. (Editor). 1951. The Chemistry and Technology of Food and Food Products, 3 Vols. Interscience Publishers, New York.

KLEIN, G. 1932. Handbook of Plant Analysis. Springer-Verlag, Berlin.

LEES, R. (Editor). 1968. The Laboratory Handbook of Methods of Food Analysis. Chemical Rubber Co., Cleveland.

LONG, C. (Editor). 1961. Biochemist's Handbook. E. and F. Spon Publishing Co., London.

McCANCE, R. A., and WIDDOWSON, E. M. 1960. Composition of Foods. Med. Res. Council, Spec. Rep. 297. Her Majesty's Stationery Office, London.

MEYER, L. H. 1960. Food Chemistry. Reinhold Publishing Corp., New York.

MILLER, J. (Editor). 1953. Organic Analysis. Interscience Publishers, New York.

MULLER, E. (Editor). 1958. Methoden der Organischen Chemie. G. Thierme, Stuttgart, Germany.

PAECH, K., and TRACEY, M. V. (Editors). 1956. Modern Methods of Plant Analysis. Springer-Verlag, Berlin.

POTTER, N. N. 1968. Food Science. AVI Publishing Co., Westport, Conn.

RADT, F. (Editor). 1940. Elseviers' Encyclopedia of Organic Chemistry. Elsevier Press, Amsterdam, The Netherlands.

RAUEN, v. H. M. 1956. Biochemical Handbook. Springer-Verlag, Berlin.

RAPPOPORT, Z. (Editor). 1968. Handbook of Tables for Organic Compound Identification. Chemical Rubber Co., Cleveland.

SOBER, H. A. (Editor). 1968. Handbook of Biochemistry. Chemical Rubber Co., Cleveland.

STANDEN, A. (Editor). 1967. Kirk-Othmer's Encyclopedia of Chemical Technology. J. Wiley & Sons, New York.

STECHER, P. G. (Editor). 1960. The Merck Index of Chemicals and Drugs. Merck and Co., Rahway, N.J.

WATT, B. K., and MERRILL, A. L. 1963. Composition of Foods, Agriculture Handbook. 8. U.S. Dept. Agr., Washington, D.C.

WEAST, R. C. (Editor). 1968. Handbook of Chemistry and Physics, 49th Edition. Chemical Rubber Co., Cleveland.

WEISSBERGER, A. (Editor). 1948. Physical Methods of Organic Chemistry. Interscience Publishers, New York.

WILLIAMS, R. J., and LANSFORD, E. M. (Editors). 1967. Encyclopedia of Biochemistry. Reinhold Publishing Corp., New York.

WINTON, A. L., and WINTON, K. B. 1935–1939. Structure and Composition of Foods. J. Wiley & Sons, New York.

Reviews
Advances in Agronomy
    Analytical Chemistry and Instrumentation
    Carbohydrate Chemistry
    Chemistry Series
    Chromatography
    Clinical Chemistry
    Colloid Science
    Comparative Biochemistry and Physiology
    Enzymology and Related Subjects of Biochemistry
    Food Research
    Lipid Research
    Protein Chemistry
Annual Reports on Progress of Chemistry
Annual Review of Biochemistry
    Microbiology
    Physiology
Bacteriological Reviews
Biological Reviews
Chemical Reviews
Chromatographic Reviews
Methods in Medical Research
Methods of Biochemical Analysis
Methods in Enzymology
Nutrition Abstracts and Reviews
Physiological Reviews
Progress in the Chemistry of Fats and Other Lipids
    Medicinal Chemistry
Recent Advances in Food Science
Reports on the Progress of Applied Chemistry
Vitamins and Hormones
Yearbook of Agriculture—U.S. Dept. Agr.

Abstracts, Bibliography, Indexes
Agricultural Index
Analytical Abstracts
Applied Science and Technology Index
Bibliographic Index
Bibliographic Current List of Papers, Reports, and Proceedings of International Meetings
Bibliography of Agriculture
Bibliography of Chemical Reviews
Biological Abstracts
Chemical Abstracts
Chemical Titles
Chemisches Zentralblatt
Current Chemical Papers
Current Contents
World Bibliography of Bibliographies

Periodicals
Acta Chemica Scandinavia
Agricultural and Biological Chemistry (Tokyo)

Agronomy Journal
*Analytical Biochem*istry
*Analytical Chem*istry
*Analytica Chim*ica *Acta*
Analyst, The
*Angewa*ndte *Chem*ie
*Anna*len der *Chem*ie
*Anna*les des *Fals*ifications et des *Fraudes*
*Anna*ls of the *N*ew York *Acad*emy of *Sci*ences
*Appl*ied *Microbiology*
*Appl*ied *Spect*roscopy
*Arch*ives of *Biochem*istry and *Biophys*ics
Australian Journal of *Biol*ogical *Sciences*
*Baker's Dig*est
*Ber*ichte der Deutschen Chemischen Geselschaft (discontinued)
*Biochem*ical and *Biophys*ical *Research Commun*ications
*Biochem*ical Journal
*Biochem*ische *Z*eitschrift (discontinued)
Biochemistry
*Biochim*ica et *Biophys*ica *Acta*
*Biotechnol*ogy and *Bioeng*ineering
*Brewers Dig*est
*Brit*ish Journal of *Nutr*ition
Brot und *Gebaeck*
*Bull*etin de la *Societe Chim*ique de *France*
*Bull*etin de la *Societe* de *Chim*ie *Biol*ogique
Canadian Journal of *Biochem*istry and *Physiol*ogy
*Carbohydrate Res*earch
*Cereal Chem*istry
*Cereal Sci*ence *Today*
*Chem*ische *Ber*ichte
*Chem*istry and *Ind*ustry (*London*)
*Chem*istry and *Phys*ics of *Lipids*
*Clin*ica *Chim*ica *Acta*
*Electroanal*ytical *Chem*istry
Endeavour
Enzymologia
Ernaehrungsforschung
*Eur*opean Journal of *Biochem*istry
Experientia
*Federation Proc*eedings
Fette, Seifen, Anstrichmittel
*Food Eng*ineering
*Food Manufa*cture
Food Science
*Food Technol*ogy
Getreide und *Mehl*
*Helv*etica *Chim*ica *Acta*
*Hoppe Seylers' Z*eitschrift fur *Physiol*ogische *Chem*ie
*Ind*ustrial and *Eng*ineering *Chem*istry
Journal of *Agr*icultural and *Food Chem*istry
Journal of the *Am*erican *Chem*ical *Society*
Journal of the *Am*erican *Oil Chemists' Society*
Journal of the *Assoc*iation of *Official Anal*ytical *Chem*ists
Journal of the *Assoc*iation of *Public Anal*ysts
Journal of *Bacteriol*ogy

Journal of *Biochemistry (Tokyo)*
Journal of *Biological Chemistry*
Journal of the *Chemical Society*
Journal of *Chromatography*
Journal of *Dairy Research*
Journal of *Dairy Science*
Journal of *Food Technology*
Journal of *Food Science*
Journal of *Gas Chromatography*
Journal of *General Microbiology*
Journal of *Histochemistry* and *Cytochemistry*
Journal of the *Institute of Brewing*
Journal of *Lipid Research*
Journal of *Milk* and *Food Technology*
Journal of *Molecular Biology*
Journal of *Nutrition*
Journal of the *Science of Food* and *Agriculture*
Laboratory Practice
Lipids
Microchemical Journal
Microchimica Acta
Milling
Mitteilungen aus dem Gebiete der Lebensmitteluntersuchung und Hygiene
Muhle, Die
Nahrung, Die
Nature
Naturwissenschaften, Die
Physiologia Plantarum
Plant Physiology
Poultry Science
Proceedings of the National Academy of Science, U.S.
Proceedings of the Nutritional Society of England and Scotland
Proceedings of the Society of Experimental Biology and Medicine
Science
Staerke, Die
Stain Technology
Transactions of the New York Academy of Science
Wallerstein Laboratories Communications
Zeitschrift fur *Analytische Chemie*
Zeitschrift fur *Ernaehrungswissenschaft*
Zeitschrift fur *Lebensmittel Untersuchung und Forschung*
Zeitschrift fur *Naturforschung*

CHAPTER 2

Sampling

The validity of the conclusions drawn from the analysis of a food depends, among other things, on the methods used in obtaining and preserving the sample. Sampling and any subsequent separations are the greatest sources of error in food analyses. An ideal sample should be identical in all of its intrinsic properties with the bulk of the material from which it is taken. In practice, a sample is satisfactory if the properties under investigation correspond to those of the bulk material within the limits set by the nature of the test.

SAMPLES

The sample should be large enough for all intended determinations. Homogenous samples of 250 gm (or ml) are generally sufficient. Samples of spices are often limited to 100 gm, and of fruits and vegetables increased to 1000 gm. The samples should be packed and stored in such a way that no significant changes occur from the moment of sampling until the analysis is completed. The container should be identified clearly. Official and legal samples must be sealed in such a way that they cannot be opened without breaking the seal.

The quality control laboratory analyzes various samples (Pearson 1958). *Raw materials* are analyzed to determine whether the delivery approximates previous deliveries or if the material from a new supplier is up to the buying sample. *Process control samples* are generally analyzed by rapid tests (i.e. refractometer, hydrometer) in the plant to make adjustments in producing an acceptable and uniform product. Periodic checks of *finished products* show whether the food meets legal requirements, is acceptable to the consumer, and has reasonable shelf-life. *Buying samples* are submitted by suppliers of raw materials prior to delivery. Most *complaint samples* are submitted by customers. In a competitive market, information about products being sold by other manufacturers is of interest to the management. The composition of *competitors' samples* is also valuable in developing new products.

11

TERMINOLOGY AND SAMPLING CONSIDERATIONS

The Analytical Commission of Terminology of IUPAC (International Union of Pure and Applied Chemistry) has proposed several definitions to be used in sampling. These include: *Sample:* A portion of material taken from the consignment and selected in such a way that it possesses the essential characteristics of the bulk. *Sampling procedure:* The succession of steps set out in the specification which ensures that the sample shall possess the essential characteristics of the bulk. *Sampling unit:* The minimum-sized packages in the consignment which the sample may represent. *Increment:* A stated amount of material to be taken from the sampling unit. *Gross sample:* The sample prepared by mixing the increments. *Subsample:* A smaller sample produced in a specified manner by subdivision of the gross sample. It has the essential characteristics of the gross sample. *Analysis sample:* The amount of the laboratory sample taken by the operator for the test. This sample should be such a size that it may be regarded as homogenous and possess the characteristics of the bulk that it represents. Alternatively, the characteristics may vary in successive analysis samples. This sample is the "item" on which a variate value is obtained for statistical purposes.

According to Kramer and Twigg (1970), factors that determine selection of a sampling procedure include: (1) *purpose of inspection:* acceptance or rejection, evaluation of average quality, and determination of uniformity; (2) *nature of lot:* size, division into sublots, and loading or stacking; (3) *nature of test material:* its homogeneity, unit size, previous history, and cost; and (4) *nature of test procedures:* significance, destructive or nondestructive assay procedures, and time and cost of analyses.

STATISTICAL CONCEPTS

Statistically-sound sampling plans and procedures have been developed to provide the most efficient technique with regard to cost and information. Details of such procedures are readily available, (Barlett and Wegener 1957; Bowman and Remmenga 1965; Bowker and Goode 1952; Bureau of Ordnance 1952; Cochran 1963; Dept. of Defence 1963; Dodge and Romig 1944; Kramer and Twigg 1966; U.S. Dept. of Agr. 1964).

The following is a summary of the basic statistical concepts involved. In a series of n observations in which x is the value of a single observation and $\bar{x}$ the mean (average) of n observations, d is the deviation of a single observation from the mean $d = x - \bar{x}$. The standard deviation is $s_0 = \sqrt{\Sigma d^2/n}$. If the total number of items is N, the true mean is $\mu = \Sigma x/N$, the true standard deviation is $\sigma = \sqrt{\Sigma d^2/N}$, and $\bar{x} = \Sigma x/n$ is an estimate of the true mean. Generally, s_0 is smaller than σ, as it is derived from a smaller number of observations and is less likely to include all extreme values. To account for this, for n smaller than 10, s is calculated as $s = \sqrt{\Sigma d^2/(n-1)}$ and gives a better estimate of σ than s_0.

The dimensions of s are the same as of x. The deviation can also be expressed as $t = d/s$. This is useful in calculating the fraction of a sampled lot that varies from the mean by a certain value. It has been found that one standard deviation ($t = 1$) represents a deviation on both sides of the mean (total range of $2s$) within which 68.2% of all observations will occur for a population with a normal distribution provided a sufficiently large (>25) number of observations is made; 95.4% of all observations fall within a range of $t = 2$ (or $4s$); and 99.7% within a range of $t = 3$ (or $6s$). This means that the chance that a single observation will not exceed $3s$ is over 99%, and the chance that the error will not exceed s is 68%. The standard error $s_{\bar{x}}$ (standard deviation of the sample mean) decreases as the number of samples increases according to the relation $s_{\bar{x}} = s/\sqrt{n} = \sqrt{\Sigma d^2/n(n-1)}$. Thus, if s is known, one can determine the number of samples (or sample increments) required for a desired standard error $s_{\bar{x}}$. As a result of the nonlinear relation between $s_{\bar{x}}$ and n, only seldom is it justified to increase n above 20. In practice, in 20 observations the range covered is about $3s$, i.e. 3 times the standard deviation.

From the equation $n = (ts/e)^2$ it is possible to calculate the number of observations, n, with the standard deviation s, required to have a confidence that the error, e of the observed mean will not exceed a certain magnitude. In that equation, t is the confidence or probability factor; when $t = 1$ the probability is 68%, when $t = 2$ it is 95%, etc. The above is true provided the analytical error is substantially smaller than the sampling error. Generally, a confidence of 95% ($t = 2$) is satisfactory; for a confidence of 99%, $t = 2.6$.

In practice, unless variability within a lot is to be determined, the lot is sampled according to a specific pattern, and equal or suitable proportions are combined into a grand composite that is analyzed.

MANUAL SAMPLING

Samples are frequently taken manually. Apparently homogenous materials such as single-phase liquids or well-mixed powders should be mixed thoroughly immediately before sampling. Small quantities of powders or solutions can be mixed by rotating and shaking in a closed container that has a volume at least twice that of the sample. Mixing may also be accomplished by pouring several times from one container to another (Assoc. Vitamin Chemists 1951). Laboratory samples of powders or ground materials may be obtained by quartering of thoroughly-mixed samples, discarding two opposite quarters, remixing the remaining material, and repeating the process till the sample is reduced to a desired size. Sample dividers that mechanically mix and divide powdered or granular materials (Fig. 2.1) may be purchased from several apparatus supply houses.

Courtesy Seedburo Equipment Co.

FIG. 2.1. BOERNER SAMPLER

A sample of grain is placed in the hopper and then released down the sides of a cone the point of which is directly under the center of the opening. Around the base of the cone are three pockets or openings. The grain falling down the sides of the cone is cut into 36 separate streams, alternating into 2 streams which empty into the pans.

Heterogenous materials or materials of unknown character are generally sampled by geometric sampling. The lot is divided, mentally, into several regular geometric units (cubes, cylinders, etc.), and a subsample is taken from several locations in the lot.

The size of the subsample must be related to size of the lot. Depending on heterogeneity, the subsamples may or may not be mixed.
If subsamples are combined into a grand composite, they are mixed
and reduced in size to prepare the final sample for the laboratory.
In the analyses of grains, gross impurities (stones, straw) are often
determined mechanically in the grand composite before reduction in
size. Data on the determination of gross impurities and the size of
the original composite and laboratory sample must be recorded.

Granular or powdered solids are generally sampled by probes
and triers (Fig. 2.2 and 2.3). Liquids require thorough mixing
before sampling. Partly or completely frozen, crystallized, or
solidified fluids must be liquified completely and mixed. If such
mixing is practically unattainable, samples must be taken at various
heights. Examples of liquid samplers are shown in Fig. 2.4. Milk
must be thoroughly mixed as the fat rises to the top and the
composition changes on standing. On the other hand, excessive
mixing of cream is inadvisable. Butter and hard cheese samples
are generally taken with a stainless steel borer; soft cheeses are

FIG. 2.2. SPIRAL PROBE FOR SAMPLING SOYBEANS

sampled by cutting out a representative segment. The greatest difficulty arises in sampling large fruits and vegetables. Often selection of a large number of individual units, to compensate for variation, is required.

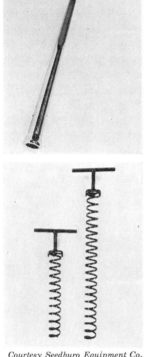

FIG. 2.3. TRIERS: TOP, FOR SACKED RICE; BOTTOM, FOR COTTONSEED

Courtesy Seedburo Equipment Co.

CONTINUOUS SAMPLING

In continuous quality control manual samplers have been replaced by mechanical samplers. There are three basic types of mechanical samplers (Johnson 1963). (1) The *riffle cutter* is composed of equally-spaced dividers designed to continuously remove a small fraction of the stream. Usually, a riffle divides the stream equally and the sample is passed through the same riffle or successive riffles for further proportional reduction of the sample to a quantity convenient for analysis. This device is commonly used in laboratories in cutting and quartering a larger sample.

(2)   The *circular* (or Vezin) sampler can be used for intermittent or continuous sampling and is applicable for both wet and dry materials.   The cutter appears as a truncated wedge of a circle that passes through the falling stream once each revolution.   Size of the segment or cutter opening determines the size of the sample. The sample is large (5 to 10% of the entire stream) and a secondary sampling is generally required.   If the feed stream is relatively homogenous, the sampler can be converted to an intermittent type engaged by a magnetic brake and timer.   (3)   The most popular and lowest-cost sampling unit is the *straight-line* sampler that can be operated either intermittently or continuously.   In this type of sampler, the cutter moves in a straight line and at uniform speed across the entire stream.

FIG. 2.4.   LIQUID SAMPLER

Graduated extension tube telescopes inside plunger and locks at any point to determine sampling distance from bottom. Valve lifts when bottom of tank is reached or valve may be lifted by cord attached to valve plunger to collect samples from intermediate levels.

Courtesy Seedburo Equipment Co.

The factors that affect the amount of sample taken by a sampler include the size of the cutter opening, the speed of the cutter as it travels through the stream, and the frequency of making a sample cut.

FIG. 2.5.   RIFFLE TYPE SAMPLER FOR DIVIDING OR HALV-
ING DRY MATERIALS

Material poured into hopper is divided into 2 equal portions
by 2 series of chutes which discharge alternately in opposite
directions into separate pans.

Generally, to allow free entrance into the cutter, the minimum width of a cutter opening should be three times the size of the maximum size of the particles to be sampled. The cutter should move through the stream at a speed that gives a smallest sample without deflecting particles that should enter the cutter. The amount of sample taken per cut (Q) can be determined by the formula $Q = P \times W/L$; P = lb/sec of feed; W = cutter width in in.; L = cutter speed in in./sec (generally about 30). Frequency of sampling depends on the uniformity and homogeneity of the sampled material. Mechanical sampling may be combined with high-speed analysis and automated processing adjustment. Sampling at high flow rates may require secondary and even tertiary subsampling to reduce the primary sample to a convenient quantity for analysis.

SAMPLING ERRORS

Sampling errors are caused by several factors. Lack of randomness in selection may result from both instrumental limitations or deficiencies or from personal bias in human selection.

Manual methods of sampling powdered or granular materials are subject to numerous errors (Quackenbush and Rund 1967). Baker *et al.* (1967) studied the factors that affect sampling bias by triers. These factors include: *particle shape*—round particles flow into the sampler compartments more readily than angular particles of similar size; *surface adhesiveness*—an uncoated hygroscopic material flows into the sampler compartment more readily than nonhygroscopic materials of similar shape and of either larger or smaller size; and *differential downward movement* of particles (on the basis of size) when disturbed during sampling. Consequently, horizontal cores contain a higher proportion of smaller-sized particles than vertical cores. The latter is probably also the main bias factor affected by diameter and opening size of sampling tube.

Changes in composition may occur during or after sampling; they include gain or loss of water, loss of volatiles, physical inclusion of gases, reaction with container material or foreign matter in container, damage to fruits or vegetables by mechanical injury and subsequent enhanced enzymatic or chemical changes.

The main problem arises, however, from nonhomogeneity of foods. We encounter both a macroheterogeneity (among various units of a lot), and microheterogeneity (within various parts of a unit). The latter is especially important in vitamin assay and in determination of minor components. Thus, nicotinic acid and thiamine are concentrated in the aleurone and scutellum tissues of the wheat kernel; epidermis cells of grapes are rich in anthocyanin pigments; and essential oils in citrus fruits are mainly in cells of the flavedo layer. But also major components are distributed unevenly. Proteins, lipids, minerals, and crude fiber are higher in the outer layers than in the endosperm of cereal grains; variations in water, sugars, and organic acids are found in various tissues of fruits and vegetables; and uneven distribution of fat in meat makes it imperative to often express some analytical values on a fat-free basis.

Most of the difficulties are overcome by fine grinding and mixing of large samples. In some instances, however, attempts at homogenization of a food sample are wrought with difficulties; in others, apparently homogenous preparations have a tendency of segregation or stratification. Failure to recognize and appraise the variations may limit or even invalidate conclusions from analytical data.

To summarize, the aim of sampling is to secure a portion of the material that satisfactorily represents the whole. The more heterogenous the material, the greater the difficulties and required efforts to obtain a truly representative sample.

20 FOOD ANALYSIS

BIBLIOGRAPHY

Assoc. Vitamin Chemists, Inc. 1951. Methods of Vitamin Assay. Interscience Publishers, New York.

Baker, W. L., Gehrke, C. W., and Krause, G. F. 1967. Mechanisms of sampler bias. J. Ass. Offic. Anal. Chemists 50, 407–413.

Barlett, R. P., and Wegener, J. B. 1957. Sampling plans developed by U.S. Dept. of Agr. for inspection of processed fruits and vegetables. Food Technol. 11, 526–532.

Bowker, A. H., and Goode, H. P. 1952. Sampling Inspection by Variables. McGraw-Hill Book Co., New York.

Bowman, F., and Remmenga, E. E. 1965. A sampling plan for determining quality characteristics of green vegetables. Food Technol. 19, 617–619.

Bur. Ordnance. 1952. Sampling Procedures and Tables for Inspection Variables. NAVORD-OSTD-80. Bur. Ordnance, Dept. Navy. Govt. Printing Office, Washington, D.C.

Cochran, W. G. 1963. Sampling Techniques, 2nd Edition John Wiley & Sons, New York.

U.S. Dept. Defense. 1963. Sampling Procedures and Tables for Inspection of Attributes. 105D. Mil. Std.

Dodge, H. F., and Romig, H. G. 1944. Sampling Inspection Tables. J. Wiley & Sons, New York.

Johnson, N. L. 1963. Sampling devices. Food Technol. 17, 1516–1520.

Kramer, A., and Twigg, B. A. 1970. Fundamentals of Quality Control for the Food Industry, 3rd Edition, Vol. 1. AVI Publishing Co., Westport, Conn.

Pearson, D. 1958. Food analysis-techniques, interpretation and legal aspects. IV. The quality control chemist in the food industry. Lab. Pract. 7, 92–94.

Quackenbush, F. W., and Rund, R. C. 1967. The continuing problem of sampling. J. Assoc. Offic. Anal. Chemists 50, 997–1006.

U.S. Dept. Agr. 1964. U.S. standards for sampling plans for inspection by attributes —single and double sampling plans. Federal Register 29-FR-5870.

CHAPTER 3

Preparation of Samples

Care, time, and effort devoted to the preparation of samples for analysis should be commensurate with the information required and the accuracy and precision of the analytical results desired. If the sample is not prepared properly for analysis, or if the components have been altered during that preparation, the results will be inaccurate regardless of the effort, the precision of the apparatus, and the techniques used in the analysis (Entenman 1961).

The purpose of sample preparation is to mix thoroughly a large sample in the laboratory. This apparently homogenous sample must then be reduced in size and amount for subsequent analysis. Grier (1966) reported results of a survey based on questionnaires submitted to 200 plant scientists concerning kinds of materials used in testing (from dry, stemmed hay to wet fruits); mills for size reduction of dry materials (Wiley mills, hammer mills, choppers); equipment for size reduction of wet material (food choppers, blenders, high-speed mixers); and ovens and lyophilizers for drying. The problems encountered by the analysts in the preparation of samples for analyses included: preparing representative small samples from large samples; loss of plant material; removal of extraneous material from plants without removal of plant constituents; enzymatic changes before and during analyses; compositional changes during grinding; metal contamination during grinding; changes in unstable components (i.e. chlorophylls, unsaturated fatty acids); and special preparation problems in analyses of oil seed materials.

Both the nature of the food and the analyses to be performed must be considered in the selection of instruments for grinding. An analysis can sometimes be made directly on fresh or dried material. Total nitrogen by the Kjeldahl method and total ash determinations require generally little disintegration, other than that necessary to provide a homogenous and representative sample (Pirie 1956). In practice, however, some disintegration is required and for this a wide range of techniques and equipment is available.

For determination of moisture, total protein, and mineral contents, dry foods are generally ground to pass a 20-mesh (openings per linear inch) sieve.   For assays involving extraction (lipids, carbohydrates, various forms of protein) samples are ground to pass a 40-mesh sieve. The advantages of using finer ground samples are outweighed by losses of total material and moisture, heating and undesirable chemical modifications, and even difficulties in assay (as in clogging of filter pores in extraction or filtration).

GRINDING DRY MATERIALS

Mechanical methods range from the simple pestle and mortar to elaborate and effective devices for grinding.   For fine grinding of dry materials, power-driven hammer mills are widely used.   To control the fineness of grinding, various screens (through which the ground material must pass) are inserted (Fig. 3.1).   Hammer mills are rugged and efficient, and not easily damaged by stones and dirt that may contaminate some samples.   They are used to grind such materials as cereals, oil meals, and most foods provided they are reasonably dry and do not contain excessively high amounts of oil or fat.

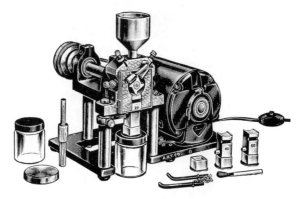

Courtesy A. H. Thomas Co.

Fig. 3.1.   Wiley Mill, Intermediate, Uses Interchangeable Sieve Top Delivery Tubes of 10 to 80 Mesh Per inch

The fineness of grinding is largely affected by the composition of the food, and especially the moisture content.   Generally, foods are ground better after drying in a desiccator or vacuum oven.   In some instances, however, pulverizing air-dried materials gives best results.

For small samples, especially if fine grinding is required, ball mills are used.   The material is ground in a container half-filled with balls. The container rotates constantly and the balls exert an impact-grinding action. The main objection to such mills is that they require many hours (or even days) for satisfactory grinding.

Grinding of oil seeds or oil-rich samples presents special problems that are described in Chap. 36.

Grinding of wet samples may result in significant losses of moisture, in heat generation, and accompanying chemical changes. Grinding, especially of hard materials, may result in serious contamination from the abrasives of the mill. This may be especially objectionable in determining mineral components. The contamination can be minimized, or at least recognized, if the working parts of the grinder are made of a resistant material, (glass, ceramic, agate), or of one that can be easily determined, metal.

Chilled ball mills can be used to grind frozen materials without preliminary drying. Grinding of frozen foods reduces undesirable chemical changes (i.e. Maillard reaction involving interaction between sugars and amino acids) at room temperature. Clemments (1959) described a grinder for frozen, resilient plant tissues. Similar grinders can be used for grinding bread crumbs and samples of foods in which heat labile components are to be studied.

GRINDING MOIST MATERIALS

For disintegration of moist materials, various fine-slicing devices are available. Some moist materials are disintegrated best by bowl

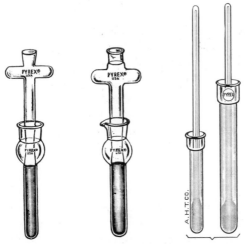

Courtesy A. H. Thomas Co.

FIG. 3.2. TISSUE GRINDERS FROM LEFT TO RIGHT: TEN-BROECK PLAIN, TEN-BROECK WITH POUR OUT, AND HOMOGENIZERS WITH PISTON-TYPE, GROUND GLASS PESTLE

cutters (for leafy vegetables and fleshy tubers and roots) or meat mincers (for fruits, roots, and meat products).   Any of the above can be either hand- or power-driven.   In addition, power-driven pestles and mortars, especially if an abrasive (such as sand) is added, can give fine subdivision.

The commercially available tissue grinders are used for small samples of soft material (Fig. 3.2).   The sample is forced through the annulus between two concentric cylinders, one of which is driven mechanically.   In addition to batch grinders, choppers, and graters for continuous food disintegration are available.

FIG. 3.3.   BLENDER-DISINTE-GRATOR APPARATUS (OSTERIZER)

Equipped with stainless steel cutting blades at speed of 17,000 or 25,000 rpm.

• **Open-Bottom Jar for Ease in Cleaning**
• **Standard Mason Jar Screw Thread**

Courtesy A. H. Thomas Co.

For grinding dilute suspensions and most soft and pasty foods, the various modifications of the Waring blendor can be used (Fig. 3.3). In those instruments, knives at up to about 25,000 rpm disintegrate a sample suspended in an extractant. Vessels to accommodate 10 to 2000 ml are available. The blenders are ideal for routine disintegration of tissues for most analytical determinations. The blenders fail whenever a thick slurry or highly texturized material is homogenized, because the rotating knives form a cavity. The blenders have been used successfully for disintegration (and sometimes simultaneous extraction) of such diverse materials as oil seeds, creamed cottage cheese (Perlmutter 1953), French dressing (Ratay 1953), and for a great variety of animal and plant products (Jones and Ferguson 1951).

In several types of colloid mills, the dilute suspension is pumped through a controlled gap between smooth or slightly serrated surfaces until the particles have been sufficiently disintegrated by shear.

Sonic and supersonic vibrations have been adapted for the dispersion of foods. Attempts have been made to disintegrate a tissue by saturating it with gas under pressure and then allowing the gas to expand by suddenly releasing the pressure. Shaking suspensions with pure sand or small glass beads for wet grinding in a ball mill is often highly efficient. This principle is the basis of the widely used Mickle disintegrator (Pirie 1956).

ENZYMIC AND CHEMICAL TREATMENT

In addition to mechanical methods of disintegration, enzymic and chemical procedures can be used. The availability of pure cellulases is particularly useful in testing materials of plant origin; proteases and carbohydrases can be used in solubilizing high molecular weight components in many foods. Dimethylformamide, urea, pyridine, phenol, dimethyl sulfoxide, synthetic detergents (that have largely replaced cholic and deoxycholic acids), and reducing agents (for disulfide bond cleavage) are some examples of chemicals that can be used effectively in dispersing or solubilizing foods or food components for analysis.

ENZYME INACTIVATION

One of the problems facing a food analyst is enzymatic modification following sampling or the preparation of samples for analyses. Generally, if total contents of a specified compound are determined (i.e. minerals, carbohydrates, nitrogen), enzyme inactivation is not essential. If, however, various forms of a compound (sugars, free and bound forms of lipids, groups of proteins) are to be determined, the tissue must be killed in such a way that potentially troublesome enzymes are immediately and completely inactivated.

Whenever possible fresh material should be analyzed. To preserve the original state of components in a living tissue, several methods of enzyme inactivation can be used. The treatment required for enzyme inactivation varies widely with the food size, consistency, and composition; the enzymes present; and the intended analytical determinations. Thus, fungal amylases are generally heat labile and can be inactivated at relatively low temperatures; some bacterial amylases are highly resistant and may survive bread baking temperatures. Extraction of chlorogenic acid from seeds or dry tissues requires heating to 90° to 100°C for 1 hr to inactivate polyphenolases. Juicy tissues should be heated for a time that will result in a temperature of 90° to 100°C for 5 to 10 min in the center of the sample (Paech 1956).

Drying should be as rapid and at as low a temperature as possible. Such drying is greatly facilitated by spreading the sample over a wide area. Generally, drying at 60°C under vacuum is recommended. If the sample contains no heat-sensitive or volatile compounds, heating for several minutes at 70° to 80°C may be advisable. Such heating inactivates most enzymes and modifies cells to enhance the rate of their drying. During drying, certain components are destroyed (enzymes, vitamins), others are almost invariably modified (proteins and lipids), and some flavor components are volatilized. If drying is not done carefully, caramelization and sugar inversion in acid foods are likely to occur. Case-drying, as a result of formation of an impervious outside layer, is often quite troublesome.

The difficulties encountered in drying plant materials were illustrated by the investigations of Gausman et al. (1952). Unripe corn kernels were dried at 3 temperatures (44°, 54°, and 83°C). Drying at the higher temperatures and especially in samples containing high moisture levels (50% or above), lowered the apparent starch contents and increased the sugar contents. Pyridoxal and pantothenic acid levels were decreased; but apparent riboflavin and nicotinic acid levels were higher after drying at elevated than at room temperatures.

Some plant materials can be stored at −20° to −30°C provided they can be cooled to the low temperature within 1 hr. However, plant acid phosphatases function at −28°C in the frozen state and in 40% methanol solution. Tissues or extracts taken for studies of phosphate metabolism can be stored in the cold after drying (Bieleski

1964). Most foods are preserved best (even for long term storage) by freeze drying. But freezing alone does not stop enzyme action. Actually, slow freezing disrupts tissues and may often enhance enzymatic changes. Fresh foods, in which the enzymes have not been inactivated prior to freezing, are especially susceptible to enzymatic attack during and after thawing.

Some enzymes can be inactivated by inorganic compounds that cause irreversible enzyme poisoning; by a shift in pH, provided other compounds are not affected by high acidity or alkalinity; or by salting out, i.e. by high concentrations of ammonium sulfate (Paech 1956). The most common methods of halting enzyme action in analyses of plant tissues include treatment with 80% methanol or ethanol, ice-cold 5 or 10% perchloric or trichloracetic acid, or a mixture of methanol-chloroform-2M formic acid (12:5:3, by volume) (Bieleski 1964). Most commonly, the plant material is cut rapidly to enhance penetration and dropped into hot redistilled 95% ethanol containing enough precipitated calcium carbonate to neutralize the organic acids in the plant tissues. The amount of ethanol is such as to give a final concentration of 80%. The treatment is most effective for preserving carbohydrates and is the standard procedure in determinations of plant gums. However, proteins may still be modified during subsequent storage.

Processing (grinding, homogenization, mixing) of fresh tissues in air is likely to oxidize labile reduced compounds. In many instances, the oxidation is caused by enzymes that show limited or no net oxidative effects in intact tissues. Such changes are often eliminated by enzyme inactivation. Often the grinding of tissues after dipping or the addition of reducing agents (i.e. bisulfite or dithiothreitol) (Cleland 1964) may reduce oxidation. In some instances, neither heating nor the addition of reducing agents eliminates undesirable oxidative changes. For extraction of ascorbic acid, 2% metaphosphoric acid or a mixture of 10% acetic acid and 5% oxalic acid have been recommended (Paech 1956). Unless proper measures are taken, losses of up to 75% ascorbic acid in the grinding stage alone may be encountered.

MINIMIZING LIPID CHANGES

Traditional methods of preparing samples may affect the composition of lipid extracts. The proportions of lipid classes in living tissue may change upon death and the samples must be chilled rapidly prior to extraction, or frozen quickly for storage (Holman 1966). Incomplete extraction procedures leave more polar lipids than nonpolar components unextracted. In determination of fatty acids, saponification of tissue minimizes losses of lipids that are extracted with difficulty, but may decrease yields of polyunsaturated fatty acids because of reaction with other tissue components in the presence of alkali. Prolonged saponification causes isomerization of polyunsaturated forms to conjugated forms.

Storage of lipids rich in polyunsaturated fatty acids presents several problems (Holman 1966). The methylene-interrupted polyunsaturated acids, because of their activated methylene groups, are easily attacked by oxygen. The rate of oxidation depends on the degree of unsaturation. The maximum rate of oxidation for linoleate is approximately 20 times that of oleate. Each additional double bond in the molecule of polyunsaturated fatty acids increases the oxidation rate by at least twice. The oxidation rate of a sample is the weighted average of the concentration and rates of oxidation of the component fatty acids.

Dry fat samples should be stored under nitrogen or dissolved in petroleum ether. Storage in ethyl ether is undesirable because it tends to form oxidative peroxides. Dilution with petroleum ether and flushing in a stream of pure nitrogen is a good practice for temporary protection. The rate of oxidation is temperature-dependent. At $-20°C$, the rate of oxidation is about $^1/_{16}$ of that at room temperature. Addition of antioxidants (0.1 to 0.05% of propyl gallate or santoquin) is effective provided it does not interfere with the analytical determinations. The antioxidant, 4-methyl-2,6-ditertiarybutylphenol is a useful antioxidant in chromatographic fractionations, as it is easily separable from the common lipids (Wren and Szczepanowska 1964). Light and especially fluorescent lighting activate oxidation. Polyunsaturated fatty acids are less damaged when stored in frozen ($-20°C$) intact tissues than after they are extracted from the tissue.

CONTROLLING OXIDATIVE AND MICROBIAL ATTACK

To minimize oxidative changes, preservation at low temperatures under nitrogen is recommended for most foods. Compositional changes (desiccation or moisture absorption. volatilization) of relatively dry foods are reduced by storage in hermetically closed containers at about 4°C. If such containers contain powdered materials and are opened frequently for removing samples, it is advisable to check whether moisture changes have taken place. Frozen foods should be wrapped in plastic material or placed in air tight containers to reduce dehydration.

To reduce or eliminate microbial attack, several methods can be used. They include freezing, drying, and the use of preservatives or a combination of any of the three. The commonly used preservatives include sorbic acid or sorbate, sodium benzoate, sodium salicylate, tylosin, formaldehyde, mercuric chloride, toluene, or thymol. Selection of a method or preservative will depend on the nature of the food; expected contamination (natural or additives), storage period and conditions; and the analyses that are to be performed.

BIBLIOGRAPHY

BIELESKI, R. L. 1964. The problem of halting enzyme action when extracting plant tissues. Anal. Biochem. 9, 431–442.

CLELAND, W. W. 1964. Dithiothreitol, a new protective reagent for SH groups. Biochemistry 3, 480–482.

CLEMMENTS, R. L. 1959. A technique for the rapid grinding of frozen plant tissue. J. Assoc. Offic. Agr. Chemists 42, 216–217.

ENTENMAN, C. 1961. The preparation of tissue lipid extracts. J. Amer. Oil Chemists Soc. 38, 534–538.

GAUSMAN, H. W. et al. 1952. Some effects of artificial drying of corn grain. Plant Physiol. 27, 794–802.

GRIER, J. D. 1966. Preparation of plant material for analysis. J. Assoc. Offic. Anal. Chemists 49, 291–298.

HOLMAN, R. T. 1966. Polyunsaturated acids; General introduction to polyunsaturated acids. Progr. Chem. Fats Lipids 9, 3–12.

JONES, A. H., and FERGUSON, W. E. 1951. Methods of preparing food and plant products with the Waring blendor and the limitations of that equipment. Food Res. 16, 281–284.

PAECH, K. 1956. General procedures and methods of preparing plant materials. In Modern Methods of Plant Analysis, Vol. 1, K. Paech, and M. V. Tracey (Editors). Springer-Verlag, Berlin (German).

PERLMUTTER, S. H. 1953. Report on preparation of samples of creamed cottage cheese with the Waring blendor. J. Assoc. Offic. Agr. Chemists 36, 187–190.

PIRIE, N. W. 1956. General methods for separation. In Modern Methods of Plant Analysis, Vol. 1, K. Paech, and M. V. Tracey (Editors). Springer-Verlag, Berlin.

RATAY, A. F. 1953. Report on preparation of sample and sampling of French dressing. J. Assoc. Offic. Agr. Chemists 36, 758–759.

WREN, J. J., and SZCZEPANOWSKA, A. D. 1964. Chromatography of lipids in the presence of an antioxidant 4-methyl-2,6-di-tert-butylphenol. J. Chromatog. 14, 405–409.

CHAPTER 4

Reporting Results and Reliability of Analyses

INTRODUCTION

Basically, analytical assays are concerned with the determination of the mass (weight) of a component in a sample. The numerical result of the assay is calculated as a weight percentage or in other modes of expression that are actually equivalent to the mass/mass ratio. The mass (weight) of a component in a food sample is calculated from a determination of a parameter whose magnitude is a function of the mass of a specific component in the sample.

Some properties are basically mass-dependent. Absorption of light or other forms of radiant energy are a function of the number of molecules, atoms, or ions in the absorbing species. For quantitative determination, a measurement is made in a solution of a given depth that contains a specified concentration of solute. Although certain properties, such as specific gravity and refractive index, are not mass-dependent, they can be used indirectly for mass determination. Thus, one can determine the concentration of ethanol in aqueous solutions by a density determination. Refractive index is used routinely to determine soluble solids (mainly sugars) in syrups and jams.

Some mass-dependent properties may be characteristic of several or even of a single component and may be used for selective and specific assays. Examples are light absorption, polarization, or radioactivity. Some properties have both a magnitude and specificity parameter (nuclear magnetic resonance and infrared spectroscopy). Such properties are of great analytical value because they provide selective determinations of a relatively large number of substances.

This chapter describes conventional ways of expressing analytical results, and defines the significance of specificity, accuracy, precision, and sensitivity in assessing the reliability of analyses.

REPORTING RESULTS

In reporting analytical results, both the reference basis and the units used to express the results must be considered. The analyses can be performed and the results reported on the edible portion only or on the whole food as purchased. Results can be reported on an as-is basis, on an air-dry basis, on a dry matter basis, or on an arbitrarily selected moisture basis (i.e., 14% in cereals).

To convert contents (%) of component Y from oven dried to an as-received basis the following formulas are used:

$$\% Y_{OD} = \frac{\% Y_{AR} \times 100}{(100 - \%_{OD} \text{ loss})}$$

$$\% Y_{AR} = \frac{\% Y_{OD} (100 - \%_{OD} \text{ loss})}{100}$$

where $_{OD}$ = Oven Dried

$_{AR}$ = As Received

Similarly, data can be converted to an air-dried basis.

If the contents are to be expressed on an arbitrary moisture basis (i.e., 14% in cereals) the following formula is used:

$$\% Y = \frac{\% Y_{AR \text{ basis}} (100 - \text{arbitrary moisture } \%)}{100 - {}_{AR} \text{ moisture } \%}$$

To weigh out a sample on an arbitrary moisture basis use the following:

$$\text{Sample weight} = \frac{\% \text{ dry matter }_{AM}}{\% \text{ dry matter}_{AR}} \times \text{required sample weight}$$

where $_{AM}$ = arbitrary moisture basis

$_{AR}$ = as received moisture basis

To obtain % dry matter, subtract percent of moisture from 100. If the moisture has been determined in two stages, air drying followed by oven drying, compute total moisture contents of sample as follows:

$$TM = A + \frac{(100 - A)B}{100}$$

where TM = % total moisture

A = % moisture loss in air drying

B = % moisture of air-dried sample as determined by oven drying

Tables, nomograms, and calculators are available to simplify calculations in expressing results on a given basis, or for weighing samples on a fixed moisture basis (i.e. 20% in dried fruit). In view of the very wide range in moisture contents in various foods, analytical results are often meaningless unless the basis of expressing the results is known.

Expression on an as-is basis is wrought with many difficulties. It is practically impossible to eliminate considerable desiccation of fresh plant material. In some instances, even if great pains are taken to reduce such losses, the results may still vary widely. Thus, the moisture contents of leafy foods may vary as much as 10% depending on the time of harvest (from early morning to late afternoon).

The moisture contents of bread crust and crumb change from the moment the bread is removed from the oven as a result of moisture migration and evaporation. Absorption of water in baked or roasted low-moisture foods (crackers, coffee) is quite substantial. In most cases, storing air-dried foods in hermetically-closed containers is least troublesome. Once the moisture contents of such foods are determined, the sample can be used for analyses over a reasonable period.

The concentrations of major components are generally expressed on a percent by weight or percent by volume basis. In liquids and beverages gm per 100 ml is often reported. Minor components are calculated as mg (or mcg) per kg or liter; vitamins in mcg or international units per 100 gm or 100 ml. Amounts of spray residues are often reported in grains per pound or ppm.

In calculating the protein contents of a food, it is generally assumed the protein contains 16% nitrogen. To convert from organic nitrogen (generally determined by the Kjeldahl method, see Chap. 37) to protein, the factor of 6.25 = 100/16 is used. In specific foods known to contain different concentrations of nitrogen in the protein, other conversion factors are used (i.e. 5.7 in cereals, 6.38 in milk).

If a food contains a mixture of carbohydrates, the sugars and starch are often expressed as dextrose. In lipid analyses (free fatty acids or total lipid contents) calculations are based on the assumption that oleic acid is the predominant component. Organic acids are calculated as citric, malic, lactic, or acetic acid depending on the main acid in the fruit or vegetable.

Mineral components can be expressed on an as-is basis or as % of total ash. In either case the results can be calculated as elements or as the highest valency oxide of the element.

Amino acid composition can be expressed in several ways; gm amino acid per 100 gm sample, or per 100 gm protein, or per 100 gm of amino acids. For the determination of molar distribution of amino acids in protein, gm moles of amino acid residue per 100 gm moles of amino acid, are computed.

In trade and industry empirical tests are often used. Thus, fat acidity of cereal grains is often expressed as mg KOH required to neutralize the fatty acids in 100 gm food. Acidity is often expressed for simplicity in ml of $N/10$ or N NaOH. The acidity of acid phosphates in baking powders is reported in industry as the number of parts of sodium bicarbonate that are required to neutralize 100 parts of the sample.

The problem of expressing results in a meaningful manner has assumed special significance in plant analyses. Several workers suggested expressing composition of growing plants on a protein or nucleic acid basis. The protein determination is simpler, more rapid, and more precise, but it is difficult to distinguish between physiologically active and storage proteins; DNA content would be a sounder basis for investigations in plant physiology. In tubers, beets, and onions, the structural cell wall material is estimated from crude fiber determination. In determining enzymatic activity, results are often expressed on a protein basis rather than on a total weight basis.

During maturation of cereal grains several changes take place. In addition to an increase in kernel size and weight as a result of incorporation of nutrients, moisture content decreases. Even if the results are compared on a uniform moisture basis (14% or dry matter), the picture may still be misleading. Thus, the protein content of maturing wheat on a weight percent basis changes little during maturation. This has led to the erroneous (and often repeated in literature) concept that proteins are synthesized and laid down in the maturing kernel at early stages, and that starch is the main component synthesized in later stages. This fallacy is disproved by calculating the results on a kernel basis. As the dry weight of a maturing wheat kernel increases about fourfold within the last three weeks, to maintain uniform protein content, the weight of protein laid down per kernel must also be quadrupled.

During storage of foods for prolonged periods, considerable losses of metabolizable organic components occur. To determine compositional changes during storage, the content of some mineral (most commonly phosphorus, sometimes calcium or magnesium) is used as the basis for calculation.

RELIABILITY OF ANALYSES

The reliability of the analytical method depends on its (1) specificity, (2) accuracy, (3) precision, and (4) sensitivity (Anastassiadis and Common 1968).

Specificity is affected primarily by the presence of interfering substances that yield a measurement of the same kind as the tested sample. In many cases, the effects of the interfering substances can be accounted for.

In calculating or measuring the contribution of several interfering substances, it is important to establish whether their effects are additive.

Accuracy of an analytical method is defined as the degree to which a mean estimate approaches a true estimate of an analyzed substance, after the effects of other substances have been allowed for—by actual determination or calculation. In determining the accuracy of a method we are basically interested in establishing the deviation of an analytical method from an ideal one. That deviation may be due to an inaccuracy inherent in the procedure; the effects of substances other than the analyzed one in the food sample; and alterations in the analyzed substance during the course of the analysis.

The accuracy of an analytical assay procedure can be determined in two ways. In the absolute method, a sample containing known amounts of the analyzed components is used. In the comparative method, results are compared with those obtained by other methods that have been established to give accurate and meaningful results.

The first approach is often difficult or practically impossible to apply, especially for naturally occurring foods. In some cases, foods can be prepared by processing mixtures of pure compounds. If the mixtures are truly comparable in composition to natural foods, meaningful information is obtained.

Several indirect methods are available to determine the accuracy of the analyses. While those methods are useful in revealing the presence of errors they cannot prove the absence of errors. When a complete analysis of a sample is made and each component is determined directly and if the sum of the components is close to 100, a certain degree of accuracy is indicated. On the other hand, an apparently good summation can result from compensation of unrelated errors in the determination of individual components. A more serious error can result from compensation of errors that are related in such a way that a negative error in one component will cancel a positive error in another component.

This may be particularly important in incomplete fractionations. Thus, the sum of proteins separated according to differences in solubility may be close to 100%, yet the separation of individual components may be incomplete or of limited accuracy.

In the recovery method, known amounts of a pure substance are added to a series of samples of the material to be analyzed and the assay procedure is applied to those samples. The recoveries of the added amounts are then calculated. A satisfactory recovery is most useful in demonstrating absence of negative errors.

If any of the procedures indicates that accuracy of an analytical method is affected by interference from substances that cannot be practically eliminated, a suitable correction can sometimes be applied. Such a correction is often quite complicated as the results may be affected by concentration of the interfering or assayed substance, or by their interaction in food processing or during the analysis.

Precision of a method is defined as the degree to which a determination of a substance yields an analytically true measurement of that substance. It is important to distinguish clearly between precision and accuracy. If numerous similar samples are to be analyzed in industrial quality control, it may often be unimportant whether the analysis yields the exact information regarding the composition of the sample. The information may be useful provided the difference between the precise and accurate determination is consistent. The analysis that gives the actual composition (or in practice the most probable composition) is said to be the most accurate. Thus, for instance, direct and accurate determination of the bran content of wheat flour is both laborious and imprecise. The bran content can be estimated indirectly from the amount of crude fiber in a flour. This estimation is based on the fairly constant ratio between crude fiber (determined by a precise empirical method) and actual bran contents. Still simpler, is the estimation of bran content from total mineral content or reflectance color assay of a flour.

Since accurate moisture determination is often time-consuming and complicated, a rapid and reproducible moisture test that is highly correlated with an official and accurate moisture assay may be used for routine quality control.

To determine the precision of an analytical procedure and the confidence that can be placed on the results obtained by that procedure, statistical methods are used. The most basic concept in statistical evaluation is that any quantity calculated from a set of data is an *estimate* of an unknown *parameter* and that the estimate is sufficiently reliable. It is common to use English letters for estimates, and Greek letters for the true parameters (see also the chapter on sampling).

If n determination x_1, x_2, ... x_n are made on a sample, the average $\bar{x} = \Sigma x/n$ is an estimate of the unknown true value, μ.

The precision of the assay, is given by the standard deviation, σ.

$$\sigma = \pm \sqrt{\frac{\Sigma(x - \mu)^2}{n}}$$

If the number of replicate determinations is small, ($<$10) an estimate of the standard deviation (s) is given by

$$s = \pm \sqrt{\frac{\Sigma(x - \bar{x})^2}{n - 1}}$$

The divisor, $n - 1$, used to estimate s, is termed the degrees of freedom and indicates that there are only $(n - 1)$ independent deviations from the mean.

The standard deviation is the most useful parameter for measuring the variability of an analytical procedure.

If s is independent of x for a given concentration range, s can be computed from the results of replicate analyses on several samples of similar materials. In that case, the sums of the squares of the deviations of the replicates of each material are added, and the resultant total is divided by the number of degrees of freedom (the sum of the total number of determinations, n, minus the number of series of replicate determinations).

A complicating factor in determining the precision arises when the standard deviation varies with the concentration of the element present. Sometimes the range of concentration can be divided into intervals and the standard deviation given for each interval. If the standard deviation is approximately proportional to the amount present, precision can be expressed in percent by using the coefficient of variation (CV).

$$CV = \frac{\text{standard deviation}}{\text{amount present}} \times 100$$

If the data show a varying standard deviation, transformation of the data into other units in which the standard deviation is constant is often useful. Two widely used transformations are square roots or logarithms.

Chemical analyses are made for various purposes and the precision required may vary over a wide range. In the determination of atomic weights, an effort is made to keep the error below 1 part in 10^4 to 10^5. In most analytical work, the allowable error lies in the range 1 to 10 parts per 1000 for components comprising more than 1% of the sample. As a rule, analyses should not be made with a precision greater than required. Up to a point, precision is a function of time, labor, and overall cost (Youden 1959).

The precision of an analytical result depends on the least exact method used in obtaining the result. In expressing the result, the number of figures given should be such that the next to the last figure is certain and the last figure is highly probable yet not certain. Thus 10% and 10.00% denote widely varying precision (Paech 1956). The following is an example of how the average result (computed from several determinations) is to be expressed. Assume the moisture content of sugar was determined in triplicate, and the following results were obtained: 1.032, 1.046, and 1.036%. The average is 1.038%. But as the difference between 1.032 and 1.046 is larger than 0.010, the results should not be expressed with more than two figures after the decimal point. Thus, 1.04% indicates that the first figure after the decimal point is certain, and the second one is probable but uncertain.

The results of weighing, buret reading, and instrumental (including automatic) reading have limitations. Replication of analyses eliminates some of the errors resulting from sampling, from heterogeneity of sampled material, and from indeterminate—accidental or random—errors in the assay. But whereas repetition of an assay generally increases the precision of the analysis, it cannot improve specificity and accuracy. If, however, reasonable specificity and accuracy have been established, the precision of the assay is an important criterion of its reliability.

Sensitivity of a method used in determining the amount of a given substance is defined as the ratio between the magnitude of instrumental response and the amount of that substance. Sensitivity is measured and expressed as the smallest measurable compositional difference between two samples. That difference becomes meaningful if it exceeds the variability of the method. In instrumental analysis, the signal to noise ratio should be at least 2:1.

Sensitivity can be increased in two ways: (1) by increasing the response per unit of analyzed substance (i.e. in colorimetric assays by the use of color reagents that have a high specific absorbance, or in gravimetric determinations by the use of organic reagents with a high molecular weight), and (2) by improving the discriminatory power of the instrument or operator (i.e. in gravimetry by using a microbalance, in spectrophotometry by using a photomultiplier with a high magnifying power) (Anastassiadis and Common 1968).

BIBLIOGRAPHY

ANASTASSIADIS, P. A., and COMMON, R. H.   1968.   Some aspects of the reliability of chemical analyses.   Anal. Biochem. 22, 409–423.
PAECH, K.   1956.   General procedures and methods of preparing plant materials.   In Modern Methods of Plant Analysis, Vol. 1, K. Paech, and M. V. Tracey (Editors).   Springer-Verlag, Berlin (German).
YOUDEN, W. J.   1959.   Accuracy and precision: Evaluation and interpretation of analytical data.   In Treatise on Analytical Chemistry, I. M. Kolthoff, and P. J. Elwing (Editors).   Interscience Encyclopedia, New York.

Theory of Spectroscopy

The next several sections will be concerned with instruments used in food analysis. A knowledge of how these instruments work is a prerequisite for their maximum utilization. When an instrument manufacturer sells a piece of equipment, he guarantees certain specification and performance levels. However, most instruments are capable of delivering far more than that, and if the analyst understands how the equipment works and what it is supposed to do, he can usually extend its practical capabilities manyfold. It is also necessary to have an understanding of how an instrument operates in order to know whether the data obtained from it are correct.

The interaction of radiation with matter is the basis for many of the common instrumental techniques. Consider the light you are reading this book by. The radiation coming from that light consists of many wavelengths, ranging from very short to very long. Physicists have studied that radiation and have found that each ray behaves as if it has an electric component and a magnetic component acting at right angles to each other; they have called it *electromagnetic radiation*. (The term *light* is reserved only for the visible region. The terms *ultraviolet light* or *infrared light* are not correct and should be *ultraviolet* and *infrared radiation*.)

There are several units and terms which are used to describe electromagnetic radiation. These are given in Table 5.1.

To get some idea of the physical significance of those units consider the interatomic distances within a molecule. Those distances are measured in angstroms (H-Cl = 1.27Å). If the distance from New York to Los Angeles is 1 in. then an angstrom on the same scale would be about the width of a dime. A nm would be about the width of your hand and a μm would be about the length of a football field.

TABLE 5.1

SPECTROSCOPY UNITS

Unit	Symbol	Length	Relation to Other Units
Micrometer (old micron, μ)	μm	10^{-6} meters	
Nanometer (old milli-micron, mμ)	nm	10^{-9} meters	$1/1000$ μm
Angstrom	Å	10^{-10} meters	$1/10$ nm
Wavenumber	cm^{-1}	c/ν	$1/\lambda$ see equation 5.1
Electron volt	ev	23.06 kcal/mole or 8066 cm^{-1}	
Erg	erg	6.24×10^{11} ev/mole	

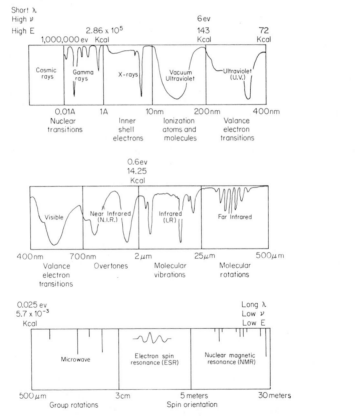

FIG. 5.1. ELECTROMAGNETIC SPECTRUM (NOT TO SCALE)

There are two basic equations useful in spectroscopy.

$$\lambda \nu = c \tag{5.1}$$

λ is the wavelength of the radiation in cm, ν is the frequency of the radiation in cycles/second, and c is a constant, the speed of light, 3×10^{10} cm/sec. The second equation is

$$E = h\nu \tag{5.2}$$

where E is the energy of the radiation, h is Planck's constant (6.62×10^{-27} erg-sec), and ν is the frequency of the radiation, as in equation 5.1.

Suppose the rays of radiation from your desk lamp plus those from any other source, are plotted in chart form as in Fig. 5.1, with the very short wavelengths to the left and the very long wavelengths to the right. Along the top of the figure are several examples of what those wavelengths mean in terms of energy; and along the bottom are some indications of what that radiation does when it interacts with a molecule.

The regions are established by the limitations of the instruments and the boundaries are continuously changing as the instruments improve. Notice that the term INSTRUMENT was used, not machine. We drive to work in machines but measurements are made with instruments.

Interaction of Radiation With Matter

Consider a ray of radiation from your desk lamp shining on your shirt. The ray can either interact with the molecules in the shirt or pass through. What allows some rays to interact and some to pass through?

There are two main requirements that must be met in order for a ray of radiation to be absorbed by a molecule. (1) The incident ray must have the same frequency as a rotational, vibrational, electronic, or nuclear frequency in the molecule, and (2) the molecule must have a permanent dipole or an induced dipole; in more common terms there must be something for the absorbed energy to do—work must be done. Refer to Fig. 5.2 and recall your physics experiment with tuning forks.

When the left tuning fork is struck with a hammer, the fork bends to the right compressing the air molecules against it.

Those molecules, in turn, compress against other molecules further to the right (compression). When the left tuning fork returns to its original position, it leaves a reduced concentration of air

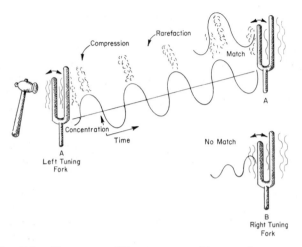

FIG. 5.2.   FREQUENCY MATCHING FOR ENERGY ABSORPTION

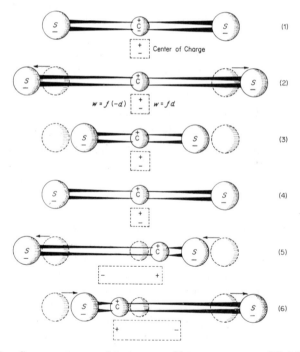

FIG. 5.3.   SYMMETRIC AND ASYMMETRIC VIBRATIONS OF A CS$_2$ MOLECULE

molecules (rarefaction). If those concentrations are plotted vs time, a wave pattern is produced.[1] When the compressed air reaches the tuning forks on the right side, both forks are bent to the right. Both forks will want to return to their original positions. The top fork (A) will return quite easily because there are less molecules to oppose it since it is in phase with the rarefaction. The bottom (B) fork will return to its original position before the entire compression wave has passed it, and the fork will meet some resistance that will slow it down. The B fork again will move to the right because of its normal vibrational frequency. This time, however, it does so as a rarefaction passes, and the fork gets little push from the weak concentration of air molecules. When the fork moves again to the left, it collides directly with the second compression coming by and is stopped from vibrating. The net effect is that the right hand A fork has the same frequency as the left fork and vibrates, while the right B fork does not vibrate because it does not have a matching frequency.

A similar effect occurs when the electric and magnetic components of the electromagnetic ray of radiation interact with the electric and magnetic components of a molecule. If they match, the energy can be transferred (absorbed), and if they do not match, the ray cannot interact and will pass on through the molecule.

The second requirement is somewhat harder to visualize. Suppose the frequencies match and the energy can be transferred. Where does it go and what does it do? It must be used up by doing work. Figure 5.3 showing diagramatically vibrations of a CS_2 molecule may help to understand what happens.

Consider (1) a CS_2 molecule with the C and S at normal distances. C is relatively more electropositive than S so the center of positive charge for the entire molecule acts as if it were at the center of the molecule. The two negative S's have the *center* of their negative charges also at the center of the molecule. Now let the molecule vibrate symmetrically, that is, the right hand sulfur will move away from the carbon the same amount the left hand sulfur moves away (2). It takes work to move the left negative sulfur away from the positive carbon, and it takes the same amount of work to move the right negative sulfur away from the positive carbon. The net effect however is that no work has been done. Remember that work (*w*) equals *force* (*f*) *times distance* (*d*). In this case the distance to the right is exactly opposite the distance to the left and the forces

[1] Neither sound waves nor radiation waves *wiggle* through the air. Many students confuse the concentration vs time plot with the actual physical process.

cancel each other.   You can see this because the center of the negative charge did not move away from the center of the molecule.
The same thing is true for (3).   Since there is no *net* work to be done
the energy of the ray can not stay in the molecule, and the ray will
pass through even though it had the same frequency as a vibration
in the molecule.

Now consider (5).   In this case the work (force times distance)
for the left sulfur does not equal the force times distance for the
right sulfur and the effect is that the center of negative charge
moves away to the left of the positive carbon.   The reverse is true
with (6).   It takes work to separate the positive and negative
center of charges so there is a place for the energy to go and the ray
of radiation will be absorbed.   Notice that the center of charges
consists of a positive and a negative end, a *dipole*.   Some molecules
have a dipole naturally;   others can have an induced dipole when
the electric and magnetic field of a ray of radiation approaches the
molecule.

Rotations, Vibrations, Electronic and Nuclear Transitions

Now that we know what is required for the absorption of radiation, what happens to molecules when rays of different wavelengths
and energies interact with them?   Refer to Fig. 5.4.

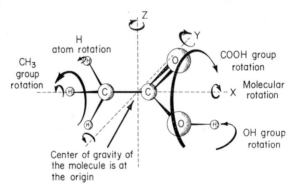

FIG. 5.4.   ATOM, GROUP, AND MOLECULAR ROTATIONS

Consider the case of 100 rays of wavelength of radiation bombarding a molecule.   If the 2 conditions for absorption are met, then
some of the 100 rays would be absorbed and a detector placed on the
other side of the molecule would read between 100 and 0.

When radiation having energies found in the NMR region strikes
the molecule there is only enough energy to cause atom rotations.

That is, the H atoms can rotate about the C—H and O—H bonds, and the O around the C=O bond, etc. Since there are 3 times as many C—H rotations possible as O—H, and since the H is bonded weaker to the C than to the O, we would expect 2 places where absorption would occur, and all other things being equal 1 band would be 3 times as strong as the other. The O rotation around the double bond is restrained and would be expected to have a different frequency and reduced intensity. Since every compound has a different arrangement and number of atoms, the wavelengths of absorption will vary in intensity and distribution in the electromagnetic spectrum. This results in a *fingerprint* that is used to determine what atoms and groups are present, the shape of the molecule, and how much of it is present.

The energies involved in the NMR region are of the order of 10^{-5} kcal/mole. Groups rotate with 0.01 to 0.1 kcal. If 0.1 to 2 kcal/mole of energy is added (far infrared region), there is enough energy to rotate the entire molecule.

There are three *fundamental* rotations for all nonlinear molecules. A molecule can rotate around the X, Y, or Z axes. Since most molecules are unsymmetrical, the rotations about each axis have a different frequency depending on the shape of the molecule. What happens if you strike a molecule at an angle and cause it to "tumble"? The "tumbling" motion can be resolved into three fundamental rotations. A "tumbling" molecule may be rotating 20% along the X axis, 70% along the Y axis, and 10% along the Z axis, the combined effect of which is "tumbling." The important idea is that regardless of the kind of "tumbling" a molecule does, it can always be resolved into a combination of the X, Y, and Z rotations. We cannot predict how each individual molecule will rotate, but we can experimentally show what the *average* of the billions of molecules in a sample will do. The spectra we see measure that average.

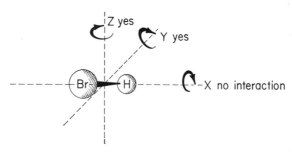

FIG. 5.5. ROTATION OF A LINEAR MOLECULE

Only two axes (Fig. 5.5) of rotation are significant for linear molecules. Hold a pencil horizontally in front of you and rotate it along its horizontal axis. The electric and magnetic fields produced by this motion are too weak to interact with a beam of radiation so there is no absorption band at that frequency.

If 1 to 20 kcal of energy per mole is added to a molecule (infrared region) there is enough energy to cause the molecule to vibrate. There are many different ways a molecule can vibrate. Referring to Fig. 5.6 the atoms can compress or stretch along the axis of a bond. This is called a *stretching* vibration. The atoms can *bend symmetrically* or *asymmetrically;* if all the atoms in the molecule move we have a *deformation* vibration. There are fundamental vibrations just as there are fundamental rotations.

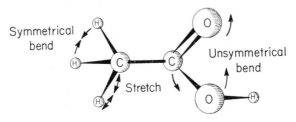

FIG. 5.6.   SOME MOLECULAR VIBRATIONS

A *fundamental vibration* is a movement of atoms such that the center of gravity of the molecule does not change. This means that when any atom of a molecule moves (mass times distance) to the left, another atom or atoms must move to the right an equal mass times distance. Refer to Fig. 5.3, part 5 and 6. Notice that when both sulfurs move to the left, the carbon compensates by moving to the right.

There are two rules that give the theoretically possible number of fundamental vibrations.

$$\text{For nonlinear molecules} \quad 3N - 6 \qquad (5.3)$$

$$\text{For linear molecules} \quad 3N - 5 \qquad (5.4)$$

N is the number of atoms in the molecule. Ethyl acetate has 14 atoms and is nonlinear, therefore it has $(3 \times 14) - 6$ or 36 fundamental vibrations. Hydrobromic acid (Fig. 5.5) has 2 atoms and is linear, therefore it has $(3 \times 2) - 5$ or 1 fundamental vibration.

The presence of a fundamental vibration does not mean that an absorption will occur. For the latter to take place work must be done. Benzene, C_6H_6, has $(3 \times 12) - 6$ or 30 fundamental vibrations, but only 4 are active in the infrared. Benzene has seven planes and a point of symmetry. Therefore if something bends or stretches on one side of the molecule exactly like a bend or stretch on the opposite side, no net work is done and no absorption can take place.

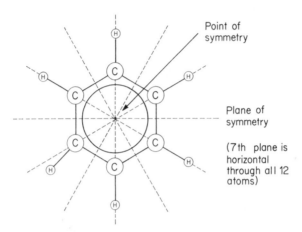

Point of symmetry

Plane of symmetry

(7th plane is horizontal through all 12 atoms)

FIG. 5.7. SYMMETRY PLANES IN BENZENE

Another factor that reduces the number of frequencies that actually appear is *degeneracy*. Consider the CO_2 molecule. It has $(3 \times 3) - 5$ or 4 fundamental vibrations, but we see only 2 absorption bands in the infrared. Figure 5.8 shows the four fundamental vibrations. The 2 bending vibrations have the same frequency and will appear as only 1 band. Since the two vibrations have the same frequency, they are said to be twofold degenerate. If four vibrations had the same frequency, they would be fourfold degenerate.

If radiation of 10 to 100 kcal/mole is applied to a molecule (near infrared, visible, ultraviolet) we can get electronic transitions. That is, an electron in the molecule raises to a higher energy level. If even more energy, generally greater than 100 kcal/mole, is added, there is not only enough energy to raise an electron to a higher energy level but there is enough energy to completely remove the electron from the molecule producing an ion. This begins to occur in the ultraviolet region and continues into the X-ray, gamma ray, and cosmic ray regions. Those ions react differently than the ions and molecules normally used in your body processes and this is why you don't stand in front of X-ray equipment very long, but

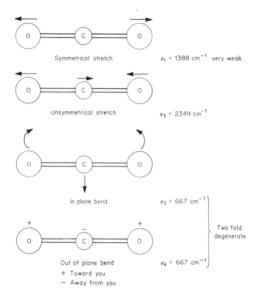

Symmetrical stretch ν_1 = 1388 cm^{-1} very weak

Unsymmetrical stretch ν_3 = 2349 cm^{-1}

In plane bend ν_2 = 667 cm^{-1}

Out of plane bend ν_4 = 667 cm^{-1}
+ Toward you
− Away from you

Two fold degenerate

FIG. 5.8.   CO$_2$ FUNDAMENTAL VIBRATIONS

why it is perfectly safe to stand in front of a radio with its low energy waves.   If several thousand kcal/mole is added to a molecule then nuclear transitions occur.   The entire nucleus changes energy states.

All atoms and molecules contain energy.   This energy takes many forms and varies greatly as to the amount.   When an atom or a molecule has its lowest energy it is said to be in the *ground state*. This is the state preferred by the molecule.   If the molecule has more energy than the lowest energy it is in an *excited state*.   When a molecule absorbs energy in the form of radiation or by a collision and is raised from the ground state to an excited state we get an *absorption spectrum*.   When the molecule gives up its excess energy and goes back to the ground state we get an *emission spectrum*.

There are many possible excited states but there is not an infinite number.   Notice in Fig. 5.9 that there is a definite amount of energy (quantum) required for each transition.   According to the classical description with a continuous number of energy levels if a molecule absorbed energy it would start to rotate at a higher frequency.   If radiation of this new frequency was then absorbed, the molecule would rotate faster.   This process could continue until the molecule was spinning so fast that it could tear itself apart. Experimentally we know that this does not happen.   A molecule has only certain energies it can absorb. If a molecule absorbs energy and rotates, it must get exactly the right energy to make the next jump before anything further will happen. Since the next jump requires twice the original energy, it is harder for the second step to take place. The third jump is even harder because this takes four times the original amount of energy. If the molecule gets an extra large amount of energy, rather than rotate at a destructive rate the molecule will start to vibrate and rotate slower (Fig. 5.9). This process can be repeated several times until the molecule vibrates with such an amplitude that it again risks destroying itself. Rather than destroy itself, an electron within the molecule moves to an excited state and uses up the excess energy.

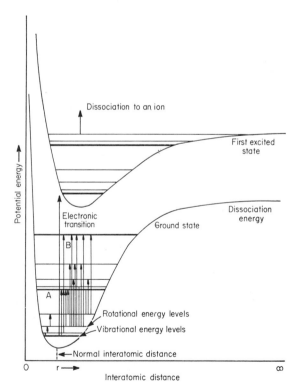

FIG. 5.9. POTENTIAL ENERGY DIAGRAM OF ROTATIONAL, VIBRATIONAL, AND ELECTRONIC TRANSITIONS

It is important to note that a molecule does not vibrate faster with an absorption of energy but with a greater amplitude. Atoms within molecules vibrate between 10^{12} to 10^{15} times a second.

Associated Problems

(6.1)   KMAN, a radio station, operates at a frequency of 1380 Kc. How long (in feet) is the wavelength associated with this frequency?   Ans. 711 ft

(6.2)   The C=O group in benzaldehyde, C_6H_5CHO, absorbs radiation at 5.9 μm. What is the vibration frequency of C=O? 1 μm = 1 $\times$ 10^{-4} cm.   Ans. 5.09 $\times$ 10^{13} sec^{-1}

(6.3)   If you were bombarded by a mole of photons from radio station KMAN (1380 Kc), how many calories of energy would you receive?   1 cal = 4.18 $\times$ 10^7 ergs.   Ans. 1.3 $\times$ 10^{-4} cal

(6.4)   If you were bombarded by a mole of X rays from a copper target (λ = 1.54 Å), how many calories of energy would you receive? Notice the difference between standing in front of a radio (problem 6.3) and in front of an X-ray tube.   Ans. 1.84 $\times$ 10^8 cal

(6.5)   How many fundamental vibration frequencies are there for the artificial sweetener, sodium cyclamate, $C_6H_{11}NHSO_3Na$?   Ans. 66

CHAPTER 6

The Visible and Ultraviolet Regions

Fear is the greatest obstacle to understanding instrumentation. We will, therefore, study the visible and ultraviolet regions first because the instruments used are quite common and more students are familiar with them.

The energies associated with these regions range from 40 to about 140 kcal/mole. This energy is sufficient to cause electronic transitions within the molecule and to ionize many substances. The electrons involved are the outer or bonding electrons, and the nomenclature used to describe the process is based on the bonds formed. When two or more atoms unite to form a molecule, the electrons involved can form sigma bonds (σ), pi bonds (π), or there may be some unused or *non*-bonding (n) electrons.

Formaldehyde will be used to illustrate the various possibilities (Fig. 6.1). Because the various electrons are involved in bonding to different extents, their energies are different and this is shown in Table 6.1. Figure 6.2 summarizes the molecular orbital picture for formaldehyde. Notice that the bonding orbitals have a large overlap with each other, whereas the antibonding orbitals have little overlap. In practice, interaction between sigma and pi electrons is small and the main transitions are $\sigma \rightarrow \sigma^*$ (vacuum UV), $\pi \rightarrow \pi^*$ (UV), $n \rightarrow \pi^*$ (near UV, visible), and $n \rightarrow \sigma^*$ (vacuum UV). It takes about 10^{-15} sec for one of those transitions to take place. The $n \rightarrow \pi^*$ transition is forbidden by selection rules, and therefore the intensity of these bands is weak, about 10^{-6} that of

TABLE 6.1

ELECTRONIC ENERGY LEVELS IN A MOLECULE

	Type	Symbol		
Least stable	Antibonding	σ	σ^*	Pronounced sigma star
Highest energy	Antibonding	π	π^*	
Energy	Nonbonding	n	n	
	Bonding	π	π	
Most stable	Bonding	σ	σ	
Lowest energy				

the $\pi \rightarrow \pi^*$. This means that most of the spectra observed in the UV and visible range is due to a π electron being raised to an excited state. The $n \rightarrow \pi^*$ transitions may be seen if nonpolar solvents are used. We can tell a $\pi \rightarrow \pi^*$ from a $n \rightarrow \pi^*$ band by using a more polar solvent. If the band is due to a $\pi \rightarrow \pi^*$ transition, the band will shift to longer wavelengths (a *bathochromic* or red shift). If the band is due to a $n \rightarrow \pi^*$ transition, the band will shift to shorter wavelengths (a *hypsochromic* or blue shift). Figure 6.3 will serve to define spectral shifts.

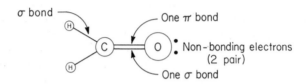

FIG. 6.1. n, σ, AND π BONDS

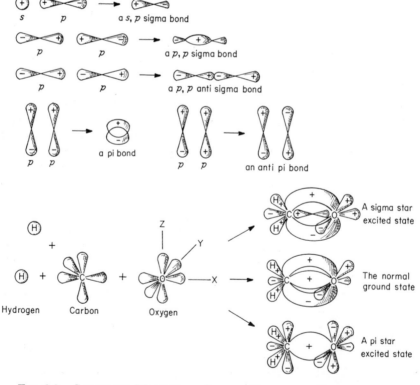

FIG. 6.2. SIMPLIFIED MOLECULAR ORBITAL PICTURE OF FORMALDEHYDE

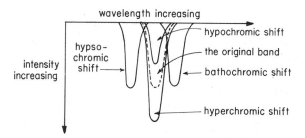

FIG. 6.3. SPECTRAL SHIFT NOMENCLATURE

Note that the geometry of a molecule in an excited state may be different from one in the ground state. Formaldehyde in the ground state is a flat molecule; in the excited state it is bent 20° and some of the bonds are longer (Fig. 6.4). This structural change can cause some rather drastic changes in chemical behavior. For example, phenol is 100,000 times more acid in the excited than in the ground state.

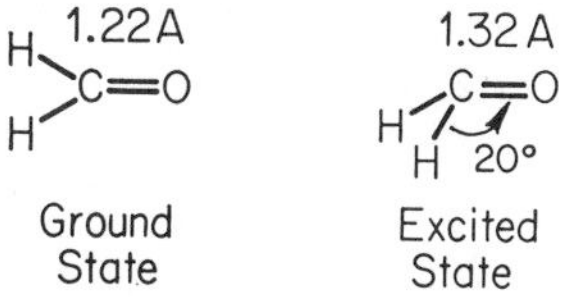

FIG. 6.4. FORMALDEHYDE IN THE GROUND AND EXCITED STATE

Ground State

Excited State

An extension of the electronic transition process is most dramatically illustrated in the visible region with molecular complexes. This is the process known as *charge transfer*. According to Rao (1967), "The main spectral feature accompanying complex formation is the broad intense absorption band in the visible or ultraviolet region due to an electronic transition. Just as the excitation of an electron in an individual molecule by a quantum of radiation may be associated with intramolecular rearrangement of charge, similarly in the complex formed by the association of two molecular or ionic species, the excitation of an electron by a photon can involve a charge rearrangement in the complex. This rearrangement, according to the Mulliken theory, involves a transfer of an electron or part of it from one component of the complex to the other. The reaction may therefore be represented as follows for the alkali halide, MX.

$$M^+ + X^- + h\nu \rightarrow (M^+, X^-)^* \rightarrow M + X \qquad (6.1)$$

where either or both the products M and X may be in the excited state. The primary excitation process is reversible and the ex-

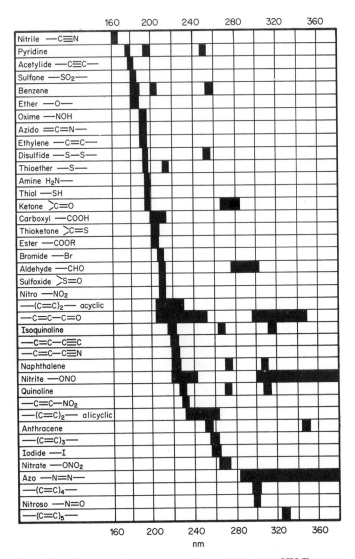

FIG. 6.5. SPECTRA CORRELATION CHART FOR THE UV REGION

cited complex $(M^+, X^-)^*$ will return to the ground level unless a secondary interaction can take place during its life span to give rise to the products M and X. It is the very short lifetime of the excited state (10^{-8}) sec which restricts the quantum efficiency of the reaction giving rise to color centers."

Figure 6.5 shows the position of the bands for the more common groups in the UV region. Because the bands are too broad (charge transfer) or too weak ($n \rightarrow \pi^*$), no similar chart in the visible region has been prepared.

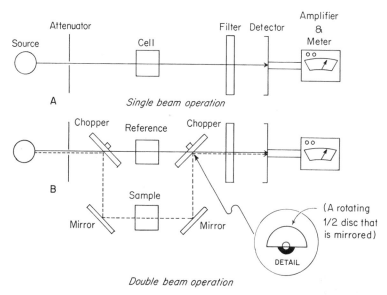

FIG. 6.6.   SINGLE AND DOUBLE BEAM OPERATION

Instrument Components

All spectroscopy instruments contain the same basic components: a source, an attenuating device, a monochromator, a cell to hold the sample, a detector and amplifier, and a meter or recorder to observe the signal.

The simplest and least expensive arrangement is called a *single beam* instrument (Fig. 6.6A). However, if the source voltage fluctuates during a scan or the solvent absorbs certain wave lengths, an error is introduced into the spectra. That error can be corrected in double beam instruments by splitting the source radiation into two beams (Fig. 6.6B). Half of the time the beam passes through the reference cell containing everything but the sample, and the other half of the time the beam passes through the sample cell. Any source or solvent effects will now be registered by both the sample and the reference, and if the detector measures the difference between the signals, the spectrum obtained is due entirely to the sample.

The device used to split the source radiation into 2 beams is called a chopper and is a rotating half disc that has a mirror on 1 side. This disc turns at about 1000 times a minute or faster.

Single beam instruments can be used for routine sample analysis, but double beam instruments are required for good, fast, research work.

Solvents

Water is the most common solvent for the visible region. Methyl iso-butyl ketone and chloroform dissolve many organic compounds. In the ultraviolet region, cyclohexane is a good solvent for aromatic compounds and if a more polar solvent is needed, 95% ethanol is satisfactory; other solvents that transmit down to 200 nm are acetonitrile, cyclopentane, hexane, heptane, methanol, 2,2,4-tri-methyl pentane, and 2-methyl butane. Almost all of those compounds are saturated (sigma bonded) aliphatic organic materials. The transition for the solvent would be a $\sigma \rightarrow \sigma^*$ and this requires more energy than is available in the UV region. Therefore those materials are transparent in the UV region and do not interfere with the sample spectra.

Sources

For the visible region a tungsten lamp is satisfactory. Only about 15% of the energy is in the visible region; yet, those bulbs provide sufficient energy for most measurements in that region. Only a few percent of the radiation from a tungsten lamp is ultraviolet radiation. The most common sources for the UV region are the H_2 discharge lamp, the Hg vapor lamp, and more recently the deuterium (D_2) discharge lamp. Xe and Kr lamps are used for the vacuum UV.

Attenuating Devices

The attenuator regulates the amount of radiation coming from the source. Since different wavelengths have different energies and because mirrors, lenses, cells, solvents, and mono-chromators all absorb a certain amount of radiation, some means must be provided to vary the intensity of the radiation coming from the source. This can be done in several ways. One method is by a sliding wedge. As the wedge covers the slit, a decreasing amount of radiation can get through.

Another method is to vary the voltage to the lamp filament. The intensity of the radiation varies as:

$$I = kV^n \tag{6.2}$$

where I = intensity
$\quad k$ = proportionality constant
$\quad V$ = voltage
$\quad n$ = 3 to 4

which means that a small line voltage change will produce a large intensity change. This is why the meter needle moves continuously in many of the low-priced instruments. In the more expensive instruments, voltage regulators are used to eliminate the fluctuation.

Cells

The cells may be either round or square, and for the visible region are usually made of glass. Glass is not satisfactory in the UV region because it absorbs most of the radiation below 360 nm. Two good materials for the UV region are quartz and fused silica. Quartz has slightly better dispersing ability than fused silica, but fused silica transmits down to 185 nm compared to 200 nm for quartz.

One problem that always occurs is that Pyrex glass cells get mixed in with quartz cells and it is hard to tell them apart. A solution of trichloroethylene has a refractive index (1.475) equal to that of Pyrex (1.474), thus a Pyrex cell placed in this solution will disappear.

Be careful when using round cells that the cell is always placed in the instrument in the same direction every time. This is necessary because round cells are seldom uniform in diameter and the glass thickness varies.

When filling cells always fill them nearly full. If they are filled only a little past the light path it is possible to have reflection and stray-light errors.

Monochromators

Before the various types of monochromators are discussed it is necessary to understand what they are for. Figure 6.7 shows the absorption spectra of a purple $KMnO_4$ solution. Suppose radiation of the wavelengths indicated in Fig. 6.8 passes through a cell full of permanganate solution. Figure 6.8 is a simplification (if not mechanically correct) of what happens. In order to obtain 100% transmittance (T), we put 20 units of each of the 5 wavelengths through the sample. From Fig. 6.7 it can be seen that the violet, blue, yellow, and red wavelengths are essentially unaffected by the permanganate solution and pass through. The green is absorbed.

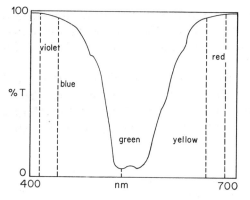

FIG. 6.7.   SPECTRA OF A PURPLE
SOLUTION

The result is that with the $KMnO_4$ the meter needle reads 75, a 25% T change.

It is apparent that only the green radiation is of any analytical value. The other wavelengths simply "clutter up" the detector. The lower part of Fig. 6.8 shows what happens when all but the green wavelengths are eliminated. Again 100 units of radiation enter the sample, but now all of the extraneous wavelengths are filtered out and only green gets through. The meter now reads 5%T, a 95%T deflection compared with a 25%T reading without the monochromator. Thus we have almost a fourfold increase in sensitivity.

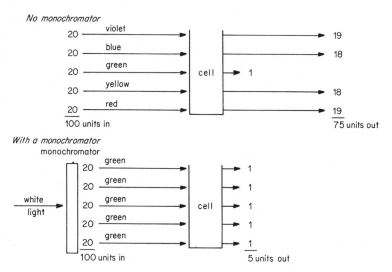

FIG. 6.8.   THE PURPOSE OF A MONOCHROMATOR

What is needed therefore for the most sensitive measurement is a device that will pass only one wavelength, that being the wavelength that interacts most intensely with the sample. Such a device is called a _monochromator_.

The least expensive instruments use as monochromators colored glass or solutions. Actually their band pass widths may be as much as 150 nm, or half of the visible region. A way of providing a much narrower band pass, 30–50 nm, is by using an interference filter shown in Fig. 6.9.

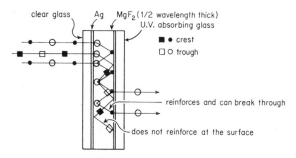

FIG. 6.9. INTERFERENCE FILTER (NOT TO SCALE)

The filter consists of 2 pieces of glass, each mirrored on 1 side and separated by a spacer. Imagine it as a jelly sandwich. The bread is the glass, the butter the silver mirror, and the jelly the spacer. When radiation strikes one glass surface and passes through it, the radiation now travels to the other side where it strikes the mirrored surface at an angle. If it does not have enough energy to get through the mirrored surface, it is reflected back to the opposite mirror, a process that may be repeated several times. Along the way the wave may be reinforced by another wave of the same wavelength. If their crests coincide at the mirrored surface, their combined energies break through the mirrored surface. This is _constructive interference_. Waves of other wavelengths cannot get in phase at the mirrored surface because of the space limitations set up by the spacer and cancel each other out, a process known as _destructive interference_. The radiation emerging from the exit side of the filter will not only be λ but $\lambda/2$, $\lambda/3$, $\lambda/4$, etc., since those wavelengths can also be in phase in the space between the glass. These are shorter wavelengths and of higher energy than the desired wavelength, and can cause a great deal of difficulty if they reach the detector. Therefore a piece of glass or other absorber is used to remove them and act as a secondary filter.

A way of producing more nearly monochromatic radiation is to use a prism, shown in Fig. 6.10. A single prism is capable of providing band widths of 10–20 nm. An advantage of a prism over an interference filter is that the prism provides a continuous spectrum over a much wider range. A disadvantage is that it has nonlinear dispersion. Figure 6.11 illustrates this effect. What this means is that 2 wavelengths separated by 5 nm at the short wavelength end of the spectrum are dispersed more than 2 wavelengths separated by 5 nm at the long wavelength end. This makes it mechanically difficult to scan the wavelengths.

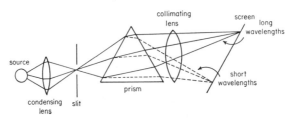

FIG. 6.10.   DISPERSION BY A PRISM

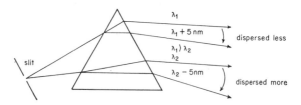

FIG. 6.11.   NONLINEAR DISPERSION

The beam of radiation emerging from the above prism can be passed through a second prism, and band widths of 3–5 nm can be obtained. If a third prism is used the radiation is highly monochromatic, but the energy is insufficient to activate a detector in a reasonable amount of time. Some Raman spectrographs use three prisms but it takes 3 to 4 days of exposure to see the spectra. This is not practical for routine analysis, and two prisms are generally the limit.

An entirely different approach for producing monochromatic radiation is to use a *diffraction grating.*

A grating is made by evaporating a thin metallic film onto a glass surface and then ruling lines in it to make openings. A good grating has 15,000 parallel smooth lines to the inch. The main disadvantage of transmission type grating is that the support

absorbs much of the radiation.   The reflection grating (Fig. 6.12)
does not have this disadvantage. λ_1 must travel further than λ_2
from the source to the viewer.   If this distance difference is equal
to one wavelength, constructive interference takes place.   Equation 6.3 gives the relationship necessary.

$$n\lambda = d \,(\sin i + \sin \theta) \tag{6.3}$$

where n = order number
The other terms are defined in Fig. 6.12.

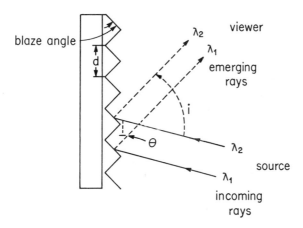

FIG. 6.12.   REFLECTION GRATING

Detectors

The main type of detector for the UV and visible regions is the
phototube.   Figure 6.13 shows a single stage phototube.

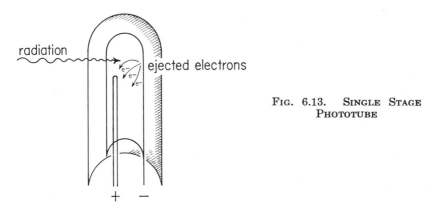

FIG. 6.13.   SINGLE STAGE
PHOTOTUBE

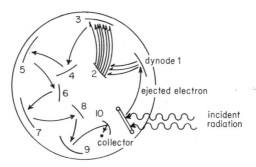

FIG. 6.14.   PHOTOMULTIPLIER TUBE

The incident radiation strikes the cathode which is coated with a mixture of cesium oxide, silver oxide, and silver.   The energy from 1 photon of radiation releases 1 to 4 electrons from the cathode surface.   Those electrons are attracted to the anode by applying a positive potential of 50–100 v.   This signal is amplified further until it is strong enough to move a recorder pen.   The detection and much of the amplification can be combined into 1 operation by a 10 stage photomultiplier shown in Fig. 6.14.   The initial process is the same as before except that now the anode is another electron emitting screen like the cathode of the single stage phototube. Each emitted electron is accelerated by a 100-v potential applied to each *dynode*.   The accelerated electron increases its energy to the point that when it strikes the 1st dynode it has enough energy to release several more electrons.   This process is repeated until dynode 10 is reached.   By now, several million electrons have been released and a corresponding amplification of the signal is obtained.

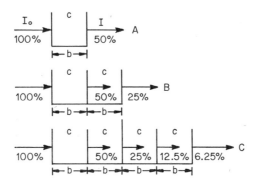

FIG. 6.15.   BEER'S LAW EXPERIMENT

Quantitative Analysis

Consider the glass cell in Fig. 6.15A having a thickness, b, and filled with a solution of concentration, c. The radiation entering the cell has an intensity I_0, and the emerging radiation has a value of I. The ratio of I/I_0 is called transmittance (T), and $I/I_0 \times 100$ = percent transmittance $(\%T)$. Since radiation I_0 has not entered the cell and none has been absorbed, it can be set arbitrarily at 100%. Assume that I is 50%. If the thickness of the cell is doubled (Fig. 6.15B), I drops to 25% or $1/2$ of 50 just as 50 was $1/2$ of 100. If the cell path is further increased the intensity keeps decreasing by the same factor.

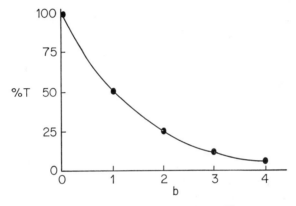

FIG. 6.16. VARIATION OF I WITH CELL THICKNESS

The curved line in Fig. 6.16 shows the plot of I vs cell thickness. An equation can be derived for this curve as follows.

$$-dI/db = kI \qquad (6.4)$$

The decrease in intensity, $-dI$, with respect to the cell thickness, db, is proportional, k, to the intensity, I. The variables can be rearranged

$$-dI/I = k\,db \qquad (6.5)$$

and give when integrated

$$-\ln I = kb + C \text{ (constant of integration)} \qquad (6.6)$$

when $b = 0$, $I = I_0$, therefore

$$C = -\ln I_0 \qquad (6.7)$$

and

$$-\ln I = kb - \ln I_0$$

This can be rewritten as

$$I/I_0 = e^{-kb} \qquad (6.8)$$

A similar set of experiments can be done with concentration as the variable, that is, going from c to $2c$ to $3c$, etc. An equation similar to (6.4) can be developed:

$$-dI/dc = k_1 I \qquad (6.9)$$

which gives an equation similar to (6.8) upon integration.

$$I/I_0 = e^{-k_1 c} \qquad (6.10)$$

The two equations, (6.8 and 6.10) can be combined:

$$I/I_0 = e^{k_{12} bc} \qquad (6.11)$$

To be useful as an analytical tool, the meter reading should double if the concentration doubles. This is not the case with equation (6.11). A new term, *absorbance*, (A) is defined as

$$\log I/T = \text{absorbance} = A = \log I_0/I \qquad (6.12)$$

Notice that this involves common logs (log), not natural logs (ln). After multiplying the log of both sides of equation (6.11) by 2.3 (to convert natural log to common log) we have

$$\log I/I_0 = -2.3\, k_2 bc \qquad (6.13)$$

Multiplying by -1 to get rid of the $-$sign and combining 2.3 k_2 nto a new constant, a, yields

$$\log I_0/I = abc \qquad \text{or } A = abc \qquad (6.14)$$

This is known as *Beer's law*.

a = absorptivity
b = cell thickness in cm
c = concentration in any convenient units. Usually this is mg/ml or μg/ml. However, if the work is to be published then c must be in moles/liter and then a is changed to ϵ (molar absorption).

How do you convert from A to $\%T$? Refer to equation (6.14) and rewrite it as

$$\log I_0 - \log I = abc = A \qquad (6.15)$$

I_0 is always set at 100%, so log I_0 is 2.   Log I is the $\%T$ read on the instrument.   Therefore:

$$2 - \log \%T = A \qquad (6.16)$$

Example

Caffeine in coffee and tea is determined at 276 nm by extracting the caffeine with dilute NH_4OH, separating it from other materials on a Celite column, and eluting it with $CHCl_3$.   A sample containing 0.5 mg caffeine/50 ml of $CHCl_3$ had a reading of $80\%T$, calculate the concentration of caffeine if the sample solution had a $\%T$ of 60.   10 mm cells were used.

For the standard

$$\begin{aligned}
A &= 2 - \log \%T \\
&= 2 - \log 80 \\
&= 2 - 1.903 \\
&= 0.097
\end{aligned}$$

For the unknown

$$\begin{aligned}
A &= 2 - \log \%T \\
&= 2 - \log 60 \\
&= 2 - 1.778 \\
&= 0.222
\end{aligned}$$

To determine the absorptivity, a, employ Beer's law using the experimental information obtained from the standard.

$$\begin{aligned}
A &= abc \\
0.097 &= a \times 1 \text{ cm} \times 0.5 \text{ mg/50 ml} \\
a &= 0.0194
\end{aligned}$$

The a thus obtained can only be used when concentrations are expressed in mg/50 ml.   For the unknown:

$$\begin{aligned}
A &= abc \\
0.222 &= 0.0194 \times 1 \text{ cm} \times c \\
c &= 1.14 \text{ mg caffeine/50 ml of } CHCl_3
\end{aligned}$$

Analysis of Mixtures

One often analyzes mixtures of two or more substances.   To avoid laborious separations, one can determine the components by using simultaneous spectrophotometric determinations.   This is based on the assumption that the absorbancies for each component can be added in a linear manner to the absorbancies of any other component.

Example

Chloral hydrate $Cl_3CCH(OH)_2$, is used as an anesthetic for cattle and is also the active component in familiar "knock-out drops" for humans.   Meat suspected of containing this drug may be examined by the method of Cabana and Gessner (1967) which determines not only the presence of chloral hydrate, but also the metabolites, trichloroacetic acid, trichloroethanol, and urochloralic acid. A 5-gm sample of steak was homogenized with 35 ml of 8% sulfosalicylic acid, centrifuged, and 1 ml of the supernatant liquid was added to 5 ml of pyridine. Two milliliters of 10 M KOH was added, the solution boiled for 5 min and a 3-ml aliquot of the pyridine layer was placed in a sample cell; 0.5 ml of water was added to clarify the solution. Standards of chloral hydrate and trichloroethanol were prepared in the same manner. The spectra obtained are shown in Fig. 6.17. Calculate the concentration of chloral hydrate and trichloroethanol in the meat.

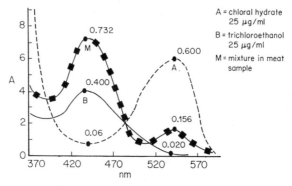

FIG. 6.17. SPECTRA OF CHLORAL HYDRATE, TRICHLORO-
ETHANOL, AND A MIXTURE OF THE TWO

We have 2 unknowns c_A and c_B, therefore, 2 equations are required in order to solve the problem. Refer to the mixture spectrum. The absorbance at 440 nm is the sum of the absorbance of A and the absorbance of $B;$ the same is true at 540 nm. Therefore:

$$A_{A(440)} + A_{B(440)} = A_{total(440)} \qquad (6.17)$$

$$A_{A(540)} + A_{B(540)} = A_{total(540)} \qquad (6.18)$$

We now have 2 equations but 4 unknowns. These equations must be rewritten in terms of concentration. This can be done by recalling the relationship, $A = abc;$ b, the cell thickness, remains constant throughout and can be dropped. The above equations can be rewritten in the following form;

$$a_{A(440)}c_{A(440)} + a_{B(440)}c_{B(440)} = A_{total(440)} \qquad (6.19)$$

$$a_{A(540)}c_{A(540)} + a_{B(540)}c_{B(540)} = A_{total(540)} \qquad (6.20)$$

Using the spectra of pure A and B, the absorptivities, a, can be obtained from $A = abc$.

For A

$$\text{at 440 nm} \quad 0.060 = a_A \times 25 \ \mu g/ml \text{ or } a_{A(440)} = 0.0024 \quad (6.21)$$

$$\text{at 540 nm} \quad 0.600 = a_A \times 25 \ \mu g/ml \text{ or } a_{A(540)} = 0.024 \quad (6.22)$$

For B

$$\text{at 440 nm} \quad 0.400 = a_B \times 25 \ \mu g/ml \text{ or } a_{B(440)} = 0.016 \quad (6.23)$$

$$\text{at 540 nm} \quad 0.020 = a_B \times 25 \ \mu g/ml \text{ or } a_{B(540)} = 0.0008 \quad (6.24)$$

These a values will remain the same for the unknown.

$$0.0024 \ c_{A(440)} + 0.016 \ c_{B(440)} = 0.732 \quad (6.25)$$

$$0.024 \ c_{A(540)} + 0.0008 \ c_{B(540)} = 0.156 \quad (6.26)$$

Multiply equation (6.26) by 20 to eliminate c_B,

$$0.0024 \ c_A + 0.016 \ c_B = \quad 0.732 \quad (6.27)$$

$$0.48 \ c_A + 0.016 \ c_B = \quad 3.120 \quad (6.28)$$

$$-0.4776 \ c_A \qquad\qquad = -2.388 \quad (6.29)$$

$$c_A = 5 \ \mu g/ml \quad (6.30)$$

Substituting this value of c_A into equation (6.25)

$$(0.0024 \times 5) + 0.016 \ c_B = 0.732$$

$$0.016 \ c_B = 0.732 - 0.012 = 0.720$$

$$c_B = 45 \ \mu g/ml$$

Since the standards were treated in the same manner as the sample, all dilution corrections are accounted for. This means that the 5-gm meat sample contained 5 μg of chloral hydrate and 45 μg of trichloroethanol, indicating that most of the drug was metabolized. A person eating a 16-oz steak would intake about 0.4 mg of the drug which is not enough to render unconsciousness.

If you have 3 unknowns in a sample, 3 equations and 3 pure spectra will be required. The method is limited to eight components. Even if a calculator is used, the spectrophotometric error is so large that though the equations are solved the accuracy is poor.

Selecting the Wavelength for Measurement

What is the optimum wavelength for a particular sample? Refer to Fig. 6.18.

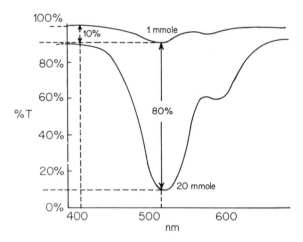

FIG. 6.18. OPTIMUM WAVELENGTH TO OBTAIN A MEASURE-
MENT

If a measurement is made at about 415 nm, a $10\%T$ change is obtained for a twentyfold concentration change, but if the measurement is made at about 515 nm, an $80\%T$ change is obtained for the same twentyfold concentration change. All other things being equal, the best wavelength to make a measurement at is the one that transmits the least radiation.

Proper Transmittance to Obtain a Measurement

Now that the best wavelength has been established in which to make a measurement, is there a best $\%T$? Figure 6.19 is a plot of $\%$ relative error vs $\%T$.

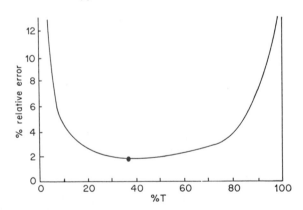

FIG. 6.19. RELATIVE ERROR VARIATION WITH TRANSMITTANCE

For a solution that obeys Beer's law the best $\%T$ is 36.7%. This is the place where the most accurate reading can be obtained. However, notice that readings between 20% and 80%T are still reasonably accurate. What is the procedure if the readings are not between 20 to 80%T? If only an occasional sample is to be analyzed, it may be easiest to either dilute or concentrate the sample. However, if the determination is to be done routinely, then the use of *precision spectrophotometry* should be considered.

Precision Spectrophotometry

A way to expand the scale so a more accurate reading can be made was developed by Hiskey (1949). If it is desired to compare two highly colored solutions using water as a blank, the water reference transmits so much more radiation than the colored solutions that small differences between the two solutions go undetected. If instead of a colorless reference, a reference solution that is colored is used, the difference between the two highly colored solutions is more pronounced and a more precise reading is obtained.

With dilute solutions, the reverse is true in that the black background of the dark current or zero reference should be replaced by a solution more transparent than the completely opaque shutter.

Examples of this are illustrated in Fig. 6.20A.

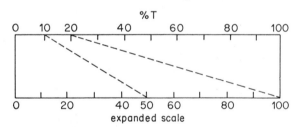

FIG. 6.20A. SCALE EXPANSION FOR SOLUTIONS HAVING A
LOW %T READING

Take for example two solutions, a standard that measures 20%T *vs* a water blank, and an unknown that measures 10%T *vs* water. In this case a small photometric error will make a large analytic error. Instead of using water as a reference for the 100%T setting, use the standard solution (20%T) and set the instrument so that it reads 100%T. The scale has been expanded fivefold (20 to 100) and now the solution that previously read 10% will read 50%. This has increased the precision fivefold.

For very dilute solutions a similar situation exists. Here the standard shows an 80%T and the unknown 90%T. In this case the scale is expanded by setting the zero of the instrument scale with a standard solution that is a little more concentrated than the unknown (Fig. 6.20B).

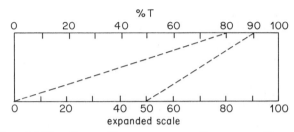

FIG. 6.20B. SCALE EXPANSION FOR SOLUTIONS HAVING HIGH % T READINGS

Some Recent Applications

Howard *et al.* (1968) have developed a method for determining benzopyrene in smoked foods. Benzopyrene is a carcinogen and is found in the curing smoke. After a cleanup procedure, a measurement at 387 nm can detect as little as 10 μg benzopyrene per kg food.

Sodium lauryl sulfate can be detected in egg whites in the 0.025% range if the material is complexed with Azure A, extracted, and measured at 637 nm (Wiskerchen 1968).

Wilderman and Shuman (1968) found that DDT residues in firm fruits, such as apples and peaches, could be recovered (95%) if they were extracted with isopropanol-hexane, 1:2, followed by chloroform-methanol, 1:1. When $AgNO_3$ and 2-phenoxyethanol is added, the blue color is measured at 450 or 510 nm.

The use of differential spectrophotometry was used by Ferren and Shane (1968) to determine caffeine in coffee, tea, and cola drinks. By this technique, separation of otherwise interfering alkaloid purines, theobromine, theophylline, and amino acids is not required and detection limits are lowered to 5 μg/gm.

Associated Problems

(Answers are to slide rule accuracy.)

(1) What is the absorbance of a solution that has a % transmittance of 82? Ans. 0.086

(2) Assume the molar absorptivity, ϵ, of bis(2,2'-biphenyl) iron (II) is 10,000. If the iron in a bread sample was determined by this method and you used 2-cm cells, what minimum concentration

of iron could you detect if the lowest absorbance you can measure is 0.005? Ans. 2.5×10^{-7} moles/liter

(3) When 3.00 μg of tris(1,10-phenanthroline)iron(II) is dissolved in 10 ml of solution the absorbance at 515 nm was 0.20; 10 mm cells were used. Calculate the $\%T$ and the molar absorptivity of this chelate. Ans. $\%T = 63$, $\epsilon = 4.3 \times 10^5$

(4) Butylated hydroxyanisole, BHA, (mw = 180) is a fat antioxidant. The current law permits no more than 200 ppm BHA. Assume that ϵ is 8000 (1 cm cells) and that the density of fat is 0.9. What size sample would you have to take to have a reading of $37\%T$? Ans. 48.5 gm

(5) The $\%T$ of separate solutions of colored substances T and Y, each at a concentration of 1×10^{-3} M, are measured at the wavelengths shown below. An unknown solution containing these two components is also measured. From the data below calculate the concentration in mg/ml of T and Y if their molecule weights are 340 and 468, respectively.

Solution	$\%T$ at 480 nm	$\%T$ at 734 nm
T	50	90
Y	90	25
Unknown	75	40

Ans. Y = 0.32 mg/ml, T = 0.105 mg/ml

BIBLIOGRAPHY

CABANA, B. E., and GESSNER, P. K. 1967. Determination of chloral hydrate, trichloroacetic acid, trichloroethanol, and urochloralic acid in the presence of each other and in tissue homogenates. Anal. Chem. 39, 1449–1452.

FERREN, W. P., and SHANE, N. A. 1968. Differential spectrophotometric determination of caffeine in soluble coffee and in drug combinations. J. Assoc. Offic. Anal. Chemists 51, 573–577.

HISKEY, C. F. 1949. Principles of precision colorimetry. Anal. Chem. 21, 1440–1446.

HOWARD, J. W., FAZID, T., and WHITE, R. H. 1968. Collaborative study of a method for benzo (γ) pyrene in smoked foods. J. Assoc. Offic. Anal. Chemists 51, 544–548.

RAO, C. N. R. 1967. Ultraviolet and Visible Spectroscopy. Butterworths, London.

WILDERMAN, M., and SHUMAN, H. 1968. Extraction procedures for DDT residues in firm fruits and in hay, colorimetric method. J. Assoc. Offic. Anal. Chemists 51, 892–895.

WISKERCHEN, J. 1968. Collaborative study of a method for sodium lauryl sulfate in egg whites. J. Assoc. Offic. Anal. Chemists 51, 540–543.

CHAPTER 7

Color of Foods

Color and discoloration of many foods are important quality attributes in marketing. Although they do not necessarily reflect nutritional, flavor, or functional values, they relate to consumer preferences based on the appearance of the product. Color characteristics of foods can result from both pigmented and originally non-pigmented compounds.

The broad area of color in foods may be divided into two general problems: (1) addition of approved synthetic colors to achieve a desired appearance, and (2) determination of natural pigments (Francis 1963). In the first, acceptability, uniformity, and reproducibility are most important. In the second we are concerned with color as an index of economic value (i.e. in grading) or with control of color in processing and storage.

Color is often used to determine the ripeness of fruit (green color in tomatoes or peaches; white color in lima beans). Color of potato chips is largely controlled by the reducing sugars content, by storage conditions of the potatoes, and by subsequent processing. The yellow color of egg yolk and of the skin of chicken is a direct function of the amount of pigment present. The color of poultry meat also can be influenced by the amount of finish put on the bird. Inclusion of xanthophyll pigments or antioxidants in poultry feed can influence skin and shank color. The brown discoloration in frozen turkeys is independent of the pigment quantity, but is related to freezing rate. Color is one of the more prominent variables apparent in raw and cured meats.

Flour color measurements are used in many countries to determine flour grade. Flour color is primarily affected by the extraction rate and the amount of bran in the flour. Additional factors are particle size, color characteristics (i.e. carotenoid pigment contents) of the endosperm and the skin of the wheat kernel from which the flour is milled, and efficiency of wheat cleaning (to remove dirt, smut, weeds, etc.) prior to milling.

Optical Aspects

Man can distinguish a multitude of colors. The main problem in

72

color determination is to develop an objective, precise, and reproducible procedure for measuring quantitatively the differences perceived by a trained observer.

Color is a characteristic of light, that is measurable in terms of intensity and wave length. It arises from the presence of light in greater intensities at some wavelengths than at others. In practice, it is limited to the band of the spectrum from 380 to 770 nm, the part of the electromagnetic spectrum that is visible to the human eye. Precise measurement of color in foods can be made and the color expressed in terms of internationally accepted units provided the measurement is not complicated by appearance factors such as gloss, mottling, and texture (Kramer and Twigg 1970). Color is an appearance property attributable to the spectral distribution of light; gloss, transparency, haziness, and turbidity are properties of materials attributable to the geometric manner in which light is reflected and transmitted.

From the point of view of optics, color is the stimulus that results from the detection of light after it has interacted with an object. Thus, three factors are involved: a light source, an object, and a receiver-detector. For standard comparisons, three standard illuminants (light sources) have been established by the International Commission on Illumination (called CIE from the initials of the French name, Commission Internationale de L'Eclairage). These are: Illuminant A—incandescent lamp at 2848°K; Illuminant B—noon-sunlight lamp at 5000°K; and Illuminant C—average daylight at 6740°K.

The light may be reflected, transmitted, absorbed, or refracted by an illuminated object. If practically all radiant energy in the visible range is reflected from an opaque surface, the object appears white. If the light through the entire visible spectrum is absorbed in part, the object appears to be gray. If it is absorbed almost completely the object is black.

The term *lightness* (or *value*) describes the relation between reflected and absorbed light, without regard to specific wavelength. *Hue* is an aspect of color which we describe by words as green, blue, yellow, or red. This perception of color results from differences in absorption of radiant energy at various wavelengths. Thus, if the shorter wavelengths of 400 to 500 nm are reflected to a greater extent than the other wavelengths, the color is described as blue. Maximum reflection in the medium wave length range results in a green or yellow color; and maximum reflectance at the longer wave lengths (600 to 700 nm) indicates red objects. *Chroma, saturation,* or *purity* describes reflection at a given wavelength and shows how much a color differs from gray. It describes, for example, how a red brick differs from a red tomato if the two are of the same lightness and hue. Visual perception of color as described by the three variables hue, value, and chroma can be arranged in a cylindrical coordinate system, as shown in Fig. 7.1. If the light is reflected from a surface evenly at all angles, we have the impression of a product with a flat, dull, or diffuse appearance. If the reflection is stronger at a specific angle or in a beam we observe gloss or sheen as a result of *specular* or directional reflectance.

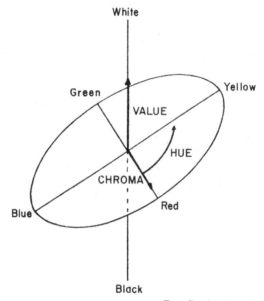

From Bernhardt (1969)

FIG. 7.1.   VISUAL PERCEPTION OF COLOR

This can be described in a simple case by the three variables hue, value, and chroma that are arranged in the figure in a cylindrical coordinate system.

The standard system against which other systems should be compared is that proposed by CIE in 1931. The system uses a chromaticity diagram representing three-dimensional space for a standard illuminant. The CIE system specifies a color by three quantities—X, Y, and Z—called tristimulus values. Those values represent the amounts of three primary colors—red, green, and violet—that are required for a standard observer to get a match. If each of the tristimulus values is divided by the sum of the three, the resulting values x, y, and z, called chromaticity coordinates, give the proportion of the total stimulus attributable to each primary. Since the sum of the three is unity, the values x and y alone can be used to describe a color, and when plotted on a two-dimensional area result in a chromaticity diagram. The third dimension, of lightness or darkness, is defined by the Y tristimulus value. The chromaticity diagram of the CIE system is shown in Fig. 7.2.

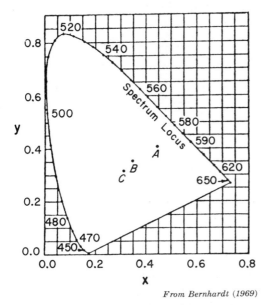

FIG. 7.2. CHROMATICITY DIAGRAM OF THE CIE SYSTEM

All real colors lie within the horseshoe-shaped locus marked with the wavelengths of the spectrum colors.

From Bernhardt (1969)

Several recording spectrophotometers draw a continuous curve of light reflected by a sample for the visible spectrum range. The essential components of a reflectance spectrophotometer are shown in Fig. 7.3. The CIE has recommended a standard method of reducing by calculation the spectrophotometric curve across the entire range of visible wave lengths to three numbers that adequately and accurately describe the color. For this purpose, CIE has established three curves, called color mixture curves, that represent the color vision of the standard observer under standard light conditions. For a color specification, the reflectance of a sample at wave length intervals of 10 nm is multiplied by the respective values of X, Y, and Z for a standard light source. This shows how much energy of each wave length of the illuminant source is reflected from the sample. Next, the products of X, Y, and Z are totaled for each wavelength. Then x, y, and z are determined from the ratios:

$$x = \frac{X}{X + Y + Z} \qquad y = \frac{Y}{X + Y + Z} \qquad z = 1 - (x + y) \quad (7.1)$$

This calculation is quite laborious and has been replaced by tristimulus integrators attached to the spectrophotometer.

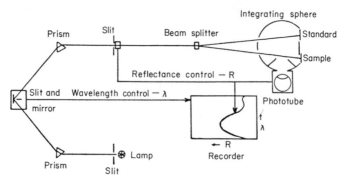

From Bernhardt (1969)

FIG. 7.3. THE ESSENTIAL COMPONENTS OF A REFLECTANCE SPECTROPHOTOMETER

A white-coated integrating sphere is used to collect the light reflected from the surface of the sample and direct it to the phototube.

The Sample

As mentioned previously, in practice it is also important to know the effect of physical characteristics of a surface on its resultant reflectance. Color, as complex as it is, is only one aspect of the appearance of an object. Gloss, texture, relative opacity, and surface uniformity are also important. Thus, colors of objects with different appearances can be matched only for a given set of viewing conditions. It is virtually impossible, for instance, to match the appearance of a flat surface to a textured one.

The preparation of a sample for color measurements is dictated by the specific requirements for which it is evaluated (Mackinney and Chichester 1954; Little and Mackinney 1969). This requirement ranges from the absolute restriction that the sample be intact and unaltered to a wide latitude in sample alteration.

In selection of raw materials, the objective is a rapid, preferably on-line evaluation of unaltered or minimally altered samples. The methods and instruments are adapted to give rapid acceptance or rejection of a single commodity. The sample can be altered provided the answer that is obtained is relevant to the evaluation.

An ideal sample for measuring reflectance is flat, homogenously pigmented, opaque, and light-diffusing. Practically all foods are irregular in shape, texture, particle size, and surface characteristics; are both light-transmitting and light-diffusing; and show inhomogenous pigment distribution. Ideal samples for transmittance measurements are clear and moderately light-absorbing. In practice, those criteria are seldom met.

Compressing a dry powdered sample (i.e. flour) to a pellet gives an approximately ideal reflecting surface. Some products, i.e. coffee beans, can be ground and compressed but the procedure is difficult to reproduce. Some dried foods (lyophilized egg whites or yolks, instant coffee, sweet potato flakes, and gelatin desserts) can be pressed into wafers between thin discs of Teflon (Berardi et al. 1966). Measurements of the surface reflectance characteristics of thin layers of high-moisture homogenized samples are affected by their translucency, depth of layer, and background. Resolution of spectral reflectance curves of homogenized tuna samples is greatly improved by treating the surface of the sample with sodium dithionate (Little and Mackinney 1969).

Simplified Tristimulus Color Systems

Color identification with a recording spectrophotometer is costly and time consuming. More economical and practical (for routine purposes), though less accurate, tristimulus systems are available.

In the *Munsell system*, color is determined by using scales of hue, value, and chroma. Hue is defined on the Munsell color tree in the circumferential direction by 5 main and 5 intermediate hues for decimal notation. Chroma (the degree of saturation) is defined by 10 or more steps in the radial direction, and value (the degree of lightness) by 10 steps in the vertical direction. In the use of the Munsell system, 3 or 4 overlapping discs are used (Nickerson 1946)—one each for hue and chroma, and 1 or 2 to provide adjustment for value. The discs are adjusted until the color obtained by rapid spinning of the discs, matches the color of the tested object. The results may be reported as such, as a ratio of the two chromatic discs, or may be transformed to the CIE system values by calculation (Essau 1958) or by the use of chromaticity diagrams on which Munsell notations are superimposed (Billmeyer 1951).

Nickerson (1946) described an instrument that used a Munsell disc colorimeter in which the angle, intensity, and type of light source are controlled. To reduce the time required for color matching, the lens is rotated instead of rotating the discs. A colorimeter with discs of specified Munsell notations is also available.

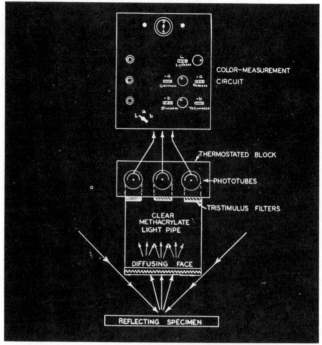

Courtesy Hunter Associates Laboratory, Inc.

FIG. 7.4.   BLOCK DIAGRAM OF THE TRISTIMULUS
HUNTER COLOR METER

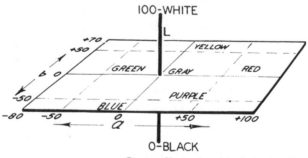

Courtesy Hunter Associates Laboratory, Inc.

FIG. 7.5.   THE AXES OF THE L, a, b COLOR VALUES OF THE
HUNTERLAB COLOR METER

The tristimulus photoelectric colorimeter developed by Hunter (1952) is relatively inexpensive, rugged, and well-adapted to routine tests.   The instrument consists of three separate circuits, filters, and photocells (Fig. 7.4).   The Hunter R_d (diffuse reflectance) or L (lightness) values are directly comparable to the Y of the CIE system or value of the Munsell system (Fig. 7.5). The Hunter a value denotes redness or greenness; the Hunter b values measure yellowness or blueness. The a values are functions of X and Y, the b values of Z and Y. Determining the a and b values provides information equivalent to that of determining the hue and chroma dimensions of the Munsell system. Hunter and CIE values are interconvertible by calculation (Hunter 1958). Similarly, Hunter values can be converted to Munsell notations mathematically (Davis and Gould 1955) or by use of charts (Billmeyer 1951).

Courtesy Lovibond af America, Inc.

FIG. 7.6. THE LOVIBOND TINTOMETER

The *Lovibond Tintometer* (Fig. 7.6) is a visual colorimeter used widely in the oil industry. The instrument has a set of permanent glass color filters in the three primary colors—red, yellow, and blue. The colors are calibrated on a decimal scale in units of equal depth throughout each scale. The oil sample is placed in a glass cell and the filters are introduced into the optical system until a color match is obtained under specified conditions of illumination and viewing. Color of oils is measured with transmitted light. Either transmitted light for molten waxes and fats, or reflected light against a white background for solid samples is used. Attachments to the Lovibond Tintometer provide means of converting the readings from Lovibond units into trichromatic coordinate values. In addition, modifications for determining color of liquid, solid, powder, or paste materials are available.

Abridged Methods

The greatest accuracy is generally obtained in a function involving all three coordinates. In some cases, however, the work and com-

putations may not be justified by the small increase in accuracy.   In
some instances, color can be determined satisfactorily by abridged
methods instead of determining three dimensions of color.

In frozen lima beans, the Hunter L value (or Y of the CIE system)
are correlated with panel scores.   With applesauce, however, all
three Hunter values (L, a, and b) contribute to consumer evaluation,
since the multiple correlation including all tristimulus values is
significantly higher than any single value or any combination of two
values.   The effect of color on grade score of lima beans can be
determined directly from a simple table that compares the Hunter L
value with scores for three grades (Kramer and Hart 1954).   For
determining U.S. grade for color of applesauce, nomographs are
available.

The color of raw tomatoes can be measured with the Agtron E
colorimeter from the reflectance ratio at 546 and 640 nm of cut surfaces of two halves of a tomato (Smith and Huggins 1952) or from
determination of a single reflectance value on Agtron F (Smith
1953).

An approximation of the color of tomato juice can be obtained
from a reading of percent reflectance at one wavelength of Agtron F
(Gould 1954).   The correlation with panel scores is improved if two
color attributes (or their ratio) such as a (redness) and b (yellowness)
readings of the Gardner color and color difference meter are determined (Kramer 1954).   A still more accurate figure representing the
color can be obtained by including a function of all three color
parameters.   Such a score developed by Yeatman et al. (1960) for
grading purposes can be described by the function

$$\frac{aL}{\sqrt{a^2L + b^2L}}/L \qquad (7.2)$$

An instrument called the "tomato colorimeter" was developed to
measure the color of tomato juice in terms of the above equation
(Hunter and Yeatman 1961).

The brightness, or tristimulus y is a useful index of the degree of
roasting, development of flavor, and the amount of extractable coffee
solids (Francis 1963).   For fine control of color development, however, roasting to a given color is inadequate.   The time-temperature
relations in roasting must be controlled, and the relations between
visual appearance and tristimulus reading must be measured.
Whatever method is used, it is always advisable to select the appropriate parameter for the abridged procedure after tristimulus color
data of the product have been correlated with panel scores.

Aqueous solutions of pure sucrose are clear and water-white. Such solutions may display a slight coloration and haze from traces of impurities. Since the coloration is low, it is difficult to express in terms of tristimulus coordinates or related systems. The color is, therefore, commonly expressed in terms of an absorption index at a specified wavelength (Bernhardt 1969). When the solution has a haze, the transmitted light is attenuated through scattering and the absorption index cannot be determined in a single measurement. The scattering (turbidity) is the result of differences in the refractive indices of the solution and the trace impurities suspended in the solution. Total attenuation index of sugar solutions measures the contributions of absorption and turbidity. Total attenuation can be determined by measuring light transmitted by the colored and scattering solution. The turbidity can be determined independently of absorption by measuring scattered light. The absorption index (light attenuation due to color) can be calculated from the difference between the total attenuation and scattering. In determining the attenuation index through transmittance measurements, serious errors may occur if some of the scattered light is permitted to enter the photo detector. The scatter error is reduced in a special photometer designed by Bernhardt *et al.* (1962).

Simple color grading equipment using predetermined fixed color ranges is available. For rapid grading of raw materials (i.e. lima beans, maple syrup, or honey) standard colored plastic blocks, visually equivalent to the desired color, are used.

Determination of Pigments

Methods for extraction of pigments and measurement of their concentration at specified wavelengths in a photoelectric colorimeter or spectrophotometer have been developed (Kramer 1965). They are satisfactory provided the color impression of a food is correlated with the extracted pigment. The limitation of the method lies in the color as seen by the eye not being closely correlated with the concentration of the extracted pigment, and an undue emphasis on total pigment contents whereas in practice, color of the outer surface is responsible for consumers' color impressions.

Thus the value of spectrophotometric determination of pigments in meat extracts is limited. The published methods measure the proportion of various forms of myoglobin in the total sample, whereas generally only the surface color is evaluated by the consumer. There are, however, many instances in which determining transmittance of light at a specified wavelength of a pigment extract is satisfactory. In practice, the pigment is extracted with appropriate solvent(s) in a blender, the extract is diluted, cleared by centrifugation, and transmittance is determined at a wavelength giving maximum absorption for the pigment.

A translucent solution transmits its own color and absorbs the complimentary color. Red colored solutions absorb most in the green, yellow solutions in the blue-violet, and blue solutions in the orange-red. The major water-soluble red pigment in beets is betanin. A purple solution of beet juice transmits the purple color but absorbs the complimentary color. Consequently, at 525 nm (green) transmittance is lowest and absorption is highest. At 700 nm (red) practically all the light is transmitted. The greater the pigment concentration, the lower the percent transmittance of the complimentary color. In most analytical assays, the transmittance of the complimentary color is measured (Kramer and Smith 1946).

Beta-carotene is important in sweet potatoes as a nutritional component (precursor of vitamin A) and as a contributor to the pleasant orange color. Correlation between beta-carotene and color is high, provided effects of moisture contents, pithy breakdown, and texture are accounted for. Generally beta-carotene comprises up to 90% of the total carotenoid pigments in sweet potatoes. If the pigment composition varies widely, both the type and quantity of each major pigment must be determined and weighted for their contribution to the color (Francis 1969).

Carotenoid pigments and their stability are important in macaroni manufacture. Extraction of the pigments from durum semolina with water-saturated butanol or naphta-alcohol and absorbance measurements at 440 nm are performed routinely.

Cranberry juice contains 4 major red anthocyanin pigments, a number of minor red pigments, and 6 yellow flavonoid pigments. Good chemical methods of estimating anthocyanin content of cranberry and cranberry products are available (Francis 1957). Simplified colorimetric methods can be used to determine the pigment content of fresh juice, but the method must be modified in assay of stored juice in which pigment degradation has taken place. Spectrophotometry can be used to determine the color of wine and pigment degradation during storage and processing.

BIBLIOGRAPHY

ANON. 1966. Color matching scores. Chem. Eng. News 50, 51-52, 55-56.
BERARDI, L. C., MARTINEZ, W. H., BONDREAUX, G. J., and FRAMPTON, V. L. 1966. Rapid reproducible procedure for wafers of dried food. Food Technol. 20, 124.

BERNHARDT, W. O.   1969.   Color and turbidity in solutions.   Food Technol. *23*, No. 1, 30–31.
BERNHARDT, W. O., EIS, F. G., and McGINNIS, R. A.   1962.   The sphere photometer.   J. Am. Soc. Sugar Beet Technol. *12*, No. 2, 106.
BILLMEYER, F. W.   1951.   Nomographs for Converting Hunter Color Values to C. I. E. Values.   E. I. duPont de Nemours and Co., Wilmington, Del.
BILLMEYER, F. W.   1967.   Optical aspects of color.   Opt. Spectra 1, No. 2, 59, 61–63.
COMMISSION INTERNATIONAL DE L'ECLAIRAGE.   1931.   Proceedings of Eighth Session.   Cambridge, England.   Sept. 19–29.   Cambridge Univ. Press.
DAVIS, R. B., and GOULD, W. A.   1955.   A proposed method for converting Hunter color difference meter readings to Munsell hue, values, and chroma notations corrected for Munsell values.   Food Technol. *9*, 536–540.
ESSAU, P.   1958.   Procedures for conversion of color data for one system into another.   Food Technol. *2*, 167–168.
FRANCIS, F. J.   1957.   Color and pigment measurement in fresh cranberries.   Proc. Am. Soc. Hort. Sci. *69*, 296.
FRANCIS, F. J.   1963.   Color control.   Food Technol. *17*, No. 5, 38–42, 44–45.
FRANCIS, F. J.   1969.   Pigment content and color in fruits and vegetables.   Food Technol. *23*, No. 1, 32–36.
GOULD, W. A.   1954.   Simplified color instrument now available to industry.   Food Packer *35*, No. 1, 33.
HUNTER, R. S.   1952.   Photoelectric tristimulus colorimetry with three filters.   U.S. Dept. Comm. Natl. Bur. Std. (U.S.).   Circ. *C 429*.
HUNTER, R. S.   1958.   Photoelectric color difference meter.   J. Opt. Soc. Am. *48*, No. 12, 985–995.
HUNTER, R. S., and YEATMAN, J. N.   1961.   Direct-reading tomato colorimeter.   J. Opt. Soc. Am. *51*, 552.
KRAMER, A.   1954.   Color dimensions of interest to the consumer.   Quartermaster Food and Container Institute, Symposium-Color in Foods.   Chicago, Ill.
KRAMER, A.   1965.   Evaluation of quality of fruits and vegetables.   *In* Food Quality, Effects of Production Practices and Processing, G. W. Irving, Jr., and S. R. Hoover (Editors).   Publ. *77*, Am. Assoc. Advan. Sci. Washington, D.C.
KRAMER, A., and HART, W. J.   1954.   Recommendations on procedures for determining grades of raw, canned, and frozen lima beans.   Food Technol. *8*, 55–62.
KRAMER, A., and SMITH, H. R.   1946.   Preliminary investigation and measurement of color in canned foods.   Food Res. *11*, 14–31.
KRAMER, A., and TWIGG, B. A.   1970.   Quality Control for the Food Industry, 3rd Edition, Vol. 1.   AVI Publishing Co., Westport, Conn.
LITTLE, A. C., and MACKINNEY, G.   1969.   Colorimetry of foods—the sample as a problem.   Food Technol. *23*, No. 1, 25–28.
MACKINNEY, G., and CHICHESTER, C. O.   1954.   The color problem in foods.   Advan. Food Res. *5*, 301–551.
MACKINNEY, G., and LITTLE, A.   1962.   Color of Foods.   AVI Publishing Co., Westport, Conn.
NICKERSON, D.   1946.   Color measurement and its application to the grading of agricultural products.   U.S. Dept. Agr. Misc. Publ. *580*.
SMITH, T. J.   1953.   Tomato grading by electronics.   Food Eng. *25*, No. 9, 53.
SMITH, T. J., and HUGGINS, R. A.   1952.   Tomato classification by spectrophotometry.   Electronics *25*, No. 1, 92.
YEATMAN, J. N., SIDWELL, A. P., and NORRIS, K. H.   1960.   Derivation of a new formula for computing raw tomato juice color from objective color measurement.   Food Technol. *14*, 16.

CHAPTER 8

Fluorescence and Phosphorescence

The phenomenon of *luminescence* was first studied by Stokes in about 1852. Stokes noticed that when fluorspar was placed in the sun it seemed to *glow*. The orange jackets of highway workers that appear very bright even on cloudy days, road signs that "glow in the night," and the afterglow when a television set is turned off are modern examples of this phenomenon. *Luminescence* is a general term to describe systems that can be made to *glow*. Those systems can be further divided according to the *glow* producing mechanism. The two major divisions of present-day analytical importance are *fluorescence* and *phosphorescence*.

In general, compounds that fluoresce or phosphoresce are those that contain either an electron donating group (amines, alcohols, and hetero atoms) or multiple conjugated double bonds (aromatic rings). Notice that these groups contain either nonbonding or pi electrons. The presence of groups that tend to withdraw electrons such as carboxyl, azo, the halides and nitro groups usually destroy fluorescence. Figure 8.1 shows the energy levels involved in fluorescence and phosphorescence.

Consider a molecule having a pair of nonbonding electrons in which one electron is spinning opposed to the other. Such a state is called a *singlet state*. It generally requires radiation energy in the ultraviolet region to raise one of these electrons to an excited state (1 to 2 in Fig. 8.1). This process takes about 10^{-15} sec and happens so fast that the molecule does not have time to change its basic geometry. The electron may immediately lose all of the energy it just acquired and return to its original energy level, a process known as *resonance fluorescence* (2 to 1 in Fig. 8.1).

Vibrational relaxation is the process by which a molecule in the excited state loses energy by dropping back to the lowest vibrational level in the excited state (2 to 3 in Fig. 8.1). In solution this excess energy can be taken up by the solvent molecules and no radiation is given off. The entire process takes from 10^{-11} to 10^{-13} sec. This process is so efficient that it can be said that molecules in solution will emit radiation from the lowest vibrational level of an excited

84

state.   The lifetime of molecules in this lowest level of the first excited state is from 10^{-7} to 10^{-9} sec.

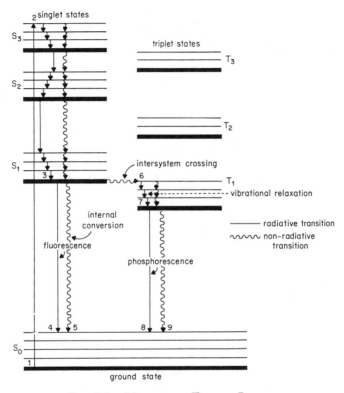

FIG. 8.1.   MOLECULAR ENERGY LEVELS

If the molecule now returns to the ground state with the emission of radiation (3 to 4 in Fig. 8.1) the process is called *fluorescence*. Note that the energy emitted is less than the amount of energy originally absorbed.   This emitted energy is usually in the visible or near infrared region.   Our eyes cannot detect the ultraviolet radiation that started the process, but they can detect the lower energy visible radiation that ends the process and this is why certain substances tend to "glow."   Physically, fluorescence is regarded as an instantaneous process, that lasts about 10^{-8} sec.

Intersystem crossing is the process of changing from a singlet state (paired electron spins) to a triplet state (unpaired electron spins) as shown in 3 to 6 in Fig. 8.1.   This can be very fast, of the order of 10^{-8} sec, and is enhanced by the presence of heavy atoms within the molecule.   Fast as it is, intersystem crossing is still much slower than vibrational relaxation and as a result, intersystem crossing usually occurs after the molecule has reached the lowest vibrational level in the singlet excited state. After the intersystem crossing has taken place the molecule immediately (10^{-13} sec) goes to the lowest vibrational level by vibrational relaxation. The lifetime of this lowest triplet state varies from 10^{-4} to 10 sec which is very long compared to the lowest excited singlet state. If the molecule drops from the triplet state to the ground state with the emission of a quantum of radiation, the process is called phosphorescence (7 to 8 in Fig. 8.1).

Physically this is a slow process and phosphorescence is usually associated with compounds that tend to have an afterglow, that is they glow after the source of exciting radiation is shut off.

Because the triplet level lifetime is relatively long, there is a greater chance for the molecule to lose its energy through a radiation-less transition either by vibrational coupling or by collision. The latter is so common that few systems will phosphoresce at room temperature and they must be cooled to liquid N_2 temperature ($77°K$) before the collisions are reduced to such a degree that phosphorescence can occur.

Fluorescence Instrumentation

The source must be an ultraviolet source so the hydrogen, deuterium, or mercury lamps discussed previously can be used. Recently, xenon arcs are becoming more popular. Quartz or fused silica optics and cells are required. Since the fluorescent radiation is in the visible region a multiplying phototube can be used for the detector. As these components are all part of an ultraviolet spectrophotometer, what then is the difference between an ultraviolet spectrophotometer and a fluorimeter? The difference is shown in Fig. 8.2. Note that all that is necessary is to move the detector so that the radiation *that comes from the side of the sample* is measured. Fluorescent radiation is emitted in all directions and advantage is taken of this fact in designing a fluorimeter. It is possible to measure the amount of fluorescence that *comes out of the sample* in the same direction as the incident radiation. This requires a very good monochromator to separate the excitation radiation from the fluorescent radiation, and can be done accurately only with great difficulty. It is comparable to weighing a ship and then the ship and the captain to determine the weight of the captain. However, by moving the detector so that it will be at a right angle to the source radiation only the fluorescent radiation is measured and much more accurate results can be obtained.

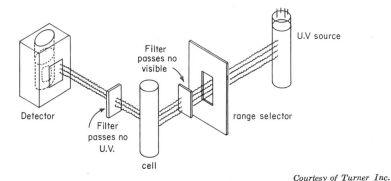

Courtesy of Turner Inc.

FIG. 8.2.   DIAGRAM OF A TURNER FLUORIMETER

Experimental Precautions

Some compounds are destroyed by the intense ultraviolet radiation that is used to excite the molecules.   Consequently, the slow analyst may find it difficult to get reproducible results.

The adjustment of pH is very important.   The nonbonding electrons can be protonated thereby making them no longer nonbonding, and the fluorescent spectra can be shifted considerably or even destroyed.

Temperature is another factor that must be controlled.   One of the difficulties with fluorescing molecules is that most of them lose their energy by collision rather than by fluorescing.   A good way to reduce molecular collision is to solidify the material by freezing it. Therefore if a system does not fluoresce or does so weakly, place the sample and the cell into a dry ice-acetone bath, or better yet, into liquid nitrogen.   The sample is frozen in a few seconds and in many cases the fluorescence intensity increases 10 to 1000 times.   The aluminum 8-hydroxyquinoline system can be changed quite noticeably just by cooling it with the refrigerated water from a drinking fountain.   With liquid nitrogen, readings must be taken within a few seconds after the cell is removed from the freezing solution because the cold cell will freeze moisture from the air onto its surface and ruin the analysis.   This can be reduced considerably by blowing dry air through the cell compartment.

Self-absorption is a problem at high sample concentrations. The emitted fluorescent radiation from one molecule has just the right energy to be absorbed by a second molecule.   This radiation absorbed by the second molecule may then be lost by further collision or it may be reradiated.   If it is reradiated, the original ray has gone through 2 molecules; yet, a detector would count it as having come from only 1 molecule and low results would be obtained. This is a very serious problem and only solutions in which the quotient abc is below 0.01 will follow Beer's law. Advantage will be taken of the efficiency of this process when atomic absorption is discussed in Chap. 10.

Quenching is a process in which foreign substances lower or eliminate the fluorescence. Usually only a trace is necessary to ruin a determination (Hercules 1966).

Quantitative Analysis

The fluorescence intensity is proportional to the initial radiation 'times abc.

$$\log \left(\frac{F_0}{F_0 - F}\right) = abc \tag{8.1}$$

where $F_0 = 100$ (usually set by a quinine standard)
F = fluorescence measured

Note that it is the LOG of $F_0/(F_0 - F)$ that is equal to abc. Equation 8.1 can be written as

$$\log F_0 - \log (F_0 - F) = abc \tag{8.2}$$

F_0 is usually set at 100 using a 1 ppm quinine standard and as log of 100 is 2

$$2 - \log (100 - F) = abc \tag{8.3}$$

For very dilute solutions the determination of an unknown is basically done as in visible spectroscopy. The method of standard addition, however, is much faster and is often preferred. The standard addition method involves measuring the fluorescence of the unknown, adding a small known amount of the material being determined to the original sample, and measuring the fluorescence again. From the increase in fluorescence due to the added standard, the original concentration can be computed. If the calibration plot is linear, using the method of standard addition is simple. If, however, the calibration plot yields a curve, some calculation is necessary. The following example will show how this calculation is done.

Example

Cholesterol is determined in egg noodles by treating the product with a base, extracting it with ether, irradiating it at 546 nm, and measuring the fluorescence at 577 nm.   Two grams of such a product (10% moisture) was treated as described before, extracting both the cholesterol and small amounts of other steroids.   The extract was treated to oxidize cholesterol to a nonfluorescing compound.   The solution was diluted to 100 ml; 25 ml was placed in a fluorometer cell (previously calibrated with quinine), and a reading of 7% was recorded.   The oxidized cholesterol in the fluorimeter cell was reduced back to a fluorescing species, producing a reading of 60%.   When to a 24-ml aliquot of the original sample solution was added to 1 ml of a standard solution of cholesterol containing 0.1 mg/ml, an instrument reading of 90% was obtained.   What is the % cholesterol in egg noodles?

Use equation 8.3.   For the standard plus the unknown,

$$2 - \log (100\text{–}90) = abc$$
$$2 - 1.0 = 1.0 = abc \text{ (uncorrected for background)}$$

In a case like this, when it is assumed that abc is greater than 0.01, the background reading cannot just simply be subtracted from the sample reading because it is the $\log F_0/(F_0 - F)$ that is equal to abc, and to subtract one fluorescence reading from another would be similar to subtracting $\%T$'s in colorimetry, and is not permissible.

Twenty-five milliliters of the sample solution produced a background fluorescence of 7%.   By using 24 ml in the standard and 25 ml for the background reading, the effect of sample size is very small.

For the background

$$2 - \log (100\text{–}7) = abc$$
$$2 - 1.97 = 0.03 \text{ background}$$

Therefore $1.0 - 0.03 = 0.97 = \log F_0/(F_0 - F)$ corrected for background.

The value of 0.97 is due to the standard plus the unknown so the unknown must now be subtracted.   The problem now becomes: what fluorescence reading will be produced by 24 ml of cholesterol if 25 ml had a reading of 60%?   Since a and b will remain constant, the log of $F_0/(F_0 - F)$ should be in the ratio of 24/25.

$$\frac{25}{\log \left(\dfrac{100}{100 - 60} \right)} = \frac{24}{\log \left(\dfrac{100}{100 - F} \right)}$$

$$\text{let } \log \left(\frac{100}{100 - F} \right) = X$$

$$\frac{25}{0.40} = \frac{24}{X} \qquad X = 0.39 \text{ (uncorrected for background)}$$

$$0.39 - 0.03 = 0.36 \text{ corrected for background.}$$

When the unknown, 0.36 is subtracted from the unknown plus the standard, 0.97, then the difference of 0.61 must be due to the 0.1 mg of cholesterol added.

$$\frac{0.61}{0.1} = \frac{0.36}{R} \qquad R = 0.059 \text{ mg of cholesterol/25 ml of sample solution}$$

$$\frac{0.000059 \text{ gm} \times 4 \times 100}{1.8 \text{ gm}} = 0.0013\% \text{ cholesterol (on dry matter basis)}$$

Phosphorescence Instrumentation

The instrument used is a fluorimeter with a rotating can or a pair of rotating blades around the sample cell. Their purpose is to separate the fast fluorescence spectra from the slower phosphorescence spectra.

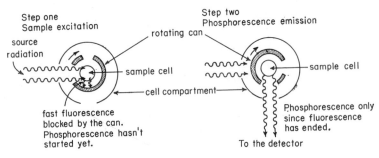

FIG. 8.3. PRINCIPLE OF THE ROTATING CAN PHOSPHOROSCOPE (TOP VIEW)

Since the phosphorescence lifetime varies from compound to compound, it is necessary to have a variable speed motor to turn either the can or the blades.

The better instruments provide a way to scan the spectra to determine the best excitation spectra and then use this information to determine the emission spectra.

Experimental Precautions

The phosphorescence lifetimes depend on the solvent, temperature, intensity of the exciting radiation, length of excitation time, and the presence of foreign compounds to effect intersystem crossing.

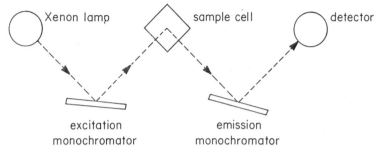

Courtesy of Aminco-Bowman Co.

FIG. 8.4. OPTICAL DIAGRAM OF THE AMINCO-BOWMAN FLUOROMETER-
PHOSPHORIMETER

A low temperature is highly desirable and the most common
system is liquid nitrogen (77°K). Liquid nitrogen can be obtained
in most cities at prices varying from 20¢ to $1.50/liter. About 40
ml fill the sample Dewar flask and suffice for several determinations.
One must use solvents that are liquid at room temperature and
freeze as a clear glass at 77°K. Cracks or snow produced by the
frozen solution scatter the radiation and make it almost impossible
to do quantitative work. The best solvents include: (1) pentane,
(2) diethyl ether:isopentane:ethanol (5:5:2) (EPA), (3) EPA:
chloroform (12:1), (4) triethylamine:diethyl ether:n-pentane (2:
5:5), (5) n-pentane : n-heptane (1:1), and (6) methyl cyclohexane:
n-pentane 4:1 to 3:2.

EPA is the most commonly used but the choice depends on the
solubility of the sample.

Another experimental difficulty is to prevent ice from getting into
the Dewar flask holding the sample. This is caused by moisture in
the air condensing and freezing on the cold surface then falling into
the Dewar. The snow does not settle to the bottom of the Dewar
because the liquid N_2 bubbles continuously as it evaporates and
stirs the snow. This makes a noisy signal, gives a water peak, and
decreases the sensitivity. The interference can be greatly reduced
by passing dry air through the sample compartment.

Qualitative and Quantitative Analysis

Each phosphorescence is unique for each molecule and may be
used to identify phosphors in a mixture. Quantitative analysis is
handled the same as fluorescence; five parameters are used for
qualitative analysis: (1) the shape and position of the phosphorescence curve; (2) the mean lifetime, τ; (3) the vibrational pattern or

band spacing $\Delta \nu$; (4) the frequency of the vibration transition, ν, the highest frequency band, the transition from the lowest triplet to the lowest singlet; and (5) the quantum efficiency, Φ (the ratio of the quanta of light phosphoresced to the quanta absorbed).

SOME RECENT APPLICATIONS

Vitamin A can be determined in concentrations as low as 0.1 $\mu g/$ ml by exciting it at 327 nm and measuring it at 510 nm (Drujan et al. 1968).

The amino acid, tryptophan, if irradiated at 280 nm and examined at 348 nm, can be detected at levels as low as 5×10^{-5} M (Tappel 1968).

Strassman et al. (1968), have shown that if alpha and beta-substituted malic acids are converted to beta keto acids by reaction with sulfuric acid and then complexed with resorcinol, fluorescing derivatives of 7-hydroxy coumarin are formed. For example, 0.2 μg of isopropylmalate can be detected.

Adrenaline and noradrenaline fluoresce if condensed with ethylenediamine dihydrochloride. An extract with 1-butanol is measured at 510 nm; 0.002 γ/ml can be detected (Viktora et al. 1968).

Aflatoxins, recently identified fungal poisons, can be separated by thin layer chromatography but are hard to detect. Pons (1968) has shown that 98% of fractions B_1, B_2, G_1, and G_2 can be determined by fluorescence after separation by thin layer chromatography; 0.0005 μg of G_2 can be detected.

Ethoxyquin can be extracted from eggs with isooctane and determined by its fluorescence band at 425 nm.

In Chap. 6 a colorimetric procedure for benzopyrene, a carcinogen, in smoked fish and ham was mentioned. By using fluorescence the detection limit can be lowered from 10 to 4 $\mu g/kg$.

It is a standard procedure now to coat thin layer plates with a fluorescing material which glows upon irradiation. The compounds, which usually quench the fluorescence, appear as black spots on a colored background. This is a very sensitive detection method; yet, the compounds can be eluted for further testing because seldom does a chemical reaction take place.

Applications of phosphorescence in food analysis are still limited.

PROBLEMS

(1) During the preparation of a food coloring dye an undesirable oxidation side reaction produces a fluorescing compound. A 5% NaOH solution was added to 4 gm of the reaction mixture and the

side reaction compound extracted. The layers were separated and the aqueous layer diluted to 50 ml. Ten milliliters of this solution was placed in a fluorimeter and a reading of 10% was obtained. A few drops of concentrated H_2SO_4 were added and a reading of 56% was obtained. A mixture of 9 ml fresh solution, 2 drops concentrated H_2SO_4, and 1 ml of a dye standard (0.5 mg/ml) produced a reading of 95%. Calculate the number of mg of oxidation product present per gm of reaction material. Ans. 1.72 mg/gm

(2) A 4-gm sample of ham is to be checked for its vitamin B_1 content. It is extracted with HCl, treated with the enzyme phosphatase, and diluted to 50 ml. A 10-ml aliquot is treated with ferricyanide before diluting it to 25 ml. A second 10-ml aliquot is diluted as a blank. A standard solution containing 0.1 μg/ml is treated as above, again preparing a blank.

Fluorescence reading	Solution
3	sample blank
5	standard blank
55	sample, oxidized
80	standard, oxidized

Calculate the amount of B_1 in 1 oz of ham. Ans. 177 μg

(3) Preliminary experiments show that thyroxine, $C_{15}H_{11}I_4NO_4$, can be determined by phosphorescence. Thyroxin, an amino acid of the thyroid gland, exerts a stimulating effect on food metabolism. A standard containing 0.5 μg/10 ml produced a phosphorescence of 70%T at 77°K. A 0.7-gm tablet, extracted with EPA and diluted to 25 ml had a phosphorescence of 40%T. What is the total amount of thyroxine in the tablet? Ans. 0.71 μg

BIBLIOGRAPHY

DRUJAN, B. D., CASTILLON, R., and GUERRERO, E. 1968. Application of fluorometry in the determination of vitamin A. Anal. Biochem. 23, 44-52.

HERCULES, D. 1966. Fluorescence and Phosphorescence Analysis, Interscience Publishers, New York.

PONS, W. A. 1968. Fluorodensitometric measurements of aflatoxins on TLC plates. J. Assoc. Offic. Anal. Chemists 51, 913-914.

STRASSMAN, M., CECI, L., and TUCCI, A. F. 1968. Fluorometric assay of malic acid and substituted derivatives. Anal. Biochem. 23, 484-491.

TAPPEL, A. L. 1968. Automated measurement of proteolytic enzymes. Anal. Biochem. 23, 466-473.

VIKTORA, J. K., BAUKAL, A., and WOLFE, F. W. 1968. New automated methods for estimation of small amounts of adrenaline and noradrenaline. Anal. Biochem. 23, 513-528.

CHAPTER 9

Infrared Spectroscopy

Infrared spectroscopy is believed to have started with the experiment reported by Herschel in 1800. Herschel placed a thermometer at successive points in a daylight spectrum and noticed an unusually large heating effect in the region immediately beyond the red region of the visible spectrum, hence—infrared. The next century was spent in developing detectors, in attempting to understand infrared sources, and in making empirical observations on structure-spectra relationships. Physicists made the early significant contributions; particularly they developed prior to World War I the molecular vibration-rotation theories. It wasn't until just prior to World War II, when it was shown that infrared could be used for functional group analysis, that chemists became interested in infrared.

In infrared work two units of wavelength are commonly used, the micrometer (old micron) and the wavenumber. Rotational frequencies are of the order of 10^{11} to 10^{13} Hertz, and vibrational frequencies are in the 10^{13} to 10^{15} Hertz region. In order to have more manageable numbers and still maintain a proportional relationship between energy and frequency the wavenumber concept was developed.

Recall that $\lambda \nu = c$. Divide ν by c (3×10^{10}) and the result, $1/\lambda$, is now 3×10^{10} smaller than the original frequency, which means we deal with numbers in the range 20 to 5,000. This is still proportional to energy, since the equation $\Delta E = h\bar{\nu}$ must only be changed to $\Delta E = h\bar{\nu}c$ to have an identity. The term, $\bar{\nu}$, is the symbol for the wavenumber and has the units cm^{-1}. Equation (9.1) is used to convert μm to cm^{-1}.

$$\text{Wavelength } (\mu m) = \frac{10,000}{\text{wavenumbers } (\bar{\nu})} \qquad (9.1)$$

Example

What is the frequency in wavenumbers corresponding to 5.40 μm?

$$\bar{\nu} = \frac{10,000}{5.40} = 1851$$

Infrared Instrumentation

Sources.—We need a source that provides a continuous spectrum at a high intensity. Recall that a molecule absorbs radiation if the radiation matches a frequency of some transition within the molecule. If a molecule can absorb radiation it can also emit it, and thereby become a source for that wavelength. We want all of the wavelengths in a continuous spectrum, so the source compound must be able to absorb all wavelengths in order to emit all wavelengths. If a compound absorbs all of the wavelengths striking its surface it will be black and we have what is known as the *ideal black body radiator*.

Figure (9.1) is a plot of intensity at various temperatures *vs* wavelength for an ideal black body radiator.

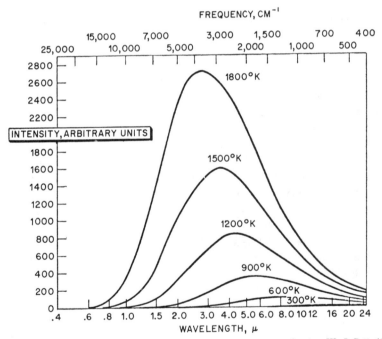

Courtesy W. J. Potts (1963)

FIG. 9.1. BLACK BODY EMITTANCE VS WAVELENGTH

Notice that as the temperature of the source increases, the intensity of the emitted radiation increases. Equation 9.2 shows the relationship.

$$I_{\lambda \max} = 1.3T^5 \times 10^{-15} \qquad (9.2)$$

where I = watts/cm^2

T = °K of the source

This equation indicates that if the absolute temperature of the source is doubled its emission intensity increases 2^5 or 32 times. Therefore one way to increase sensitivity would be to increase the source temperature. However, this cannot be done at random since it can also be observed from Fig. 9.1 that as the temperature of the source increases, the wavelength of maximum intensity shifts to a shorter wavelength. Rayleigh has shown that short wavelength radiation can be scattered easily.

$$T = k(cld^3)/(d^4 + \alpha\lambda^4) \qquad (9.3)$$

Here T = the amount of scattering. Notice that with λ^4 in the denominator, a very small shift in radiation to a shorter wavelength will cause a very large increase in the amount of radiation that can be scattered. This short wavelength radiation has a high energy since ($\Delta E = h\nu$). Even if a small amount of this high energy radiation reaches the detector, it will overwhelm the low energy signal that should be detected. Great care must be taken with infrared instruments to ensure that this stray radiation does not reach the detector. This is done by controlling the source temperature and/or by using filters and baffles.

A source temperature of about 1500°K is generally used in commercial equipment although experimental research equipment may use 6000°K carbon arcs.

Nernst Glower.—A hollow rod is made of a fused mixture of the oxides of zirconium (90 parts), yttrium (7 parts), and either erbium or thorium (3 parts). The rod is usually about 3 mm in diameter and about 5 to 6 cm long. Sometimes a molybdenum strip is attached to each end of the electrode as an expansion spring to prevent the joint from cracking when the instrument is repeatedly turned on and off. The glower has a negative coefficient of resistance and must be preheated before it will conduct. If the glower is water-jacketed, 1 to 2% stability can be maintained; and if in addition it is also enclosed, 0.1% stability can be achieved.

The glower has an intense radiation up to about 10 μm and its normal lifetime is up to several hundred hours. This source produces about 200 times the radiation at 1 to 2 μm that it does between 14 to 15 μm.

Globar.—A globar is made from carborundum (SiC) and is about $1/4$ in. in diameter and 2 in. long. The usual operating temperature is about 1400°C with 2000°C, the point where (SiC) vaporizes, being the upper limit. This source conducts at room temperature and requires no preheating. A globar has many advantages: it is sturdy, it is a more stable source than a Nernst glower, it is better than a Nernst glower beyond 10 μm, and its radiation matches that of a black body.

Disadvantages of a globar are that it has a relatively short life, it overheats at temperatures in excess of 1200°C, and must be cooled. It requires a lot of power, and the binding material evaporates gradually and absorbs some of the radiation. In addition, the operating voltage gradually changes because the binding material is evaporating and this requires a more complex electronic control of the source.

Nichrome Wire.—A closely wound spiral of nichrome wire, heated to 1500°K by the passage of an electric current. The main advantage is a positive temperature coefficient of resistance. Both the glower and the globar had negative coefficients which meant that their temperature was lowest and the resistance highest at the ends of the source where the electrical connections are made. This results in a high dissipation of energy at the connections, and causes their eventual arcing and subsequent burnout. The image of the wire spiral is not uniform and this is eliminated by enclosing it in a ceramic cylinder.

Cell and Prism Materials.—If a material is to be transparent in the infrared region and thus serve as cell and prism material, its molecules must have frequencies of motion different from that of the incident radiation and the material to be examined. Most materials examined by infrared are organic compounds which are held together mainly by covalent bonds. Cell materials should then be made from something different than covalently bonded compounds such as ionic bonded systems. This means that most cell materials are alkali or alkaline earth halides. Several of the materials used are shown in Table 9.1.

The entire infrared region was originally presented as being from 700 nm to about 500 μm. Why then is there interest in materials that only transmit to 5–7 μm? The reason is that much of the information that can be obtained in the far infrared and microwave regions can be obtained between the 2–7 μm region if the monochromator has sufficient dispersion to resolve the vibrational-rotational interactions in this region. If this can be done, the very expensive and difficult to operate far infrared and microwave instruments are not necessary.

TABLE 9.1

INFRARED TRANSMITTING MATERIALS COMMONLY USED

Material	Usual Cutoff Wavelength, (μm)	Refractive Index
Glass	2.5	
SiO$_2$ (quartz)	3.5	1.45
TiO$_2$ (rutile)	5.0	2.6
LiF	5.5	1.38
Al$_2$O$_3$ (ruby)	5.5	1.77
MgO (periclase)	7.0	1.74
Irtran 1	7.5	1.36
CaF$_2$	8.5	1.43
Irtran 3	10	1.39
BaF$_2$	11	1.6
Si	11	3.4
Irtran 2 (ZnS)	14	2.28
NaCl (halite)	15	1.54
Irtran 4	20	2.42
KCl (sylvite)	20	1.49
AgCl	22	2.07
KBr	25	1.53
KRS-6 (TlCl-TlBr 40%)	30	2.18
CsBr	35	1.69
KRS-5 (TlI-TlBr 42%)	39	2.63
KI	40	1.67
CsI	50	
Diamond	250	2.41
Polyethylene	>250	

In general, the alkali halides have several disadvantages. They are water soluble, cleave and scratch easily, and have a high-temperature coefficient which means that close temperature control is necessary to secure mechanical stability and constant dispersion. Silver chloride is not water soluble but is quite sensitive to visible radiation. Probably the best salt material at the present time is the eutectic mixture of thalium bromide (42%) and thalium iodide (58%), known as KRS-5. It is easily recognized because of its orange color. It is toxic, hard to polish, and stains readily; but its solubility in water is low and its high refractive index (2.63) make it a favorite for multiple internal reflection attachments which will be discussed in a later section.

Although diffraction gratings are replacing prisms almost completely in infrared spectrophotometers, the need for a good cell material still remains. Today, the *Irtran* series is the best solution of the problems of water solubility, high refractive index, and the ability to withstand large temperature changes. For example, Irtran-2 shows no change when exposed to water at 23°C for 336 hr, a 6% NaCl solution for 168 hr, 0.5 N HNO$_3$ for 24 hr, 0.5 N NH$_4$OH for 24 hr, and ethyl ether, acetone, chloroform, or benzene for 24 hr. It can withstand temperatures of more than 800°C and can be polished to good optical tolerances. Samples of 0.08-in. material have survived water quench tests from 165°C to ambient temperature.

Cells.—The purpose of the cell is to hold the compounds while they are being examined with the instrument. The cell windows are made from the materials just discussed and extreme care must be used to keep the cells in good condition. Cells are usually stored in desiccators to reduce moisture problems, and they should not be exposed to large differences in temperature lest they crack. Cells are generally carried downward, that is, with the hand on top so body heat does not warm the cell thus changing its optical properties. The high solubility of most cell materials makes it mandatory that the solvents and compounds placed in the cells are as dry as possible. Carbon tetrachloride and carbon disulfide are favorite solvents (care—CS_2 reacts with primary amines). Cell thicknesses vary from a few hundredths of a millimeter for liquid samples to several meters for those used with gases. As a result, a variety of cells are required for the normal range of problems encountered.

Sandwich Cells.—This cell is probably the most common type. A thickness varying from 0.01 to 1.0 mm is normal. Regardless of the care exercised, the cell windows eventually fog due to traces of moisture in the sample. This reduces the sensitivity of the analysis and the windows must be polished. Polish the plates clear and flat. One way of doing this is by the alcohol-felt method. Stretch a piece of felt over a glass plate and hold it in place with a spring clip or rubber bands. Pour a small amount of the polishing compound (400 mesh Fe_2O_3) on the top part of the felt and moisten about $1/3$ of the pad with 95% alcohol. Rub the plate rapidly over the polishing compound and alcohol, gradually working down to the dry felt. About 30 passes over the alcohol and 10 over the dry felt are satisfactory.

Cavity Cells.—These are made of one piece by an ultrasonic cavitation process, and are much cheaper than sandwich cells. They work quite well for exploratory problems and are relatively inexpensive.

Wedge Cell.—In double beam operation when solutions of solids are to be determined, the spectra of the solvent interferes unless it is blanked out. This is difficult to do in the infrared region because the cell thickness is so small that it is hard to find two perfectly matched cells, and even then the amount of solvent in the reference would be more than in the sample cell. What is needed is a continuously variable cell thickness for the reference side. The wedge cell (contains a wedge-shaped spacer) is used for this purpose. For example, for a sample cell of 0.025 mm thickness, a wedge varying from 0.01 to 0.08 mm is used. This cell is moved horizontally until the major solvent band is just cancelled out. It can be done quite quickly.

Gas Cells.—For fairly pure compounds a 2- to 10-cm gas cell is used. For trace analysis it is customary to use a multiple pass cell, since cells larger than about 10 cm do not fit into most instruments.

KBr Pellet Methods

The infrared spectra of solids are generally obtained from a solution of the compound. Solvent bands may interfere and many compounds are not soluble in the solvents commonly used. To simply place small crystals between the cell windows does not work satisfactorily because too much radiation is scattered. In order to reduce scattering, the sample particles should be about 2 μm in diameter, or if the particles are larger, they should be surrounded by a medium of similar refractive index.

A technique previously used was the mull method. The procedure involved placing 2–5 mg of the sample into a small mortar, add 4-5 drops of Nujol (mineral oil) and grind the mixture into a paste. The paste was spread between the cell windows. The mineral oil reduces the scattering considerably. The main disadvantage is that the —CH$_2$— and —CH$_3$ bands of the mineral oil interfere with the sample, and that solvents, including mineral oil, often interact with the sample to distort the spectra.

A technique which solves most of the forementioned problems is the KBr pellet method. This method was developed by two sisters, Stinson and O'Donnell (1952), during the course of determining both UV and IR on the same sample of cytosine and isocytosine without reflection losses. Essentially, the compound is mixed with optically pure KBr and pressed into a disc-shaped pellet which is placed in the sample beam of the instrument. Normal size pellets are about the size of a dime, with micropellets being 1–0.5 mm in diameter.

The die originally designed is expensive and requires an elaborate technique to prepare a clear pellet and to remove it without breaking. However, if only qualitative results are desired, or if an internal standard is used, then a much more rapid and less expensive procedure is available. Figure 9.2 shows the setup for the rapid method.

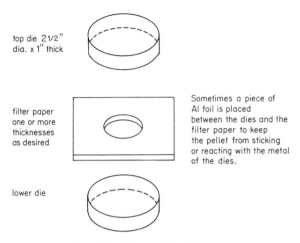

top die 2 1/2" dia. x 1" thick

filter paper one or more thicknesses as desired

Sometimes a piece of Al foil is placed between the dies and the filter paper to keep the pellet from sticking or reacting with the metal of the dies.

lower die

FIG. 9.2. RAPID KBr METHOD

The following is a description of how pellets can be made by the rapid method.

Grind 0.1 gm of the sample in a small agate mortar until it is very fine. Add 2 gm of DRY KBr and grind the two together until they appear to be completely mixed. (A "wiggle bug" is convenient for this.) With a cork borer, cut a hole about 1 cm in diameter in a piece of filter paper or stiff cardboard. The pellet thickness can be varied by the number of pieces of cardboard used. Place a 2-in. square of aluminum foil on the lower die, place the cardboard on top, add about 0.2 to 0.4 gm of the pellet mixture, place another piece of aluminum foil on top, and finally add the top die. Place this in a press (any machinist can make an A frame about 18 in. high and a heavy duty car or light truck hydraulic jack can be used), and apply about 80,000 psig for about 1 min and then release the pressure. The pellet will usually crack in several places at this time because the trapped air is being released. Now reapply the pressure and remelt the KBr. The pellet formed will be imbedded in the paper and can easily be handled; in fact, it can be mounted directly in the sample holder if the original piece of cardboard was cut correctly. The pellet should be transparent. A milky pellet indicates not enough pressure was applied, while a brownish pellet indicates the KBr is old and needs replacing. A wet pellet indicates the KBr is old and needs replacing. A wet pellet is indicated by 30–40% transmittance at the 2 μm part of the spectra.

An even simpler and less expensive method is to use the minipress shown in Fig. 9.3. It consists of a big nut with a bolt in each end.

The ends of the bolts are polished and hardened.   Pressure is applied by 2 wrenches and the sample is left in the center of the nut
when the 2 bolts are removed.   The nut is then placed in the beam
of the instrument.   The only disadvantage is that the sample must
be destroyed before a new sample can be made thus preventing a
"file" of samples to be built-up for a particular project.

Courtesy Wilks Scientific Co.

FIG. 9.3.   MINI-PRESS

When using KBr pellets it is necessary to reduce the reference
beam intensity to compensate for the KBr.   Rather than make
another pellet, which is hard to duplicate, various attenuation
devices have been made.   The two shown in Fig. 9.4 are the most
common commercial attenuators, the right one being quite easy
for the individual to make.   It requires a small piece of metal and
a scrap of venetian blind type window screen.

Detectors

The detectors used for the visible and ultraviolet regions require
too much energy to respond to infrared radiation.   The problem

Courtesy Wilks Scientific Co.

FIG. 9.4.   EXTERNAL BEAM ATTENUATORS

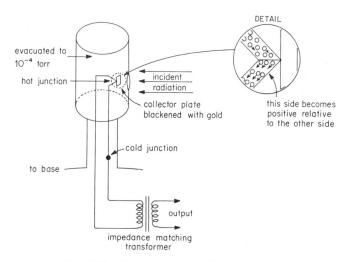

FIG. 9.5.   THERMOCOUPLE CONSTRUCTION

may be put in its proper perspective by the following example.   Suppose someone lit a match on a dark night several blocks away from you.   Your eye would immediately detect the visible radiation but your body would not feel the heat given off.   The infrared detector must detect this heat.   If 0.5% analytical accuracy is to be achieved, then temperature differences of $5 \times 10^{-5}°C$ must be measured. Infrared detectors are therefore constructed differently than those for the visible and ultraviolet regions.   The three most common detectors found in commercial instruments are thermocouples, bolometers, and pneumatic devices.

Thermocouples.—Figure 9.5 shows the basic idea of how a thermocouple works. A thermocouple is formed when two different metals are placed together. Two such junctions are necessary, one to serve as the detector and one to serve as a reference. Radiant energy is focused on the hot junction and because of the different characteristics of the two metals, one side of the junction warms up more than the other side. This temperature (energy) difference causes electrons to leave one metal faster than the other metal and the net result is that at the junction a small voltage is produced which is proportional to the amount of heat striking the junction. The cold junction serves as a reference.

Blackening the absorbing surface of the hot junction improves the efficiency of the absorption of the radiation. If gas molecules from the surrounding air strike the junction, they can cool it, thereby decreasing its sensitivity. Therefore, the junction is placed in a small cavity and evacuated. The combination of surface blackening and evacuation can improve the sensitivity 10 to 20 times.

Bolometers.—While a thermocouple works fairly well as an infrared detector, more sensitive devices are needed. A bolometer, which measures a difference in current, is a more sensitive device than the thermocouple. The familiar Wheatstone bridge circuit is shown in Fig. 9.6.

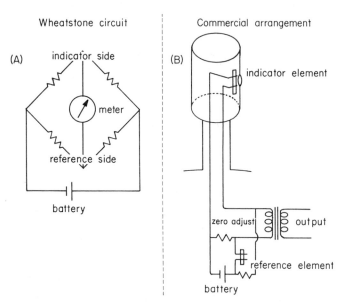

FIG. 9.6. BOLOMETER CONSTRUCTION

The bolometer is a ribbon of electrical conducting material (thermocouple, thermistor, wire filament) through which a steady current is passed. When radiation strikes the sensing element the electrical resistance changes and the corresponding current change is measured. By using the Wheatstone bridge circuit any infrared radiation changes due to variations in external conditions that change the temperature are canceled by the reference side of the detector, thus only the radiation from the sample is measured.

The sensitive element in the bolometer may be a thermocouple, a piece of metal (nickel is popular), a thermistor (a thin piece of Mn, Ni or Co oxides), or a superconductor.

A superconductor is a substance that loses its normal resistance when cooled to very low temperatures (Fig. 9.7).

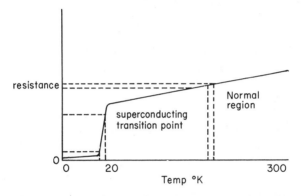

FIG. 9.7. NORMAL AND SUPERCONDUCTING REGIONS

Notice that the resistance change per unit of temperature change is much greater in the superconducting transition region than in the normal region. Superconducting bolometers are 20–30 times more sensitive than normal bolometers. Tantalum nitride and niobium nitride are two compounds that exhibit this effect. However, liquid helium and/or liquid hydrogen must be used to attain sufficiently low temperatures.

Pneumatic Detector.—This detector developed by Golay (1947) uses an entirely different approach than other detectors. It was mentioned earlier that in striking a match at night visibility detection is more sensitive than infrared detection. What Golay did in effect was to make a visible detector that was controlled by infrared radiation. Figure 9.8 shows schematically this detector.

The detector consists of a pneumatic chamber with a capillary tube covered with a flexible cover that has a mirrored surface on the

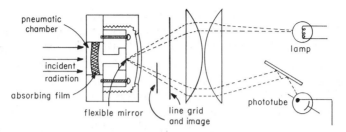

FIG. 9.8.   GOLAY DETECTOR

outside.   This chamber is filled with xenon.   Inside the chamber is a metallized membrane of very low thermal capacity (Al on collodion) to absorb the incident radiation.

The infrared radiation is absorbed by the inner membrane which in turn heats the xenon.   As the gas in the chamber expands and contracts it moves the membrane at the end of the capillary.   A faint beam of light is directed at the membrane and the amount the light moves is detected with a photocell.

A Golay detector is very sensitive, usually being surpassed only by superconducting bolometers.   It is capable of covering the infrared region out to 700 μm.   Its short wavelength sensitivity falls when the wavelength is comparable to the film thickness, and its long wavelength sensitivity falls when the wavelength is comparable to the diameter of the receiving element.

Multiple Internal Reflection

Multiple internal reflection (MIR) or attenuated total reflection (ATR), developed by Fahrenfort (1961), is used particularly in the infrared spectral region.   It is used for the analysis of a variety of materials.   Not since the introduction of the KBr technique has there been a development of equal importance in the field of infrared spectroscopy.   The technique is particularly valuable if it is undesirable or inconvenient to impose on the specimen the pre-

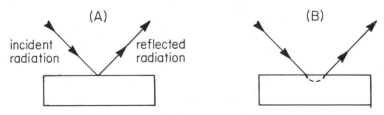

FIG. 9.9.   REFLECTED RADIATION

parative procedures required for normal transmission work, or where the region of interest is a thin layer adjacent to a surface. Figure 9.9A shows how we commonly draw a diagram illustrating the reflection of radiation.

Under certain conditions this has been found to be incorrect and what really happens is that the incident beam penetrates a few molecular layers before it is reflected (Fig. 9.9B). The amount of reflection that occurs is roughly inversely proportional to the ratio of the index of refraction of the plate and that of the medium in contact with it. Since air has a low refractive index (1.0), a radiation beam being totally reflected from the face of the prism in contact with air suffers practically no change. If a material with an index equal to or greater than that of the plate is placed in contact with a plate, all of the energy will pass into the second medium.

Further, the index of refraction of a material undergoes a rapid change at wavelengths where the material absorbs. Hence, when such a material is brought in contact with a prism (see Fig. 9.10), the internally reflected beam will lose energy at these absorbed wavelengths (where the index is high), so that a plot of the reflected energy will produce a curve that is very similar to, but not identical to, a conventional transmission spectrum. The amount of apparent penetration of energy into the sample depends on the prism's refractive index (high index—low penetration) and the angle of incidence of the beam of radiation (high angle—high penetration). The actual thickness of the sample has no effect on the spectrum —and herein lies the real importance of the MIR technique as the spectroscopist no longer must prepare a thin film for the infrared energy to pass through. He can obtain an equivalent spectrum by merely placing his sample in direct contact with the reflecting face of the prism. Figure 9.10 shows how a multiple internal reflection (MIR) spectrum is obtained.

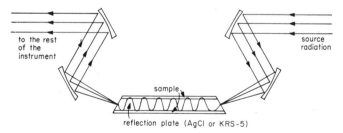

to the rest of the instrument

source radiation

sample

reflection plate (AgCl or KRS-5)

Courtesy Wilks Scientific Co.

FIG. 9.10.   MULTIPLE INTERNAL REFECTION

KRS-5 is usually preferred because it has a high refractive index (2.63) which provides a better refractive index ratio.

Qualitative Analysis

Bands in the shorter wavelength region (2–10 μm) are thought to result mainly from the stretching and bending vibrations of individual bonds, and are therefore considered characterististic of the diatomic structural units of functional groups. Bands in the longer wavelength region (7–25 μm) appear to be caused by more complex vibrations of polyatomic units and of the molecule as a whole.

It is to be noted that spectra determined in different solvents are not always identical. If a compound is pure, it will have a moderate number of bands and they will be sharp. Generally, impurities of greater than 1% must be present before they can be detected successfully.

A quick scanning of the spectrum for strong absorption bands or characteristic patterns will often provide a useful preliminary diagnosis. Table 9.2 and Fig. 9.11 show a recommended procedure.

TABLE 9.2

PRELIMINARY SCANNING PROGRAM

(1) Absorption at 2.5 to 3.2 μm	OH, NH compounds
check a. 5.7 to 6.1	acids
b. 5.9 to 6.7	amides (usually two bands)
c. 7.5 to 10.0	—O— compounds
d. about 15.0	primary amines (broad)
(2) Sharp absorption at 3.2 to 3.33	Olefins, aromatics
check a. 5.0 to 6.0	benzenoid patterns (weak)
b. 5.9 to 6.1	olefins
c. 6.1 to 6.9	aromatics (two bands)
d. 11.0 to 15.0	aromatics (several strong bands)
(3) Sharp absorption band at 3.33 to 3.55	Aliphatics
check a. 6.7 to 7.0	—CH$_2$—, —CH$_3$
b. 7.1 to 7.4	—CH$_3$
c. 13.3 to 13.9	—(CH$_2$)$_4$—
(4) Two weak bands at 3.4 to 3.7	Aldehydes
check a. 5.7 to 6.1	aldehydes and ketones
(5) Absorption at 4.0 to 5.0	Acetylenes, nitriles
(6) Strong sharp bands at 5.4 to 5.8	Esters, acyl halides (1 peak) anhydrides (2 peaks)
check a. 7.5 to 10.0	—O— compounds
(7) Strong sharp bands at 5.7 to 6.1	Aldehydes, ketones, acids
(8) Strong bands at 7.5 to 10.0	—O— compounds
(9) Strong bands at 11.0 to 15.0	Aromatics, chlorides

The spectra correlation chart by Colthup (1950) is reproduced as Fig. 9.11 for a more detailed examination of spectra.

FIG. 9.11. SPECTRA-STRUCTURE CORRELATORS ⟶

Quantitative Analysis

Quantitative analysis using infrared radiation is extremely difficult on an absolute basis, because it is hard to determine cell thickness accurately and because wavelength settings are hard to reproduce. However, some good empirical methods have been developed which make quantitative analysis quite practical.

Since most infrared work is done in cells 0.01–0.25 mm thick, this thickness must be known to 0.0001–0.00025 mm in order to get even 1% accuracy. There are two ways of doing this. One method is based on an interference pattern similar to that of the interference filter, and the other is to use a standard absorber. If you take a freshly polished cell and determine a spectrum of the empty cell, you get a spectrum like that shown in Fig. 9.12 if your instrument is linear in micrometers.

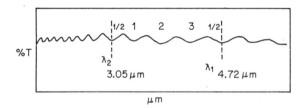

FIG. 9.12. INTERFERENCE PATTERN OF AN EMPTY CELL

Notice the manner in which the number of bands is counted. By measuring the space between the bands, the cell thickness can be determined from:

$$b = \frac{n(\lambda_1 \times \lambda_2)}{2(\lambda_1 - \lambda_2)}$$ (9.2)

where b = spacing in micrometers

n = number of bands between λ_1 and λ_2

λ_1 and λ_2 are the wavelengths in μm between the bands counted

Example

Calculate the spacing of a cell in millimeters if the empty cell produces a spectrum between 3 and 5 μm with peaks at 3.05, 3.44, 3.90, 4.31, and 4.72 μm.

$$b = \frac{4(4.72 \times 3.05)}{2(4.72 - 3.05)} = 17.24 \ \mu\text{m} = 0.0172 \ \text{mm}$$

The following procedure is taken from Barnes Engineering Bulletin Ba-5. Benzene is an excellent standard absorber for calibrating infrared cells which are not smooth enough to give interference fringes. The 1960 cm^{-1} or 5.05 μm band may be used for calibrating cells which are less than 0.1 mm in pathlength and the 845 cm^{-1} band or 11.8 μm band may be used for calibrating cells 0.1 mm or greater in pathlength.

(1) For the 1960 cm^{-1} band

$$\text{cell thickness in mm} = 0.1 \times \text{absorbance} \qquad (9.3)$$

(2) For the 845 cm^{-1} band

$$\text{cell thickness in mm} = \text{absorbance}/0.24 \qquad (9.4)$$

The second problem is determining the absorptivity. This is difficult because it is almost impossible to reproduce slit widths and wavelength settings with sufficient accuracy to ensure that an a determined by one operator will be the same as an a determined by another, especially if different instruments are used. Because of these difficulties, several empirical methods have been developed. The two useful ones are the *base line method* and the *empirical ratio method*. These methods are illustrated in Fig. 9.13. Plot the log of I_0/I vs concentration to establish a calibration curve.

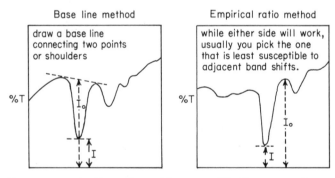

FIG. 9.13.   (A) BASE LINE METHOD, (B) EMPIRICAL RATIO METHOD

Infrared as a Gas Chromatographic Detector

An infrared spectrophotometer is an excellent detector in combination with a gas chromatograph to identify the resolved components. Because materials can emerge from a gas chromatograph very rapidly, it is necessary to have either a few second scan rate or a means of bypassing the unwanted peaks, keeping the sample hot and providing sufficient detection limits to be able to see the small amount of sample which is even further diluted by the carrier gas. Infrared instruments with a few second scan rate are available but are quite expensive. Figure 9.14 shows a method to do this, as developed by the Wilks Co. The idea is to provide a long path in the infrared beam for the sample to pass through. The tube is gold-plated to provide low absorption losses, yet be chemically inert.

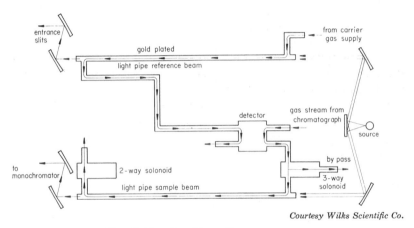

Courtesy Wilks Scientific Co.

FIG. 9.14. GAS FLOW SCHEMATIC

Some Recent Applications

Ethylene oxide is used as a fumigant in various food products. Ethylene oxide itself is moderately toxic, but chlorohydrin, which is formed from it, is quite toxic, 18 ppm being all that is required for poisoning.

Helling and Bollag (1968) have found a way to determine cate-chols in the presence of phenols, resorcinols, and hydroquinones by precipitating them with lead and identifying them by their infrared patterns. Catechols are formed during the metabolism of aromatic compounds. As little as 25 μg gives an excellent spectrum.

On the lighter side, it is possible to tell smokers from nonsmokers by sampling their breath and looking at the intensity of the CO band at 4.65 μm.

One of the main problems in using infrared in the analysis of foods results from water either absorbing all of the radiation or else ruining the sample cells.   Silica gel will absorb almost all organic vapors, which then may be quantitatively desorbed for subsequent analysis.

Infrared Reflectance

Investigations conducted by K. H. Norris and co-workers at the Instrumentation Research Laboratory, Agricultural Research Service, USDA, have led to the development of an instrument which utilizes infrared reflectance to measure moisture, protein, and oil contents of foods and agricultural products.

Most of the basic near-infrared reflectance spectra were recorded on a computerized spectrophotometer, built around a Cary Model 14 prism-grating monochromator with optics optimized for the infrared.   The monochromator was coupled to a digital computer for collection and analysis

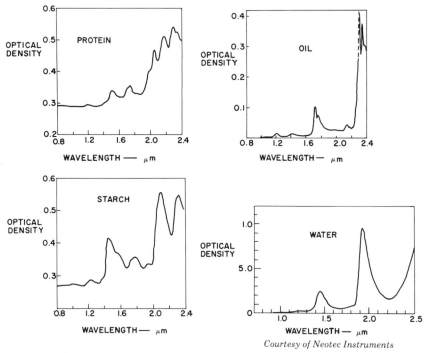

Courtesy of Neotec Instruments

FIG. 9.15.   INFRARED REFLECTANCE SPECTRA OF PROTEIN, OIL, STARCH AND WATER

of data. The sample, packed into a cell, was illuminated through a quartz window and diffusely reflected radiation was collected with four lead sulfide cells. The signal from the lead-sulfide detector was amplified with a logarithmic-response amplifier, digitized, and fed to the digital computer. Data were recorded as log (R_1/R_λ) where R_1 is the reflectance of the reference and R_λ is the reflectance of the sample at each wavelength (λ). Log (R_1) is a constant and log (R_1/R_λ) may be expressed as log $(1/R_\lambda)$ + log R_1 or log $(1/R_\lambda)$ + c, where c is a constant. Plotting log $(1/R_\lambda)$ as a function of wavelength gives a curve that is comparable with an absorption curve having peak readings at wavelengths that correspond to absorption bands in the sample. Typical spectra for protein, oil, starch, and water are given in Fig. 9.15.

The reflectance spectra of a sample of ground beef shows the summation of the spectra of water, carbohydrates, proteins, and oil (Fig. 9.16). Several methods of treating the data can be used to predict the composition from reflectance spectra:

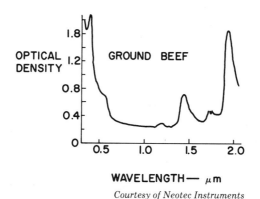

Courtesy of Neotec Instruments

FIG. 9.16. INFRARED REFLECTANCE SPECTRUM OF GROUND BEEF

(A) The log $(1/R)$ values are determined at up to eight selected wavelengths for each component.

(B) The difference in log $(1/R)$ values at up to four selected pairs of wavelength points is measured. The wavelengths are selected at a maximum absorption wavelength for the measured component and at a nearby wavelength at which the absorption is a minimum.

(C) The function $(R_{\lambda_1} - R_{\lambda_2})/(R_{\lambda_1} + R_{\lambda_2})$, i.e., the slope of the reflectance curve in the region of an absorption band varies linearly with concentration of a component; λ_1 and λ_2 are close wavelengths, near to, but not at a peak absorption point for that component.

(D) The second derivative of the log $(1/R)$ at a peak absorption point.

All four data treatments involve developing multiple correlations for calibration and use of those calibrations to determine protein, oil, or water content. New improved methods for data treatment are developed.

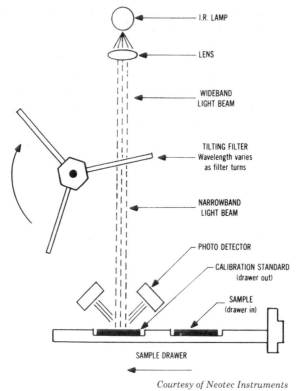

Courtesy of Neotec Instruments

FIG. 9.17. DIAGRAM OF THE GRAIN QUALITY ANA-LYZER

Numerous investigations have shown that there is a good linear relationship between the chemical composition and the reflectance in the near infrared. However, the reflectance spectra are sensitive to extremes in particle size (reflectance increases as particle size decreases). A reflectance change of up to 50% can occur as particle size is reduced; the second derivative technique provides least sensitivity to particle size effects.

As it is undesirable to extrapolate beyond the compositional limits of samples used for calibration and standardization, it is necessary to obtain samples which cover the entire expected range of composition. Finally, temperature changes of the instrument can cause both wavelength and reflectance sensitivity changes.

Two companies—Technicon Instruments, Corp., Tarrytown, N.Y. and Neotec Instruments, Inc., Silver Springs, Md.—now market near infrared instruments to measure the composition of foods and agricultural products.

The following are from the descriptions of the two instruments:

Neotec Grain Quality Analyzer.—In the GQA, the light beam passes through a *tilting* filter system which changes angle in relation to a tungsten light source. In this manner, more than 300 measurement points are scanned 10 times every second using only 3 narrowbandpass infrared filters. At all 300 points, the amount of IR light reflected by the grain sample is measured by a photodetector. Data is automatically fed into a built-in computer which solves third order equations and displays protein, oil and water content on the digital readout.

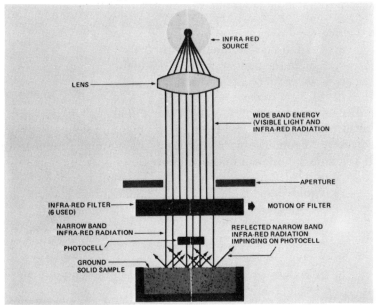

Courtesy of Technicon Instruments

FIG. 9.18. DIAGRAM OF THE INFRA-ANALYZER

Each grain type—soy, corn, milo, wheat, etc.—has its own "coefficients" or "influence factors" in the equations. These are determined via computer calibration and programmed into the GQA on a flat printed circuit "Q" card. To read constituent quantities of any grain type, simply insert the appropriate "Q" card into the unit. Answers can be obtained on a dry basis, wet basis, 14% moisture basis, or any other standard simply by changing the "Q" card.

The Dickey-john Instrument.—The infra-analyzer utilizes the difference in reflectance of six narrow wave bands of energy in the near-infrared spectrum from a ground sample. The reflected energy is detected by a sensitive photocell, sampling the reflected energy level from each of the six different wave bands many times per second. This signal output is amplified and channeled thru a synchronizing device and then is processed thru the computing electronic circuitry. In the computation process a series of mathematical equations are solved resulting in a presentation of direct readings on a digital readout device. The display on the readout device is numerical and shows percentage content of the sample for moisture, protein, and oil.

For additional information see Norris (1974), Trevis (1974), Williams (1975), and Pomeranz and Moore (1975).

PROBLEMS

(1) Convert:
    a. 4 μm to wavenumbers   Ans. 2500
    b. 2.5 μm to angstroms   Ans. 25,000
    c. 2354 cm^{-1} to μm   Ans. 4.25
    d. 13.2 μm to cm^{-1} Ans. 759

(2) Binary mixtures of A and B were analyzed in the infrared. The absorptions at 3.25 and 5.98 μm are characteristic of A and B, respectively. From the following data calculate the amount of A and B (in mole %) present in the unknown.

Composition (Mole %)	Transmittance (%) 3.25 μm	5.98 μm
20	66	69
40	46	56
60	38	48
80	31	41.5
unknown	40	53.5

Ans. A = 53, B = 47

(3) An empty absorption cell (freshly polished) was placed in an infrared spectrophotometer and a spectrum was recorded over the range 7 to 9 μm. The spectrum showed maxima at 7.12, 7.43, 8.09, 8.40, and 8.72 μm. Calculate the spacing of the cell in mm. Ans. 0.077 mm

(4) If an infrared cell is 0.025 mm thick, how many fringes would you expect to find between 2.0 and 6.0 μm? Ans. 16.6

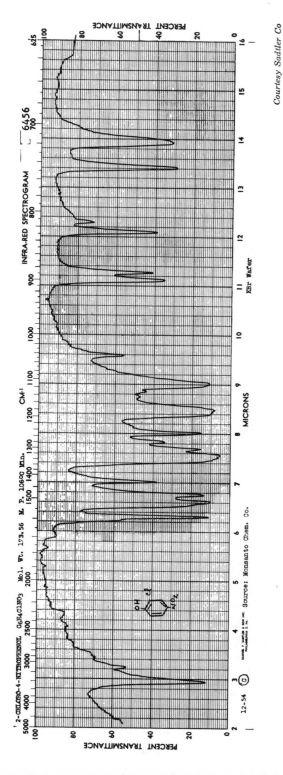

FIG. 9.19. UNKNOWN SPECTRA

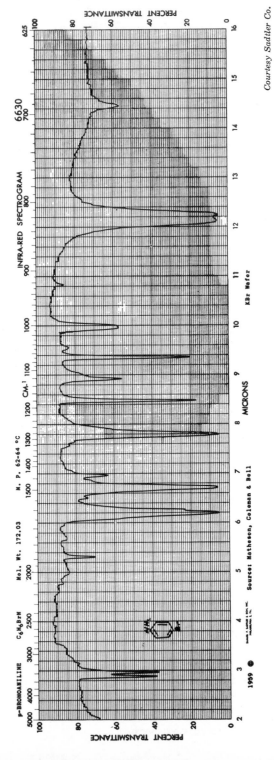

Courtesy Sadtler Co.

Fig. 9.20. Unknown Spectra

(5) Using the data in Fig. 9.19 and an organic qualitative analysis book, i.e., Shriner and Fuson, identify the following compound. It produces a brown color with ceric nitrate and forms a brown precipitate with ferric hydroxide. Its empirical formula is C_6H_4-$ClNO_3$. Ans. 2-chloro-4-nitrophenol (Sadtler Spectra No. 6456)

(6) A compound gives a positive Hinsberg test and the resulting compound is soluble in alkali. It is known to contain Br but gives a negative NaI test. Its density is 1.8 gm/ml. Use Fig. 9.20 and an organic qualitative analysis book (such as Shriner and Fuson) and identify this material.

Ans. *p*-bromoaniline (Sadtler Spectra No. 6630)

BIBLIOGRAPHY

CHEN, T., and GOULD, J. 1968. Infrared studies of ethylene chlorohydrin. J. Assoc. Offic. Anal. Chemists *51*, 878–883.

COLTHUP, N. B. 1950. Spectra-structure correlations in the infrared region. J. Opt. Soc. Am. *40*, 397–400.

FAHRENFORT, J. 1961. Attenuated total reflectance. Spectrochim. Acta *17*, 698–709.

GOLAY, M. 1947. A pneumatic infrared detector. Rev. Sci. Instr. *18*, 357–362.

HELLING, C. S., and BOLLAG, J. M. 1968. Microanalysis of catechols as lead salts by infrared spectroscopy. Anal. Biochem. *24*, 34–43.

NORRIS, K. H. 1974. Reports on design and development of a new moisture meter. Agric. Eng. *45*, No. 7, 370.

POMERANZ, Y., and MOORE, R. B. 1975. Reliability of several methods for protein determination in wheat. Baker's Dig. *49*, No. 1, 44–48, 58.

STINSON, M. M., and O'DONNELL, M. J. 1952. The infrared and ultraviolet absorption spectra of cytosine and isocytosine in the solid state. J. Am. Chem. Soc. *74*, 1805–1808.

TREVIS, J. E. 1974. Seven automated instruments. Cereal Sci. Today. *19*, 180.

WILLIAMS, P. C. 1975. Application of near infrared reflectance spectroscopy to analysis of cereal grains and oilseeds. Cereal Chem. *52*, 561.

CHAPTER 10

Flame Photometry and Atomic Absorption

Flame photometry will be discussed only to the extent that it serves as a means of comparison with atomic absorption. Almost everyone that has taken a course in chemistry has placed various salts on a spatula and held them in the flame of a Bunsen burner and watched the colors form. If you did this systematically you find that only the alkalies, alkaline earths, and a few other elements would give off radiation that appears colored. Why do some elements give off radiation that appears colored and others do not? Figure 10.1 is a simplified energy diagram which may help explain the phenomenon.

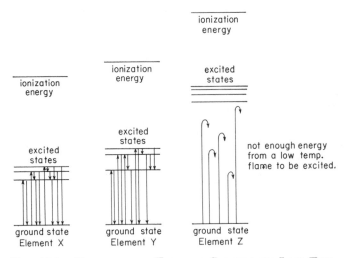

FIG 10.1. DIAGRAM FOR EMISSION SPECTRA AT LOW TEMPERATURES (NOT TO SCALE)

The heat energy from the flame can raise an electron in some atoms (elements X and Y) from the ground state to an excited state. If the excited electron drops back to the ground state from its original level in one jump the radiation given off is called a *resonance line*. If the electron loses its energy in steps, some of the energy may be given off in the visible and infrared regions or it may lose its

120

energy entirely by radiationless collisions. With the alkali and alkaline earth metals the energy required to excite the atoms is relatively low and there is enough energy in a gas-air flame (900°– 1200°C) to do this. However, with other elements the energy required to get to the excited state is greater than a Bunsen burner can provide (element Z).

Each metal has a different set of energy levels since each metal has a different nuclear charge and a different number of electrons. Therefore the wavelengths of emitted radiation will be different for each element, and a measure of the position of these wavelengths can be used to identify what atoms are present. If the intensity of this radiation is measured, then quantitative analysis is also possible.

The purpose of a flame photometer is to provide a flame whose temperature is hot enough to excite as many elements as possible, to determine which wavelengths are given off, and what their intensities are. It is possible to determine between 50 and 60 elements using present-day instruments. Figure 10.2 is a block diagram of a flame photometer.

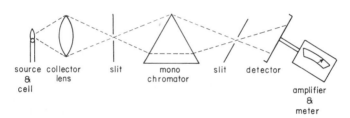

source collector   slit   mono   slit   detector
&   lens   chromator
cell
                                                              amplifier
                                                              &
                                                              meter

FIG. 10.2. BLOCK DIAGRAM OF A FLAME PHOTOMETER

A flame photometer can be considered as a UV-visible spectrophotometer in which the cell and source have been combined.

Interferences

Flame photometry interferences can generally be considered to be due to ionization, background emission, self-absorption, chemical interactions, and spectral overlap.

Ionization causes a decrease in intensity and occurs generally at low concentrations. At low concentrations there are few atoms present, they absorb all the energy, and they can be ionized. As the concentration increases, the energy is spread over more atoms, and less ions are produced. If ionization is a problem, add a small amount of an easily ionizable metal such as the alkalies. This will take up excess energy and the material sought will be ionized less.

Background emission comes from the flame and from other elements in the sample. This can be handled as follows. The emission due to the flame can be measured without any sample present and the instrument set at zero. If other elements are present, a synthetic blank is prepared and the radiation at the peak intensity is determined. Then move off to one side of the peak emission line and take another (the lowest) reading. If it is not zero, then background emission is present. The example problem in the quantitative analysis section shows how this calculation is actually made.

Self-absorption occurs when radiation emitted from one atom is just the right energy to be absorbed by another atom. The net

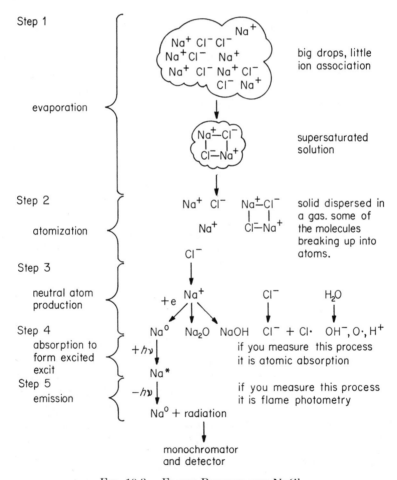

FIG. 10.3. FLAME PROCESS FOR NaCl

effect is that several atoms can be involved in absorbing and emitting, yet the detector sees only the final process, and the results indicate a much lower concentration than is actually present. This is a very serious problem, and in fact it is so efficient that advantage is taken of it in atomic absorption spectroscopy.

Spectral interferences are caused by two or more elements having emitted radiation whose wavelengths overlap. A good monochromator will generally reduce this considerably. Most of the remaining spectral interference is due to the emission of compounds. However, compound emission is generally in the green-red region, whereas atom spectra is in the violet-ultraviolet region. Therefore, if spectral interference is a problem, switch to another wavelength, preferably toward the UV region.

Chemical interference is caused by anions being present that complex the cations; thus, effectively lowering the metal concentration. Phosphates and aluminates are particularly troublesome. Chelating agents are generally used to reduce this type of interference. The chelate, although it ties up the metal also, is an organic material which burns away in the flame releasing the metal.

Processes Within the Flame

The process involved in flame photometry and atomic absorption is actually a very inefficient one and those attractive colors you see are due to a relatively few atoms. Figure 10.3 shows a representation of some of the steps involved.

Step 1.—The sodium chloride solution is aspirated into the flame by flame gases such as oxygen-hydrogen, oxygen-acetylene, air-acetylene, nitrous oxide-acetylene, and nitric oxide-acetylene. The solution is broken into drops, about 95% of which are too big to be useful because the solvent cannot be evaporated fast enough and the sample passes through the flame without further action.

Step 2.—The solvent is removed and the solid salt is dispersed in the flame gases. The heat energy from the combustion and the collisions between molecules produces atoms, ions, and free radicals.

Step 3.—Neutral atoms require the smallest amount of energy to raise them to an excited state, so any process that interferes with this process decreases the overall sensitivity of the measurement. Compounds may form with the combustion gases and their products, as well as with other elements in the sample. Although the amount of neutral atoms produced varies considerably, efficiencies of 10–15% are more likely than 80-90%.

Step 4.—To raise a neutral atom to an excited state may not take very much energy but it is a very inefficient process. Equation 10.1 can be solved to find out how many of the neutral atoms are eventually excited.

$$N_{ex} = N_g \frac{g_{ex}}{g_g} e^{-\Delta E/kT} \qquad (10.1)$$

where N_{ex} = number of neutral atoms excited
  N_g = number of neutral atoms in the ground state
  g_{ex}/g_g = statistical weight factor for the particular energy levels. For Na it is 2, and 3 for zinc
  ΔE = the energy to excite the atom
  k = Boltzman constant
  T = Flame temperature in °K

Table 10.1 shows the results of this calculation for sodium and zinc. Notice that at 3000°K only 1 out of 58,800 neutral sodium atoms is in the excited state and only 1 out of 558 trillion zinc atoms is excited. Zinc cannot be determined with a flame photometer, but can be determined easily by atomic absorption. Sodium can be determined in the 0.01 ppm region by flame photometry.

TABLE 10.1

VALUES OF N_{ex}/N_g FOR SODIUM AND ZINC

Element	2000°K	3000°K	4000°K
Na	9.86×10^{-6}	5.88×10^{-4}	4.44×10^{-3}
Zn	7.29×10^{-15}	5.58×10^{-10}	1.48×10^{-7}

Step 5.—The excited atom can return to the ground state either by losing its energy by molecular collision or by giving off a ray of radiation. It is unfortunate that we cannot control the direction in which the ray of radiation will come off. Since it can be radiated in all directions, most of it is lost and only that amount that can be collected by mirrors and lenses can be used.

In addition to the above losses, there is a high dilution loss. In order to vaporize 2 ml of a solution, about 10 liters of gas are required. If only 0.1 ml (95% is lost because of large drops) of the sample is used then this represents a dilution of 10,000. The original 10 liters of gas are further expanded to about 80 liters when heated to the flame temperature. The net result is that there are many places where an improvement in sensitivity can be made.

Our equation 10.1 gives some clues as to how to improve the sensitivity. One is to lower the ΔE and the other is to raise the temperature. Neutral atoms have the lowest ΔE so it would be logical to make more neutral atoms. It has been found that if alkali metals are added in small amounts to the sample, more neutral atoms of the element desired are formed. An electrical discharge across the flame also has a similar effect.

If the temperature can be raised to 4000°K then 1 atom of sodium in 4000 is excited, a substantial improvement over the 1 in 58,000 at 3000°K. The problem is how to produce a hotter flame. Theory predicts that the flame temperature for an oxygen-hydrogen flame should be much hotter than the normally measured 2800°K. The combustion product, water, dissociates appreciably at 2800°K and dissipates much of the energy.

A combustible mixture is needed that will produce products that will not dissociate. An oxygen-cyanogen $(CN)_2$ flame produces CO_2 and N_2, both of which dissociate little. As a result, a flame temperature of 5000°K is possible. However, cyanogen is quite expensive and very toxic so its use has been limited.

We have seen that there are many interferences in flame photometry, and that it is also very inefficient when utilizing the emission spectra. Those limitations led to a new approach.

The new approach was provided by Walsh (1955). Recall from the previous discussion on NaCl that at 3000°K only 1 out of 58,800 neutral atoms were excited. Recall also that a major interference in fluorescence and flame photometry was self-absorption, that is, wavelengths emitted from one atom were easily absorbed by another atom of the same kind.

What Walsh did in effect, was to provide a source of radiation whose wavelengths were exactly the same as the atoms in the flame required, thereby providing a much more efficient way for the neutral atoms to be excited. Figure 10.4 shows a diagram of an

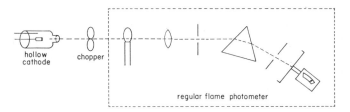

Fig. 10.4. A Schematic Diagram of an Atomic Absorption Instrument

atomic absorption instrument. Notice that a chopper and a hollow cathode have been added to a regular flame photometer. The hollow cathode, which will be discussed in more detail later, is used to produce a high intensity of radiation similar to an element to be determined in the flame. The chopper is used to produce a pulsating signal for easier amplification and to provide a means to distinguish hollow cathode radiation from flame radiation.

In flame photometry, the heat energy from the flame and the energy from molecular collisions is absorbed. This is not very efficient because many "energies" are available and only a few will excite the neutral atoms. This is where the hollow cathode lamp is effective. Its purpose is to provide the "right kind" of radiation. If magnesium is to be determined, then a magnesium hollow cathode is used; if calcium, then a calcium hollow cathode is adequate. Figure 10.5 is a diagram of a hollow cathode.

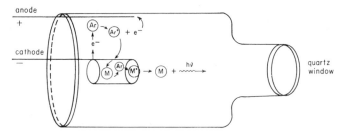

FIG. 10.5. NORMAL HOLLOW CATHODE

Electrons given off by the cathode accelerate toward the anode. Along the way they ionize argon atoms which are then attracted to the cathode. When the relatively large, highly accelerated argon atoms smash into the cathode, their energy is sufficient to knock several metal atoms off of the cathode surface. These atoms may then be excited by collision with the filler gas ions. When the excited atoms give off their energy, the radiation emitted is characteristic of the metal the cathode was made from. As an example, a copper cathode will give off radiation that copper atoms in the flame can readily absorb and that no other element will absorb. Therefore atomic absorption spectroscopy is almost completely free of spectral interference.

Why a hollow cathode? One reason is the high intensity of radiation it provides and another is to decrease the *Doppler effect*. If an atom is moving in the same direction as the ray of radiation it emits, then the atom's velocity is added to the velocity of the emitted ray changing its frequency.   The reverse is true when the atom moves away from the emitted ray.   This is the Doppler effect and it is undesirable because the atoms now give off a band of energies rather than the single energy desired.   This decreases the efficiency of absorption and also causes greater spectral interference.   The shape of the hollow cathode provides an electric and magnetic field that reduces the Doppler effect.

The main difference between flame photometry and atomic absorption is that in flame photometry the radiation emitted from the flame is measured, and in atomic absorption the decrease in the intensity of the radiation from the hollow cathode due to the absorption by the atoms in the flame is measured.

Chopper.—The atoms in the flame still emit radiation and since this is the same wavelength as that being absorbed, an error is introduced.   This is eliminated by the use of a chopper.   The chopper cuts the beam from the hollow cathode producing an alternating pulse of energy, while the energy emitted by the atoms in the flame is continuous.   It is possible electronically to distinguish these different signals and this is done by the detector system.

Burners.—The first atomic absorption instruments used the same type of burners as flame photometers.   These are seldom used now because being the turbulent flow type they gave high background noise and the number of atoms in the light path is small.   In order to increase the sensitivity of the system, mirrors can successfully be used with some elements.

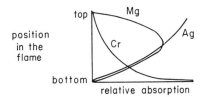

FIG. 10.6.   SOME FLAME PROFILES

Changing the flame profile is sometimes useful even if a mirror system is not used, because it may be possible to move the burner a few millimeters and improve the sensitivity severalfold.

To increase sensitivity without mirrors, several burners have been placed in a row thereby providing a longer sample path.   This increases sensitivity, but it is difficult to align all the burners and the resultant noise is disturbing.   A uniform flame that burns quietly and that has a long path length is attained in the laminar flow, elongated flame, or slot burner.

The long flame provides a longer path for absorbing radiation. It is a quiet flame because the big solvent droplets that produce the turbulence are eliminated by draining. A preheater is sometimes used to evaporate large amounts of solvent. Care must be taken with this type of burner because the fuel is premixed before it gets to the flame, and when the sample aspiration is stopped, a slight negative pressure develops, drawing the flame inside the chamber, producing an explosion. Air is used rather than oxygen for combustions because the flame speed is slower and the explosion hazard is reduced.

With an air-acetylene mixture it is possible to aspirate 70% perchloric acid without an explosion. The burner head should be coated with Desicote to keep it from dissolving.

TABLE 10.2

COMPARISON OF FLAME MIXTURES, COMBUSTION TEMPERATURES
AND FLAME SPEED IN ATOMIC ABSORPTION

Mixture	Combustion Temperature (°C)	Flame Speed (cm/sec)
Air-propane	1925	82
Air-acetylene	2300	160
50% O_2, 50% N_2	2815	640
O_2-acetylene	3060	1130
N_2O-acetylene	2955	180
NO-acetylene	3080	90

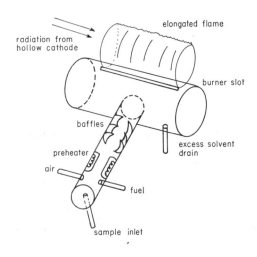

FIG. 10.7.　SLOT BURNER

Recently, nitrous oxide was used instead of air.   Nitrous oxide does not form oxides with the aspirated metals nearly as readily as air or oxygen so more atoms are reduced.   In addition, the nitrous oxide flame is several hundred degrees hotter than an air-acetylene flame so flame compounds are destroyed to a greater extent yet the flame speed is slow.   Table 10.2 compares flame mixtures, temperatures, and speeds.

Nitrous oxide is an anesthetic and a good exhaust system must be provided.

Pumps

A burner of the type shown in Fig. 10.7 can handle much more sample than it can itself aspirate into the chamber.   In addition, there is a pronounced viscosity effect on aspiration rate.   To overcome both problems, the sample is pumped into the burner.   A fivefold increase in sensitivity is generally attained and the results are much more reproducible.

Solid Samples

If a dilute solution is to be examined, the peak is often too broad and sensitivity is lowered.   With the normal aspiration rate of 3 to 5 ml per min, a width of 15–20 sec is usual.   However, if the sample could have the solvent removed and then aspirated, a much more intense peak should be obtained.   This has been done by using a tantalum boat about 3 in. long and about $1/4$ in. wide.   One milliliter of sample is placed into the boat, and the boat placed next to the flame, which heats it very rapidly and evaporates the solvent immediately.   The boat is then placed in the flame and the sample is aspirated in about 4 sec, so fast in fact that electronic recording devices are necessary to record the results.   Two samples per minute can be run and sensitivities can be increased fiftyfold.

Another way of handling solid samples is to mix the sample with a sort of gun powder into a pellet.   When the pellet is placed in the flame and ignited, or placed on a metal plate and ignited, the sample can be assayed.   One problem is that the combustion mixtures to date contain trace quantities of almost all of the common metals.

Determinable Elements

At the present time the following elements *cannot* be determined directly; Fr, Ra, Ac, Tc, Os, C, N, P, O, S, Po, all halogens and inert gases, plus Pm in the rare earths.   However, it is believed that in the near future it should be possible to determine Ra, Os, C, N, P, S, and the halogens except At.

Detection Limits

A distinction is made here between detection limits and sensitivity. Sensitivity is usually defined as the amount of the element that will produce a 1% change in the absorption reading, whereas detection limits are defined as the amount of material that will produce a signal twice as large as the background noise. Atomic absorption is capable of detecting 0.001 μg/ml for some systems with 1 μg/ml being most common.

Quantitative Analysis

If many samples are to be determined it is usually simpler to prepare a calibration curve. If only a few samples are to be determined at irregular intervals, the method of standard addition is preferred. The simple procedures given below were developed for flame photometry, but they are also adaptable for atomic absorption. Two solutions are required. The first solution contains an aliquot of the unknown, the second solution contains an aliquot of the unknown plus a measured amount of a known solution of the element to be determined. The increase in intensity (flame photometry) of the second solution over the first solution is due to the added standard. Using this value, the concentration of the unknown can be calculated.

Example

Solution 1 containing 25 ml of an unknown wine solution, which contains copper, was diluted to 50 ml and had an absorbance of 0.50. Solution 2, containing 25 ml of the wine and 5 ml of a 1 mg/ml copper standard, was diluted to 50 ml and had an absorbance reading of 0.60. What is the concentration of the copper in the wine?

The standard was diluted tenfold (5 ml to 50 ml), therefore 0.1 mg/ml of Cu had an absorbance of 0.1 (0.60–0.50). The unknown was 5 times as strong (0.50), but it had been diluted twofold (25 ml to 50 ml), so the original solution would have had an absorbance of 1.0 (2 × 0.50). From the standard, therefore, the solution contains 1 mg Cu per ml.

The above problem assumed that Beer's law was obeyed completely. How do you correct the calculation if the system deviates from Beer's law and there is a significant background enhancement? Figure 10.8 shows a plot of what happens and what measurements are needed.   Using the above values a proportionality factor can be obtained and a corrected and more accurate concentration determined.

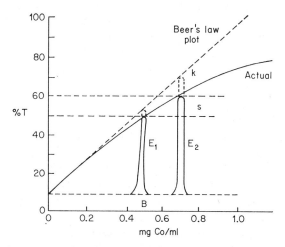

E_1 = emission reading of the unknown (In order to determine
        k for the first time this value will have to be calculated
        from a known solution).
E_2 = emission reading for the unknown plus the standard.
B = background
S = concentration of standard added, corrected for dilution.
U = unknown concentration
k = deviation constant

FIG. 10.8.   STANDARD ADDITION SPECTRA FOR THE DETER-
MINATION OF TRACES OF COBALT IN BEER

Example

Refer to Fig. 10.8.   Suppose 25 ml of a 0.80 mg/ml cobalt solution had been diluted to 50 ml and gave a reading E_1 and a second sample to which had been added 10 ml of a 0.08 mg Co/ml solution gave a reading E_2.   What is the k value, and what would be the concentration of an unknown if it had the same E_1 reading?

The equations necessary to solve the problem are:

$$(E_2 - B) = k(U + S) \qquad (10.2)$$

$$(E_1 - B) = kU_{\text{found}} \qquad (10.3)$$

$$(E_2 - E_1) = kS_{\text{found}} \qquad (10.4)$$

$$U_{\text{found}} \times \frac{S_{\text{added}}}{S_{\text{found}}} = U_{\text{actual}} \tag{10.5}$$

Using equation 10.2

$(60 - 10) = k(0.4 + 0.16)$ (corrected for dilution)

$k = 50/0.56 = 89.3$

Using equation 10.4

$(60 - 50) = 89.3 S_{\text{found}}$ $S_{\text{found}} = 0.112$

Using equation 10.3

$(50 - 10) = 89.3\ U_{\text{found}}$ $U_{\text{found}} = 0.448$

Using equation 10.5

$0.448 \times (0.16/0.112) = 0.64$ mg Co/ml

Some Recent Applications

The atomic absorption technique can be used to determine almost any metal in foods or food additives. Certain organic materials that react or form complexes with metal ions may also be determined indirectly. Meranger and Somers (1968) have shown that if one first evaporates the alcohol, Cu, Pb, Zn, Ni, Co, Cr, and Cd can be determined in wines and ciders. An example of an indirect method was developed by McCready (Potter *et al.* 1968) who determined sugar in dried parsley and frozen strawberries by adding copper to an extract. In a basic solution, sugar reacts with copper to form insoluble Cu_2O which is centrifuged from the remaining solution and the unreduced copper is then determined.

PROBLEMS

(1) While fruits and vegetables contain between 0.2 and 8.0 ppm Zn, herring can contain as much as 1200 ppm. The following data were obtained by means of an atomic absorption technique. What is the Zn content of the herring? Ans. 650 ppm

Zn Conc, Ppm	% Absorbance Reading
1100	87
900	60
700	35
500	17
300	4.5
100	2
Unknown	32

(2)   The normal bloodstream contains about 0.2 ppm Pb.   If this is increased to about 0.6 ppm, lead poisoning may occur.   Ten grams of sardines were digested with perchloric acid.   The Pb was complexed with diethyldithiocarbamate and extracted into 20 ml of chloroform.   A 5-ml aliquot was diluted with 5 ml of pure chloroform, and the solution gave an atomic absorption reading of 50%. A second 5-ml aliquot diluted with 5 ml of 0.01 mg Pb/ml chloroform produced a reading of 80% T.   Calculate the ppm Pb in the sardines.   Ans. 12

(3)   The quality of mustard can be determined by distilling the mustard oil into ammonia, forming thiosinamine.

$$C_3H_5NCS + NH_3 = S{=}C \underset{\textstyle NH_2}{\overset{\textstyle NHC_3H_5}{\big\backslash}}$$

allyl isothiocyanate   thiosinamine

Thiosinamine when treated with $AgNO_3$ precipitates an amount of Ag_2S corresponding to the content of S.   Black mustard seed must contain 0.6% to meet Federal standards.

To 5 gm of powdered mustard, 100 ml of water and 20 ml of ethanol are added.   After extraction, the mustard oil is distilled into 10 ml of 10.0% NH_3 and 20 ml of 0.2320 N $AgNO_3$ is added. The Ag_2S was filtered off and the resulting solution (1) had a reading of 60% by atomic absorption analysis.   A 10-ml aliquot of solution 1 had 0.2 ml of 0.064 gm Ag/ml added to it and this produced a reading of 70%.   Calculate the % mustard oil in the sample. Ans. 0.8%

BIBLIOGRAPHY

MERANGER, J. C., and SOMERS, E.   1968.   Determination of heavy metals in wines by atomic absorption spectrometry.   J. Assoc. Offic. Anal. Chemists *51*, 922–925.

POTTER, A. L., DUCAY, E. D. and McCREADY, R. M.   1968.   Determination of sugar in plant materials; measurement of unreduced copper by atomic absorption spectrophotometry.   J. Assoc. Offic. Anal. Chemists *51*, 748–750.

WALSH, A.   1955.   The application of atomic absorption spectra to chemical analysis.   Spectrochim. Acta *7*, 108–112.

CHAPTER 11

X Rays

Roentgen discovered X rays in 1895 and within a few years reported several basic facts on their properties. Among those properties are: (1) X rays travel in straight lines and with the speed of light; (2) X rays are not charged particles; (3) the generation of X rays may be accomplished by impinging a beam of high energy electrons onto a target, the higher atomic weight targets being the most efficient sources for X rays; (4) X rays effect a change in photographic emulsions; (5) electrical charges are dissipated when exposed to X rays; and (6) fluorescence is induced in many materials such as calcium tungstate and zinc sulfide. It wasn't until 1912, when it was found that copper sulfate and zinc blend would diffract X rays, that the analytical significance of X rays was realized.

Today, many areas of X-ray analysis have been developed such as X-ray absorption, diffraction, fluorescence, emission, absorption edge, low angle scattering, k-capture, soft X rays, and most recently, the electron probe. In direct quantitative analysis, X-ray fluorescence is now the most important technique with X-ray absorption quite valuable in many cases.

Why should X rays be considered for use as an analytical technique? X-ray absorption and fluorescence are atomic properties that are only slightly affected by the way an atom is bonded to another atom; the relationship between the mass absorption coefficient and wavelength is simple, elaborate separation schemes are not necessary; and they provide a means to determine which cation is associated with which anion. These factors combine to make X-ray techniques fast, simple, accurate, and in many cases, quite sensitive.

Figure 11.1 will be used to help explain how X rays are produced. Suppose an electron traveling at a very high velocity, strikes an atom. Since it has a high energy it is possible to overcome the negative charge, due to the electrons around the nucleus, and get into the inner levels of the atom. During this process, an electron will lose some of its energy because it requires work to push through the negatively charged field. However, if the electron has sufficient energy so that it is not repelled, it may transfer a part of its remaining energy to an electron in the K-shell around the nucleus causing it to be ejected. This creates a vacancy in a very low energy level and electrons from other orbitals will try to fill it. Since L-shell electrons are the most readily available, they usually succeed in filling this vacancy. When the L electron drops from its higher energy (less stable position) to the lower energy (more stable position) it loses

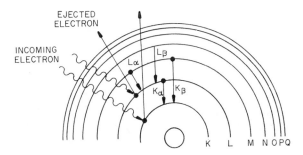

FIG. 11.1. PRODUCTION OF X-RAYS

energy. This yields a large amount of energy. When calculated in terms of wavelengths by equations $E = h\nu$ and $\lambda\nu = c$, the wavelengths are only a few angstroms in length.

The vacancy in the K-shell could have been filled by an electron from the M-shell, a less likely possibility but a more energetic one if it happens. Since in both cases a K-shell vacancy was filled, we call the X rays produced, K X rays, and the one filled from the closest shell a K_α and the other a K_β X ray.

If the original entering electron penetrated only as far as the L-shell, then the X ray would be called an L X ray and there would be L_α and L_β X rays.

Suppose a K electron is ejected and the vacancy is filled by an L electron. The L vacancy is in turn filled by an M electron, etc. When does the process stop? Theoretically it can go on until the outermost orbital is reached. But in practice it stops rather quickly. The efficiency of X-ray production is quite low. Only about 0.01 to 0.1% of the electron beam current produces X rays. This means that there is a very large number of electrons coming into the orbitals from the outside to fill the vacancies and stop the process.

Each element has a different arrangement of electrons about its nucleus and, since this means a different set of energies, we get X rays that are characteristic of each element. Figure 11.2 is a plot of X-ray intensity vs wavelength.

The K_β intensity is about 10 to 15% that of the K_α intensity. Notice that there is a large amount of radiation that is continuous.

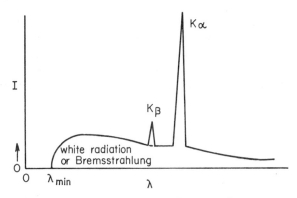

FIG. 11.2.   TYPICAL X-RAY EMISSION SPECTRUM

This is caused by the other 99.9% of the electrons that did not produce X rays, but still entered the electrical field of the atom and were slowed down (braking or Bremsstrahlung) by radiative collision with the orbital electrons.

X-ray Tubes

The purpose of the X-ray tube is to provide a beam of electrons with sufficient energy to produce X rays.   Figure 11.3 is a diagram of an X-ray tube.   A high voltage is placed across the filament and an electron cloud is formed around it.   When 5000 to 100,000 volts is applied to the target (anode) the electrons leave the cathode and hit the target, producing X rays characteristic of the target material. These X rays are emitted out of the tube through a thin Be window. Much of the energy is expended as heat and the target must be cooled, usually by water.

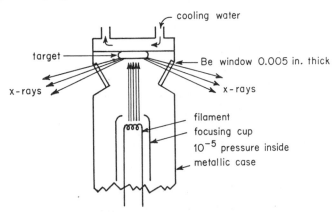

FIG. 11.3.   SCHEMATIC DIAGRAM OF AN X-RAY TUBE

Detectors

Figure 11.4 illustrates the ranges of detectors usually employed. The detailed operation of these will be explained in the section on radioactivity

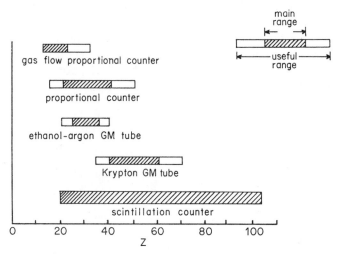

FIG. 11.4. X-RAY DETECTOR RANGES

X-ray Absorption

As with other radiations comprising the electromagnetic spectrum X rays are absorbed in a first-order process, just as visible light, and X ray absorption follows a Beer's law type equation.

$$I = I_o e^{-(\mu/\rho)\rho x} \tag{11.1}$$

where I_o = the initial intensity of the X-ray beams

I = the intensity of the beam after it has passed through the sample

x = the thickness of the sample material

ρ = the density of the sample

(μ/ρ) = mass absorption coefficient. Empirically it has been found that this can be approximated by

$$(\mu/\rho) = \frac{\alpha N_o Z^4 \lambda^{5/2}}{W} \tag{11.2}$$

where α = a constant, the same for all elements having Z greater than 13

Z = atomic number

N_0 = Avogadros number

W = atomic weight of the absorbing element

μ = linear absorption coefficient

From equation 11.2 it can be seen that absorption increases as the fourth power of the atomic number increases. Light elements are not very good absorbers of X rays and it is for this reason that a bone (rich in calcium, at. no. 20) shows up in an X-ray photograph, whereas skin and muscle (carbon, at. no. 6, hydrogen, at. no. 1) can hardly be seen. A dental filling (silver, at. no. 47) and a bullet (lead, at. no. 82) show up quite clearly. Table 11.1 shows the (μ/ρ) for several elements absorbing a copper K_α X ray.

TABLE 11.1

MASS ABSORPTION COEFFICIENTS (μ/ρ) CM^2/GM FOR A CU K_α X RAY AT 1.539 Å

Element	(μ/ρ)	Element	(μ/ρ)	Element	(μ/ρ)	Element	(μ/ρ)
H	0.48	Na	32.1	Ni	48	W	176
Li	1.10	Mg	40.8	Cu	50.9	Pt	202
Be	1.60	Al	49.0	Zn	58.6	Au	213
B	2.45	S	91	Br	89	Pb	230
C	4.52	Cl	103	Ag	217	Ne	24.0
N	7.45	Ca	163	Sn	247	A	114
O	11.1	Fe	325	I	290		

Example

DDT($C_{14}H_9Cl_5$, mol. wt 354.50) was extracted from 400 gm of whole green beans with 600 ml of acetone. This extract was concentrated to 3 ml and placed in an X-ray absorption cell having a cross-sectional area of 8.0 cm². The density of this solution was 0.792 gm/cm³. Using a X-ray beam from a tube with a copper target (K_α = 1.539Å), it was found that the intensity of the X-ray beam was decreased to 14.4% of the incident beam. Determine the amount of DDT in ppm on the green beans.

The first step is to calculate the (μ/ρ) for DDT and acetone. Refer to Table 11.1 for (μ/ρ) values for C, H, O, and Cl. For DDT, mol. wt = 354.50.

$$X_C = \frac{168.14}{354.50} = 0.474$$

$$X_H = \frac{9.072}{354.50} = 0.023$$

$$X_{Cl} = \frac{177.28}{354.50} = 0.500$$

So that for DDT,
$(\mu/\rho) = (0.474 \times 4.52) + (0.026 \times 0.48) + (0.500 \times 103) =$
      carbon   hydrogen   chlorine

53.95 cm^2/gm
    DDT

Similarly for acetone, mol. wt = 58.08

$$X_C = \frac{36.03}{58.08} = 0.621$$

$$X_H = \frac{6.048}{58.08} = 0.103$$

$$X_{Cl} = \frac{16.00}{58.08} = 0.276$$

So that for acetone,
$(\mu/\rho) = (0.621 \times 4.52) + (0.103 \times 0.48) + (0.276 \times 11.1) =$
      carbon   hydrogen   oxygen

6.50 cm^2/gm
  acetone

Now from equation 11.1 for our example,

$$\ln(I/I_o) = -(\mu/\rho) \times 0.792 \times \frac{3.00}{8.00}$$

$$= -(\mu/\rho) \times 0.296$$

where $I = 0.144\, I_0$ so that $I/I_0 = 0.144$ and $\ln 0.144 = -1.94$
    Thus:

$$(\mu/\rho) = \frac{-1.94}{-0.296} = 6.64 \text{ cm}^2/\text{gm}$$

Now it is necessary to determine what combination of the DDT and acetone (μ/ρ) values will give 6.64. Let X equal that fraction of DDT in the sample.

$$(X \times 53.95) + (1\text{-}X)(6.50) = 6.64$$
      DDT   acetone   unknown

where $X = 0.0034$ or 0.34% DDT in the 3-ml solution examined.

$$0.0034 \times 3 \text{ ml} \times 0.792 \text{ gm/ml}$$

$$= 0.0080 \text{ gm of DDT in 400 gm of beans}$$

Recall that 1 ppm = 1 mg/kg; 8 mg/400 gm = Y/1000 gm. Y = 20 ppm, an extremely high concentration.

This doesn't look like a simple assay method, but you must remember that if you were doing this routinely the (μ/ρ) values need only be calculated the first time, and only the last few steps need to be computed. If a calibration curve is prepared, the process is even faster and should be used for a complex system to avoid clean-up extractions.

The generally advantageous range for determinations by X-ray absorption methods is between 10 and 100% transmission. As little as 10^{-10} gm of calcium and phosphorus can be determined in biological specimens.

X-ray Diffraction

X-ray diffraction is not a routine analytical technique, but a knowledge of the principle involved is necessary for an understanding of X-ray fluorescence.

One of the properties of X rays is that they travel in straight lines and they are very penetrating. In addition, as electromagnetic radiation, X rays may be diffracted. In the case of X rays, the right order of spacings for diffraction exists in crystals where the atomic or ionic distances are of the order of a few angstroms.

When an X ray strikes a layer of atoms in a crystal it can be diffracted as shown in Fig. 11.5. If another X ray (λ_2) strikes the layer below the first layer, and the total distance it travels is an even number of wavelengths behind the first wavelength, λ_1, then λ_2 can add to λ_1, since they are both in phase. If this is repeated for several

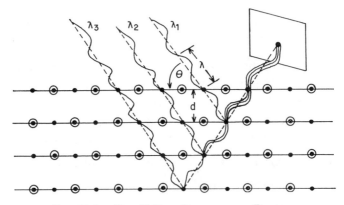

FIG. 11.5. THE X-RAY DIFFRACTION PROCESS

layers, the emerging X-ray beam is strong enough to be detected by a photographic film or other X-ray detector.

X-ray diffraction is now used primarily to examine the crystalline structure of materials, and to determine bond lengths and angles, expansion coefficients, and atomic and molecular weights. In food analysis it has found only limited application.

X-ray Fluorescence

Figure 11.6 shows one arrangement for X-ray fluorescence. The X rays from an X-ray tube are used as the source of high energy to produce secondary X rays in the sample. The sample is placed in an aluminum or plastic holder. The secondary X rays from the sample are characteristic of the elements in the sample. These X rays are collimated with a set of capillary tubes, usually made from nickel, and directed upon a crystal of known d spacing such as LiF. Recall the Bragg equation, $n\lambda = 2d \sin\theta$. We know d, and if θ is measured, then λ, the characteristic X rays and therefore the elements present in the sample can be determined. Since the angle θ depends on the wavelength, the analyzer crystal must be rotated to determine the various elements. As the crystal rotates through an angle θ, the detector must rotate through an angle 2θ. The crystal, the detector, and the mechanical parts necessary to provide the rotation is known as a goniometer.

The collimator greatly reduces the intensity of the final signal at the detector. One arrangement to eliminate the collimator is to use

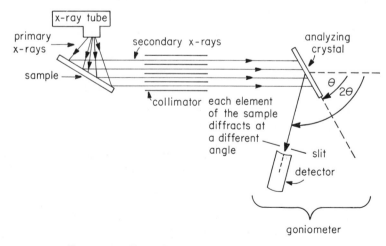

FIG. 11.6. FLAT CRYSTAL-REFLECTION METHOD

a curved crystal. The crystal used is a thin piece of mica. By bending the crystal, there is a focusing of the wavelength to a point in front of the detector.

X-ray Sources

The choice of source is determined by the elements being examined. This is somewhat of a peculiarity limited to X rays and may be explained by referring to Fig. 11.7.

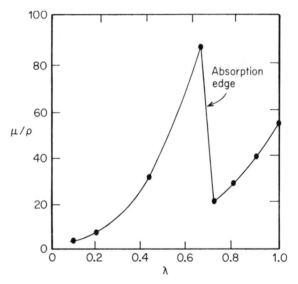

FIG. 11.7. ABSORPTION OF Mo AS A FUNCTION OF WAVE-LENGTH

As the wavelength decreases, the μ/ρ decreases. When the wavelength slightly exceeds the wavelength of a characteristic emission line for the sample, there is a rapid increase in the absorption of X rays. "As a general rule, the closer the characteristic tube radiation approaches the absorption edge value on the short wavelength side, the more efficient is the X-ray excitation process" (Birks 1960). Each element has 1 K, 3 L, and 5M absorption edges.

For biochemical assays, Cr, Mo, and W are the tubes of choice. Most are operated at 50 kv and 30 to 45 ma. At 60 kv there is barely enough energy to excite the Ba K_α ($Z = 56$) line, and usually less intense L lines are used for the heavier elements.

Collimators

Collimators may be either capillary tubes or parallel plates, with the latter the most common at the present. Generally they are made out of nickel. They range in size from 1 to 4 in. in length and from 5 to 20 mils separation. The closer the spacing and the longer the collimator, the higher the resolution but the poorer the detection limits.

Analyzing Crystals

The Bragg equation places some severe restrictions on d. As a result, a variety of different spacings must be available if all elements are to be examined with maximum sensitivity. Table 11.2 lists several crystal materials and the lowest atomic number element that can be determined.

TABLE 11.2

SELECTED ANALYZING CRYSTALS

Material	$2d$ Spacing (Å)	K_α	L_α
LiF	4.028	K(19)	In(49)
NaCl	5.039	S(16)	Mo(42)
Quartz	8.50	Si(14)	Rb(37)
Ethylene diamine tartrate (EDDT)	8.808	Al(13)	Br(35)
Calcium sulfate	15.12	Na(11)	Ni(28)
Mica	19.8	F(9)	Mn(25)
Potassium acid phthalate (KAP)	26.0	0(8)	V(23)
Half K salt of cyclo- hexane, 1,2-di acid	31.2		
Tetradecano amide	54		
Dioctadecyltere- phthalate	84		
Dioctadecyladipate	90		

The reflectivity of the last four materials in Table 11.2 is extremely high which means that the detection limits for the lighter elements can be lowered. Those materials are still being evaluated.

The use of oriented soap films that can be made easily into curved surfaces is a recent innovation that permits d values from 25 to 100 Å.

Detectors

Detectors are usually gas-flow proportional counters, thallium activated NaI scintillation counters, or the Xe proportional counter. They are discussed in detail in the chapter on radioactivity.

It should be noted that whereas the characteristic spectrum of an X-ray tube has superimposed on it the white radiation, the fluorescent spectrum does not have this disadvantage. This is shown in Fig. 11.8.

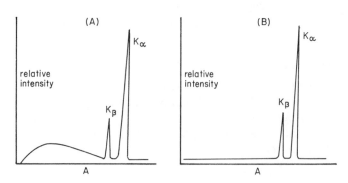

FIG. 11.8.   (A) CHROMIUM TARGET AT 50 Kv, (B) CHROMIUM EXCITED BY X RAYS FROM A W TUBE (FLUORESCENCE) OPERATED AT 50 Kv

Quantitative Analysis

For quantitative analysis, the intensity of the radiation from a given element is nearly proportional to the amount of the element present in the sample.   Since there are small deviations from linearity, it is the usual practice to prepare calibration curves.

In addition, one must correct the intensities, or intensity ratios, for absorption of the particular radiation by the sample.   This is done by using the equation 11.1.   A method recently developed (Cuttitta and Rose 1968) using a slope-ratio is quite effective in reducing matrix problems.

Some Recent Applications

The lighter elements, Ca, P, Cl, S, in biological tissue can be determined by using a W source, a EDDT crystal, and a gas flow proportional counter (Alexander 1965).

P, S, Cl, Ca, and K have been determined in rice flour, sugar, sweet potatoes, cucumbers, citrus fruit, and naval stores using a Cr tube and a KAP crystal.   The systems were found to be linear from 0.03 to 10% (Piccolo et al. 1968).

Champion et al. (1966) used X-ray fluorescence to determine Sr in oysters, milk, and grasses.   They could detect 3 ppm with 3%

precision.   They used a W target at 40 Kv and 20 ma, a LiF crystal, and a NaI scintillation counter.

Determination of elements important in food toxicology was described by Natelson and Whitford (1964).

Associated Problems

(1)   It was suspected that gasoline was spilled onto a shipment of potatoes.   The tetraethyl lead residue on the potatoes was extracted with n-octane, reduced to 2 ml, and placed in a cell with a cross-sectional area of 8.55 cm^2.   The sample, containing a small amount of lead tetraethyl in n-octane, had a density of 0.710 gm/cc.   Using an X-ray beam from a tube with a copper target, it was found that the intensity of the X-ray beam was decreased to only 38% of the incident beam.   Determine the percentage of lead tetraethyl in the n-octane extract.   Ans. 1.32%

(2)   The mass absorption coefficient (μ/ρ), for iron is 1.82 cm^2/gm at 50 kev.   What is (μ/ρ) for Ni at this same energy?   Ans. 2.33 cm^2/gm.

(3)   A first order diffraction of Cu K$_\alpha$ X rays from the 200 plane of rock salt is observed as strong radiation at an angle of 15.9°.   What is the distance between the 200 planes in NaCl?   Ans. 2.82 Å

(4)   Using a tungsten X-ray tube and a calcite analyzing crystal $(d = 3.029$ Å), a very strong line was observed at $\theta = 17.19°$ for an unknown metallic specimen.   Assuming that the spectrum is first order, calculate the wavelength of the observed X ray.   Ans. 1.79 (Co K)

(5)   Suppose you were interested in determining vanadium, iron, and nickel in cooking oil by an X-ray fluorescence method.   You have a tungsten target tube operating at 50 kev and 40 ma, and a LiF crystal.   At what values of 2θ would you look for these elements if you wanted to measure the K$_\alpha$ radiation?   K$_\alpha$ for V, Fe, and Ni is 2.504, 1.938, and 1.659 Å, respectively.   Ans. V = 76.93°, Fe = 57.6°, and Ni = 48.76°

BIBLIOGRAPHY

ALEXANDER, G.   1965.   An X-ray fluorescence method for the determination of calcium, potassium, chlorine, sulfur, and phosphorus in biological tissues. Anal. Chem. *37*, 1671–1674.

BIRKS, L. S.   1960.   The electron probe: an added dimension in chemical analysis.   Anal. Chem. *32*, No. 9, 19A–28A.

CHAMPION, K. P., TAYLOR, J. C., and WHITTEM, R. N.   1966.   Rapid X-ray fluorescence determination of traces of strontium in samples of biological and geological origin.   Anal. Chem. *38*, 109–112.

CUTTITTA, F., and ROSE, H. J.   1968.   Slope-ratio technique for the determination of trace elements by X-ray spectroscopy.   A new approach to matrix
problems.   Appl. Spectry. 22, 321–324.
NATELSON, S., and WHITFORD, W. R.   1964.   Determination of elements by
X-ray emission spectroscopy.   Methods Biochem. Anal. 12, 1–68.
PICCOLO, B., MITCHAM, D., and O'CONNER, R. T.   1968.   X-ray fluorescence
method for the quantitative analysis of light elements in chemically treated
cottons and other agricultural products.   Appl. Spectry. 22, 502–505.

CHAPTER 12

Potentiometry

Modern electroanalytical techniques other than potentiometry have been used little in food analysis because most people interested in food analysis have little familiarity with electroanalytical techniques and instruments. It is the purpose of this chapter to explain the major principles involved and to give some examples of potential and useful applications.

The heart of any electroanalytical technique is the electrode. When an electrode is placed in a solution it tends to send its ions into the solution (electrolytic solution pressure), and the ions in solution tend to react with the electrode (activity). Those two factors, the electrode pressure and the activity of the solution ions, combine to produce what is called an *electrode potential*. At a given external pressure and temperature, the potential is a constant that is characteristic of the metal.

We distinguish two classes of electrodes, *reference electrodes* and *indicator electrodes*. The potential of a reference electrode does not change significantly with changes in the surrounding solution, whereas the potential of an indicator electrode changes as the concentration of one of the ions in the solution changes.

No one has succeeded in devising a method for measuring an absolute single electrode potential, and a second electrode is always necessary to complete the circuit. However, relative electrode potentials can be obtained quite easily. By international agreement the potential of the reaction

$$2\,H^+ + 2\,e^- \rightleftarrows H_2 \tag{12.1}$$

is given a value of 0.00 v and all other electrodes are compared to it. Table 12.1 shows several of the common systems.

The term, $E°$, signifies that this is the potential of the system when all the species present are at unit activity.

The terms *activity*, *activity coefficient*, and *concentration* are quite confusing to many students. An example to illustrate the difference may help. Suppose there are 20 students in class (the concentration, C) and that 10 of them did today's assignment (the activity, a). The fraction (activity coefficient, γ) that are working is 0.5.

TABLE 12.1

REDUCTION POTENTIALS

Oxidized Form	Reduced Form	$E°$
$F_2 + 2\ e^- =$	$2\ F^-$	$+2.85$
$Pb^{4+} + 2\ e^- =$	Pb^{2+}	$+1.69$
$MnO_4^- + 8\ H^+ + 5\ e^- =$	$Mn^{2+} + 4\ H_2O$	$+1.52$
$ClO_3^- + 6\ H^+ + 6\ e^- =$	$Cl^- + 3\ H_2O$	$+1.45$
$Cr_2O_7^- + 14\ H^+ + 6\ e^- =$	$2\ Cr^{3+} + 7\ H_2O$	$+1.36$
$ClO_4^- + 8\ H^+ + 8\ e^- =$	$Cl^- + 4\ H_2O$	$+1.34$
$O_2 + 4\ H^+ + 4\ e^- =$	$2\ H_2O$	$+1.229$
$NO_3^- + H_2O + 2\ e^- =$	$NO_2^- + 2\ OH^-$	$+1.01$
$Ag^+ + e^- =$	Ag	$+0.7995$
$Fe^{3+} + e^- =$	Fe^{2+}	$+0.771$
$O_2 + 2\ H^+ + 2\ e^- =$	H_2O_2	$+0.682$
$Hg_2SO_4 + 2\ e^- =$	$2\ Hg + SO_4^-$	$+0.615$
$Fe(CN)_6^{3-} + e^- =$	$Fe(CN)_6^{4-}$	$+0.36$
$Cu^{2+} + 2\ e^- =$	Cu	$+0.345$
$Hg_2Cl_2 + 2\ e^- =$	$2\ Hg + 2\ Cl^-$	$+0.268$
$AgCl + e^- =$	$Ag + Cl^-$	$+0.2222$
$Sb_2O_3 + 6\ H^+ + 6\ e^- =$	$2\ Sb + 3\ H_2O$	$+0.152$
$Sn^{4+} + 2\ e^- =$	Sn^{2+}	$+0.15$
$2\ H^+ + 2\ e^- =$	H_2	0.000
$Fe^{3+} + 3\ e^- =$	Fe	-0.036
$Pb^{2+} + 2\ e^- =$	Pb	-0.126
$Sn^{2+} + 2\ e^- =$	Sn	-0.140
$Ni^{2+} + 2\ e^- =$	Ni	-0.25
$Tl^+ + e^- =$	Tl	-0.336
$Cd^{2+} + 2\ e^- =$	Cd	-0.4020
$U^{4+} + e^- =$	U^{3+}	-0.50
$Cr^{3+} + 3\ e^- =$	Cr	-0.71
$Zn^{2+} + 2\ e^- =$	Zn	-0.762
$Al^{3+} + 3\ e^- =$	Al	-1.67
$Th^{4+} + 4\ e^- =$	Th	-2.06
$Na^+ + e^- =$	Na	-2.71
$K^+ + e^- =$	K	-2.92
$Li^+ + e^- =$	Li	-3.06

$$a = \gamma C \tag{12.2}$$

$E°$ therefore is an ideal situation and is not likely to be found often in practice. The $E°$ must be corrected for the actual system and this can be done by applying the Nernst equation.

$$E_{actual} = E° + \frac{2.3RT}{nF} \log \frac{(a_{ox})}{(a_{red})} \tag{12.3}$$

where R = 8.316 joules/mole-degree

T = temperature in °K

n = the number of electrons involved

F = the Faraday; 96,500 coulombs

(a_{ox}) = actual activity of the oxidized species

(a_{red}) = actual activity of the reduced species

(At 25°C, $2.3RT/F$ = 0.0591).

Concentrations are much easier to work with than activities, and fortunately at low concentrations the activity so closely approaches the concentration that actual concentrations are used in the Nernst equation. Using the permanaganate system in Table 12.1 as an example;

$$E_{actual} = 1.52 + \frac{0.0591}{5} \log \frac{[MnO_4{}^-][H^+]^8}{[Mn^{2+}][H_2O]^4}$$

Sign Convention

If an electrode reaction is written as a reduction such as:

$$Zn^{2+} + 2\,e^- = Zn \qquad E^\circ = -0.762$$

$$Cu^{2+} + 2\,e^- = Cu \qquad E^\circ = +0.345$$

A − sign indicates the metal is a better reducing agent than H_2, and a + sign indicates that the metal is a better oxidizing agent than H_2. The signs also indicate the polarity of the electrode if it is used in a battery.

Overvoltage

The difference between the equilibrium electrode potential and the actual electrode potential is called *overvoltage*. This occurs because some electrodes, because of temperature, current density, electrode crystal structure, or surface area, do not arrive at an equilibrium value rapidly. Excess energy required to force the equilibrium is provided by an increase in the applied voltage; it is more negative for the cathode and more positive for the anode. A few examples are shown in Table 12.2.

TABLE 12.2

HYDROGEN OVERVOLTAGE ON SEVERAL METALS

Metal	Current Density (Amp/Cm²)	
	0.01	0.10
Platinized platinum	0.035	0.055
Smooth platinum		0.39
Gold	0.56	0.77
Iron	0.56	0.82
Silver	0.76	0.90
Mercury	1.10	1.18

Source: Hammer and Wood (1958).

Overvoltage is very important in analytical procedures. Water will form hydrogen at 1.23 v. This means that under ideal conditions, a metal that would require 1.3 v to be reduced could not be determined, because the water in the system would decompose first. However, as a result of electrode overvoltage, it may be possible to plate out the metal providing its overvoltage is significantly different than that of hydrogen.

Reference Electrodes

The major reference electrodes are the standard hydrogen electrode (SHE), the saturated calomel (SCE), and the silver–silver chloride electrode. The standard hydrogen electrode is the international standard, but is seldom used for routine work because more convenient electrodes and reliable calibration buffers are available. The SHE will not be discussed here.

Calomel Electrode.—The calomel electrode is the most widely used reference electrode. Its operation is based upon the reversible reaction:

$$Hg_2Cl_{2(s)} + 2\ e^- = 2\ Hg + 2\ Cl^- \qquad (12.4)$$

where E_{25}° = 0.2444 satd. KCl (common U. S. electrode)

= 0.2501 3.5N KCl (common European electrode)

= 0.3356 0.1N KCl

Figure 12.1 shows one arrangement of this electrode.

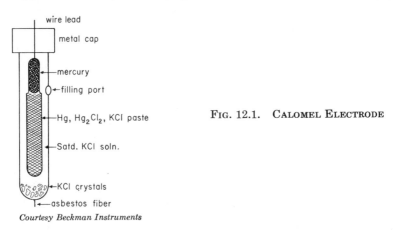

wire lead

metal cap

mercury

filling port

Hg, Hg$_2$Cl$_2$, KCl paste

Satd. KCl soln.

KCl crystals

asbestos fiber

Courtesy Beckman Instruments

FIG. 12.1. CALOMEL ELECTRODE

This electrode is unstable above 80°C and should be replaced by a Ag-AgCl electrode at high temperature.

In order to make an electrical connection between the electrode and the sample solution, some means must be provided that will produce a negligible potential and be reproducible. This can be done by a *salt bridge*. The saturated KCl salt bridge is the most common type. Why KCl? Why saturated? Refer to Fig. 12.2 which is a schematic diagram of an enlarged section of a liquid-liquid junction.

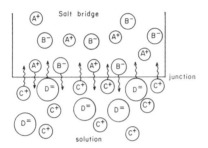

FIG. 12.2. ORIGIN OF LIQUID-LIQUID JUNCTION POTENTIALS

Consider the A^+ and B^- ions. Initially there are none of these in the sample solution and a large concentration in the salt bridge. These ions, following the laws of thermodynamics, will diffuse from the region of greater concentration (more order, lower entropy) to the lower concentration (less order, more entropy). However, the smaller A^+ ions can diffuse faster than the larger B^- ions with the result that a small charge difference or potential is established. The same thing can happen with the C^+ and $D^=$ ions from the sample diffusing into the salt bridge. The net result is called the *liquid-liquid junction potential* and can amount to 20–30 mv. This is a large value compared to the few tenths of a mv variation that is necessary for good electrode potential measurements.

If the hydrated $+$ ion and the $-$ ion were the same size, they would diffuse at the same rate and no potential would be established. K^+ and Cl^- are two ions with approximately equal mobilities, and KCl is the usual salt bridge material. This choice reduces considerably half of the junction potential. We are not able to choose our sample solutions so that the ions diffuse at the same rate. However, there is a way of significantly reducing the diffusion by the sample ions. If the salt solution in the bridge is saturated with KCl so that there are millions of ions coming out of the bridge compared to a few sample ions, then the sample ions find it hard to diffuse against this stream of ions. Although the solution diffusion is not eliminated, it is reduced to the point where the junction potential is less than 1 mv.

A solution $3N$ in KCl and $1N$ in KNO$_3$ has been found to be the best salt bridge concentration. For ions that have Cl^- interference

(Pb, Hg, Ag), a NaNO₃ salt bridge is preferred.   The necessity of
having a stream of ions coming from the salt bridge makes the depth
of immersion of the electrode into the sample solution quite important.

A pressure head of about 2 cm is required to ensure that the KCl
salt solution will stream out of the fiber or porous disc at the rate of
0.1 ml/day.

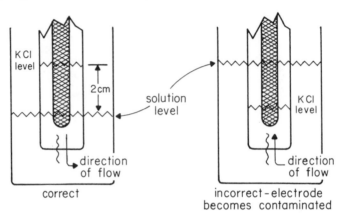

FIG. 12.3.   DEPTH OF IMMERSION FOR CALOMEL ELECTRODES

Silver-Silver Chloride Electrode.—The Ag-AgCl electrode is a
very reproducible electrode.   Its operation is based on the reversible
reaction:

$$AgCl_{(s)} + e^- = Ag_{(s)} + Cl^- \qquad (12.5)$$

where $E^\circ + E_j$ = 0.1992 (0.1 N HCl + satd. KCl)

$$= 0.1981 \text{ (in buffer solutions)}$$

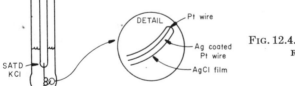

FIG. 12.4.   SILVER-SILVER CHLO-
RIDE ELECTRODE

Figure 12.4 shows a Ag-AgCl electrode.

The electrode is prepared by electroplating a layer of Ag on a Pt
wire, then converting the surface Ag to AgCl by electrolysis in HCl.
An alternate method uses thermal decomposition of Ag₂O and
AgClO₄.

The solution surrounding the electrode should be saturated with KCl and AgCl. The electrode is very sensitive to Br$^-$ and O_2 interference, and requires about 30 hr to reach equilibrium after it is initially prepared. For most determinations this electrode is relatively free of interfering side reactions and is the internal reference electrode in glass electrodes.

Indicator Electrodes

Glass Electrode.—The most common electrode in use today for the measurement of hydrogen ion concentration is the glass electrode. The relationship between pH and potential can be described as follows. The half cell reaction $H_2 = 2 H^+ + 2 e^-$ can be substituted into equation 12.3.

$$E = E° + \frac{0.0591}{2} \log \frac{(a_{H^+})^2}{(a_{H_2})} \tag{12.6}$$

Log $(a_{H^+})^2 = 2 \log (a_{H^+})$ and (a_{H_2}) is 1 at atmospheric pressure, so

$$E = E° + \frac{0.0591}{2} \times 2 \times \log (a_{H^+}) \tag{12.7}$$

Recalling that

$$pH = -\log (a_{H^+}) \tag{12.8}$$

and substituting this into equation 12.7,

$$E = E° - 0.0591 \text{ pH} \tag{12.9}$$

The result is that pH is linearly related to the electrode potential. This is why most pH meters have both a millivolt and a pH scale. A schematic diagram of a glass electrode is shown in Fig. 12.5.

The following discussion on the operation of the glass electrode is from a presentation by Rechnitz (1967).

"If a cross section is taken through a membrane of a functioning glass electrode, one finds that the membranes structure is distinctive and consists of several discrete regions and interfaces. This structure can be represented by the simplified schematic:

internal solution	hydrated gel layer	dry glass layer	hydrated gel layer	external solution

The operation of glass electrodes must be examined in terms of these various layers and of the processes that take place at the inter-

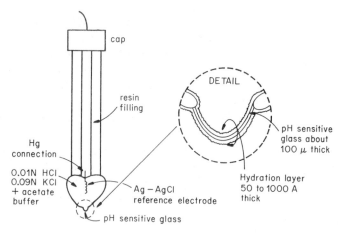

FIG. 12.5.   GLASS ELECTRODE

faces.   Additional "intermediate" layers may also exist on either side of the dry glass layer.

In practice, the dry glass layer constitutes the bulk of the membrane thickness, and the hydrated layers vary in depth from 50 to 1000 Å for the most useful electrodes, depending on the hygroscopicity of the glass.   When the dry electrode is first immersed in an aqueous medium, the formation of the hydrated (external) layer causes some swelling of the membrane.   Thereafter, a constant dissolution of the hydrated layer takes place with the accompanying further hydration of dry glass so as to maintain the thickness of the hydrated layer at some roughly constant, steady state value.   The rate of dissolution of the hydrated layer depends on the composition of the glass and also on the nature of the sample solution.   The rate of dissolution largely determines the practical lifetime of the electrode; lifetimes vary from a few weeks to several years.   (Storing the electrode in aqueous solution when not in use increases its life about 30%.   Repeated leaching and drying hasten the destruction of the silicon-oxygen network by the extracted lye.)

The hydrated layer must be present, at least on the external electrode surface, for glass electrodes to function properly in solutions containing water, probably because the movement of ions in glass is aided by the hydration of the glass.   For example, the diffusion coefficients of univalent cations in hydrated glass, about 5×10^{-11} cm^2/sec, are about 1000 times greater than in dry glass.

Schwabe and Dahms using radioactive tracers to study the hydrated layer found that:

"(1) The hydrated glass surface undergoes cation exchange according to $Na^+_{soln} + H^+_{glass} = Na^+_{glass} + H^+_{soln}$ when the membranes are in contact with a solution containing Na^+. Anions are not exchanged. The pH dependence and mass balance of the cation exchange process rules out the self exchange of Na^+ alone.

"(2) Greater quantities of metal ion are taken up than are required for simple monolayer coverage of the glass, indicating that metal ions diffuse into the hydrated layer.

"(3) The takeup of metal ions by the glass as a function of pH can be correlated with the electrode properties of the glass. Furthermore, the selectivity order of ion exchange for a series of alkali metal ions is the same as the potentiometric selectivity order of the glass electrode."

The overall potential at the glass electrode consists of the algebraic sum of a phase boundary (ion exchange) contribution and a diffusion potential contribution.

An understanding of the true mode of the functioning of glass electrodes was delayed for many years by the uncritical acceptance by chemists of the attractive, but erroneous, hypothesis that hydrogen ions selectively penetrate the glass membrane to yield the electrode potential. Although it is true that hydrogen ions (not hydronium ions) undergo exchange across the solution-hydrated layer interface, these ions do not penetrate the glass membrane under normal circumstances. Schwabe and Dahms elegantly demonstrated this with their coulometric experiment involving prolonged electrolyses on glass electrode bulbs filled with tritium labeled sample solutions. If hydrogen ions penetrate the membrane some appreciable fraction of the total quantity of electricity passed during the electrolysis should be accounted for by the transport of H^+; that is, tritium should be found on the nonlabeled side of the membrane. In fact, the quantity of tritium found on the nonlabeled side of the membrane never exceeded the natural tritium content of the outside solution even after 20 hr of intensive electrolysis at elevated temperatures.

Of course, some small but finite current must flow during the potentiometric measurement and, thus, there must be transport across the glass membrane system. Charge can be transferred across the solution-hydrated layer interface by ion exchange and, within the hydrated layer, by diffusion. But, how then is charge carried through the dry glass portion of the membrane to complete the circuit? The absence of a Hall effect in electrode glasses indicates that electronic conduction can be ruled out (even in glass electrodes with semiconductor additives), and all available experimental evidence points to the fact that the current is carried by an ionic mechanism. When an Na_2O-Al_2O_3-SiO_2 glass is electrolyzed under completely anhydrous conditions, sufficient amounts of metallic sodium are always formed at the cathode to account for all of the current passed in the electrolysis. The current is, in fact, carried entirely by the cationic species of lowest charge available in any given glass. No single sodium (for sodium silicate glass) moves through the entire thickness of the dry glass membrane but, rather, the charge is transported by an interstitial mechanism where each charge carrier needs to move only a few atomic diameters before passing on its energy to another carrier."

Ion Selective Electrodes

The glass electrode is generally thought of as being only of value in measuring pH but if the composition of the glass membrane is changed, this electrode can be quite sensitive for other cations. Electrodes for Na, K, NH_4, Ag, Li, Cs, and Rb (all univalent ions) are already commercially available.

For example, a glass consisting of 72.2% SiO_2, 6.4% CaO, 21.4% Na_2O (mole %) is the usual glass for H^+ determinations, but if the glass composition is changed to 71% SiO_2, 11% Na_2O and 18% Al_2O_3 the electrode is 2800 times more sensitive to Na than K. If this is further changed to 68% SiO_2, 27% Na_2O and 5% Al_2O_3, the electrode then becomes K sensitive ($K_{K^+/Na^+} = 20$). An electrode made from a glass containing 52.1% SiO_2, 19.1% Al_2O_3, and 28.8% Na_2O is 100,000 times more sensitive to Ag^+ than H^+.

Precipitate Impregnated Membrane Electrodes.—According to Rechnitz, "the success of glass membranes as cation selective electrodes rests largely on the fact that the hydrated glass lattice contains anionic "sites" that are attractive to cations of appropriate charge to size ratio." If a similar exchange process could be set up at a membrane material having cationic sites, it would in principle be possible to prepare anion-selective electrodes. Actually, it is quite easy to construct ion exchange membranes that permit exchange of either cations or anions; the difficulty is that those membranes show insufficient selectivity among anions and cations of a given charge to be satisfactory as practical electrodes. The problem then is one of finding an anion exchange material that will display appreciable selectivity among anions of the same charge, and also have suitable properties to permit its processing into a membrane electrode. Thus far, silicone rubber has been found to be the most effective.

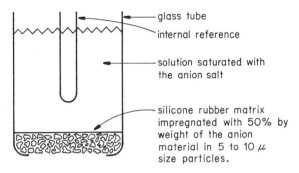

glass tube
internal reference
solution saturated with the anion salt
silicone rubber matrix impregnated with 50% by weight of the anion material in 5 to 10 μ size particles.

FIG. 12.6. PRECIPITATE IMPREGNATED ELECTRODE

An inert, semiflexible matrix (silicone rubber) is used to hold an active precipitate phase (AgI for an I $^-$ electrode) in place. Such membranes are called heterogeneous or precipitate impregnated membranes. Fisher and Babcock (1958) used radioactive tracer materials to show that the electrode potential is determined by the electrical charge on the surface of the inorganic precipitate particles, and that the current is carried by the transport of the counter ion through the membrane.

The main advantage of the membrane type electrodes over the older metal-metal halide electrodes is their insensitivity to red-ox interferences and surface poisoning. Figure 12.6 is a schematic presentation of a precipitate impregnated electrode.

At the present time electrodes sensitive to chloride, bromide, iodide, sulfide, sulfate, phosphate, and hydroxide ion have been developed.

Solid-State Electrodes.—The active membrane portion of this electrode consists of an inorganic single crystal doped with a rare earth (Fig. 12.7). For example the Orion fluoride electrode is ap-

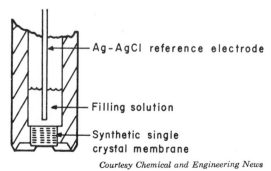

Ag-AgCl reference electrode

Filling solution

Synthetic single crystal membrane

Courtesy Chemical and Engineering News

FIG. 12.7. SOLID STATE ELECTRODE

parently crystalline lanthanum fluoride that has been doped with europium(II) to lower its electrical resistance and facilitate ionic charge transport.

The electrode has approximately a 10-fold selectivity for fluoride over hydroxide, and at least a 1000-fold selectivity for fluoride over chloride, bromide, iodide, hydrogen carbonate, nitrate, sulfate, and monohydrogen phosphate. The electrode does not require pre-conditioning or soaking prior to use. Chloride, bromide, iodide, and sulfide electrodes are also available.

Liquid-liquid Membrane Electrodes.—The range of selective ion-exchange materials could be greatly extended if such materials could be used in electrodes in their liquid state. Liquid ion exchange materials that possess high selectivity for specific ions may be tailored by appropriate chemical adjustment of the exchanger on the molecular level.

The main problems hindering the development of successful liquid-liquid electrodes are mechanical. It is necessary, for example, that the liquid ion exchanger be in electrolytic contact with the sample solution, yet actual mixing of the liquid phases must be at a minimum.

At the present time electrodes are available for calcium, magnesium, cupric copper, chloride, perchlorate, and nitrate. Figure 12.8 shows a liquid-liquid membrane electrode.

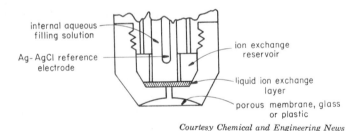

internal aqueous filling solution

Ag-AgCl reference electrode

ion exchange reservoir

liquid ion exchange layer

porous membrane, glass or plastic

Courtesy Chemical and Engineering News

FIG. 12.8. LIQUID-LIQUID MEMBRANE ELECTRODE

Bimetallic Electrodes.—A very sensitive, yet inexpensive and simple electrode pair consists of two wires, usually of different composition. Their performance is based on the fact that one of the electrodes reaches equilibrium with the solution before the other one does. The main advantages of electrodes of this type are that they can be made very tiny and they can be bent into any shape desired. Nickel and platinum are preferred for organic oxidation-reduction titrations. A platinum-platinum rhodium pair is satisfactory for

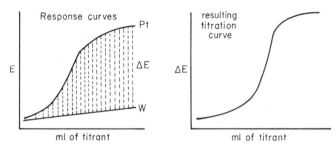

FIG. 12.9. A Pt-W ELECTRODE PAIR

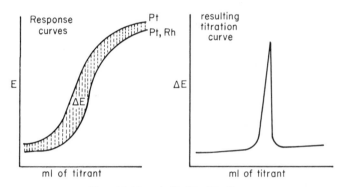

FIG. 12.10. A Pt-Pt, Rh PAIR

acid base titrations. Two cases and the resulting titration curves are shown in Fig. 12.9 and 12.10.

Potentiometry

The finest electrodes in the world are of no value unless the potentials they have generated are measured and recorded accurately. The methods of potentiometry consist of measuring the potential difference between the indicator electrode and a reference electrode at various intervals during the progress of a titration. An electrode potential is an equilibrium process and during its measurement the equilibrium must not be disturbed or the potential obtained will be incorrect. Most potential measuring devices, such as voltmeters, require considerable current for operation. When this current is passed through the electrode, either metal is plated out from the solution or gas is evolved. In either case the equilibrium at the electrode has changed and you are not measuring what is in the solution, but what is next to the electrode and this may be quite different and usually is continuously changing. The net result is that the measurement is erratic and a more refined method of measuring potential is necessary.   A null method is often used in which an equal but opposite potential is applied to the electrolytic cell. This does not alter the electrode process, but it does reduce the current flow to zero thus maintaining the equilibrium.   When the solution potential balances the external potential, a galvanometer placed in the circuit will show no deflection.   The cell potential can then be determined by measuring the external potential.   The apparatus for doing this is called a *potentiometer*.   A diagram of a student potentiometer is shown in Fig. 12.11.

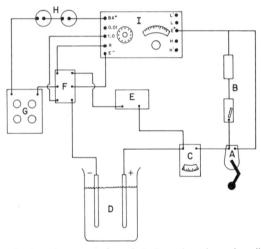

A.   Tap key, depress only for an instant as a long depression will
      change the cell equilibrium.
B.   Resistances for course and fine adjustment.
C.   Galvanometer-This can be used with metal electrodes but not
      with glass electrodes.
D.   Titration cell
E.   Standard cell-This cell is used to calibrate the instrument.
F.   Double pole double throw switch
G.   Decade resistance box-for a coarse adjustment to balance the
      potential of the bucking batteries(H).
H.   1.5 v batteries -To provide the potential in opposition to the
      cell voltage
I.   Potentiometer-contains a precision slide wire resistor and a
      series of 0.1 v steps to cover the range 0 to 1.5 v.

FIG. 12.11.   A STUDENT POTENTIOMETER

The Direction of a Reaction.—A problem often encountered is in what direction does an oxidation-reduction reaction proceed? Suppose you had a solution of U^{+3} and wanted to oxidize it to U^{+4} with Sn^{+4}.   Is the reaction feasible?   There are two rules that must be followed: (1) Write the half cell *reduction* reactions (Table 12.1 or equivalent) with the most positive reaction on top. (2) Proceed in a clockwise direction *around* the reactions to get the reaction direction. This is shown below for the Sn^{+4}, U^{+3} system.

$$Sn^{+4} + 2e^- = Sn^{+2} +0.04 \text{ v} \qquad \text{direction of}$$

$$U^{+4} + 1e^- = U^{+3} -0.05 \text{ v} \qquad \text{the reaction}$$

Notice that the system will go as desired. A solution of Sn^{+2} could not be oxidized to Sn^{+4} with a solution of U^{+4}- the reverse reaction- unless some additional external potential was applied.

Dissociation Constants of Complex Ions.—The following paragraphs are included to make you aware of an apparent difficulty encountered in making measurements of metal ions in the presence of complexing agents. The net effect of the complexing agent is to lower the concentration of the metal ion and completely alter the measured potential. This is of real importance in food analysis when trace metals are to be determined and complexing agents are present.

Example

A food sample was suspected of containing cyanide. The sample was extracted with water, made 0.01 M in $ZnCl_2$, and diluted to one liter. The potential of a metallic zinc electrode and a 1 M calomel electrode pair dipping into the solution was -1.342 v. Knowing that the dissociation constant of $Zn(CN)_4^=$ is 8.9×10^{-17}, calculate the concentration of CN^- in the solution.

$$Zn^{2+} + 2 e^- = Zn \qquad E_1^\circ = -0.763$$

$$Hg_2Cl_2 + 2 e^- = 2 Hg + 2 Cl^- \qquad E_2^\circ = +0.282 \text{ v}$$

The reaction is:

$$Zn + Hg_2Cl_2 = Zn^{2+} + 2 Hg + 2 Cl^-$$

$$E_{cell} = E_2^\circ - E_1^\circ + \frac{0.0591}{2} \log \frac{[Zn^{2+}][Hg]^2[Cl^-]^2}{[Zn][Hg_2Cl_2]}$$

All but Zn^{2+} are in their standard states, and thus their activities are equal to unity, which gives

$$E_{cell} = E_2^\circ - E_1^\circ + \frac{0.0591}{2} \log [Zn^{2+}]$$

$$-1.342 = 0.282 - [-0.763] + \frac{0.0591}{2} \log [Zn^{2+}]$$

$$\log[Zn^{2+}] = -(0.594)/(0.0591) = -10.05$$

$$[Zn^{2+}] = 8.9 \times 10^{-11} \text{ moles/liter}$$

Since we started with 0.01 M Zn^{2+} this means that the Zn^{2+} concentration has been greatly reduced by the presence of a complexing agent. We are assuming the following reaction,

$$Zn(CN)_4^= = Zn^{2+} + 4 \; CN^-$$

$$K_d = \frac{[Zn^{2+}][CN^-]^4}{[Zn(CN)_4^=]} = 8.9 \times 10^{-17}$$

$$8.9 \times 10^{-17} = \frac{[8.9 \times 10^{-11}][CN^-]^4}{[0.01 - 8.9 \times 10^{-11}]}$$

(ignore because it is small
compared to 0.01)

Therefore

$$[CN^-] = 0.01 \; M$$

However, you must remember that the 0.01 M $Zn(CN)_4^=$ would require 0.04 M CN to form the complex. The $[CN^-]$ measured is actually that which is in excess of the amount necessary to make the complex. The total CN^- present is therefore 0.04 + 0.01 or 0.05 M.

pH Meters.—The electrical resistance of the glass electrode is very high, being of the order of 500,000 ohms in very thin-walled and fragile membranes, and as high as 200,000,000 ohms in rugged commercial products. The measurement of the potential of the glass cannot be made with an ordinary potentiometer because the current which passes through the glass membrane under the very low potential of the cell is insufficient to move the galvanometer needle. Vacuum tube amplifying devices capable of registering minute amounts of current can be substituted for the galvanometer, and the measurement of the potential across the glass membrane can be made. Essentially, the commercial pH meter is an ordinary potentiometer in which the galvanometer has been replaced by a vacuum tube device for indicating the potential difference. Vacuum tube amplifiers or transistors require only minute current for their operation and they may be left connected to the cell.

Nonaqueous Titrations.—Water has some major limitations as a titrating medium for organic compounds. Not only are there relatively few organic compounds that are soluble in water, but most of those that are soluble are too weakly basic or acidic to be titrated accurately. It has long been known that the strength of basic compounds, such as amines, is much greater in glacial acetic acid than in water and that acidic compounds such as sulfonamides and phenols are more acidic in a basic solvent such as dimethylformamide. Consider a system (A) of a base in water as shown in Fig. 12.12.

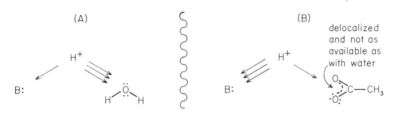

FIG. 12.12.   ACID-BASE SOLVENT SYSTEMS

The H^+ from the titrant can react with the base to form BH^+ or with the water to form H_3O^+. For weak bases, such as amines, the electrons are not as readily available as the electrons in water ($K_a = 10^{-7}$), and the proton will react with the water, not completely, but sufficiently to ruin the stoichiometry of the base titration. However, if the base, B, is placed in acetic acid ($K_a = 1.8 \times 10^{-5}$), the H^+ from the titrant will prefer to react with the more readily available electrons of B: and the titration will now go to completion.

A convenient electrode pair for nonaqueous titrations is silver-silver chloride and glass, the latter being the indicator electrode as it is used when pH determinations are made in aqueous solutions. The silver-silver chloride electrode is preferred to a saturated calomel electrode or a mercury-mercurous sulfate electrode, because the silver-silver chloride electrode eliminates the necessity of an aqueous salt bridge. The glass electrode must be soaked in water every hour or so when used in nonaqueous solvents in order to provide the hydrated layer necessary for the electrode to work.

The voltage measurement should be done with a pH meter rather than the student type potentiometer if a glass electrode is used. However, the pH meter is calibrated for the glass and calomel electrodes in aqueous solution; the EMF produced by the glass and silver-silver chloride electrode in either acidic or basic media differ markedly from the true value, and if left this way should be reported as *apparent pH*. The following example problem will show how pH in a nonaqueous system can be converted to pH in an aqueous system.

Example

An ethanol solution of benzoic acid has a pH of 10.25 and an EMF of 0.6135 v. When the same amount of benzoic acid was dissolved in water it had a potential of 0.2065 v. What is the pH of this concentration of benzoic acid in water?

$$pH = pH_s + \frac{[E - E_s)}{0.0591} \tag{12.10}$$

where pH = the pH determined in the nonaqueous solvent
pH_s = the pH in the water solvent
E = the potential of the nonaqueous system.
E_s = the potential of the water reference system.

Equation 12.10 has been recommended by both the American and British standardization agencies as a practical definition of pH. Therefore,

$$10.25 = pH_s + \frac{(0.6135 - 0.2065)}{0.0591}$$

where pH_s = 3.36 in water (quite a difference from the 10.25 in ethanol)

Some Recent Applications

The 1968 review of potentiometric titrations presented in *Analytical Chemistry* listed 288 references and the author of the review stated that he left out most of the application papers. This is a fair indication of the utility of potentiometry. Only one paper, which is an excellent example of how an apparently complex problem can be solved by a simple method, will be discussed.

This is a method for the determination of phosphine (PH_3) which is a fumigant in cereal grains and which can be a food contaminant. PH_3 is removed from the sample by flushing with dry nitrogen and passing the gases over an alcoholic solution of mercuric chloride. The phosphine is trapped [$PH_3 + 3HgCl_2 = P(HgCl)_3 + 3 HCl$] and the liberated HCl is determined by a potentiometric titration with NaOH solution. 10 μg of PH_3 can be determined if a pH meter sensitive to ± 0.001 pH unit is used.

Goldstein (1968) developed a method for the analysis of catalase using the oxygen electrode. The rate of the O_2 evolution was measured when catalase was allowed to react with H_2O_2.

The nitrate electrode, Orion 92-07 liquid ion exchange type, was used by Paul and Carlson (1968) to determine nitrate nitrogen in potatoes, tomatoes, grapes, and sugar beets. The electrode was linear from 1 ppm to the 15,000 ppm found in potatoes. An aluminum resin was used to eliminate the HCO_3^- and organic acid (citrate) interference, and silver nitrate was added to remove excess chloride. The results in the 1 to 50 ppm region had a standard deviation of 3.1%. The use of a calcium ion sensitive electrode for milk analysis is discussed by Demott (1968). Sodium sensitive electrodes were used by Halliday and Wood (1966) to measure salt in bacon.

Associated Problems

(1) What is the pH of a solution at 25°C if a potential of 0.703 v is observed between a hydrogen electrode and a SCE?   Ans. 7.72

(2) What potential would be shown by an antimony electrode (Table 12.1) relative to a SCE in a solution at pH 5.5?   Ans. 0.722 v

(3) Calculate the potential in the following cell. A Pt wire dips into a solution that is 0.05 M in Sn^{+4} and 0.025 M in Sn^{+2}. This solution is connected by a salt bridge to another solution that is 0.086 M in Hg^{+2} and 0.107 M in Hg^+. A Pt electrode dips into this latter solution.   Ans. 0.124 v

(4) A 0.2 M solution of HF is placed in a polyethylene beaker. To this is added an equal volume of 0.03 M Al^{+3}. A calomel electrode and an Al wire, connected to a potentiometer, were placed in the solution. What is the potential?   Ans. 1.519

$$AlF_6^{3-} = Al^{3+} + 6 F^- \qquad K_d = 1.4 \times 10^{-20}$$

BIBLIOGRAPHY

DEMOTT, B. J. 1968. Ionic calcium in milk and whey. J. Dairy Sci. 51, 1008.

FISHER, R. B., and BABCOCK, R. F. 1958. Effects on aging of reagent solutions on the particle size of precipitates—electrodes consisting of membranes of precipitates. Dissert. Abstr. 19, 428.

GOLDSTEIN, D. 1968. A method for assay of catalase with the oxygen electrode. Anal. Biochem. 24, 431.

HALLIDAY, J. H., and WOOD, F. W. 1966. The determination of salt in bacon by using a sodium ion responsive glass electrode. Analyst 91, 802–805.

HAMMER, W. J., and WOOD, R. E. 1958. Handbook of Physics. McGraw-Hill Book Co., New York.

PAUL, J. L., and CARLSON, R. M. 1968. Nitrate determination in plant extracts by the nitrate electrode. J. Agr. Food Chem. 16, 766–768.

RECHNITZ, G. A. 1967. Ion selective electrodes. Chem. Eng. News, June 12, 146–158.

CHAPTER 13

Coulometry

Coulometric titrations are determinations in which the titrant or reagent is electrically generated at the surface of an electrode (working electrode) immersed in a solution with a second electrode (counter electrode). Because the titrant is used almost at the instant it is generated at the electrode, coulometric methods can be designed to employ titrants which cannot be easily used in burets, such as chlorine, bromine, iodine, Ti(II), Cu(I), and Cr(II). Submicrogram quantities of materials can be determined. The instrumentation is generally quite simple and of low to moderate cost. The analyses are rapid and there are no dilution corrections. The electrodes can be placed in remote places and the system is easy to automate. These advantages combine to make coulometry a potentially powerful analytical technique.

Coulometric titrations are based on Faraday's law, which simply states that the passage of 1 Faraday of electricity (96,494 coulombs) will oxidize or reduce 1 gm equivalent to the substance under consideration. Mathematically this can be stated as,

$$g = \frac{itM}{nF} \tag{13.1}$$

where g = weight of the substance in gm
$\quad i$ = current in amperes
$\quad t$ = time in sec
$\quad M$ = formula weight of the substance
$\quad n$ = number of electrons transferred per atom
$\quad F$ = 96,494 (Faraday's constant)

It is necessary that only 1 reaction take place at the electrodes and that it proceeds with 100% current efficiency; that is, every electron reacts only with the system being investigated. In order to maintain 100% current efficiency, low currents are generally used, and the electrolysis products formed at the counter electrode must not be allowed to diffuse to the working electrode. A porous barrier is generally used to obstruct diffusion, yet allow the passage of current by the supporting electrolyte.

An alternate method (Lingane 1958) is to add a material to the system that *can be electrolyzed* at the counter-electrode, and whose reaction products are inert to the electrolysis at the working-electrode should they migrate there. Hydroxylamine and hydrazine are two such compounds.

$$2NH_2OH = N_2 + 2H^+ + 2H_2O + 2\ e^- \tag{13.2}$$

$$NH_2NH_2 = N_2 + 4H^+ + 4\ e^- \tag{13.3}$$

The H^+ formed will not be reduced if a mercury electrode is used because of its overvoltage and of course N_2 is inert.

Example

If a current of 24.75 ma flows for 284.9 sec through a solution of copper nitrate, how many milligrams of copper will be deposited at the cathode?

$$\frac{24.75 \times 10^{-3}\ (amp) \times 284.9\ (sec) \times 63.54\ (gm)}{2\ (electrons) \times 96{,}494} \times \frac{1000\ mg}{gm}$$

$$= 2.322\ mg$$

From the knowledge of the time of the current flow and the magnitude of the current, one may determine the number of gram-equivalents of titrant generated and therefore the number of gram-equivalents of the substance being titrated.

Two processes are normally employed: constant current or constant potential. The constant current process will be discussed first.

Constant Current Coulometry

In this method the current, i, is maintained constant and the time, t, required for the material in the sample to completely react is determined. Since $it = q$ (coulombs), and 96,494 coulombs = 1 eq; t, the percent composition of the sample can be quickly determined. For simple systems this is by far the easiest method. However, if more than one species is present, extreme care must be taken because of *concentration overvoltage* and *activation overvoltage*, both of which reduce the current efficiency to less than 100%.

Concentration overvoltage arises in the following manner. Recall the Nernst equation,

$$E_{actual} = E^\circ + \frac{0.0591}{n} \log \frac{[ox]}{[red]}$$

The E_{actual} depends on the concentration of the ionic species *at the surface* of the electrode. This may be considerably different than the bulk concentration as shown in Fig. 13.1. Suppose 1000 electrons per second reach the electrode surface from a constant current supply and 2000 sample ions arrive each second at the electrode surface to react.   Since there are more sample ions than electrons, all of the electrons are used up and there is 100% current efficiency. Now suppose that after a while the concentration of sample ions drops to 500 ions per second.   The constant current supply, however, will continue putting 1000 electrons per second on the electrode surface; these electrons must be used up.   Since there aren't enough sample ions for the electrons provided, the electrons must be used somewhere else.   Other ions present in the sample will react and we can tell from the Nernst equation what happens.   From the Nernst equation it can be seen that the potential will change as the concentration changes.   Different elements react at different potentials and other elements will react as well as the sample ions.   This means that 100% current efficiency is no longer obtained and the analysis is in error.   What is needed to prevent this is vigorous stirring and low currents.

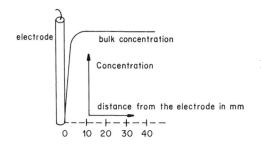

electrode

bulk concentration

Concentration

distance from the electrode in mm

0   10   20   30   40

Fig. 13.1. Variation of Concentration with Distance

Activation overvoltage is a deviation of the electrode potential from its Nernst equation value due to the passage of *electrolytic current* (not the electrode exchange current) in the solution itself. This shifts the cathode to a more negative value and the anode to a more positive value.   For small currents the effect is linear with an increase in current; for large currents, the shift is linear with the logarithm of the current.   Again low currents are required.

Example

5.00 ml of platinum (IV), added to an electrolyte of sodium bromide containing stannic chloride, was titrated with electrogenerated stannous ion, using platinum and gold electrodes.   The endpoint was detected potentiometrically after 297.3 sec at a constant current. The iR drop across a standard 10.00 ohm resistor was 0.0965 volts. What is the normality of the platinum(IV) solution? $E = iR$ or $i = E/R = 0.0965/10.00 = 0.00965$ amp.

$$\frac{0.00965 \times 297.3}{96,494} = 2.973 \times 10^{-5} \text{ gm-eq}$$

$$\frac{2.973 \times 10^{-5} \text{ gm-eq}}{5.00 \text{ ml}} = \frac{5.946 \times 10^{-3} \text{ gm-eq}}{1000 \text{ ml}} = 0.00595N$$

Lingane (1954) reported a circuit for a constant current source for all types of coulometric titrations which maintains the current constant to within ±0.01% for many hours.

A simple inexpensive circuit which is suitable for many titrations is shown in Fig. 13.2.

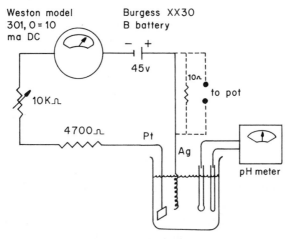

FIG. 13.2. SCHEMATIC DIAGRAM OF A SIMPLE COULO-
METRIC TITRATION SYSTEM

The electrodes shown are used conveniently in the coulometric titration of a number of acids; however, other electrodes could be used as well. If the current measured does not have to be known too accurately the 0–10 ma ammeter is suitable. Alternatively, if more accurate current values are needed, a 10.00 ohm standard resistor may be inserted, in series, into the circuit and the potential drop across this resistance determined by employing a potentiometer and using Ohm's law, $E = iR$.

Controlled Potential Coulometry

The main problem with constant current coulometry in complex systems is that the currents required to maintain 100% current efficiency are so small that the analysis can take several hours.

An improvement in this respect is accomplished by *controlled potential coulometry*. The system requires 3 electrodes; 2 that are the same as in constant current coulometry and a 3rd electrode to monitor the cathode potential. The third electrode is placed near the cathode and draws very little current so it does not disturb the system appreciably. If the cathode potential varies from a predetermined value, then the system is automatically corrected back to the proper value. The device to maintain this constant potential is known as a *potentiostat*. Since the potential is controlled, only one element will be determined at a time. However, the current will decrease as the concentration of the element decreases. Figure 13.3 compares the constant current and constant potential processes.

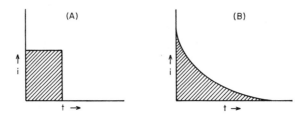

FIG. 13.3. CURRENT-TIME PLOTS FOR CONSTANT CURRENT (A), AND CONSTANT POTENTIAL (B) COULOMETRY

When the current has decreased to 0.1% of the initial current, the deposition may be regarded as completed. In some cases the current decreases to a level above 0.1% of the original value, and remains unchanged. This residual current is due to oxygen or to slow evolution of hydrogen. A usual method for determining the time for electrolysis is the *10 half life* concept. Table 13.1 shows the extent of deposition.

TABLE 13.1

FRACTION DEPOSITED PER NUMBER OF HALF TIMES

Fraction Deposited	Number of Half Times
0.50	1
0.90	3.33
0.99	6.67
0.999	10.0
0.9999	13.33

The half time is determined by noting the current at the start, and then measuring the amount of time it takes to reduce this value to *1/2*.

Controlled Potential Calculations.—A common case might be the exponential decrease of current with time, such as shown in Fig. 13.3.

$$q = \int_0^t idt \qquad (13.4)$$

if

$$i = i_o \exp(-kt) \qquad (13.5)$$

then

$$q = i_o \int_0^t \exp(-kt)dt \qquad (13.6)$$

or

$$q = \left(\frac{i_o}{k}\right) \exp\,(-kt)_0^t \qquad (13.7)$$

Example

If the initial current is 100 ma and decreases exponentially, with $k = 0.0062$ sec^{-1}, and the time of titration is 683 sec, how many equivalents of bromine are liberated from a bromine solution?

$$q = \frac{-0.100}{0.0062} \times e^{-0.0062 \times 683} + \frac{0.100}{0.0062}$$

$$= \frac{-0.100}{0.0062} [1 - e^{-0.0062 \times 683}]$$

$$= 16.13 \times (1 - e^{-4.2345})$$

$$= 16.13 \times (1 - 10^{-1.839})$$

$$q = 16.13 (1 - 0.0145)$$

$$q = 16.13 \times 0.9845 = 15.90 \text{ coulombs}$$

The reaction to liberate bromine is

$$2\ Br^- = Br_2 + 2\ e^-$$

Thus 1 equivalent of Br_2 is $159.83/2 = 79.92$ gm. $15.90/96,494 = 0.000165$ equivalents of Br_2 liberated; and in mg of Br_2 liberated this is, $0.165 \times 79.92 = 13.17$ mg Br_2.

The area under the curve (Fig. 13.3B) can be calculated mathematically as in the above example or it can be integrated graphically, electrically, or mechanically.

For systems that behave reasonably well a graphical method may be used. Plot the log current *vs* time and then

$$A = \frac{i_o}{2.303\ k} \qquad (3.8)$$

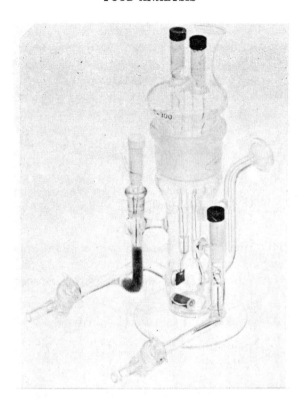

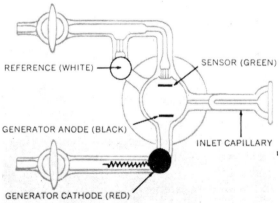

Courtesy Dohrman Instrument Co.

FIG. 13.4.   DOHRMAN SULFUR CELL

where A = the area
  i_o = the intercept
  k = slope

A chemical means of determining q is a silver coulometer. The current used in the analysis is passed through a silver solution. A pair of platinum electrodes is placed in the solution and the cathode weighed before and after the analysis. 1.117972 mg of Ag will be deposited for each coulomb that passes. Thus, major current fluctuations that can occur with inexpensive equipment can be accounted for.

The Coulometer as a Gas Chromatographic Detector.—One of the most common applications of coulometry in food analysis is as a detector for a gas chromatograph such as that manufactured by the Dohrman Co. Basically the idea is as follows. The volatile food components, or extracts, are separated by the chromatograph and passed through a quartz combustion tube heated to 800°C. The material is decomposed to simple compounds such as CO_2, H_2O, NO_2, SO_2, Cl_2, and Br_2. Those materials are then passed into a coulometric titration cell and titrated by a constant current process, the selectivity depending on the type of electrodes and the supporting solution. Figure 13.4 is a diagram and a photograph of the Dohrman sulfur cell.

The following formulae taken from the Dohrman technical bulletin show how the calculations are made.

$$\text{microequivalents} = \frac{q \times 10^6}{96{,}494 \text{ coulombs/eq}} \qquad (13.9)$$

$$q = it = \left(\frac{\text{mv/in.}}{R(\Omega)}\right) 10^{-3} \times \frac{60 \text{ sec}}{\text{min}} \times \frac{\text{min}}{\text{in.}} \times A \qquad (13.10)$$

$$= V/R \times 6.0 \times 10^{-2} \times S \times A \qquad (13.11)$$

where V = mv/in. recorder sensitivity
  R = ohms resistance of the range ohms switch
  S = chart speed of the recorder in min/in.
  A = area under the peak in square inches

$$\text{microequivalents} = \frac{V \times S \times 6.0 \times 10^{-2} \times 10^6 \times A}{R \times 96{,}494} \qquad (13.12)$$

The quantity of a chlorinated compound can be determined by applying the following formula:

μgm of compound =

$$\frac{A \times \dfrac{\text{min}}{\text{in.}} \times \dfrac{\text{mv}}{\text{in.}} \times \dfrac{35.45 \text{ gm}}{\text{eq}} \times \dfrac{60 \text{ sec}}{\text{min}} \times \dfrac{10^6 \, \mu\text{gm}}{\text{gm}} \times \dfrac{10^{-3} \text{ v}}{\text{mv}} \times 10^2}{\left(\text{Recorder input resistance}\left(\text{range} \times \% \text{ Cl in the compound}\right.\right.}$$

$$\left.\left. \times \, 96,494 \, \frac{\text{coulombs}}{\text{eq}} \text{ in ohms}\right)\right)$$

$$(13.13)$$

For a sensitivity of 0.1 mv/in. and a chart speed of $^1/_2$ in./min, equation 13.13 becomes

$$\text{Cl} \quad \mu\text{gm} = \frac{A \times 442}{\text{range} \times \% \text{ Cl}} \qquad (13.14)$$

$$\text{S} \quad \mu\text{gm} = \frac{A \times 199.2}{\text{range} \times \% \text{ S}} \qquad (13.15)$$

$$\text{N} \quad \mu\text{gm} = \frac{A \times 174.2}{\text{range} \times \% \text{ N}} \qquad (13.16)$$

Determination of Halogen.—The halogenated compound is converted to Cl_2, Br_2, or I_2 during the combustion process. The electrolytic cell contains 70% acetic acid in water. The reference electrode is Ag-AgOAC (satd).

A typical cell reaction, using chloride as an example, is as follows,

$$Ag^+ + Cl^- = AgCl$$

Depletion of silver in the electrolyte is detected by the reference/sensor pair and is replaced electrically at the generator electrodes by a current from the microcoulometer. The reaction at the generator electrode is $Ag = Ag^+ + e^-$

Determination of Sulfur.—The sulfur compound is converted to SO_2 at temperatures above 800°C. The electrolyte is 0.04% acetic acid, 0.05% KI. The generator and sensor electrodes are Pt and the reference electrode is $Pt:I_3^-$.

A typical cell reaction, using SO_2 as an example, is as follows,

$$I_3^- + SO_2 + H_2O = SO_3 + 3\,I^- + 2H^+$$

The triiodide depletion in the electrolyte is detected by the reference/sensor electrodes and is replaced at the generator electrodes. The reaction at the generator anode is, $3\,I^- = I_3^- + 2\,e^-$. Any system that will react with iodide (I_3^-) can be determined.

Determination of Nitrogen.—The nitrogen is converted to ammonia by combustion with H_2 over a Ni catalyst. The electrolyte in the cell is 0.04% Na_2SO_4, the generator and sensor electrodes are Pt, and the reference electrode is Pb in saturated $PbSO_4$. Any acid or base can be titrated with this system. BaO removes acid gases, so bases can be determined.

A typical cell reaction, using ammonia as an example, is as follows,

$$NH_3 + H_2O = NH_4OH$$
$$NH_4OH + H^+ = NH_4^+ + H_2O$$

Depletion of the hydrogen ion in the electrolyte is detected by the reference/sensor electrodes and is replaced as follows,

$$^1/_2H_2 = H^+ + e^-$$
$$H_2O + e^- = OH^- + ^1/_2H_2$$

As little as 500 pg of N_2 can be determined.

Determination of Phosphorus.—The phosphorus is converted to PH_3 by combustion at 950°C with H_2 over a Ni catalyst. Any H_2S or HCl that is formed is removed by a small section of Al_2O_3. The cell is the same as that for halogens. The reaction product is not known although the reaction proceeds as if Ag_2PH were formed.

Some Recent Applications

The simultaneous and selective detection of P, S, and Cl in pesticides has been developed by Burchfield et al. (1965). The compound is combusted in H_2 and the elements converted to PH_3, H_2S, and HCl. The normal halogen cell determines all three at the same time. Al_2O_3 is added to remove H_2S and HCl; PH_3 remains in solution. Silica gel absorbs HCl and separates PH_3 and H_2S.

Pease (1968) determined terbacil (Cl) and bromacil (Br) down to 0.04 ppm on apples, peaches, and citrus fruit.

Pease and Kirkland (1968) determined methomyl in cabbage and sweet corn down to 0.02 ppm by extracting the methomyl with ethyl acetate, purifying the extract by a further extraction into chloroform, and hydrolyzing the product to the corresponding oxime by a base, extracting this into ethyl acetate, and determining the S coulometrically.

S,S,S,-tributyl phosphorotrithioate was determined in cotton seed by Thomas and Harris (1965).

The triazine herbicides, atrazine, propazine, simazine, ametryne, prometryne, and simetryne, are detected in the 20 to 100 ppb region on peanuts, sugar beets, Brussel sprouts, pineapples, and potatoes by the coulometric method (Mattson *et al.* 1965). Neither visible nor UV spectrophotometry was able to distinguish the various compounds, nor was spectroscopy as sensitive.

The possibilities for the future are almost unlimited. Table 13.2 shows some of the recent systems for electrogenerating various titrants. Almost all of these can be applied to food analyses either directly or indirectly.

TABLE 13.2

SYSTEMS FOR ELECTROGENERATING VARIOUS TITRANTS

Substance Determined	Electrogenerated Titrant
Phenols	Bromine
Phosphites	Bromine
Olefins	Bromine
Cyclic β diketones	Bromine
S compounds	Iodine
Proteins	Hypobromite
Organic chlorides and bromides	Silver(I)
Alkaloids and organic bases	Silver(I)
Chromate, dichromate, oxalate, molybdate, tungstate, ferrocyanide	Lead(II)
Phosphate	Bismuth(III)
Organic acids in sugars	OH^-
Organic nitrogen compounds	H^+
Aliphatic amides	Oxidation at Pt electrode
Catechol	Oxidation at Pt electrode
Nitroso compounds	Reduction to amines
Nitro compounds	Reduction in dimethyl sulfoxide

Associated Problems

(1)   The following measurements were made in a coulometric titration of arsenic (III) ions extracted from 50 lb of Jonathan apples. The titrant was electrogenerated bromine.

iR drop across the resistance:	0.6748 v
calibrated resistance:	89.52 ohms
generation time:	6.36 min

Calculate the amount of arsenic (ppm) present on the apple skins. Ans. 0.16 ppm

(2)   In an electrolytic determination of bromide ion in 50.00 ml of a solution, the quantity of electricity, as determined from a mechanical current-time integrator, was 131.5 coulombs. Calculate the weight of bromide ions in the original solution. Calculate the potential of the silver electrode that should be employed throughout the electrolysis.

$$K_{sp} \text{ AgBr} = 5.0 \times 10^{-13}, \text{ AgBr} + e^- = \text{Ag} + \text{Br}^-$$
$$E° = +0.073 \text{ v}$$

Ans. a. 5.45×10^{-2} gm; b. -0.653 v

(3) A particular timer employed in constant current coulometric titrations is marked in units of 0.01 min. If it is desired to have each unit (0.01 min) correspond conveniently to one microequivalent, what value of the current in milliamperes, would be required from the power supply? Ans. 160 ma

(4) In the determination of copper in a brass cooking pan, the copper was coulometrically determined by means of the method of Miller and Hume (1960) using thioglycollic acid. The endpoint was detected amperometrically using two mercury electrodes. A 2.705-mg aliquot of the brass was titrated at a constant current of 10.66 ma. The time of the analysis was 247.3 sec. Determine the copper content, in weight percent. Ans. 31.4%

(5) Atrazine, $C_2H_{14}ClN_5$, 16.41% Cl, is extracted from 2 lb of cranberries with 100 ml of hexane. A $10\mu l$ sample is injected into the chromatograph and the Cl determined coulometrically. The chart speed is 0.5 in./min, the range resistance is 256 ohms, the peak area is 1.8 in.^{2}, and the recorder sensitivity is 0.1 mv/in. Calculate the concentration of atrazine on the cranberries in ppm. Ans. 2.08 ppm

(6) Simetryne, $C_8H_{15}N_5S$, 15.03% S, is extracted from 100 lb of potatoes with 2 liters of hexane which is then concentrated to 100 ml. A 10 μl sample is injected into a chromatograph and the sulfur determined coulometrically. The chart speed is 1 in./min., the recorder sensitivity is 0.05 mv/in. and the range resistance is 1280 ohms. The peak area is 0.15 in.^{2}. Calculate the ppb simetryne in the potatoes. Ans. 0.34 ppb

BIBLIOGRAPHY

BURCHFIELD, H. P., RHOADES, J. W., and WHEELER, R. J. 1965. Simultaneous and selective determination of phosphorus, sulfur and halogen in pesticides by microcoulometric gas chromatography. J. Agr. Food Chem. *13*, 511–516.

LINGANE, J. J. 1954. Constant current source for coulometric titrations. Anal. Chem. *26*, 1021–1023.

LINGANE, J. J. 1958. Electroanalytical Chemistry, 2nd Edition. Interscience Publishers, New York.

MATTSON, A. M., KAHRS, R. A., and SCHNELLEN, J. 1965. Use of micro-coulometric gas chromatograph for triazine herbicides. J. Agr. Food Chem. *13*, 120–122.

MILLER, B., and HUME, D. N. 1960. Coulometric titrations with electrically generated sulfhydryl compounds—applications of thioglycollic acid. Anal. Chem. *32*, 524–528.

PEASE, H. L. 1968. Determination of terbacil residues using microcoulometric gas chromatography. J. Agr. Food Chem. *16*, 54–56.

PEASE, H. L., and KIRKLAND, J. J. 1968. Determination of methomyl residues using microcoulometric gas chromatography. J. Agr. Food Chem. *16*, 554–557.

THOMAS, R. and HARRIS, T. H. 1965. The microcoulometric determinations of S,S,S-tributyl phosphorotrithioate in cottonseed. J. Agr. Food Chem. *13*, 505–508.

CHAPTER 14

Conductivity

In conductivity use is made of two parallel platinum foil (1 cm²) electrodes, about 1 cm apart.   The amount of current that will pass between these two electrodes is a function of the concentration, temperature, and kind of ions.   The polarity of the electrodes is usually changed about 1000 times a second in order to prevent electroplating and electrode polarization.

The equivalent conductance is defined as

$$\Lambda = \frac{1000K}{C} \tag{14.1}$$

where C is the concentration expressed as normality, K, the *specific conductance* is expressed in mho's (the reciprocal of the resistance). Λ varies with concentration, and the value of Λ at infinite dilution is Λ_0.   Equivalent conductivities are more commonly broken into *ionic conductivities* which can be added together to form *molar conductivities*.

$$\Lambda_0 = \sum(\lambda_+) + \sum(\lambda_-) \tag{14.2}$$

TABLE 14.1

IONIC CONDUCTANCES AT 25 °C IN WATER

Ion	λ_+	Ion	λ_-
H_3O^+	350	OH^-	197.5
Li^+	39	NH_4^-	80
Na^+	50	Cl^-	76
K^+	74	Br^-	78
NH_4^+	74	I^-	76.5
Ag^+	62.5	NO_3^-	71
$1/2\ Mg^{2+}$	55	HCO_3^-	44
$1/2\ Ca^{2+}$	60	CH_3COO^-	41
$1/2\ Sr^{2+}$	60	ClO_4^-	67
$1/2\ Ba^{2+}$	64	$1/2\ CO_3^{2-}$	70
$1/2\ Cu^{2+}$	56	$1/2\ C_2O_4^{2-}$	24
$1/2\ Zn^{2+}$	55	$1/2\ SO_4^{2-}$	80
$1/2\ Pb^{2+}$	73	$1/2\ CrO_4^{2-}$	82
$1/2\ Fe^{2+}$	54	$1/3\ PO_4^{3-}$	80
$1/3\ Fe^{3+}$	68	$1/3\ Fe(CN)_6^{3-}$	101
$1/3\ La^{3+}$	70	$1/2\ Fe(CN)_6^{4-}$	111

Values of ionic conductivities at infinite dilution at 25°C for a number of ions are given in Table 14.1.

The instrument generally used is a Wheatstone bridge circuit arranged as shown in Fig. 14.1.

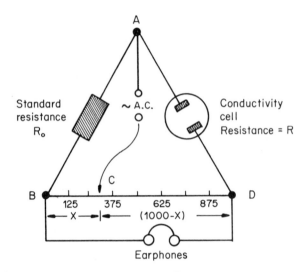

FIG. 14.1. SCHEMATIC CIRCUIT DIAGRAM FOR THE MEAS-
UREMENT OF CONDUCTANCE

A standard resistance, R_0, is set to approximately match the resistance of the conductivity cell, R. By movement of the slide wire contact, C, a point will be found where the noise heard in the earphones is minimized. Then,

$$\frac{\text{cell resistance}}{\text{std resistance}} = \frac{R}{R_0} = \frac{\overline{CD}}{\overline{BC}} = \frac{(1000 - X)}{X} \tag{14.3}$$

where $\overline{BD} = 1000$ divisions, and X, is the number of divisions. Thus, the measurement of X as a function of the volume of titrant added permits us to prepare conductometric titration plots.

Acid-base Titrations

Consider the case of the titration of 100 ml of 0.10 N NaOH with 1.0 N HCl. From Table 14.1 we see that the OH$^-$ has a large conductance, the Na$^+$ a fairly small conductance. As the HCl titrant is added, the reaction

$$\text{OH}^- + \text{H}^+ = \text{H}_2\text{O}$$

proceeds, so that the OH$^-$ concentration decreases, and hence the conductance of the solution decreases (Fig. 14.2).

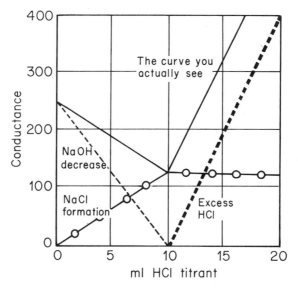

FIG. 14.2.   TITRATION OF NaOH WITH HCl

After the equivalence point is reached, the added H$^+$ is in excess, and as this H$^+$ has a very large conductance value, the conductivity of the solution again rises sharply.

Note that the NaOH contribution to the conductance decreases extremely rapidly while the NaCl contribution increases slowly, thereby decreasing the slope of total conductance.   Following the equivalence point, the NaCl contribution to the conductance remains constant while the HCl contribution rises sharply.   Thus the total conductance of the solution rises sharply following the end-point.

With weak acids, weak bases, or both, other shapes of titration curves may be obtained.   However, the principle in constructing these curves is the same as with strong acids and bases.

Figure 14.3 shows a few of the major types.   Note that weak acids and bases are only slightly ionized so that they do not themselves contribute much to the total conductance of the solution, although the salt formed still exerts an influence.   Also readily noticeable from the figure is that a weak base is a better titrating agent for a weak acid than a strong base.

182   FOOD ANALYSIS

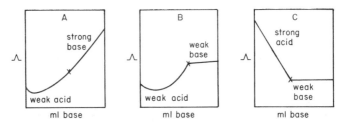

FIG. 14.3. CONDUCTOMETRIC TITRATION CURVES INVOLVING
WEAK ACIDS AND BASES

Conductance depends on the concentration of the ions present.
As a titration proceeds and the solution is diluted, the conductance
decreases due to this dilution effect and the endpoint is less distinct
than it could be.   To remedy this, two approaches are generally
used.   The first is to use a titrant ten times stronger than the titrated
material so the dilution is held to a minimum.   If this fails, then a
correction for the dilution is made using the following equation,

$$G_{\text{actual}} = \left(\frac{V + v}{V}\right) G_{\text{obs}} \qquad (14.4)$$

where V = the initial volume
      v = the volume of titrant added

Example

Twenty milliliters of NaOH was titrated to 50% of the equivalence
point by adding 5.85 ml of 1.00 N HCl.   The observed conductance
was 29.0 ohm^{-1} cm^2.   What is the actual conductance?

$$G_{\text{actual}} = \left(\frac{20 + 5.85}{20}\right) \times 29.0$$

$$= 37.4 \text{ ohms}^{-1} \text{ cm}^2$$

Oxidation Reduction Systems

Conductometric titrations involving ox-red systems are very
difficult because the equivalent conductances of the ions involved are
quite similar ($Fe^{+2} = 54$, $Fe^{+3} = 68$).

Precipitation Titrations

In the same way in which acid-base titrations were carried out by
the conductometric method, precipitation titrations can be performed.   Figure 14.4 illustrates the titration of 100 ml of 0.01 N
AgNO$_3$ with 1.0 N KCl at 25°C.   Note that the equivalent con-

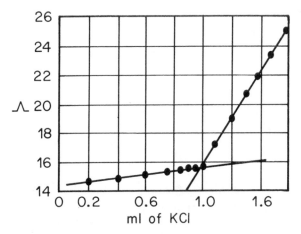

FIG. 14.4. TITRATION OF $AgNO_3$ WITH KCl

ductance of the K^+ is slightly greater than that of the Ag^+ so that a slight but steady increase in the conductance of the solution is observed prior to the equivalence points.

This titration curve could be sharpened by choosing a cation in the titrant with a lower conductance than that of Ag^+. From Table 14.1 we see that such a cation is Li^+, so the titration of $AgNO_3$ with LiCl should give a sharper break in the titration curve than that observed with KCl. Figure 14.5 shows this to be the case.

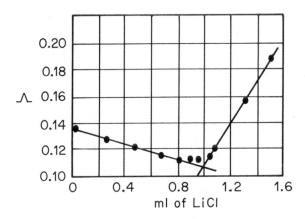

FIG. 14.5. TITRATION OF $AgNO_3$ WITH LiCl

From Fig. 14.5 one also notes that there is a rounding-off effect in the region near the equivalence point. This is due to the solubility

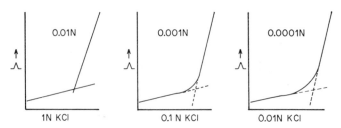

FIG. 14.6.   EFFECT OF CONCENTRATION ON THE SHARPNESS OF
THE ENDPOINT

product of the precipitate formed in this titration.   Figure 14.6
shows somewhat more dramatically the effect of concentration on
the types of endpoints observed, the case being the titration of
$AgNO_3$ with various concentrations of KCl.

A Conductivity Detector for a Gas Chromatograph

Probably the most successful detector of this type is the one
developed by Coulson (1965).   The sample is separated by a gas
chromatograph, combusted, the gases mixed with a stream of deion-
ized water, and the conductivity of this solution is measured.   The
detector works well with halogens, sulfur, and nitrogen (convert to
NH_3 with H_2), and is as sensitive as thermal conductivity cells for
carbon.

According to Coulson, electrolytic conductivity gas chromatography requires less clean up of food extracts, soil extracts, and
water extracts, than is required for microcoulometric gas chromatography, because smaller aliquots of samples may be analyzed.
Smaller samples injected into the chromatograph result in longer
column life and sharper peaks due to more rapid volatilization of the
pesticide residues from other plant extractives.   Sharper peaks result in better resolution and more positive identification of unknown
peaks.   The greater sensitivity of electrolytic conductivity detection
makes possible routine analysis at the low ppb level.

BIBLIOGRAPHY

COULSON, D. M.   1965.   Electrolytic conductivity detector for gas chromatography.   J. Gas Chromatog. *3*, 134.

CHAPTER 15

Electrophoresis

In a restricted sense, electrophoresis refers to the movement of charged colloidal particles and macromolecular ions under the influence of an electric field. But recent usage includes viruses, biological cells and subcellular organelles, and small organic molecules such as amino acids (Cann 1968). Electrophoresis in food analysis is, generally, restricted to proteins. The charge on the protein depends on the pH of the solution. Any protein has an isoelectric point (pI) at which the net charge is zero and at which it has a zero mobility. At pH values below pI, the protein migrates as a cation, and the mobility increases with decreasing pH. At pH values above the pI, the protein migrates as an anion, the mobility increasing with increasing pH.

Differences in migration velocities provide a sensitive means of separating components of a mixture that is otherwise difficult to fractionate. It is often the only method available for the quantitative analysis and fractionation of biological fluids, and for the characterization of their purified components. In addition it is a powerful tool for the detection and characterization of macromolecular interactions.

Electrophoretic methods can be divided into moving boundary electrophoresis and zone electrophoresis.

Moving Boundary Electrophoresis

Moving boundary electrophoresis is the transport of charged molecules dissolved in a buffer in a U tube, both branches of which are in contact with a buffer connected to electrodes which provide a constant voltage gradient. Modern technics of moving boundary electrophoresis began with the development of the Tiselius (1931) tube or cell that has remained till today the prototype of most instruments in this area. See Fig. 15.1. The cell is immersed in a water bath maintained at about 1°C so that the temperature inside the cell is 4°C, at which the density of water is maximum and at which small temperature variations cause little convection. The optics most commonly used for visualizing the protein concentration gradients caused by the electric field are Schlieren optics. They show as a

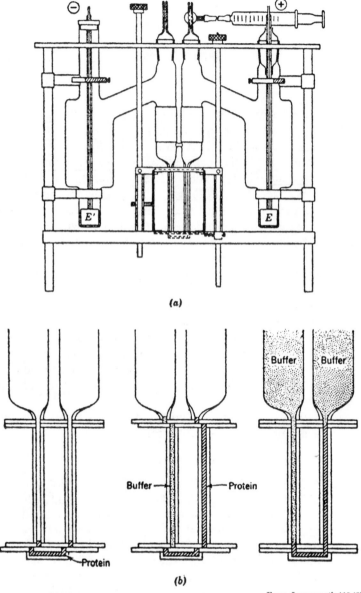

(a)

(b)

From Longsworth (1942)

FIG. 15.1. DIAGRAMS OF TISELIUS ELECTROPHORESIS APPARATUS
(a) The cells, electrode, vessels, and support (b) Initial formation of
boundaries in the electrophoresis cell.

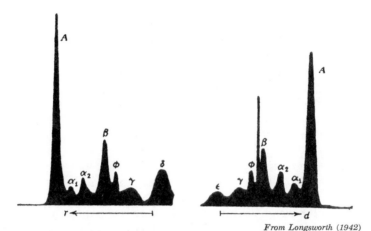

From Longsworth (1942)

FIG. 15.2. ELECTROPHORETIC PATTERNS OF HUMAN PLASMA PROTEINS

continuous curve the change in protein concentration versus
migrated distance.   The various fractions appear as Gaussian curves
(Fig. 15.2).   Some Tiselius electrophoretic instruments are also
equipped with Raleigh interference optics which show as a series of
parallel vertical interference bands the relation of total protein concentration to migrated distance.   Those patterns combined with the
Schlieren pattern are useful in determining the relative protein concentration of each peak (Van Oss 1968).

The use of moving boundary electrophoresis is limited today to
the determination of diffusion coefficients, to obtain small amounts
of pure samples of the fastest and slowest migrating proteins that can
be removed with special syringes, and to study interactions between
proteins from irregularities in the Schlieren patterns.

Zone Electrophoresis

Zone electrophoresis on solid supports and in gels (Fig. 15.3) is the
method of choice in situations where separations, rather than precise
physicochemical measurements, are desired.   In addition, zonal
methods are suitable for the analysis of small quantities of material
by fairly simple procedures.

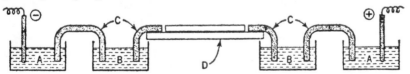

FIG. 15.3.   DIAGRAM OF APPARATUS FOR ZONE ELECTROPHORESIS

(A) salt solution (B) buffer (C) filter paper soaked in buffer, and
(D) supported gel or paper

A bewildering number of papers appear each year describing new apparatus, procedures, and applications. A number of review articles and books (see for example Block *et al.* 1958; Bloemendal 1963; Nerenberg 1966) with extensive bibliographies are available.

The important advantages of zone electrophoresis are: (1) simple and inexpensive apparatus which permit simultaneous analysis of several samples in a relatively routine procedure, (2) simple procedures for visualization of zones and for isolation of fractions, (3) improved resolution by combining electromigration with molecular sieving (i.e. in starch and polyacrylamide gels), and (4) adaptability to either large scale preparative or microanalytical separations. In addition, zone electrophoresis can be used for the investigation of low molecular weight substances that are difficult to analyze by the moving boundary methods. Those advantages are gained, however, at the sacrifice of accuracy and precision, particularly with respect to the determination of mobilities from migration rates in solid media. This loss is important if one wishes to use zone electrophoresis for identification and characterization of macromolecules in the same way as in moving boundary electrophoresis (Cann 1968).

Among the many supporting media which have been used for zone electrophoresis, the most important ones are paper and cellulose acetate, and agar, starch, and polyacrylamide gels. In starch gel and polyacrylamide gel electrophoresis, molecular sieving is utilized to great advantage. Because size, shape, and electrophoretic mobility determine relative mobility in the gels, the degree of resolution is much improved. In addition, strongly dissociating agents (such as concentrated urea solutions) can be incorporated into the starch gel to permit separation of protein subunits which tend to aggregate in aqueous solutions.

Paper electrophoresis was first used for protein separations by von Klobusitzki and Konig (1939). In paper electrophoresis, a narrow band of protein solution is deposited on a support strip which is soaked in buffer. The constant voltage gradient, applied through the two ends of the paper separates the proteins into bands according to their electrophoretic mobilities. To overcome the partial sorption of the proteins on the paper, the cellulose acetate strips were developed. The latter have an additional advantage in that they can be made transparent by treatment with paraffin oil or 5 to 15% acetic acid in ethanol at the end of electrophoresis, and then staining the proteins. The dark stained protein bands stand out better against the transparent background.

Instead of paper, some investigators prefer open-pored gel-like agar or agarose (at a concentration of about 1%) (Wieme 1963).

With the gel methods, the sample is deposited by pipetting a measured amount of protein in a slot in the gel. Agar gels are negatively charged and show a strong electroosmotic transport toward the cathode. Consequently, the sample should not be deposited too far from the middle of the gel plate. Agarose gels are neutral, and the problem does not arise. In addition, the migration of many proteins, and particularly lipoproteins, is limited by their interaction with agar but not with agarose. The use of starch as a support was introduced by Smithies (1955), and of polyacrylamide by Barka (1961) and by Ornstein and Davis (1962).

After the electrophoretic run, the supporting medium (paper, cellulose acetate, or gel) is "fixed" in alcohol or acid, or heated to insolubilize the proteins. The supporting medium is then stained with a protein stain. Cellulose acetate strips are generally stained with a Ponceau-S solution in 5% trichloracetic acid; thus, the fixing and staining operations are combined in 1 step. The strips are washed with 4% acetic acid containing tap water and can be stored for reference, photographed, or scanned with a densitometer. Strips stained with Amido Black are washed for several hours in 10% acetic acid or methanol. For lipoprotein staining, Sudan Black is used most. The strips are generally washed with 50% ethanol-water mixtures. Glycoproteins can be stained with PAS (periodic acid-Schiff reagent containing fuchsin). Replacing fuchsin by pararosaniline improves considerably the stability of the reagent. In addition, the separated bands can be eluted and the intensity of the stain measured spectrophotometrically.

For scanning the separated bands, several instruments are available. Measuring the absorbance of unstained bands at 205 nm provides a measure of the peptide bond, but the determination at the low wavelength requires rather expensive equipment. Ultraviolet scanning at 280 nm has the drawback that the absorbance of various proteins at that wavelength depends on their tryptophan and tyrosine contents, and varies widely with the separated proteins.

Scanning of separated and stained bands, followed by integration, provides information on the relative distribution of the proteins. The interpretation of the quantitative results must be made with caution. Generally, there is some overlapping of proteins and the extent of staining of individual protein bands (on an equiprotein basis) varies widely.

Methods of Paper Electrophoresis

The basic process of paper electrophoresis is as follows: the material to be separated into its components is either spotted or streaked onto the center of a strip of paper which is saturated with a buffer solution. Each end of the paper dips into a container holding several hundred milliliters of additional buffer solution. An electrode is placed in each container, one being connected to the positive terminal of a DC source and the other to the negative terminal. A potential of about 5–10 v/cm is applied, and if any charged particles are present in the material, they will move toward the terminal of opposite charge.

There are four main procedures involving the use of paper as a stabilizer. These are the sandwich, ridgepole, solvent immersion, and the horizontal strip techniques. Schematic diagrams of the last two of these are shown in Fig. 15.4 and 15.5.

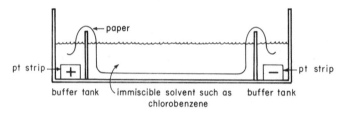

FIG. 15.4.   SOLVENT IMMERSION TYPE ELECTROPHORESIS

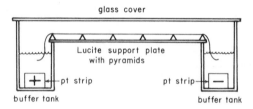

FIG. 15.5.   HORIZONTAL STRIP TYPE ELECTROPHORESIS

In the sandwich technique, a paper strip, saturated with buffer and containing a spot of the specimen to be analyzed, is placed between glass plates and connected to electrode vessels. The edges of the glass plates are usually closed with silicone grease to retard evaporation. If clamps are used to press the plates together, great care must be taken to ensure that the pressure is uniform, otherwise erratic migration will occur due to potential changes in the buffer system.

The solvent immersion type consists of immersing a paper strip, soaked in buffer solution, into a nonconducting, immiscible solvent such as carbon tetrachloride or chlorobenzene. Chlorobenzene is preferred because its density is nearly the same as the buffer saturated paper, and this keeps the paper from floating or sinking. Thus, more reproducible results can be obtained. Two major advantages of this technique are that the buffer evaporation is held to a minimum, and the immersion solvent acts as a coolant to reduce thermal migration within the paper, which can be a serious problem if high currents are used. The electrode vessels are made quite large so that they will hold a large amount of buffer. This is necessary to reduce pH changes caused by the electrode reaction at the ends of the filter paper.

The horizontal suspension technique requires that the paper strip be held in a taut horizontal position. The ends of the paper dip into electrode vessels. The entire system is placed in a box or chamber to maintain constant temperature. In some cases helium is placed in the chamber to act as a heat sink. This is one of the best techniques for mobility studies because the paper is horizontal and the effects of movement of the supporting solutions are minimized.

Calculations for Solution and Paper Electrophoresis

In order for a particle to migrate in an electric field, it must possess a net electrostatic charge. This charge is an integral multiple of 4 $\times 10^{-10}$ esu. Many compounds that do not normally have a charge can still be separated electrophoretically provided some charging process is used. Charging processes that have been found to be successful are reactions with acids and bases, dissociation into ions by polar solvents, hydrogen bonding, chemical reactions, polarization, and ion pair formation.

The widest application of electrophoresis is in separations for the isolation and purification of small amounts of compounds. The best method of referring to these separations and comparing one system to another has been the *mobility* of the compounds involved. Correct mobilities are therefore important and anyone using electrophoresis should be familiar with the various factors influencing the mobility and know what corrections should be made.

Consider a particle placed on a strip of paper which is saturated with a buffer and has a potential applied to it. The force (F) exerted on the particle is equal to the charge of the particle (Q) times the field strength (E).

$$F = QE \tag{15.1}$$

At first glance it would appear that the particle would move toward one end of the paper with an increasing velocity. However, it turns out that the velocity is constant. This is apparently because as the particle moves through the buffer it meets a retarding force caused by the viscosity of the solvent. This retarding force increases linearly with the particle acceleration, thus maintaining the velocity of the particle essentially constant. Stokes has shown that this opposing force for a sphere moving in a viscous medium can be expressed as,

$$F_s = 6\pi r\eta v \tag{15.2}$$

where F_s = viscous retarding force
      r = radius of the particle in cm
      η = viscosity of the medium in poises
      v = electrophoretic velocity in cm/sec

When these forces are equal

$$E = Q = 6\pi r\eta v \tag{15.3}$$

The mobility of a particle is defined as

$$U = v/E \tag{15.4}$$

where U is in cm^2/sec and E is in volts/cm. Substituting equation 15.3 into equation 15.4,

$$U = Q/6\pi\eta vr \tag{15.5}$$

Example

Calculate the mobility of the Ba^{2+} ion in a 0.0625 M solution of $BaCl_2$. The radius of Ba^{2+} in solution is 2.78 Å and the viscosity of this solution is 10.310 millipoises.

Referring to equation 15.5:

$$Q = 2 \times 4.8 \times 10^{-10}$$

$$r = 2.78 \times 10^{-8} \text{ cm}$$

$$\eta = 1.03 \times 10^{-2} \text{ poises}$$

Since the mobility has the units cm^2 $volt^{-1}$ sec^{-1}, the factor 300 is necessary to convert from practical volts to esu's.

$$U = \frac{2 \times 4.8 \times 10^{-10}}{6 \times 3.14 \times 2.78 \times 10^{-8} \times 1.03 \times 10^{-2} \times 3 \times 10^2}$$

$$= 5.9 \times 10^{-4} \text{ } cm^2 \text{ } volt^{-1} \text{ } sec^{-1}$$

The above mobility will not be verified experimentally because there is an additional retarding effect on the mobility due to the negative chloride ions moving in the opposite direction from the Ba^{2+} ions. This has a tendency to decrease the mobility from the Ba^{2+}, and is related directly to the ionic strength of the solution. By incorporating the Debye-Huckel equation the following correction term is obtained.

$$\frac{1}{1 + rA\sqrt{\mu}} \tag{15.6}$$

where μ = ionic strength = $1/2 \ \Sigma \ cZ^2$
c = concentration in moles/liter
Z = valence of the ion
A = constant = 0.233 × 10⁸ for water at 25°C
r = radius of the particle in cm

Example

What is the mobility of the Ba^{2+} ion when the other ions are considered?

$$U = \frac{Q}{6\pi r\eta} \times \frac{1}{1 + rA\sqrt{\mu}}$$

$$\sqrt{\mu} = \sqrt{1/2 \ (0.0625 \times 2^2) + (2 \times 0.0625 \times 1^2)}$$

$$\mu = 0.433$$

The correction term then is

$$\frac{1}{1 + 2.78 \times 10^{-8} \times 0.233 \times 10^8 \times 0.433} = 0.781$$

Multiplying the mobility obtained in the previous example by 0.781

$$U = 5.9 \times 10^{-4} \times 0.781 = 4.62 \times 10^{-4} \ cm^2volt^{-1}sec^{-1}$$

The preceding equations were developed from theory; however, in practice a relationship for calculating mobilities is desired that involves more readily obtainable measurements. Recalling that the units involved in equation 15.4 were volts/cm and cm/sec, the equation can be rewritten as equation 15.7, in which needed parameters are easily obtained.

$$U = \frac{p_m l_m}{Vt} \tag{15.7}$$

where U = mobility in $cm^2volt^{-1}sec^{-1}$
p_m = the distance the ion moves in cm
l_m = distance between the electrodes in cm
V = voltage in volts (not to be confused with E which is volts/cm)
t = time in sec

For some electrophoretic cells, the distance l_m is hard to obtain accurately so that volts/cm is determined from

$$E = i/qk \qquad (15.8)$$

where E = volts/cm
  i = current in amperes
  q = cross-sectional area of the cell in cm^2
  k = specific conductivity of the solution in ohm^{-1}cm^{-1}

k is determined from

$$k = C/R \qquad (15.9)$$

where R = resistance of the solution in the cell in ohms
  C = the cell constant in cm^{-1}. The cell constant is determined by the classical method of Jones and Bradshaw (1933) in which they found that if 7.457 gm of dry KCl were added to 1 liter of conductivity water at 20°C the resulting conductivity was 0.007138 ohm^{-1}cm^{-1} at 0°C.

Example

A 3% solution of egg albumin buffered at a pH of 7.2 was placed in an electrophoresis cell having a diameter of 1.2 cm. A current of 20 ma was used and it took 208 min for the egg albumin to move 5.87 cm toward the positive electrode. If the cell constant was 18.3 cm^{-1} and the resistance of the system was 7690 ohms at 0°C, calculate the mobility of the material.

Step (1). Calculate the specific conductivity using equation 15.9.

$$k = \frac{18.3 \text{ cm}^{-2}}{7690 \text{ ohms}} = 0.00238 \text{ ohm}^{-2}\text{cm}^{-2}$$

Step (2). Calculate E using equation 15.8.

$$E = \frac{0.020}{(3.14 \times 0.62)\,(0.00238)} = 7.43 \text{ volts/cm}$$

Step (3). Calculate U using equation 15.7. Remember that the l_m has been incorporated into the E value since $E = V/l_m$.

$$U = \frac{-5.87}{(208 \times 60)\,(7.43)}$$

$$= -6.3 \times 10^{-5} \text{ cm}^2\text{volt}^{-1}\text{sec}^{-1}$$

The sign is negative because the ion was negative as indicated by motion toward the positive electrode.

Heat Production.—A problem that often arises is that of heat production. The current passing through the cell develops heat and this not only causes convection currents which change the mobilities but it increases the rate of evaporation of the buffer solution when paper electrophoresis is employed. This loss of buffer causes a change in mobility due to changes in potential, and current and it also causes a capillary action in the paper since the drier paper will take up solution from the buffer tank. The usual maximum amount of heat that can be tolerated is 0.15 watts/cm^3. The heat produced can be calculated from equation 15.10.

$$H = i^2/q^2k \qquad (15.10)$$

where H = heat developed in watts/cm^3 and i, q, and k are as previously defined.

Example

Determine the heat developed in the solution described in the previous example.

$$H = \frac{(0.020)^2}{(1.13)^2 (0.00238)} = 0.132 \text{ watts/cm}^3$$

Electro-osmosis.—It has been found that if water is placed in a capillary tube and positive and negative electrodes are placed in contact with the water, the water will migrate toward the negative electrode indicating that the water is positively charged. This migration can be altered by placing other ions in the system, and in fact can be reversed. This movement of the solution in an electric field is called *electro-osmosis.* Consequently, the mobilities must be further corrected because the compound migrating in the electrophoretic system can be held back or speeded up by the movement of the buffer solution. This is similar to the action of a tail wind or head wind on an airplane. To make that correction, a material with no inherent mobility is selected. Therefore, if it moves, it does so because of the solution movement carrying it along. This movement can either be added or subtracted from the mobility of the compound under investigation to correct for the electro-osmosis. One such compound that is used with proteins is dextran. It has a mobility of only -0.16×10^{-5} and can be used as a correction factor as follows:

$$U_{cpd} = \frac{P_{cpd} \pm P_{dex}}{Et} \qquad (15.11)$$

where P_{cpd} = distance the compound moved
$\quad\;\; P_{\text{dex}}$ = distance the dextran moved
$\quad\;\; \pm$ = + if the dextran moves opposite to the compound and
$\qquad\quad$ − if it moves in the same direction.

Migration Path Length Correction.—It has been found that the mobility of a compound determined in free solution is different from that determined using paper as a support. This is believed to be due to the fact that the particle in the paper must wind in and out among the fibers of the paper, thus traveling a longer path than that which is actually measured. Figure 15.6 illustrates the problem involved.

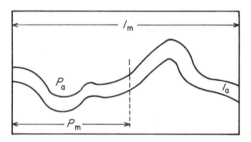

FIG. 15.6.   PARTICLE MIGRATION IN PAPER

Referring to Fig. 15.6, equations 15.12 and 15.13 can be obtained.

$$l_a = l_m(l_a/l_m) \qquad\qquad (15.12)$$

$$p_a = p_m(l_a/l_m) \qquad\qquad (15.13)$$

Rearranging equation 15.7

$$p_m = UVt/l_m \qquad\qquad (15.14)$$

However, we know this is not correct and should be expressed in terms of p_a and l_a as

$$p_a = UVt/l_a \qquad\qquad (15.15)$$

In order to get this into a workable equation, the relationships shown in equations 15.12 and 15.13 are substituted for p_a and l_a.

$$p_m = \frac{UVt}{l_m}\left(\frac{l_m}{l_a}\right)^2 \qquad\qquad (15.16)$$

The value of $(l_m/l_a)^2$ can be obtained by conductivity measurements.

$$\left(\frac{l_m}{l_a}\right)^2 = \frac{R_o v_t}{R v_b} \qquad\qquad (15.17)$$

where R_o = resistance, in ohms, of the electrolyte occupying a volume equal to the buffer in the paper

R = resistance, in ohms, of the paper saturated with buffer

v_t = total volume of the paper and the buffer in it

v_b = volume of the buffer in the paper

Another way of determining the ratio is to obtain the mobility in the paper and also in free solution. Any difference will be due to a different path length. Typical ratio values of (l_m/l_a) are 0.5 to 0.7 for most papers.

Isoelectric Point.—Consider the amino acid NH_2—R—COOH which in neutral solution would exist as $NH_3{}^+$—R—COO$^-$. It may react to acid or base as follows:

$$NH_3{}^+—R—COO^- + H^+ = NH_3{}^+—R—COOH$$
(a positive ion)

$$NH_3{}^+—R—COO^- + OH^- = NH_2—R—COO^- + H_2O$$
(a negative ion)

If placed in an electrophoresis cell, the amino acid would migrate toward the anode in basic systems and toward the cathode in acid systems. As mentioned before, there should be a pH at which the net charge is zero, and the amino acid would show no movement. This point where no movement occurs is called the *isoelectric point*; it is important as it affects separations. Consider the situation where the pH is such that part of the amino acid ions are positive and part are negative. Here we should get a very diffuse band and probably no separation. What is desired, therefore, to get sharp resolution is to have the compound we want separated as far away from the isoelectric point as possible so that all of the ions are of the same charge.

Disc Electrophoresis

This is a relatively new development of a general technique called gel electrophoresis. In moving boundary electrophoresis it is difficult to remove a particular fraction without stirring and remixing the solution. To overcome this problem many different types of supporting materials (i.e. paper) were tried. By using starch gel, the separated components could be removed easily by slicing the gel.

Gel electrophoresis, however, in many cases, takes a long time, one reason being because of the nature of the gel the sample is added in a broad band.

The technique of disc electrophoresis is a way of concentrating the sample in a very narrow "disc" before the actual separation occurs.

(The separated components are also disc-shaped.)   This means that
the separation can be done much faster and better.   For example,
where a filter paper technique yields 7 protein fractions of a serum,
a gel gives 30 components.   Although it is true that a narrow initial
sample zone is one reason for good separations, another property of
gels is that it acts like a sieve.   The theory of this technique was
developed primarily by Ornstein (1964), and the actual analytical
technique was developed by Davis (1964).

In cases where the ionic mobilities are quite similar, paper electrophoresis may not resolve the components.   However, a gel is a
latticed network whose pores can be made of molecular size and act
as a sieve to further separate compounds based on differences in size
and shape.

The gel usually used in disc electrophoresis is polyacrylamide,
made by polymerizing acrylamide with N,N-methylene bis-acryl-
amide (Bis).   This is catalyzed by riboflavin and or N,N,N,N-
tetramethylenediamine (Temed) and by light; a lamp is used to
hasten the process once the solutions are placed in the cell.   The
pore size can be varied by the ratios of the chemicals used.   In a
recent technique, a continuous gradient of pores down the column
was used.   These polyacrylamide gels are relatively inert and have
almost no ionic side groups.

A disc cell is filled with three sections (Fig. 15.7).   These include
a large-pore sample concentration gel, a large pore spacer gel to remove the materials used to concentrate the sample, and a small pore
gel used to separate the components.

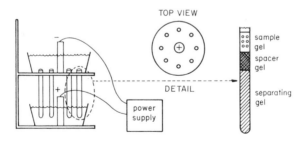

FIG. 15.7.   DISC ELECTROPHORESIS APPARATUS

According to Davis (1964), "The concentration step is achieved by
introducing a mixture of sample ions into an electrophoretic column
near the boundary of two ions whose sign is like that of the sample
ions at a given pH.   One ion is faster (chloride), the other, an ion
of a weak acid or base (glycine), slower than all of the sample ions at

this pH. The counter ion (Tris, hydroxymethyl aminomethane), the ion of sign opposite that of the slow, fast, and sample ions, serves as a buffer to set initially and then to regulate the various pH's behind the moving anion boundary. Tris is used with proteins. The electrical polarity is set so that the fast ion is situated ahead (in the direction of migration) of the sample and slow ions. Application of a voltage results in the segregation from one another and stacking of the constituent ions of the sample into contiguous zones in order of their relative mobilities, the entire sample being sandwiched between the fast and slow ions." The sample thickness can be very thin. Bromphenol blue is added as a marker dye to show when the migration down the entire cell is complete.

The next step is to remove the slow ion from the sample. The spacer segment is used for this by employing a different pore size and a different pH, so that the mobility of the ion of the weak acid or base now continuously overtake and pass through the sample species, establishing a comparatively uniform voltage gradient in which electrophoretic separation of the sample occurs.

The third step is to actually electrophoretically separate the sample components. A small pore gel is used for this. After the separation, the gel is removed from the cell, transferred to a larger diameter cell, and dye added to react with the separated compounds. The unbound dye is removed from the gel by electrophoresis.

Electrofocusing

Electrofocusing or isoelectric focusing is the name given to the phenomenon occurring to ampholytes in a pH gradient influenced by an electric field (Haglund 1967). The gradient is obtained by imposing a dc potential on an electrolyte system in which the pH steadily increases from the anode to the cathode. Provided the pH gradient is stable, ampholytes (such as proteins or peptides) present in the electrolyte system are repelled by both electrodes, and each ampholyte species collects at a place in the gradient where the pH of the gradient is equal to the isoelectric point of that species. Though the theory of electrofocusing has been known for a long time, it has only recently become a practical method of protein fractionation. This has been mainly due to commercial availability of carrier ampholytes for use in electrofocusing, and of improved instruments for separation.

The carrier ampholytes are mixtures of aliphatic polyamino-polycarboxylic acids with molecular weights of 300–600. The pK values of the amino acid carboxylic groups determine the pI of each ampholyte. When preparing a column, the carrier ampholytes are dissolved in a density gradient of a nonionic compound (i.e. sucrose) to obtain a convection-free medium. The sample can be layered at some chosen level in the column or evenly distributed throughout the whole volume. The pH gradient is formed by the carrier ampholytes when a voltage (200–1200 v, dc, yielding a current up to 10 ma) is applied across the mixture. The proteins in the sample migrate to the point where they are electrically neutral (i.e. pH = pI). To reduce convection, the column is thermostated during the operation. The time required for electrofocusing, normally, ranges from 24 to 72 hr. At the end of the run the contents of the column are drained (without remixing the gradient-stabilized and separated substances) through a capillary to a fraction collector. The effluent is detected by a flow analyzer to identify the separated sharp zones.

The two main applications of isoelectric focusing in biochemistry are (1) the analytical separation of high molecular weight ampholytes, especially proteins, according to their isoelectric points (a separation of 2 species requires a difference in isoelectric point of only 0.02 pH units); (2) characterization of proteins by their isoelectric points in a single, simple experiment, compared to 16 runs required in free electrophoresis. In the pH range 5 to 8, the precision and reproducibility of isoelectric point determination in electrofocusing are about 0.01 pH unit.

As any other method, electrofocusing has several limitations. Thus, lipoproteins are denatured at the isoelectric point, and some proteins are only slightly soluble at their isoelectric point, especially in a salt-free system. Adding urea increases solubility and may be used to an advantage in some cases. Increasing the ampholyte concentration, and thus the ionic strength, can also increase solubility of some proteins.

Gel electrofocusing is a microanalytical modification of isoelectric focusing in a sucrose density gradient (Wrigley 1968). Carrier ampholytes and protein sample are set in a polyacrylamide gel using conventional disc electrophoresis apparatus. Fractionation takes 1–3 hr. Protein zones are detected by precipitation in the gel with 5% trichloracetic acid or with protein dyes after removal of carrier ampholytes.

Preparative Electrophoresis

Electrophoresis in horizontal thick slabs of buffer soaked into an appropriate medium (mainly potato starch) has been for many years the simplest method of preparative electrophoresis. By this method,

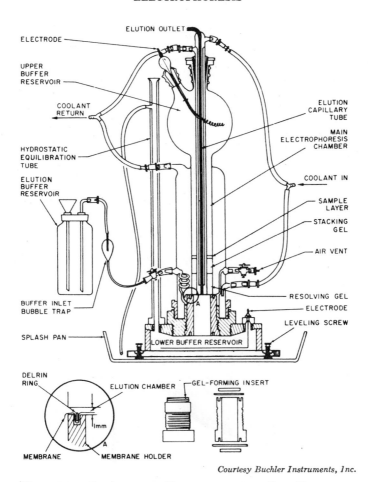

ELUTION OUTLET

ELECTRODE

UPPER
BUFFER
RESERVOIR

COOLANT
RETURN

ELUTION
CAPILLARY
TUBE

MAIN
ELECTROPHORESIS
CHAMBER

HYDROSTATIC
EQUILIBRATION
TUBE

ELUTION
BUFFER
RESERVOIR

COOLANT IN

SAMPLE
LAYER

STACKING
GEL

AIR VENT

RESOLVING GEL

ELECTRODE

LEVELING SCREW

BUFFER INLET
BUBBLE TRAP

SPLASH PAN

LOWER BUFFER RESERVOIR

DELRIN
RING

ELUTION CHAMBER

GEL-FORMING INSERT

1mm

MEMBRANE

MEMBRANE HOLDER

Courtesy Buchler Instruments, Inc.

FIG. 15.8. PREPARATIVE POLYACRYLAMIDE GEL ELECTRO-
PHORESIS

batches of 5–50 ml liquid (or more in cooled columns) can be separated in 10 to 20 hr. A commercial, continuous fractionation and
elution electrophoresis apparatus (Fig. 15.8) is adapted from the
design of Jovin *et al.* (1964). Samples containing up to 200 mg proteins in 160 ml can be separated within 10 hr. The instrument is
capable of utilizing several supporting media, though it has been used
primarily for polyacrylamide gel electrophoresis.

Associated Problems

(1) A 0.2 M solution of $MgCl_2$ is prepared. What is the mobility
of the chloride ion if its solution radius is 1.93 Å, and the viscosity of

the solution is 8.54 millipoises? What is the mobility if ionic strength is considered? Ans. -1.24×10^{-5} cm^2 volt^{-1} sec^{-1}; -1.07×10^{-5} cm^2 volt^{-1} sec^{-1}

(2) What is the solution radius of the acetate ion if a 0.125 M solution of sodium acetate showed a mobility of -2.49×10^{-4} cm^2 volt^{-1} sec^{-1}. The viscosity is 9.633 millipoises. Ans. 1.69Å

(3) The electrophoretic fractionation of a pig embryo plasma produced a spot 58 mm away from the origin 13 hr and 36 min after a current of 9 ma had been applied. The cell constant is 15.2 cm^{-1} for the 8 mm cell, and a resistance of 2550 ohms was obtained when a 0.02 M sodium phosphate 0.15 M NaCl buffer was used. Calculate the mobility and the heat evolved. Ans. 2.0×10^{-5} cm^2 volt^{-1} sec^{-1}; 5.3×10^{-2} watts/cm^3

(4) The red blood cells of a guinea pig were found to migrate 38 mm in 86.5 min when placed in a 0.15 M phosphate buffer. A current of 12 ma was used. Calculate the mobility and the heat evolved. The cell was 1.0 cm in diameter, had a constant of 18.9 cm^{-1}, and 8140 ohms was the resistance when the buffer was in the cell. Ans. 1.12×10^{-4} cm^2 volt^{-1} sec^{-1}; 1.0×10^{-1} watts/cm^3

Some Recent Applications

Gelling and thickening agents have been detected by Padmoyo and Miserez (1967), using electrophoresis on cellulose acetate film. Paper electrophoresis has been used by Ney (1967) for the detection of egg white in liquid yolk, and albumin protein changes in stored eggs were examined by Croisier and Sauveur (1967). High voltage paper electrophoresis of water soluble coal tar food dyes was done by Niitsu (1964). Improved techniques for enzyme resolution by starch gel electrophoresis have been described by White and Kushnir (1966).

Cheeses, such as Cheddar, blue, Camembert, etc. may be distinguished from each other by their peptide patterns on polyacrylamide gels according to Ney et al. (1966).

Thin-slab polyacrylamide techniques have been described by Cowie (1968) and agar-gel by Hill et al. (1966) for the identification of fish. An enzyme method which uses starch gel electrophoresis to locate esterase bands has been proposed by Thompson (1968) as a further means of animal and fish species identification.

Electrophoresis on silica gel has been used by Maier and Diemair (1966) to separate histamine and histidine.

BIBLIOGRAPHY

BARKA, T. 1961. Studies on acid phosphates. II. Chromatographic separation of acid phosphates of rat liver. J. Histochem. Cytochem. 9, 564–571.

BLOCK, R. J., DURRUM, E. L., and ZWEIG, G. 1958. A Manual of Paper Chromatography and Paper Electrophoresis, 2nd Edition. Academic Press, New York.

BLOEMENDAL, H. 1963. Zone Electrophoresis in Blocks and Columns. Elsevier Publishing Co., Amsterdam, The Netherlands.

CANN, J. R. 1968. Recent advances in the theory and practice of electrophoresis. Immunochem. 5, 107–134.

COWIE, W. P. 1968. Identification of fish species by thin-slab polyacrylamide gel electrophoresis of the muscle myogens. J. Sci. Food Agr. 19, 226–229.

CROISIER, G., and SAUVEUR, B. 1967. Alterations in the electrophoretic characteristics of albumen proteins induced by the storage of eggs. Ann. Biol. Animale, Biochim. Biophys. 7, 317–324.

DAVIS, B. 1964. Disc electrophoresis. II. Method and application to human serum proteins. Ann. N.Y. Acad. Sci. 121, 404–427.

HAGLUND, H. 1967. Isoelectric focusing in natural pH gradients—a technique of growing importance for fractionation and characterization of proteins. Sci. Tools 14, No. 2, 17–24.

HILL, W. S., LEARSON, R. J., and LANE, J. P. 1966. Fish and other marine products: Identification of fish species by agar gel electrophoresis. J. Assoc. Offic. Anal. Chemists 49, 1245–1247.

JONES, G., and BRADSHAW, B. C. 1933. The measurement of the conductance of electrolytes. V. A redetermination of the conductance of standard KCl solutions in absolute units. J. Am. Chem. Soc., 55, 1780–1800.

JOVIN, J., CHRAMBACH, A., and NAUGHTON, M. A. 1964. An apparatus for preparative temperature regulated polyacrylamide gel electrophoresis. Anal. Biochem. 9, 351–369.

LONGWORTH, L. G. 1942. Recent advances in the study of proteins by electrophoresis. Chem. Rev. 30, 323–340.

MAIER, H. G., and DIEMAIR, W. 1966. Zum Nachweis von Histamin neben Histidin in Bohnenkaffee. Z. Anal. Chem. 223, 263.

NERENBERG, S. T. 1966. Electrophoresis, A Practical Laboratory Manual. F. A. Davis, Philadelphia.

NEY, K. H. 1967. Detection of eggwhite in commercial liquid yolk. Fette, Seifen, Anstrichmittel 69, 794–795.

NEY, K. H., WIROTAMA, I. P. G., and SEITZ, I. 1966. Study of cheese by means of polyacrylamide gel electrophoresis. Intern. Dairy Congr., Proc., 17th, Munich 4, 283.

NIITSU, Y. 1964. High voltage paper electrophoretic analysis of water soluble coal tar dye for food. Japan Analyst 13, 1239–1942.

ORNSTEIN, L., and DAVIS, B. J. 1962. Disc electrophoresis. Eastman Kodak Co., Rochester, N.Y.

ORNSTEIN, L. 1964. Disc electrophoresis. I. Background and theory. Ann. N.Y. Acad. Sci. 121, 321–349.

PADMOYO, M., and MISEREZ, A. 1967. Identification of gelling and thickening agents by electrophoresis and staining on cellulose acetate strips. Mitt. Gebiete Lebensm. Hyg. 58, 31–49.

SMITHIES, O. 1955. Zone electrophoresis in starch gels; group variations in the serum of normal human adults. Biochem. J. 61, 629–638.

THOMPSON, R. R. 1968. An enzymatic (esterase) method for identification of animal and fish species. J. Assoc. Offic. Anal. Chemists 51, 746–748.

TISELIUS, A. 1931. Moving Boundary Method of Studying Free Electrophoresis of Proteins. Thesis, Upsala, Sweden.

VAN OSS, C. 1968. Separation and purification of plasma proteins: analytical and preparative separation, purification and concentration methods. In Progress in Separation and Purification, E. S. Perry (Editor). John Wiley & Sons, New York.

VON KLOBUSITZKI, D., and KONIG, P. 1939. Biochemische Studien uber die Gifte der Schlangengattung. Arch. Exp. Path. Pharmakol. *192*, 271–278.

WHITE, J. W., JR., and KUSHNIR, I. 1966. Enzyme resolution in starch gel electrophoresis. Anal. Biochem. *16*, 302–312.

WIEME, R. J. 1963. Agar Gel Electrophoresis. Elsevier Publishing Co., New York.

WRIGLEY, C. 1968. Gel electrofocusing—a technique for analyzing multiple protein samples by electrofocusing. Sci. Tools *15*, No. 2, 17–23.

CHAPTER 16

Voltammetry (Polarography)

If the world of electroanalytical chemistry were limited to potentiometry, most of the measurements and information we are now
able to obtain could not have been done.

In a potentiometric system, the system controls us because we
merely measure what is going on in the solution naturally. However, if we force the issue; that is, make something happen that
wouldn't ordinarily happen, then we gain control of the situation
and can obtain a wealth of new information. Remember— *Voltage*
controls what reacts and *Current* controls how much reacts.

At the turn of this century, although it was well known that
voltammetry had great possibilities, it was also found that reliable
measurements could seldom be made. The problem was that the
electrodes were not reproducible and the results depended to a large
extent on the past history of the electrode.

This problem was solved in 1922 by Heyrovsky who used a dropping mercury electrode (Fig. 16.1). This consists of a very fine
glass capillary which is filled with extremely pure mercury. As the
mercury comes from the capillary it forms a small drop which is continuously expanding, exposing a new and reproducible surface to the
solution it is immersed in.

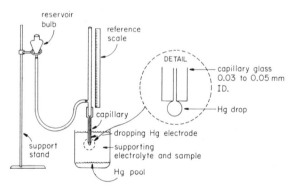

FIG. 16.1. DROPPING MERCURY ELECTRODE ARRANGE-
MENT

205

The advantages of the dropping mercury electrode are (1) the surface area is reproducible, (2) the new surface eliminates poisoning effects, (3) the high overpotential of hydrogen on Hg renders possible the deposition of substances difficult to reduce, (4) mercury forms amalgams with many elements and this reduces their potential, (5) diffusion currents assume a steady current immediately and are reproducible, (6) the range is from $+0.4$ to -1.8 v referred to a normal H_2 electrode (above $+0.4$ v, Hg dissolves and at -1.8 v H_2 is evolved), and (7) the lower limit of detection with conventional polarography is 10^{-5} M/liter.

Heyrovsky coined the term *polarography* as the method is based on the polarization of the cathode. We now prefer to call the process *volt-ammetry*, since the resulting curves are actually a plot of current *vs* voltage.

The dropping mercury electrode opened the door to modern electroanalytical chemistry and we now have such instruments as ac, dc, pulse, cyclic, and derivative polarography, chronoamperometry, chronopotentiometry, chronocoulometry, amperometry, and coulostatic analysis—to mention only the major techniques. However, if a basic understanding of dc polarography is obtained, the rest of the techniques can be readily understood. For this reason the section will be limited to dc polarography.

A polarograph works as follows. The dropping mercury electrode (DME) is placed in the sample solution and adjusted so that about 20 drops a minute are obtained. This electrode is usually made the cathode $(-)$. The anode $(+)$ may be the mercury pool which forms at the bottom of the cell (not very reliable), a saturated calomel electrode (quite reliable), or any other type of reference electrode. The cathode is made the controlling electrode by making its area much smaller than that of the anode. A small electrode is more easily polarized and is the controlling electrode. Since the anode has a larger area and the current is small (microvolts at most), the concentration polarization at this electrode is usually negligible and its potential is regarded as constant. THE SOLUTION IS NOT STIRRED. The reason for this will be discussed later. A voltage, usually starting at zero and increasing negatively, is applied to the drop. When the drop voltage reaches the value necessary for a species in the sample to be reduced, a current will begin to flow due to the transfer of electrons from the mercury drop to the metal ions (the polarized electrode is depolarized). The current that results is then plotted *vs* the voltage applied. Figure 16.2 indicates the basic electrical diagram. Figure 16.3 shows what we might expect to get.

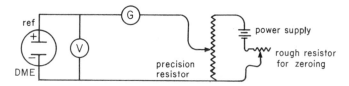

FIG. 16.2.   A SIMPLE DC POLAROGRAPH

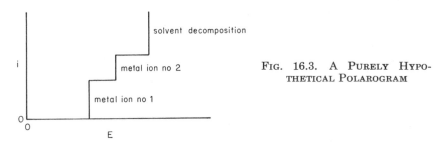

FIG. 16.3.   A PURELY HYPO-
THETICAL POLAROGRAM

When the voltage is zero, the electrode has no attraction for any of the ions present, and no current should flow.   As we increase the voltage we reach a point where the most easily reduced ion will react, and a current begins to flow.   Now the current will reach a limit depending on how many ions are in the solution that can diffuse to the electrode.   The greater the concentration the greater the current; this is the basis of quantitative voltammetry.   If the voltage is again increased, we reach the point where the next ion can be reduced.   However, the first ion is still being reduced so the current produced is the sum of both ions.   The height of this wave will be proportional to the number of ions of the second type in the solution. If the voltage is further increased, a point will be reached at which the water solvent starts to decompose and to liberate hydrogen. Since the water concentration is large it will produce a wave that will go off scale and completely obscure any metal ions that might have been reduced at a more negative voltage.

Figure 16.3 showed what we might expect; but what do we actually get?   This is shown in Fig. 16.4 for the reduction of lead using a very poor technique.

In region 1 we had expected no current, yet there is a slight current that increases slightly with voltage.   This is called the *residual current* and it is the current necessary to charge the drop, forming what is known as the *Helmholtz double layer*, more electrons being necessary as the voltage increases.

In region 2 we might expect a straight line but we get a sawtoothed signal.   This is because the mercury drop continuously in-

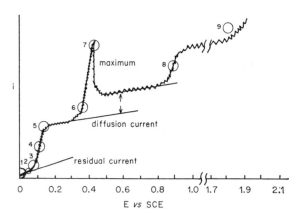

FIG. 16.4.   A POLAROGRAM OF LEAD USING A POOR TECH-
NIQUE

creases in size as it leaves the capillary.   At the start the drop is
small and few ions can diffuse into it so the current is small.   As the
drop increases in size, more ions can diffuse into it and the current
increases.   When the drop falls off, the current will not go back to
zero because there is still a small mercury surface exposed, and also
the speed of the pen is slow enough to dampen out part of the oscilla-
tion.   The net effect is a saw-toothed pattern.

In the regions 3 and 5 we get a rounding of the curve rather than a
sharp break.   Remember that the energies of ions are not all the
same but follow a Gaussian distribution, some are of low energy and
some are of high energy.   The high energy ions will react with the
electrode a bit sooner than the main bulk of the ions so an *early* current is obtained.   The wave increases normally (region 4) until we
approach the upper limit where the lesser energetic ions are not reduced as easy as the *normal* ions, and more voltage must be applied.

What is the source of the wave in region 4?   Lead is supposed to
be reduced at about -0.4 v vs SCE and this is only -0.1 v.   It
turns out that oxygen in the solution is very easily reduced and the
reduction occurs in two steps:

$$O_2 + 2\ H^+ + 2\ e^- = H_2O_2 \qquad E° = -0.05\ v \qquad 16.1$$

$$H_2O_2 + 2\ H^+ + 2\ e^- = 2\ H_2O \qquad E° = -0.9\ v \qquad 16.2$$

This is a serious problem.   Notice that the oxygen reduction produces most of the current and takes up most of the chart paper,
thereby reducing the utility of the method.   The oxygen must be
removed.   This is done by bubbling nitrogen through the sample
solution for about 5 min before an analysis is made.

Region 6.   This is the normal lead wave.

Region 7.   This is called a *maximum*.   Its exact cause is not completely known, but it can be eliminated by adding a *suppressor* to the sample.   Suppressors are surface active agents such as gelatin, Triton X-100 (a synthetic detergent), or such compounds as bromcresol green.   Only a few drops of dilute solution (0.002–0.01%) of the suppressor should be added or the entire wave can be eliminated.

Region 8.   The second oxygen wave.

Region 9.   The water solvent reduction.

Figure 16.5 shows a voltamogram of a mixture of four metals in an aqueous solution using $(NH_4)_2SO_4$ as the supporting electrolyte.

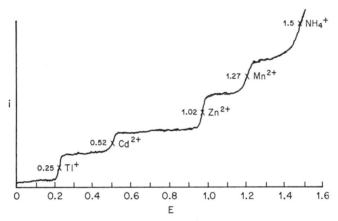

FIG. 16.5.   A VOLTAMOGRAM OF A MIXTURE OF FOUR COMPONENTS

Migration Current

It was noted earlier that the sample solution should not be stirred. This is because the wave height is proportional to the sample concentration only if ions arrive at the cathode by one process, in this case by diffusion; stirring would upset the diffusion process.   However, there are many many ions in solution and the cations, seeing the negative potential of the cathode, will move toward the cathode and the anions will move toward the anode.   This current is not due to diffusion and is called the *migration current*.   Can you see that a species such as $Cd(CN)_4^{2-}$ would actually cause a decrease in the Cd^{2+} current under these conditions?

To reduce that effect a *supporting electrolyte* is added.   This is a solution about 1000 times more concentrated than the solution to be measured, and it is made out of a salt that is very difficult to reduce (Na, K, Li). Refer to Fig. 16.6. Since there are so many more K+ ions than C+ ions, when a potential is applied to the DME, the K+ ions migrate to the electrode, and surround it, but cannot be reduced because the voltage (−0.60 is not high enough to reduce the K+ ions). The C+ ions other than those in proximity of the electrode are prevented from migration and reach the electrode because of normal diffusion.

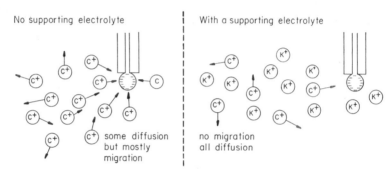

FIG. 16.6. THE EFFECT OF A SUPPORTING ELECTROLYTE

Diffusion Current

The experimental results of polarography were obtained long before a mathematical interpretation was forthcoming. The first successful equation that accounted for the reduction of the ions, the diffusion toward a spherical drop whose area was continuously changing and whose surface is moving toward the bulk solution, was developed by Ilkovic in 1934.

$$i_d = 607 n D^{1/2} C m^{2/3} t^{1/6} \qquad (16.3)$$

where i_d = average current in μa (708 is used when i_d is the maximum current)

t = drop time in sec

m = flow rate of Hg in mg/sec

D = diffusion coefficient of the ion under study (cm²/sec)

C = concentration in millimoles/ liter

n = number of electrons taking place in the reaction at the dropping mercury electrode

Temperature does not enter into this equation directly, but the diffusion coefficient changes 1–2%/°C at room temperature. Therefore for accurate work, temperature control to a few tenths of a degree is necessary.

Example

A 5×10^{-4} M solution of $BaCl_2$ in 0.1 M $(CH_3)_4NCl$ has a $E_{1/2}$ of -1.94 v vs SCE and shows an average diffusion current of 4.0 μa. The dropping rate was found to be 24 drops per min and when 20 drops were collected they weighed 0.0750 gm. Calculate the diffusion coefficient.

$n = 2$, since $Ba^{2+} + 2e^- = Ba$

$C = 0.5$ mM/liter

$$t = \frac{60 \text{ sec}}{24 \text{ drops}} = 2.5 \text{ sec/drop}$$

$$m = \frac{0.0750 \text{ gm}}{20 \text{ drops}} = 3.75 \text{ mg/drop}; \times 24 \text{ drops}/60 \text{ sec} = 1.5 \text{ mg/sec}$$

Substituting these values into the Ilkovic equation,

$4.0 = 607 \times 2 \times D^{1/2} \times 0.5 \times (1.5)^{2/3} \times (2.5)^{1/6}$

Using logarithms to evaluate the fractional powers

$\log (1.5)^{2/3} = 0.1761 \times 2/3 = 0.1174$

anti-log is 1.31

$\log (2.5)^{1/6} = 0.3979 \times 1/6 = 0.0663$

anti-log is 1.165

Thus $D^{1/2} = \dfrac{4.0}{607 \times 2 \times 0.5 \times 1.31 \times 1.165} = 0.0043$

$$D = 1.9 \times 10^{-5} \text{ cm}^2/\text{sec}$$

Calibration Curve Method

This is the most convenient method when a large number of similar samples are to be analyzed. Voltamograms are obtained for 3 to 5 solutions of varying concentrations and the diffusion current, corrected for residual current, is plotted vs concentration. Figures 16.7 and 16.8 indicate how the data are obtained and the calibration curve prepared; E_{de} is the potential of the dropping electrode.

Once such a calibration curve is prepared, the various unknowns can readily be determined. Care must be taken to ensure that the temperature and drop time are the same in all cases.

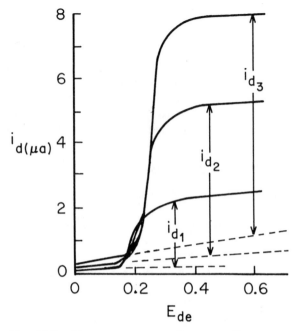

FIG. 16.7. VOLTAMOGRAMS OF STANDARD SOLUTIONS

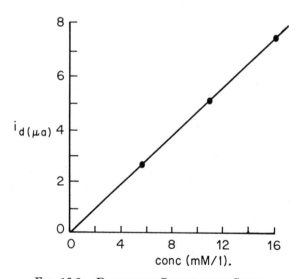

FIG. 16.8. RESULTING CALIBRATION CURVE

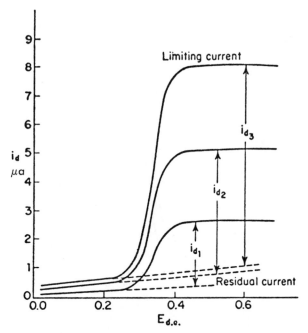

FIG. 16.9. PREPARATION OF A CALIBRATION CURVE
Diffusion current change with concentration.

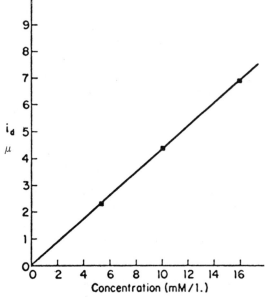

FIG. 16.10. CALIBRATION CURVE

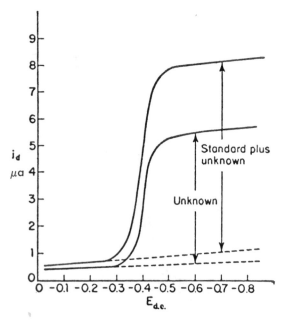

FIG. 16.11. THE STANDARD ADDITION METHOD

Standard Addition Method

When only a few samples are to be analyzed, the method of standard addition appears to be the most convenient. Two voltamograms are required: one for the unknown, and one for the unknown plus an added known amount of the same material. Figures 16.9, 16.10 and 16.11 indicate how the experimental data are utilized. The volume of the solution changes upon the addition of the standard, and this produces a corresponding change in the concentration. The result is that a correction for this dilution must be made. Equation 16.4 shows how this is done.

$$C_{\text{unk}} = \frac{i_d \, v \, C_{\text{std}}}{(\Delta i(V + v) + i_d v)} \qquad (16.4)$$

where $i =$ the increase in the diffusion current due to the added standard
      $V =$ the original volume
      $v =$ the added volume

Example

A 30.0-gm sample of grapefruit sections was stored in a new experimental tin can.   This sample was homogenized, diluted to 250 ml, and centrifuged.   A 25-ml aliquot was taken for polarographic analysis and a current of 24.9 μa was observed.   Five milliliters of a standard solution containing 6.0×10^{-4} M Sn was added and the diffusion current rose to 28.3 μa.   Calculate the % tin in the grapefruit.   The atomic weight of tin is 118.70:

$$C_{unk} = \frac{24.9 \times 5 \times 6.0 \times 10^{-4}}{(28.3 - 24.9)(25 + 5) + 24.9 \times 5}$$

$$= 3.3 \times 10^{-4} \text{ M}$$

$$118.70 \times 3.3 \times 10^{-4} \times \frac{250 \text{ ml}}{1000 \text{ ml}} = 0.0098 \text{ gm of Sn}$$

$$\frac{0.0098 \text{ gm} \times 100}{30.0 \text{ gm}} = 0.032\%$$

Some Recent Applications

Nonprotein sulfhydryl in biological tissues can be determined by ac polarography in concentrations as low as 1.45 μg/ml (Charles and Knevel 1968).

Hartley and Bly (1963) found that 2,6-xylenol reacts with nitrite to produce 4-nitroso-2,6-xylenol which is polarographically reducible. Nitrate also reacts but is reduced at a different potential, thus mixtures of the two can be determined.

Acrylonitrile monomers from packaging wrappers can contaminate food products.   These monomers can be determined polarographically by a method developed by Compton and Buckley (1965).

Fumaric acid is used as an acidulant in gelatin desserts and pie fillings; as little as 1 μgm/ml can be determined.   With an NH_4Cl support, the reduction takes place at -1.65 v vs Hg pool.   If tetramethylammonium bromide is the support, better results are obtained and the reduction takes place at -1.15 v.   The details are presented by Smith (1966) and Smith and Gajan (1965).

Associated Problems

(1)   A method is to be developed for determining traces of a thallium chloride rat poison on grapes.   A 5×10^{-3} M solution had a diffusion current of 2.28 μa.   If 25 sec were required for 8 drops of Hg to fall and the total weight of the Hg is 0.2453 gm, calculate the diffusion constant for this system.   Ans. 1.85×10^{-4}

(2)   Pyrazine (2 e$^-$ change) in coffee aroma is to be determined. If the diffusion coefficient is 0.98×10^{-5} cm^2 sec^{-1} and the same

capillary was used as in problem (1), calculate the concentration of pyrazine using the Ilkovic equation. Ans. 0.356 mM/liter

(3) Arsenic (III) produces reduction waves when 1 N H_2SO_4 is used as the supporting electrolyte. A 20-gm sample of a banana peel was examined for arsenic and produced a wave of 41.7 μa. Arsenic in concentrations of 150, 250, 350, and 500 μgm produced wave heights of 19.3, 32.1, 45.0, and 64.3 μa, respectively when treated in a manner similar to the sample. Calculate the % As in the sample. Ans. 0.0016%

(4) Potassium antimonyl tartrate in 0.4 N sodium tartrate has a half wave potential of −0.8v vs SCE. Several solutions of this compound were prepared and the following results obtained:

Conc. (mM/liter)	μa
1.0	0.94
3.2	3.10
5.6	5.40
7.4	7.12

A 0.0407 gm sample dissolved in 25 ml of solution, containing Sb had an i_d of 5.72 μa. Calculate the % Sb in the sample. Ans. 43.7%

(5) Potassium can be determined polarographically by forming potassium tetraphenyl borate(III) which is dissolved in N,N-dimethylformamide. Tetrabutylammonium iodide is used as the supporting electrolyte. The potassium impurities of the sodium salt were investigated by dissolving 0.3 gm of the salt in 20 ml of the solvent, adding the tetraphenyl borate and the supporting electrolyte, and diluting to 50 ml. A diffusion current of 5.32 μa was obtained. Ten milliliters of a standard solution containing 6.0 × 10^{-4} M K was added and a diffusion current of 6.4 μa was produced. Calculate the % K in the sample. Ans. 0.0017%

BIBLIOGRAPHY

CHARLES, R., and KNEVEL, A. M. 1968. An alternating current polarographic determination of nonprotein sulfhydryl in biological systems. Anal. Biochem. 22, 179–186.

COMPTON, T. R., and BUCKLEY, D. 1965. Polarographic determination of residual acrylonitrile and styrene monomers in styrene-acrylonitrile copolymers. Analyst 90, 76–82.

HARTLEY, A. M., and BLY, R. M. 1963. Polarographic determination of nitrite as 4-nitroso-2,6-xylenol. Anal. Chem. 35, 2094–2100.

HEYROVSKY, J. 1923. Electrolysis with a dropping Hg electrode. I. Phil. Mag. 45, 305–315.

ILKOVIC, P. 1934. Collection Czech. Chem. Commun. 6, 498–513.

SMITH, H. R. 1966. Fumaric acid in foods. J. Assoc. Offic. Anal. Chemists 49, 701–702.

SMITH, H. R., and GAJAN, R. J. 1965. Detection and determination of fumaric acid in foods. J. Assoc. Offic. Anal. Chemists 48, 699–700.

CHAPTER 17

Mass Spectroscopy

Mass spectrometry has long been useful in identification and determination of the composition of mixtures, particularly those organic in nature. Techniques are now available to use a mass spectrometer directly as a detector for the components separated by a gas chromatograph. Thus, an understanding of the basic principles of mass spectroscopy is essential for all analysts.

The basic idea of mass spectroscopy is to produce ions (only positive ions will be discussed here) by bombarding organic molecules with high energy electrons, then accelerating these ions in a definite direction so that they can be separated according to their mass or velocity. The separated ions are then detected and their intensity measured.

There are several types of mass spectrometers available commercially. The three most common types are the electromagnetic, Fig. 17.1, the time of flight, Fig. 17.2, and the quadrupole, Fig. 17.3.

Each spectrometer has three major components; the ion source, the analyzer, and the detector.

Ion Source

Consider the molecule $CH_3CH_2CH_2SH$. Suppose that a high energy electron strikes the molecule and transfers its energy to it. The diagram below shows what can happen.

$$C_3H_8S + e^-$$

- $\rightarrow C_3H_5{}^+$ $(m/e = 41)$ 4.7% $(m/e = \text{mass/charge})$
- $\rightarrow CHS^+$ $(m/e = 45)$ 11.2%
- $\rightarrow CH_2S^+$ $(m/e = 46)$ 42.8%
- $\rightarrow CH_3S^+$ $(m/e = 47)$ 3.4%
- $\rightarrow C_3H_5S^+$ $(m/e = 73)$ 0.9%
- $\rightarrow C_3H_6S^+$ $(m/e = 74)$ 21.2% (parent ion molecule)
- $\rightarrow$ negative neutral and all other $+$ ions 15.9%

The percent composition obtained depends upon the energy of the impinging electrons. If a very low energy electron strikes a molecule, it may break only one bond. If the energy of the impinging electron is increased, different bonds can be broken. A plot of the fraction of the ions produced vs the electron energy is called a clastogram, an example of which is shown in Fig. 17.4.

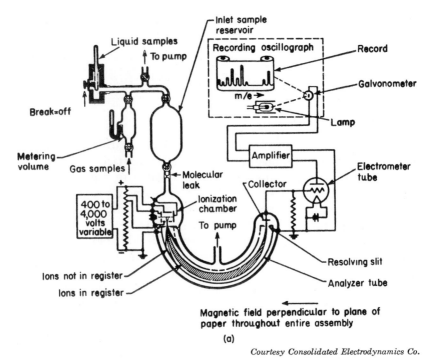

(a)

Courtesy Consolidated Electrodynamics Co.

FIG. 17.1.   ELECTROMAGNETIC MASS SPECTROMETER

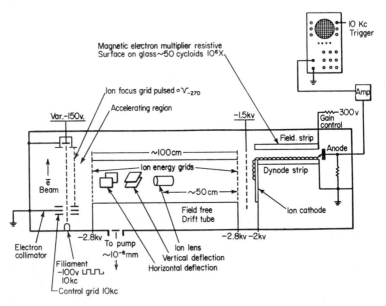

FIG. 17.2.   TIME OF FLIGHT MASS SPECTROMETER

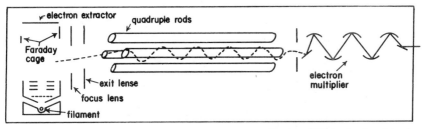

FIG. 17.3. QUADRUPOLE MASS SPECTROMETER

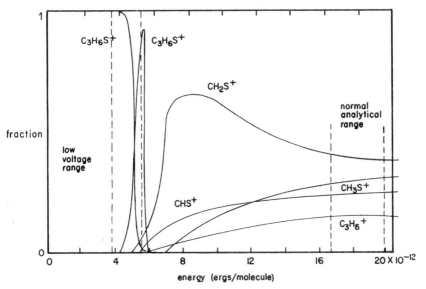

FIG. 17.4. CLASTOGRAM OF *N*-PROPYL MERCAPTAN

Notice that the ratio of the ions reaches a fairly constant value on the right of the diagram. This corresponds to 50 to 70 v applied to the ion source, and is the useful range for analysis. Figure 17.5 shows an ion source in more detail.

Electrons emitted from the filament are accelerated between E_1 and E_2 by a potential of about 70 v. When an electron strikes a molecule, M, positive ions, +, negative ions, −, electrons, e⁻, and neutral molecules may be formed. The positive species are separated from the negative species by E_3 and E_4 which have a small potential of a few volts across them. The positive ions are then accelerated down the tube E_5, E_6, etc., each having several hundred volts applied to them. A total accelerating voltage of 2000 v is normal.

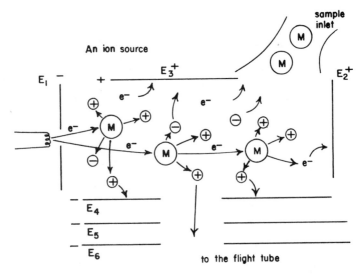

FIG. 17.5. An Ion Source

Analyzers

The ions, accelerated in the ion source, are then separated. In the electromagnetic instrument, the separation of ions of different mass to charge (m/e) ratios is accomplished by a magnetic field at right angles to the flight path of the ions. The lighter ions are deviated more than the heavier ions, so separation occurs. (Note: a vacuum of about 10^{-6} torr is always applied. The reason for this is that once the positive ions are produced, you do not want to lose them prematurely by collision with the other molecules and becoming neutralized.)

After the formation of ions in an electron beam, the ions are accelerated through a potential drop of V. In so doing, the ions receive kinetic energy, T, equal to

$$T = mv^2/2 = eV \tag{17.1}$$

where m = mass of the ion in gm
       v = velocity of the ion in cm/sec
       e = charge on the ion, 4.8×10^{-10} esu
       V = the voltage used to accelerate the ion

Upon entering a magnetic field, H, the ions will take paths that are arcs of a circle with a radius, R, equal to

$$R = mv/eH \tag{17.2}$$

Combining 17.1 and 17.2 and rearranging gives

$$m/e = H^2R^2/2V \qquad (17.3)$$

The angle of deflection (radius) is usually fixed for a given analyzer tube, so that to focus ions of given (m/e) values on the detector system, either H or V must be varied.

Example

Consider an electromagnet type mass spectrometer, having $V = 2000$ and $R = 7.00$ in. What must be the magnetic field to focus the CO_2^+ ion $(m/e = 44)$ on the detector? Use equation 17.3. Here, R is in cm and V is in erg/esu. But 300 practical volts $= 1$ erg/esu. Thus,

$$V = 2000/300 = 6.667 \text{ erg/esu}$$

and $\quad R = 7.00 \times 2.54 = 15.78 \text{ cm}$

$$H^2 = 2V(m/e)/R^2$$

$$H^2 = \frac{2 \times 6.667 \times (44/6.02 \times 10^{23})}{(15.78)^2 \times 4.80 \times 10^{-10}} \times 9 \times 10^{20}$$

$$= 7.33 \times 10^6$$

$$H = 2708 \text{ gauss}$$

$$(9 \times 10^{20} \text{ is the factor to convert } esu^2 \text{ to } emu^2)$$

The time of flight technique is a bit different. The magnets are removed and the flight path is straight instead of curved. Even without a magnetic analyzer it is still possible to analyze the various ions on the basis of equation 17.1. Since all ions fall through the same potential drop, V, all ions have the same kinetic energy, T, but different velocities which are inversely proportional to the square root of the mass (see equation 17.1, $v = \sqrt{2T/m}$).

$$t = d(m/e2V)^{1/2} \qquad (17.4)$$

where t = the flight time in sec

$\quad d$ = the flight distance in cm

Example

A time of flight mass spectrometer has a flight length of 93.0 cm. The accelerating potential is 2530 v. How long would it take for a CH_3OH^+ ion to traverse the spectrometer and how far behind a CH_2OH^+ ion would it be?

CH_2OH^+

$t = 93.0 \times (31 \times 300 \times 10^{-16}/2 \times 2.530 \times 4.80 \times 6.02)^{1/2}$
$= 7.415\ \mu sec$

CH_3OH^+

$t = 93.0 \times (32 \times 300 \times 10^{-16}/2 \times 2.530 \times 4.80 \times 6.02)^{1/2}$
$= 7.534\ \mu sec$

$\Delta t = 0.12\ \mu sec$

The electric quadrupole mass spectrometer is basically different from the previous instruments. A schematic diagram of the positioning of the four long, parallel, electrodes is shown in Fig. 17.6. Although the electrodes are shown to have a uniform hyperbolic cross section, cylindrical rods can be used if they are carefully spaced. The opposite pairs of electrodes are connected electrically.

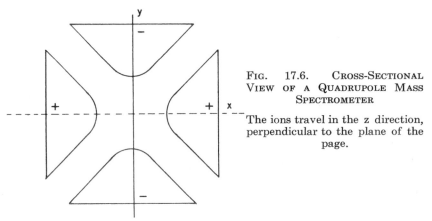

FIG. 17.6. CROSS-SECTIONAL VIEW OF A QUADRUPOLE MASS SPECTROMETER

The ions travel in the z direction, perpendicular to the plane of the page.

Both a dc voltage, U, and an rf voltage, $V_0 \cos\omega t$, are applied to the quadrupole array. Ions are injected from a conventional ion source into the rf field in the z direction. The ions travel with a constant velocity. Figure 17.7 shows what happens to ions that are too heavy or too light for the particular field strength.

Detectors

The detector is an electrode onto which the positive ions fall. Connected in series between this electrode and ground is a resistor. Electrons from ground rush to neutralize the positive charge on the collector. Current across the resistor causes a potential drop that is proportional to the current flow (Ohm's law), so that by measuring the voltage the number of ions of each type can be determined.

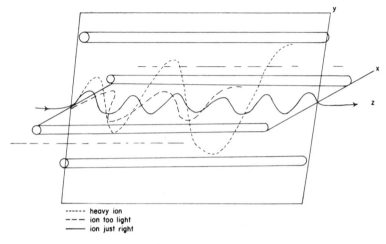

------ heavy ion
--- ion too light
——— ion just right

Fig. 17.7.   Quadrupole Oscillations

As early as 1943, Cohen, used an electron multiplier in a mass spectrometer.   The principle is the same as was described for photomultiplier tubes in a previous chapter.   The detector has a high sensitivity and a rapid response.   Multiplication factors of 10^5 to 10^6 are usual.

The Wiley magnetic electron multiplier, 1956, employs crossed magnetic and electric fields to control the electron trajectories.   As the individual groups of ions arrive at the end of the field free flight tube (time of flight mass spectrometer), they collide with the plane ion cathode of the magnetic electron multiplier.   The plane ion cathode is used because it eliminates ion transit time variations encountered with a curved ion cathode (Fig. 17.9).   Each ionization produces a group of electrons, and because of the crossed magnetic and electric fields present, the electrons follow a cycloidal path down the dynode strips of the multiplier.   In this manner a current gain of the order of 10^6 is obtained.

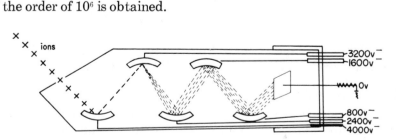

Fig. 17.8.   Cohen Type Electron Multiplier

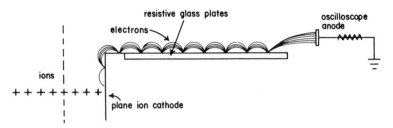

FIG. 17.9.   WILEY ELECTRON MULTIPLIER

Low Voltage Mass Spectrometry

At present, the primary use for a mass spectrometer by the food analyst is as a detector for a gas chromatograph.   The cracking pattern will tell what is present, and the height of the peaks can be used to determine how much is present.   In order to simplify the analysis, a low ionization voltage is sometimes used.   From Fig. 17.10 it can be seen that as the voltage of the electron beam is decreased, the amount of fragmentation is also decreased, particularly below about 25 v.   However, there is also an accompanying reduction in sensitivity of ion detection.   In practice, an intermediate value of the ionizing voltage is usually selected such that the fragmentation will be held to a minimum commensurate with the best sensitivity possible.

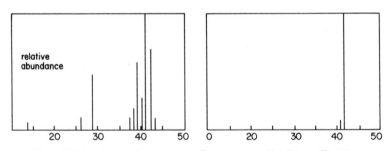

FIG. 17.10.   MASS SPECTRA OF PROPYLENE (A) 70 v, (B) 12 v

Figure 17.10, a spectra of propylene, gives some indication of how much simpler the spectra becomes at a lower voltage.

Quantitative Analysis

There are times when the single peak coming from the gas chromatograph is really a mixture of two or more components.   The height of the peaks in a mass spectrum is directly proportional to the pressure of the sample in the ion source.   If the sample is a mixture, then the heights will be proportional to the partial pressures.

There are two methods by which the solutions may be made: (1) the linear simultaneous equations method, and (2) the technique of subtractions.   An example of the subtraction technique, which is the simpler of the two, will be given here.

Subtraction Technique.—The subtraction technique is similar to the analysis of mixtures in the visible region discussed earlier. In visible spectroscopy, the absorbances of the components in the mixture were additive; in mass spectroscopy, the partial pressures are additive.

The basic idea is to take the unknown spectrum and then determine the spectrum for each compound known or suspected to be in the mixture.   Since all of these spectra may have been determined at different pressures, it is necessary to convert them all to one common pressure, which is the pressure of the unknown.   Each known spectrum is then subtracted from the unknown spectrum, in turn, until the unknown spectrum is reduced to zero.

Example

The data in Table 17.1 will be used to explain the problem. Notice that in each spectrum there is one peak that is arbitrarily set at 100.   All other peaks are proportional to the reference peak and the numbers in the tables show these ratios.

Let 1 be ethyl alcohol, 2, methyl alcohol, 3, methyl aldehyde, and 4, ethyl aldehyde.   From Dalton's law, $P_1 + P_2 + P_3 + P_4 = P_t$, and we know that the total pressure P_t is related to the number of moles of gas.   We may write this in terms of mole fraction of the components: $N_1 + N_2 + N_3 + N_4 = N_t = 1$.   Let a be a factor that correlates P_t and N_t, i.e., $P_t a = N_t = 1$.   Thus $a = 1/P_t$ and $P_1 a = N_1$, $P_2 a = N_2$, etc.

TABLE 17.1

A MIXTURE OF UNKNOWN COMPOSITION

m/e	Ion Abundance	m/e	Ion Abundance
24	0.71	33	0.14
25	2.39	40	0.43
26	5.61	41	1.81
27	7.51	42	4.36
28	17.7	43	12.5
29	100.0	44	18.3
30	43.4	45	9.13
31	37.9	46	3.96
32	8.51	47	0.11

TABLE 17.2

SPECTRA OF KNOWN COMPOUNDS

m/e	Ethyl Alcohol Ion Abundance	Methyl Aldehyde Ion Abundance	Methyl Alcohol Ion Abundance	Ethyl Aldehyde Ion Abundance
24	0.41			1.55
25	2.05			4.77
26	8.31			9.08
27	23.2		0.15	4.50
28	5.61	30.9	6.35	2.67
29	23.9	100.0	64.7	100.0
30	6.05	88.5	0.80	1.14
31	100.00	1.91	100.0	
32	1.27		66.7	
33	0.07		1.02	
40	0.23			0.95
41	1.14			3.88
42	3.03			9.16
43	7.98			26.7
44	1.76			45.6
45	5.1			1.24
46	16.12			
47	0.47			

TABLE 17.3

ETHYL ALCOHOL SPECTRUM $\times$ 0.2455

m/e	Ion Abundance	m/e	Ion Abundance
24	0.10	33	0.02
25	0.50	40	0.055
26	2.04	41	0.28
27	5.70	42	0.75
28	1.38	43	1.96
29	5.87	44	0.43
30	1.48	45	8.62
31	24.6	46	3.96
32	0.31	47	0.11

TABLE 17.4

THE ORIGINAL UNKNOWN SPECTRUM MINUS THE ETHYL ALCOHOL SPECTRUM

m/e	Ion Abundance	m/e	Ion Abundance
24	0.61	33	0.12
25	1.89	40	0.37
26	3.57	41	1.53
27	1.81	42	3.62
28	16.3	43	10.5
29	94.1	44	17.9
30	41.9	45	0.51
31	13.3	46	0.00 out
32	8.20	47	0.00 out

Let us first select the largest peak in the mass spectrum of the unknown which can arise from only one component of the mixture. In our example mixture, $m/e = 46$ can arise only from ethyl alcohol; $m/e = 47$ is too small to be reliable.

We see that the pure ethyl alcohol has an intensity of 16.12 at $m/e = 46$, whereas the unknown has a value of 3.96 at the same m/e. (Note: it is not necessary for you to determine the spectra of each compound you suspect in the sample. Tables of spectra are commercially available and they can be used in exactly the same way as shown in this example.) This means that the two are at different pressures and we must express them on a common basis. This can be done by multiplying 16.12 by a factor to convert it to 3.96. Then $16.12 \times P_1 = 3.96$. $P_1 = 0.2455$ in this case. Now, multiply all the peaks in the ethyl alcohol spectrum by 0.2455 to reduce it to the same pressure as it would have in the unknown. This is shown in Table 17.3.

This new reduced spectrum is now subtracted from the original unknown spectrum to give a new unknown spectrum without the ethyl alcohol component (Table 17.4).

Notice that the $m/e = 46$ and 47 peaks have been reduced to zero.

Now let us repeat this process, eliminating this time ethyl aldehyde, since it is the only component that has any peaks above $m/e =$

TABLE 17.5

ETHYL ALDEHYDE SPECTRUM $\times$ 0.393

m/e	Ion Abundance	m/e	Ion Abundance
24	0.61	40	0.37
25	1.88	41	1.53
26	3.57	42	3.60
27	1.77	43	10.5
28	1.05	44	17.9
29	39.3	45	0.49
30	0.45		

TABLE 17.6

THE ORIGINAL SPECTRUM MINUS ETHYL ALCOHOL AND ETHYL ALDEHYDE

m/e	Ion Abundance	m/e	Ion Abundance
24	0.00 out	33	0.12
25	0.02 out	40	0.00 out
26	0.00 out	41	0.01 out
27	0.04 out	42	0.02 out
28	15.3	43	0.00 out
29	54.8	44	0.00 out
30	41.5	45	0.02 out
31	13.3		
32	8.20		

TABLE 17.7

METHYL ALCOHOL SPECTRUM $\times$ 0.123

m/e	Ion Abundance	m/e	Ion Abundance
27	0.02	31	12.3
28	0.78	32	8.20
29	7.96	33	0.12
30	0.10		

TABLE 17.8

THE ORIGINAL SPECTRUM MINUS THREE COMPONENTS

m/e	Ion Abundance	m/e	Ion Abundance
28	14.5	31	1.0
29	46.8	32	0.0 out
30	41.4	33	0.0 out

TABLE 17.9

METHYL ALDEHYDE SPECTRUM $\times$ 0.468

m/e	Ion Abundance	m/e	Ion Abundance
28	14.5	30	41.4
29	46.8	31	0.9

40. $m/e = 44$ is the most reliable since it is the largest and there are no other compounds that have a similar m/e. The factor this time is $45.6 \times P_4 = 17.9$ (use the value in the new spectrum), and $P_4 = 0.393$. The new ethyl aldehyde spectrum is shown in Table 17.5.

And now the new unknown spectrum, minus both the ethyl alcohol and the ethyl aldehyde is Table 17.6.

There are now only six peaks left and they must be due to methyl alcohol and methyl aldehyde. Since methyl alcohol does not have an $m/e = 32$ or 33 peak, it is eliminated next using the $m/e = 32$ peak. $66.17 \times P_2 = 8.2$, and $P_2 = 0.1230$. The new methyl alcohol spectrum is Table 17.7.

The new unknown spectrum minus ethyl alcohol, methyl alcohol, and ethyl aldehyde is Table 17.8.

Finally, $100 \times P_3 = 46.8$ so $P_3 = 0.468$. The new methyl aldehyde spectrum is Table 17.9.

The final unknown spectrum is Table 17.10.

Since all of the m/e values have been reduced to zero or <0.1, this means that all of the sample has been accounted for and there is no other component left.   Now $P_1 + P_2 + P_3 + P_4 = P_t$.

TABLE 17.10

FINAL UNKNOWN SPECTRUM

m/e	Ion Abundance	m/e	Ion Abundance
28	0.0 out	30	0.0 out
29	0.0 out	31	0.1 out

$$P_1 = 0.2455$$
$$P_2 = 0.393$$
$$P_3 = 0.123$$
$$P_4 = 0.468$$
$$\overline{P_t = 1.2295}$$

Therefore, $a = 1/P_t = 1/1.230 = 0.813$

$N_1 = 0.2455 \times 0.813 = 0.200 = 20.0$ mole $\%$ ethyl alcohol

$N_2 = 0.123 \times 0.813 = 0.100 = 10.0$ mole $\%$ methyl alcohol

$N_3 = 0.468 \times 0.813 = 0.381 = 38.1$ mole $\%$ methyl aldehyde

$N_4 = 0.393 \times 0.813 = 0.319 = 31.9$ mole $\%$ ethyl aldehyde

Interfacing the Mass Spectrometer to a Gas Chromatograph

A gas chromatograph can separate complex mixtures quantitatively into individual unidentified components, and the mass spectrometer is excellent for the identification of those compounds. However, combining these two instruments poses some problems. The components from a chromatograph are at atmospheric pressure and diluted with a large amount of carrier gas flowing at 50–100 ml/min. The mass spectrometer operates at 10^{-5} to 10^{-6} torr, and requires a relatively small sample.

Early attempts consisted of placing a cold trap at the chromatograph outlet, and transferring the trapped components to the mass spectrometer. This procedure is quite slow and the entire process must be repeated for each component.

What is needed is a means of removing the carrier gas, thereby concentrating the sample, reducing the pressure 100,000,000 fold in less than a minute, and scanning each component as it emerges from the chromatograph. Rapid scan spectrometers are available, the interfacing is a problem. Several ways have been proposed but two that are the easiest for the ordinary analyst to make and use are the fritted glass tube (Watson and Bieman 1964, 1965), and the heated porous Teflon tube (Lipsky et al. 1966).

Fritted Glass.—This consists of an ultrafine fritted glass tube, Fig. 17.11, surrounded by a metal chamber vacuum jacket. A

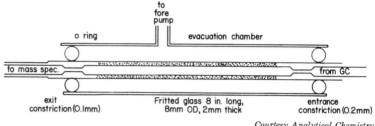

Courtesy Analytical Chemistry

FIG. 17.11.   FRITTED GLASS INTERFUCE

vacuum pump is attached to the metal chamber and the entire system is heated to 275°–300°C.   When the sample and the carrier gas emerge from the chromatograph, the low molecular weight helium atoms effuse through the fritted glass much faster than the higher molecular weight sample.   The separated helium is pumped out of the system, and the sample plus some residual helium go to the mass spectrometer.   The entrance constriction is required to maintain effusion.   The mean free path of the gas must be ten times the diameter of the pores through which it passes.   Since the pores in the fritted tube are approximately 1 μ wide, the pressure inside the tube should be a few torr.   The entrance constriction maintains this pressure drop.   The exit constriction controls the sensitivity of the apparatus.

Porous Teflon Tube.—This interface was developed to overcome the large sample losses encountered with the fritted glass tube. Organic polymer films have varying degrees of permeability to gases. Teflon FEP, a copolymer of tetrafluoroethylene and hexafluoropropylene, is used because it can stand temperatures up to 300°C without decomposing.   A 7-ft length of thin walled (0.005 in.) capillary tubing (0.020 in. od × 0.010 in. id) is used.   A 4-ft coil of 0.01 in. id stainless steel tubing is used as the inlet and exit pressure reducer.   An additional exit constriction is necessary in some cases. Efficiencies of 40–70% are obtained and no memory effects have been observed.

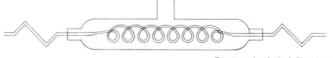

Courtesy Analytical Chemistry

FIG. 17.12.   POROUS TEFLON INTERFACE

Some Recent Applications

A mass spectrometer attached to a gas chromatograph has been used by Nonaka *et al.* (1967) to determine components of cooked chicken meat volatiles; by Anderson and Von Sydow (1964) to identify the higher boiling compounds of black currants; by Schultz *et al.* (1964) to identify orange juice volatiles; and 87 compounds in hops by Buttery *et al.* (1965).

Associated Problems

(1) Assume that in a time of flight mass spectrometer, with a flight path of 0.850 m, the accelerating voltage is 2800 v. What is the flight time in microseconds for the $m/e = 84$ and 86 of chloroform? Ans. 10.60 and 10.33 μ sec

(2) In an observation of the n-butane spectrum with a conventional electromagnetic spectrometer, it was observed that the $m/e = 29$ peak was focused on the collector when an accelerating voltage of 2850 v and a magnetic field of 2430 gauss were employed. (a) Determine if the magnetic field is held at 2430 gauss, what must be the accelerating voltage to focus the $m/e = 27$ ion on the collector? (b) If the accelerating voltage is held at 2950 v, what must be the magnetic field to focus $m/e = 25$ on the collector? Ans. 3065 v, 2095 gauss

(3) From the following mass spectrometric information, determine the mole fraction composition of the unknown using the subtraction technique. A = 1-butanol, B = 2-butanone, C = cyclobutane, and D = 3-methyl-1-butyne.

m/e	A	B	C	D	Unknown	
26	7.9	5.3	22.6	5.8	18.0	
27	55.7	16.2	42.0	36.8	90.0	
28	17.7	3.0	100.0	2.5	35.5	
31	100.0	0.6	. . .	1.2	48.0	
39	18.9	2.4	19.8	29.3	42.0	Ans. D = 40
41	62.8	1.7	89.3	17.1	52.5	B = 30
43	59.6	100.0	0.1	0.8	97.0	A = 20
53	1.2	0.6	4.7	100.0	100.0	C = 10
56	86.2	0.2	62.2	. . .	54.5	
57	6.6	6.1	2.7	. . .	7.9	
67	. . .	. . .	. . .	61.6	57.4	
72	0.16	16.5	. . .	. . .	12.2	

(4) From the following mass spectrometric information, determine the mole fraction composition of the unknown using the sub-

232   FOOD ANALYSIS

traction technique, A = 1-propanol, B = 2-propanol, C = propanol, D = ethane thiol.

m/e	A	B	C	D	Unknown	
28	6.1	1.8	68.9	41.5	36.1	
29	15.9	11.3	100.0	93.5	75.1	
31	100.0	6.4	3.2	0.1	100.0	
34	...	...	...	24.1	9.3	Ans. A = 50
45	4.2	100.0	0.1	22.4	51.4	B = 20
47	...	0.2	...	78.6	30.5	C = 10
58	0.5	0.2	38.9	10.8	12.2	D = 20
59	9.2	3.5	1.4	7.7	13.5	
60	6.4	0.4	...	1.5	7.0	
62	...	...	...	100.0	39.7	

BIBLIOGRAPHY

ANDERSON, J., and VON SYDOW, E. 1964. The aroma of black currants. Acta Chem. Scand. *18*, 1105–1114.

BUTTERY, R. G., BLACK, D. R., and KERLY, M. P. 1965. Volatile oxygenated constituents of hops. Identification by combined gas chromatography and mass spectrometry. J. Chromatog. *18*, 399–402.

COHEN, A. A. 1943. The isotopes of cerium and rhodium. Phys. Rev. *63*, 219–223.

LIPSKY, S. R., HORWATH, C. G., and MCMURRAY, V. J. 1966. Utilization of system employing the selective permeation of helium through a unique membrame of teflon as an interface for gas chromatograph and mass spectrometer. Anal. Chem. *38*, 1585–1587.

NONAKA, M., BLACK, D. R., and PIPPEN, E. L. 1967. Gas chromatographic and mass spectral analysis of cooked chicken meat volatiles. J. Agr. Food Chem. *15*, 713–717.

SCHULTZ, T. H., TERANISHI, R., MCFADDEN, W. H., KILPATRICK, P. W., and CORSE, S. 1964. Volatiles from oranges. II. Constituents of the juice identified by mass spectra. J. Food Sci. *29*, 790–795.

WATSON, J. T., and BIEMAN, K. 1964. High resolution mass spectra of compounds emerging from a gas chromatograph. Anal. Chem. *36*, 1135–1137.

WATSON, J. T., and BIEMAN, K. 1965. Direct reading of high resolution mass spectra of gas chromatographic effluents. Anal. Chem. *37*, 844–851.

WILEY, W. C. 1956. Bendix time-of-flight mass spectrometer. Science *124*, 817–820.

CHAPTER 18

Nuclear Magnetic Resonance

INTRODUCTION

Nuclear magnetic resonance (NMR) was first observed in 1945, although the concepts of nuclear spin and magnetic moments upon which the techniques are based go back to about 1925.

The NMR spectrometer is composed of the following basic units: magnet, radio frequency (rf) oscillator, and rf detector. A simple schematic diagram of the equipment is shown in Fig. 18.1.

Three fields, all at right angles to each other are required for NMR. The first, a stationary magnetic field (X axis), is required to produce a strong external field which will cause the axis of the spinning nuclei to tilt as they spin. This can be visualized by considering a spinning nucleus as a toy gyroscope. A gyroscope will spin about its axis of rotation, but when placed in an external field such as the earth's magnetic field, it will tilt and start a second rotation. This second rotation is called precession. In NMR, the magnet employed depends upon the type of analysis desired, but is usually of the order of 10,000 gauss, and must provide a very stable, homogeneous field.

The second field (Z axis), is an alternating radio-frequency field (60 Megahertz for hydrogen). The purpose of the second field in NMR is to oscillate at the same frequency as the precessing nuclei. If this occurs, then energy can be transferred from the oscillator to the nuclei, and it does work by changing the angle of precession. If the nuclei get out of phase with the oscillator, they can lose their energy and return to a lower angle of precession (see Fig. 18.2). This causes the nuclei to "wobble." It must be understood that any one nucleus does not wobble continuously, but that some nuclei are being raised to high precession levels while other nuclei are dropping back to lower precession levels, the net effect being that about 7 out of 1,000,000 nuclei "wobble."

The third field (Y axis) is used to detect this "wobble." It consists of a small coil of wire which has a voltage induced in it when the "wobbling" nuclei cut its lines of force. The small voltage generated is amplified, and the signal is displayed by a recorder.

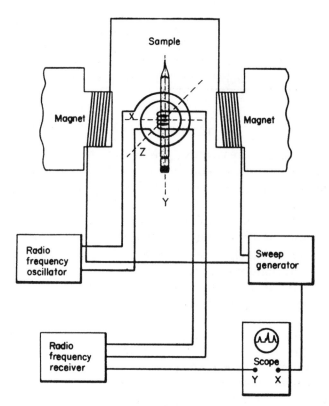

FIG. 18.1.　SCHEMATIC OF AN NMR SPECTROMETER

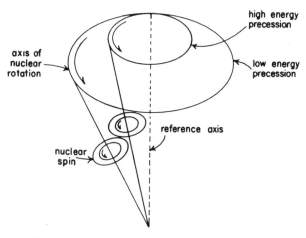

FIG. 18.2.　PRECESSION-ENERGY RELATIONSHIP

The rf oscillator frequency can be varied to match the existing frequency of the precessing nuclei or the rf frequency can be fixed and the magnetic field varied thereby changing the precessional frequency to match the rf oscillator. In practice it is easier to change the magnetic field strength. The field may be swept at different rates, but usually the rate will be of the order of 5 to 10 milligauss per minute. The spectra obtained will be a plot of signal strength on the Y axis against the magnetic field on the X axis.

The sample volume used may be varied but will be of the order of 0.01–0.5 ml. This small sample volume is a great advantage, as is the fact that this method of NMR is totally nondestructive. For solids, about 50 mg is usually dissolved in about 0.3 ml of a solvent. Carbon tetrachloride or carbon disulfide is tried first because they have no interfering hydrogens. Deuterated solvents are then used, with water the last choice. The samples are placed in small diameter tubes, of the order of 5 mm od.

It is also usual to add directly to the sample, or place in a concentric sample tube, a known material which will serve as the reference for the analysis, such as tetramethyl silane.

An additional technique which is helpful in obtaining well-resolved spectra is that of spinning the sample while in the magnetic field. This averages out the field inhomogeneities perpendicular to the direction of spinning.

NMR Spectra and Their Interpretation

Since the nuclei of atoms in different compounds are usually in different magnetic environments, they will necessitate different

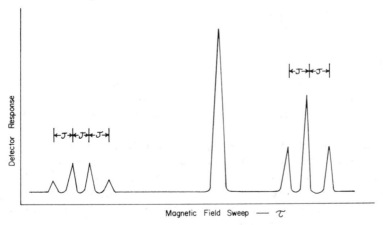

FIG. 18.3. SPECTRA OF ETHYL ACETATE

applied magnetic fields to reach resonance with an oscillator at a fixed frequency. The NMR spectra thus become fingerprints of the molecules. A typical NMR spectrum of ethyl acetate is shown in Fig. 18.3.

To illustrate some of the information available from a proton spectrum as in Fig. 18.3, let us examine the ethyl acetate case in somewhat more detail. The spectrum had the following areas, in arbitrary units: 93:140:145 for the quadruplet: singlet: triplet: Since the structural formula of ethyl acetate is

$$H_3C-C\underset{O-CH_2CH_3}{\overset{O}{\diagup}}$$

we would expect 3 groups of peaks, due to the 3 nonequivalent groups of protons: the CH_2 in the ethyl group, the CH_3 in the ethyl group, and the CH_3 in the acetate. The areas, proportional to the number of protons, would be in the ratio of $2:3:3$. If we let the average of 140 and 145, 142.5, be equivalent to 3, we find our experimentally measured areas to be in the ratio of $1.96:2.95:3.05$; or very nearly $2:3:3$, as anticipated. To identify the given lines within the groups, we see that the singlet and the triplet must be the CH_3 groups, and that the quadruplet must be the CH_2 group using the above ratios.

Looking now at the spin-spin interactions of the protons in the CH_3 of the ethyl group of the ester with the neighboring CH_2 group we see that:

and thus we could expect a triplet, wherein we have three lines due to ↑↑ ↓↓ and ↓↑ coupling, the last of these occurring twice and therefore expected to be twice as intense (i.e., an intensity ratio of $1:2:1$). Thus we readily identify the triplet as the CH_3 from the ethyl group.

The nearby singlet is now expected also to be a CH_3 group. But there is only one line. This is because there are no nearby neighboring protons in the molecule to allow spin interactions with the protons of this CH_3 group; therefore, both from the intensity and spin considerations, we conclude the singlet is the CH_3 group from the acetate portion of the molecule.

We are left with the fact that the quadruplet must be due to the CH_2 group. Again looking at the spin-spin interactions with the protons of the neighboring CH_2 group, we see that

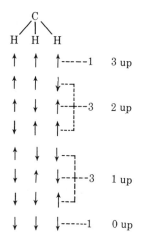

and therefore a quadruplet is confirmed, even to the prediction that the intensity ratios in the quadruplet be $1:3:3:1$.

There are some general rules about spin-spin interactions which are helpful in calculating the number of lines expected and their approximate intensities. These follow.

(1)   Nuclei of the same group do not interact to cause observable splitting.   (2) The multiplicity of the band from a group of equivalent nuclei is determined by the neighboring groups of equivalent nuclei. The neighboring group will cause a multiplicity of $s = (2nI + 1)$ in that group, where $n =$ the number of equivalent nuclei of spin I.   (3) The intensities are symmetric about the midpoint of the group and for the case of nuclei of $I = {}^1/_2$, their intensities are given by the coefficients in the expansion of

$$(r + 1)^{s-1}$$

Coupling Constants And Chemical Shifts

In Fig. 18.3, we observed a number of peaks in the NMR spectrum of ethyl acetate.   We have already discussed above that these could be considered as three groups of peaks, one a singlet, another a triplet, and the third a quadruplet.   It is of interest now to observe that the spin-spin splitting of a peak such as the triplet or quadruplet is a fixed amount.   In Fig. 18.3, we see that these are in fact equal in both the CH_2 and CH_3 groups of the ethyl portion of ethyl acetate.

TABLE 18.1

COUPLING CONSTANTS

Type of Compound	J (in Cps)
CH_3X	12
CH_3CH_2X	5–8
$\underset{H}{\overset{H}{\diagdown}}C{=}C\underset{H}{\diagup}$	17.5–18.5
$\underset{H}{\diagup}C{=}C\underset{H}{\diagdown}$	8.5–10.5
$C{=}C\overset{H}{\underset{H}{\diagdown}}$	1.5–2.0
$HC{\equiv}CH$	9
benzene ring with two H	5–8

They have been designated by J. J is the *coupling constant*, and is independent of the magnetic field.   Some representative values of J are given in Table 18.1.   The coupling constant in ethyl acetate is about 7 cps, and falls within the range of values given for CH_3CH_2X in Table 18.1.   Thus, one quickly sees how values of coupling constants may aid in structure determinations.

Another significant aid in structure determination is the chemical shift.   The chemical shift, δ, is a function of electron density about the nucleus.

The chemical shift is related to the distance from the center of a group of lines to a reference line.   Some common reference substances used, which can be used also as solvents and which produce only a single reference line, are water, benzene, and cyclohexane. The separation of the sample and reference lines, is ΔH, and thus the separation of the lines is field-dependent.   There are three ways commonly employed to determine values of δ.

(1)   Use simply a mixture of the sample and the reference. Usually this is not very satisfactory, since the value may be concentration dependent.   (2) Determine δ as a function of the concentration in the reference, and extrapolate to infinite dilution.   (3) Place the sample into one tube and then place the reference into another tube which is then inserted concentrically into the sample tube.

Approximate values of δ are given in Table 18.2 for a number of different types of compounds.

TABLE 18.2

CHEMICAL SHIFT VALUES

Group	δ	Group	δ
—SO₃H	-6.5	Ar\ \nN—H \nAr/	$+2.0$
—COOH	-6.3	Ar—SH	$+2.0$
—C—C$\diagdown^{O}_{H}$	-4.7		
—CHO	-3.3	≡C—H	$+2.3$
		CH₃—N	$+2.5$
—C$\diagup^{O}_{NH_2}$	-2.9	=C—CH₃	$+3.3$
Ar—OH	-2.4	Ar—CH₃	$+3.4$
Ar—H	-2.0	=C—CH₃	$+3.4$
=CH₂	-0.5	—CH₂—(cyclic)	$+3.5$
C=CH—C	-0.5	R—SH	$+3.5$
R—OH	-0.1	R—NH₂	$+3.7$
HOH	0.0	—C—CH₃	$+4.0$
AR—NH₂	$+1.5$	R\ \nN—H \nR/	$+4.4$
—O—CH₃	$+1.5$		
C—CH₃—X	$+2.0$		

For NMR proton work, δ has both $+$ and $-$ values. The use of δ values is somewhat inconvenient. The use of τ values,

$$\tau = 10.00 = \frac{(H_{\text{SiMe}_4} - H_S) \times 10^6}{H_{\text{SiMe}_4}} \tag{18.1}$$

where $(H_{\text{SiMe}_4} - H_S)$ is the shift using tetramethylsilane as the reference, and is expressed in ppm, was proposed by Tiers (1958). This allows us to express chemical shifts as positive values for all but the most acidic of protons. Increasing values of τ signify greater shielding of the proton. δ values, such as given in Table 18.2, relative to water, can be readily converted to τ values by adding a constant of about 5.2 ppm.

Wide Line NMR

In the previous discussion, concerning solutions, line widths were a few milligauss in width because the molecules were free to tumble in all directions and the fields produced by hydrogen nuclei, for example, tended to average out very rapidly. As a result the line width was determined by the inhomogeneity of the applied magnetic field.

However, in a solid material, the nuclei are more or less fixed in position with respect to their neighbors and because of this fixed spatial orientation, any given nucleus may be in a completely different magnetic field compared to its nearest neighbor. This means

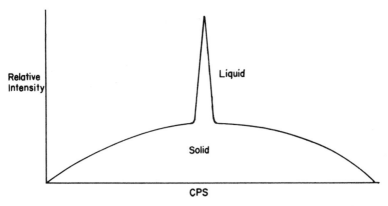

FIG. 18.4. A COMPARISON OF SOLID AND LIQUID HYDROGEN NMR
BANDS; WATER IN STARCH

that different resonance frequencies are needed for each nucleus and
the absorption band, that is the line width, will be quite large, in
some cases several hundred gauss.   Figure 18.4 shows hydrogen in a
solid (starch), and in the water surrounding it.

This difference in band widths can be used to determine moisture
in the presence of solids containing hydrogen, and the analysis can be
done in about 30 sec with an absolute error of 0.1%.

Some Recent Applications

Nuclear magnetic resonance has been used by Purcell *et al.* (1966)
to distinguish unsaturated fatty acids.   Johnston *et al.* (1964) used NMR
to investigate the hydrogenation of linolenate.   A quantitative procedure
for the beta olefinic group in fatty acid esters has been developed by Glass
and Dutton (1964).

Infrared and NMR are given by Coenen *et al.* (1967) for detecting
and quantitating the occurrence of aromatic fatty acids formed during hydrogenation of polyunsaturated fatty acids in oils.   Wide line
NMR has been applied to the rapid determination of the oil content
of soybeans by Collins *et al.* (1967), and Pohle and Gregory (1967)
determined solids in fats and shortenings by NMR.   Kung *et al.*
(1967) make use of NMR and gas chromatography to determine the
volatile acids of coffee beverages.

Toledo *et al.* (1968) reported a quantitative method for determining bound water in wheat flour and dough, and Nakanishi *et al.*
(1967) identified the structure of isochlorogenic acid from Brazilian
coffee beans.

Fats in chocolate as well as the liquid to solid ratios in this material have been determined by Oref (1965). The oil content of seeds has been determined by Conway and Earle (1963) using wide line NMR. Pohle *et al.* (1965) compared NMR and dilatometric determination of solids content of fats and shortenings.

BIBLIOGRAPHY

COENEN, J. W. E., WIESKE, T., CROSS, R. S., and RINKE, H. 1967. Occurrence, detection, and prevention of cyclization during hydrogenation of fatty oils. J. Am. Oil Chemists' Soc. *44*, 344–349.

COLLINS, F. I., ALEXANDER, D. E., RODGERS, R. C., and SILVELA, L. S. 1967. Analysis of oil content of soybeans by wide-line NMR. J. Am. Oil Chemists' Soc. *44*, 708–710.

CONWAY, T. F., and EARLE, F. R. 1963. Nuclear magnetic resonance for determining oil content of seeds. J. Am. Oil Chemists' Soc. *40*, 365–368.

GLASS, C. A., and DUTTON, H. J. 1964. Determination of beta-olefinic methyl groups in esters of fatty acids by nuclear magnetic resonance. Anal. Chem. *36*, 2401–2404.

JOHNSTON, A. E., GLASS, C. A., and DUTTON, H. J. 1964. Hydrogenation of linolenate. XI. Nuclear magnetic resonance investigation. J. Am. Oil Chemists' Soc. *41*, 788–790.

KUNG, J. T., McNAUGHT, R. P., YERANSIAN, J. A. 1967. Determining volatile acids in coffee beverages by NMR and gas chromatography. J. Food Sci. *32*, 455–458.

NAKANISHI, K., SOYAGI, S., and INOVE, Y. 1967. Reexamination of the structure of the so-called iso-chlorogenic acid. Bull. Natl. Inst. Sci. *31*, 158–164. Chem. Abstr. *66*, 28464.

OREF, I. 1965. Fat content and liquid-to-solid ratio of chocolate by wide-line nuclear magnetic resonance. J. Am. Oil Chemists' Soc. *42*, 425–427.

POHLE, W. D., and GREGORY, R. L. 1967. Standardization of nuclear magnetic resonance measurement of solids in fats and shortenings. J. Am. Oil Chemists' Soc. *44*, 397–399.

POHLE, W. D., TAYLOR, J. R., and GREGORY, R. L. 1965. A comparison of nuclear magnetic resonance and dilatometry for estimating solids content of fat and shortenings. J. Am. Oil Chemists' Soc. *42*, 1075–1078.

PURCELL, J. M., MORRIS, S. G., and SUSI, H. 1966. Proton magnetic resonance spectra of unsaturated fatty acids. Anal. Chem. *38*, 588–592.

SARAF, D. N., and FATT, I. 1967. Effect of electrolytes on moisture determination by nuclear magnetic resonance. Nature *214*, 1219–1220.

TIERS, G. 1958. Proton nuclear resonance spectroscopy. I. Reliable shielding values by "internal referencing" with tetramethylsilane. J. Phys. Chem. *62*, 1151–1152.

TOLEDO, R., STEINBERG, M. P., NELSON, A. I. 1968. Quantitative determination of bound water by NMR. J. Food Sci. *33*, 315–317.

CHAPTER 19

Radioactivity and Counting Techniques

Radioactivity is a general term applied to the emission of high energy particles and electromagnetic radiation emanating from the unstable and excited nuclei of atoms. Table 19.1 summarizes the characteristics of some of the different radiations.

Alpha particles are helium nuclei moving at high speeds (avg of 1×10^9 cm/sec) emitted from unstable nuclei having large atomic numbers. All alpha particles from a given isotope have the same energy and nearly identical penetration ranges which are very short, but producing about 25,000 ion pairs/cm while they last.

Beta particles are distinguishable from simple electrons only by the fact that they originate in the nucleus and are usually moving at high speed. A beta spectrum is continuous, having energies varying from a few thousand electron volts to several million electron volts. An average beta particle will produce about 60 ion pairs/cm.

Gamma rays originate in a nucleus which has been left in an excited state because of some previous disintegration or interaction. The fact that gamma rays have discrete energies is evidence that the nucleus exists in various energy levels. Gamma rays have high penetrating power but $1/10$ to $1/100$ of the ionizing power of beta rays.

TABLE 19.1

RADIATION CHARACTERISTICS

Radiation	Type	Charge	Typical Energy Range	Path Length Air	Path Length Solid	Primary Mechanism of Energy Loss
Alpha	Particles	+2	5–9 Mev	3–5 cm	25–40 μ	Ionization excitation
Beta	Particles	−1	0–4 Mev	0–10 cm	0–1 mm	Ionization excitation
Neutron	Particles	0	0–10 Mev	0–100 m	cm	Elastic collision with nuclei
X ray	Electromagnetic radiation	0	ev–100 Kev	μ–10 m	cm	Photoelectric effect
Gamma ray	Electromagnetic radiation	0	10 Kev–30 Mev	cm–100 m	mm–10 cm	Photoelectric effect Compton effect Pair production

TABLE 19.2

UNITS OF MEASUREMENT

Name	Symbol	Magnitude
Curie	c	3.7×10^{10} dis/sec
Rutherford	rd	1.0×10^6 dis/sec
Roentgen	r	1.61×10^{12} ion pairs/gm of air
Specific activity		dis/gm/sec
m = milli	M = mega	

Table 19.2 summarizes the common units of radioactivity measurement.

Radioactive Decay

Many equations have been developed to describe the various parts of the radioactive process but the one equation that most people come in contact with is the half-life equation.

$$t_{1/2} = \frac{0.693}{\lambda}$$ 19.1

where λ = radioactive decay constant $(\sec^{-1})$

If A is the count rate and A_0 is the number of counts at time zero, then equation 19.1 becomes

$$A = A_0 e^{\frac{-0.693t}{t_{1/2}}}$$ 19.2

Example

A sample of irradiated meat containing ^{35}S was known to contain 9.50 millicuries originally. After 1 yr and 237 days how many disintegrations per minute occur in the sample?

$$t_{1/2} = 87.1 \text{ days for } [35]S$$

$A = 9.50 \exp(-0.693 \times (365 + 237)/87.1)$

$A = 9.50 \times 0.00835$ millicuries

$A = 0.0794$ millicuries.   1 Curie $= 3.7 \times 10^{10}$ dis/sec $\times 60$ sec/min $\times 1$ curie/1×10^6 μcuries $= 2.2 \times 10^6$ dpm/μcurie

$A = 0.0794 \times 2.2 \times 10^6$ dpm $= 176,000$ dpm

Counting Devices

Although the radiations mentioned previously have high energies in bulk, individually these energies are not sufficient, by a factor of about one million, to permit direct observation and measurement.

TABLE 19.3

COUNTER CHARACTERISTICS

Detector	Sensitive Medium	Detector Multiplication	Output Signal (Volts)	Resolving Time (Sec)	Efficiency Electron	Efficiency X	Efficiency Gamma
Ionization chamber	gas	1	10^{-6} to 10^{-3}	10^{-6} to 10^{-3}	Low	Low	Low
Proportional counter	gas	10^2 to 10^4	10^{-4} to 10	10^{-6}	High	Medium	Low
Geiger counter	gas	10^7	0.1 to 10	10^{-4} to 10^{-3}	High	Medium	Low
Scintillation counter	solid liquid	10^6	10^{-2} to 10	10^{-9} to 10^{-6}	Medium	High	Very high

Detection and measurements are therefore done indirectly by utilizing the effects or interactions produced by these radiations as they traverse matter.

The four most common detectors in use today and their general characteristics are shown in Table 19.3.

Only the Geiger-Muller counter will be discussed in detail.

Geiger-Muller Counter.—The GM tube is shown schematically in Fig. 19.1. The GM tube is very sensitive to alpha and beta particles (98% efficient) compared to gamma rays (2% efficient). The GM counter is relatively inexpensive and simple to operate, but it does not discriminate between types of radiation and it has a finite lifetime. It is steadily being replaced by proportional and scintillation counters.

Suppose a ray of radiation comes through the mica window and strikes an argon atom. The argon atom is ionized to produce a positive argon ion and an electron. The positive ion moves toward the cathode about 1000 times slower than the electron moves toward the anode. The electron, attracted by the high potential of the anode is rapidly accelerated. In fact, it has sufficient energy so that if it collides with an argon atom another ion and electron can be produced. Now there are two electrons accelerating toward the anode. These can produce 4, 8, 16, etc. electrons. The net result, called the *Townsend avalanche*, is that thousands of electrons reach the anode. When these electrons reach the anode, a small current is produced and the pulse signal is measured.

In addition, some of the electrons striking the anode may have a high enough energy to knock electrons from the anode. These electrons will immediately be reattracted to the anode and they in turn can knock other electrons loose. This is known as the *photon spread*.

The total time it takes for this signal to build up is known as the rise time, t_r, and is usually 2–5 μsec.

What happened to the positive ions during this time? They are slowly moving toward the cathode as a positive space charge. If they strike the cathode with their full energy, then more photoelectrons will be generated, more than the tube can handle. The net result is that the counter will burn out. What is needed is something to dissipate this energy and that from the photon spread. Molecules, with their many energy levels, are used for this purpose, and ethanol and chlorine are favorites. The ionized argon atoms will transfer their energy to these molecules. The molecules may then form ions or free radicals or simply absorb the energy, but the

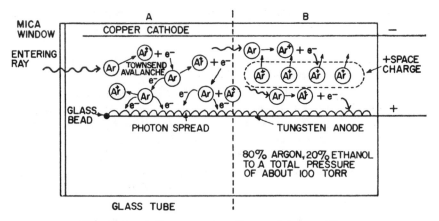

FIG. 19.1.   A DIAGRAM OF A GEIGER MULLER TUBE

net effect is that the energy is now so spread out that the cathode has little affinity for these particles, and no photoelectrons are produced.   Since there is a limit to the amount of quenching gas that can be added to this type of counter, the counter will work only as long as quenching gas is present.

What happens if a second ray of radiation enters the counter before the first ray has completed its reaction?   See Section B of Fig. 19.1.   If the ray ionized an argon atom at a point between the cathode and the positive space charge, then the electron produced will not see the anode but will recombine with the argon ion instead. The net result is that a ray of radiation entered the counter but was not counted so the counter was dead.

Now consider a ray entering the counter between the positive space charge and the anode.   The electron produced will see the anode and be attracted to it.   However, it doesn't have as much room to operate in as the original pulse and may not be detected. Again the counter is dead.   Dead times, t_d, vary but usually are between 80 to 100 μsec.   This does not mean that after 100 μsec the counter is completely ready to go again.   It means only that the next signal produced can be detected, but its amplitude will be very weak.   The time it takes for the counter to completely recover is known as the recovery time, t_{rec}, and this varies from 200 to 300 μsec.   The counter dead time affects the counting rate and as such introduces an error into the measurement.

Proportional Counters.—GM tubes are limited to about 15,000 cpm because of their long dead time.   If the voltage applied to the

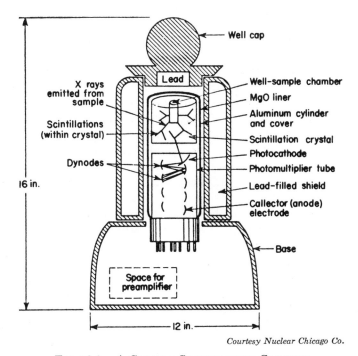

Well cap

X rays emitted from sample

Scintillations (within crystal)

Dynodes

Lead

Well-sample chamber

MgO liner

Aluminum cylinder and cover

Scintillation crystal

Photocathode

Photomultiplier tube

Lead-filled shield

Collector (anode) electrode

16 in.

Base

Space for preamplifier

12 in.

Courtesy Nuclear Chicago Co.

FIG. 19.2. A CRYSTAL SCINTILLATION COUNTER

anode of a GM tube is reduced to the place where it can collect electrons but not form the photon spread, then the output pulse is proportional to the energy of the initial ionization, since the number of secondary electrons now depends only on the number of primary ion pairs produced initially. A device operated in this manner is called a proportional counter.

The dead times are very short, of the order of 1 μsec, and therefore proportional counters can count up to 200,000 cpm. The pulse signal is much weaker than with the GM tube so a much better amplification system is needed.

You cannot make a proportional counter out of a GM tube by simply lowering the anode voltage.

Proportional counters can operate at atmospheric pressures so the quenching gas can be added continuously. Since this can be endless, a proportional counter can count indefinitely. Proportional counters are very good for alpha and beta particles and because of their ionization efficiency, alpha particles are easily distinguished from beta particles.

Scintillation Counters.—The basic principle behind the operation of the scintillation counter (Fig. 19.2) is that an energetic particle incident upon a luminescent material (a phosphor) excites the material; the photons which are created in the process of de-excitation are collected at the photocathode of the photomultiplier tube where the photons cause the ejection of electrons. The electrons ejected from the photocathode are then caused to impinge upon other electrodes, each approximately 100 v higher in potential. In the acceleration from dynode to dynode, more electrons are ejected and a large amplification is obtained.

It would be expected that the greater the energy of the incident particle, the greater the number of electrons which would be produced. This indicates that the scintillation counter could be used to obtain the energy of the particle. In the scintillation counter, advantage is not taken of the proportional properties; however, the scintillation spectrometer uses these proportional characteristics of the scintillation process to good advantage.

The *liquid scintillation counter* is particularly convenient for the counting of very low energy beta emitters, such as tritium, carbon-14, and sulfur-35, although it is not restricted to this use. As the name implies, the scintillator is not a crystal, such as NaI(Tl) nor a plastic, but a liquid solution. The solution is prepared by combining a good solvent, a scintillator, and primary solute.

The pulses of light emitted by the scintillating solution (caused by the particles in the radioactive decay process) are observed with a photomultiplier tube and counted. It is common practice to cool the photomultipliers to lower their noise; caused in part by thermal emission from the photocathode.

Many materials have been tried as liquid scintillators. One of these, contains 100 gm of naphthalene, 7 gm of 2,5-diphenyloxazole (PPO), and 50 mg of 1,4-bis(2,5-phenyloxazoyl)benzene (POPOP) per liter of pure 1,4-dioxane. The advantage of this liquid scintillator is its ability to dissolve many aqueous solutions. Here the PPO is the primary solute and POPOP is the secondary solute. A number of other good scintillator systems are known. In general, the alkylbenzenes are the best solvents.

Isotopic Dilution

The technique of isotopic dilution is a simple and extremely useful method for quantitative analysis. The method involves adding a small amount of a radio-labeled compound (of known specific activity), chemically identical to the unknown component of the unknown mixture being studied, separating a small but pure fraction of the compound, and determining its specific activity.

It is then only a matter of calculation to find the amount of the component originally present in the unknown mixture.

Let us look at this technique in a more quantitative manner. Let A = the number of grams of unknown present. There is originally no activity in this material, so the specific activity of A, S_A is zero. Now we shall add A^* grams of the radio-labeled component, chemically identical to that being analyzed. The total activity added is Q^* and the specific activity of $S_{A*} = Q^*/A^*$.

After the addition is complete, and thorough mixing has occurred, the total number of grams present is $(A + A^*)$, and the total activity is still Q^*.

Now let us separate out A' grams of the pure component. Its total activity would be $S_A' = Q'/A'$. The fraction separated comprises $A'/(A + A^*)$ of the total, and the activity is Q'/Q^* of the total.

$$A = A^* \left(\frac{S_{A*}}{S_{A'}} - 1 \right) \qquad 19.3$$

Example

A sample of 15.0 gm of liver contained an unknown amount of glycine. Ten milligrams of glycine of specific activity of 300,000 dps/gm was added to the sample, and throughly mixed. Then 12.5 mg of pure glycine was separated and counted. The count rate was 250 dps. What was the percent by weight of glycine in the original sample?

$$S_{A'} = \frac{250 \text{ dps}}{0.0125 \text{ gm}} = 20,000 \text{ dps/gm}$$

$$S_{A*} = 300,000 \text{ dps/gm}$$

$$R = S_{A*}/S_{A'} = 300,000/20,000 = 15$$

$$A = A^*(15 - 1) = 10(14) = 140 \text{ mg}$$

and therefore

$$\frac{0.140 \times 100}{15.0} = 0.93\% \text{ of glycine present in the original sample}$$

Example

To 200 ml of orange juice containing ethanol was added 5 lambda of ethanol of specific activity = 5×10^5 cpm/ml. The radiotracer was tritium. After thorough mixing, a 1.00-ml sample of ethanol was isolated from the mixture. Fifty lambda of the separated ethanol was found to have a counting rate of 200 cpm. What was the percent by volume of ethanol in the orange juice?

$$S_{A'} = \frac{200 \text{ cpm}}{0.050 \text{ ml}} = 4000 \text{ cpm/ml}$$

$$S_{A*} = 500,000 \text{ cpm/ml}$$

$$R = 500,000/4,000 = 125$$

$$125 \gg 1, \text{ so that}$$

$$A = RA^* = 125 \times 0.005 \text{ ml} = 0.625 \text{ ml}$$

and therefore

$$\frac{0.625 \times 100}{200} = 0.31\% \text{ by vol ethanol}$$

It is to be emphasized in this discussion that there are three "musts" to be observed when using isotopic dilution: (1) the tracer must be identical to the component of the mixture being investigated, (2) there must be no exchange of the tracer with any component of the mixture, and (3) there must be no side reactions with other components of the mixture.

Neutron Activation Analysis

The previous section dealt with the use of naturally occurring radioactive materials, but what about those systems that do not have available natural radioactive isotopes, or the sample is one in which a natural radioactive isotope cannot be added.

Nuclear transmutations must be produced by the interaction of the nuclei of matter with highly energetic charged particles (protons, deuterons, alpha particles) by neutrons, or by energetic photons. The production of radioactive isotopes generally is dependent upon such artificial nuclear transmutations.

A polonium-beryllium source is commonly used in the laboratory for low level irradiations. The neutrons are produced by having an alpha particle that comes from polonium bombard a beryllium nucleus to produce a neutron and carbon-12.

The reaction results in the production of neutrons of an average energy of about 4 Mev, which is far too great to produce the desired (n, γ) reaction with any reasonable efficiency. It is necessary therefore to slow down the neutrons to thermal energies (ca 0.025

mev) using water as a moderator. Collisions of the energetic neutrons with the hydrogen atoms in the water molecules result in a rapid decrease in the energy of the neutrons. One can consider then that the water moderator becomes filled with a neutron gas, the maximum density of which occurs at the source. The material to be irradiated is suspended in the water as close as is feasible to the neutron source.

Using cadmium in meat as an example, neutron activation analysis is simply to irradiate the meat, producing radioactive cadmium which will then be counted. An additional problem is that while you are preparing the sample and counting it, which takes a bit of time, the radioactive material may decay so rapidly that unless a correction is made a sizable error is present. How long do you have to irradiate meat to make it radioactive enough to count? The rest of this chapter is designed to show you how these measurements and corrections are made.

The radioactive decay obeys a statistical law identical with that encountered in monomolecular and first order chemical kinetics, and in optical absorption phenomena. Such a law states that the rate of decay is proportional to the total number of radioactive atoms present.

The half-life of the radioisotope being produced in a nuclear transformation has a definite bearing on the total time over which one can efficiently produce transmutations. The rate of decay depends only upon the number of radioactive atoms present. As one produces transmutations, the rate of decay of the produced isotope will gradually increase. The number of radioactive atoms will stop increasing when their rate of production at any instant is equal to their rate of decay. If R is the rate of production at any instant by the source, the rate of growth of the number of radioactive atoms (N^*) is given by

$$\frac{dN}{dt} = R - \lambda N^* \qquad\qquad 19.4$$

Integration of equation 19.4 leads to

$$N^* = (Rt_{1/2}/0.693)(1 - 2^{-t/t_{1/2}}) \qquad\qquad 19.5$$

Two special cases need now be considered: production of isotopes with short and with long half-lives, relative to the time of bombardment. From equation 19.5 it is seen that for an irradiation time $t = t_{1/2}$, 50% saturation is reached, for $t = 2t_{1/2}$, 75% saturation, and for $t = 5t_{1/2}$, 97% saturation. Thus bombardment times greater than 4 or 5 half-lives of the radioisotope being produced are a waste of time.

A word should be said concerning the rate of production, R, of the radioisotope at any instant by the source. This production rate is dependent upon the neutron flux, Φ, in neutrons/cm²/sec, upon the number of target nuclei, N_t, and upon the nuclear cross section for the activation, σ_{act}, in cm².

Thus R is given by

$$R = \Phi\, \sigma_{act} N_t \qquad\qquad 19.6$$

Equation 19.5 is then given as

$$N^* = \frac{\Phi\, \sigma_{act} N_t t}{0.693}\,[1-2^{-t/t_{1/2}}] \qquad\qquad 19.7$$

For the two cases discussed above, then we obtain the following forms of equation 19.4.   If $t \gg t_{1/2}$, we get

$$N^* = \frac{\Phi\, \sigma_{act} N_t t_{1/2}}{0.693} \qquad\qquad 19.8$$

If $t_{1/2} \gg t$, we get,

$$N^* = \Phi\, \sigma_{act} N_t t \qquad\qquad 19.9$$

Example

Let us assume that a sample containing 1.50 gm of phosphorus is bombarded in a nuclear pile at a flux of 5×10^{11} neutrons/cm²/sec for 1 week. The (n,γ) reaction produces phosphorus-32 with a half-life of 14.3 days. The cross section for the (n,γ) reaction is 0.19 barns (1 barn $= 10^{-24}$ cm²). Phosphorus-31 occurs to the extent of 100% in nature.   What is the resultant activity?

Using equation 19.4 we have

$$N^* = \frac{\Phi\, \sigma N_t t_{1/2}}{0.693}\,[1 - \exp(-0.693\, t/t_{1/2})]$$

but

$$-dN^*/dt = \lambda N^*$$

so that

$$-dN^*/dt = \Phi\, \sigma N_t [1 - \exp(0.693/t_{1/2})]$$

$$\frac{-dN^*}{dt} =$$

$$\frac{5 \times 10^{11} \times 0.19 \times 10^{-24} \times 1.50 \times 6.02 \text{ (no. of atoms/mole)}}{31.0} \times$$

$$10^{23} \times [1 - 2^{-7.0/14.3}]$$

$$= 2.76 \times 10^9 \times (1 - 0.71)$$

$$= 8.0 \times 10^8 \text{ dps}$$

$$\frac{8.0 \times 10^8}{3.7 \times 10^7} \text{ dps/millicurie} = 22 \text{ millicuries}$$

Example

If a sample of 10.0 gm of ammonium nitrate is irradiated for 14.0 days in a flux of 1×10^{12} neutrons/cm^2/sec, how many curies of carbon-14 will be produced? $\sigma_{act} = 1.75$ barns for the (n,p) reaction. $t_{1/2}$ of ^{14}C $= 5770$ yr. ^{14}N is 99.635% abundant in nature. Since $t_{1/2} \gg t$, we may use equation (19.6).

$$\frac{-dN^*}{dt} = \lambda N^* = 0.693 \ \Phi \ \sigma_{act} N_t (t/t_{1/2})$$

$$\frac{-dN^*}{dt} =$$

$$\frac{0.693 \times 1 \times 10^{12} \times 1.75 \times 10^{-24} \times 10.0 \times 2 \times 6.02 \times 10^{23} \times 14.0 \times 0.996}{80 \times 365 \times 5770}$$

$$= 1.21 \times 10^6 \text{ dps}$$

$$\frac{1.21 \times 10^6 \text{ dps}}{3.7 \times 10^4 \text{ cps/}\mu\text{curie}} = 33 \ \mu\text{curies}$$

Example

A sample of a metal (at. wt—100) was found to contain a very small amount of copper impurity. A 1.00-gm sample of the metal was subjected to a neutron bombardment with a flux of 5×10^{11} neutrons/cm^2/sec for 3 hr. ^{63}Cu is 69.1%, and ^{65}Cu is 30.9% in nature. The cross section for the (n,γ) reaction on ^{63}Cu is 3.7 barns. ^{66}Cu is also produced. ^{64}Cu $t_{1/2} = 12.8$ hr, and ^{66}Cu $t_{1/2} = 5$ min. After the sample was removed from the pile it was allowed to stand for 24 hr while being worked up. A 0.10-gm equivalent sample of the irradiated material was counted with a counter for

which 25 cpm corresponded to 1000 dpm.   The sample was found
to count at the rate of 3750 cpm.   What is the percent Cu impurity
in the original metal?   Assume no other radioisotopes are produced.

$$3{,}760 \times \frac{1000}{25} = 150{,}000 \text{ dpm}/0.10 \text{ gm metal}$$

or 1,500,000 dpm/gm.

But 24 hr lapsed from the irradiation so the activity actually produced was (equation 19.2):

$$A = 1.50 \times 10^6 = A\, e^{-0.693 \times 24.0/12.8}$$

$$A = 1.50 \times 10^6\, e^{1.30} = 1.5 \times 10^6 \times 3.67$$

$$= 5.50 \times 10^6 \text{ dpm}$$

or

$$\frac{5.50 \times 10^6}{60} = 9.18 \times 10^4 \text{ dps}$$

That is

$$\frac{-dN^*}{dt} = 9.18 \times 10^4 \text{ dps}$$

but

$$\frac{-dN^*}{dt} = \Phi\, \sigma_{act} N_t [1 - \exp(0.693\, t/t_{1/2})]$$

$$9.18 \times 10^4 = 5 \times 10^{11} \times 3.7 \times 10^{-24} N_t \times$$
$$[1 - \exp(-0.693 \times 3.00/ 12.8)]$$

$$= 1.85 \times 10^{-12} \times N_t(0.15)$$

$$N_t = 3.3 \times 10^{17}$$

That is, the number of ^{63}Cu target atoms was 3.3×10^{17}.   The total
number of copper atoms was

$$\frac{3.3 \times 10^{17}}{0.691} = 4.8 \times 10^{17}$$

$$4.8 \times 10^{17} \times \frac{1}{6 \times 10^{23}} \times \frac{63.54}{1} = 5.05 \times 10^{-5} \text{ gm Cu}$$

$$\frac{5.05 \times 10^{-5} \text{ gm}}{1.00 \text{ gm total}} \times 100 = 0.005\% \text{ Cu impurity}$$

Some Recent Applications

Sensitive radiotracer techniques were applied by Kristoffersen and Harper (1968) to determine volatile sulfur components of cheddar cheese made from milk obtained for a cow fed ^{35}S. Instrumental sensitivities for more than 65 elements in milk among other matrices were given by Yule (1966). Underdal (1968) used activation analysis to determine Hg in eggs.

Radioactive iodine in milk is precipitated and counted after organic decomposition in Tanaka's method (1968). Spontaneous deposition of lead-210 and polonium-210 on nickel discs before counting provided the basis for the analysis of foods for Blanchard (1966). Liquid scintillation counting after solvent extraction permitted phosphorus-32 measurements in foods by Ellis *et al.* (1966). Hamilton *et al.* (1967) used radioactive tracers to establish ashing losses for As, Na, Sn, and Zn.

Al has been determined in ascorbic acid along with Na, Cu, Cr, and Sb using neutron activation analysis by Nagy *et al.* (1968). This latter method was used by Kirchmann and Roderbourg (1965) to measure As in vegetables.

Associated Problems

(1) A breakfast cereal is known to contain thiamine. 0.500 mg of carbon-14 labeled thiamine, having a specific activity of 1.00 μcurie per gram, was added to a 500-gm sample of the cereal and mixed thoroughly by crushing and blending. A 0.40-mg sample of pure thiamine isolated from the mixture was found to have a disintegration rate of 5750 disintegrations per hour after all corrections for background, etc., were made. What is the milligram thiamine content per ounce of this cereal? Ans. 0.187 mg/oz

(2) It is desired to determine the aureomycin content of a fermentation broth. A 1.00-kg sample of the broth was subjected to isotope dilution analysis by adding 2.00 mg of carbon-14 labeled aureomycin with a specific activity of 850 cpm/mg. After thorough mixing, a sample of aureomycin was separated and purified. A 0.65-mg sample of the purified aureomycin had a count rate of 118 cpm after correction for background. What is the percent by weight of aureomycin in the fermentation broth? Ans. 0.156%

(3) Following irradiation in a nuclear pile, 14.2 μcuries of carbon-14 activity had been induced in a 5.003-gm sample of ammonium nitrate. 2.51 gm of methylamine hydrochloride was added with thorough mixing to 0.1047 gm of the irradiated NH_4NO_3 and dis-

solved in 50.0 ml of distilled water. Following a standard procedure, the phenyl isothiocyanate derivative of methylamine (ϕ—NH—CS—NHCH$_3$) was prepared from the mixture. 0.1528 gm of this derivative was placed in a liquid scintillator and the carbon-14 was counted. A count rate of 122.3 cpm was observed, and since the counting efficiency was 4.9%, the true count rate was 2508 dpm. What percent of the produced carbon-14 activity is found in the form of methylamine? Ans. 1.4%

BIBLIOGRAPHY

BLANCHARD, R. L. 1966. Rapid determination of lead-210 and polonium-210 in environmental samples by deposition on nickel. Anal. Chem. *38*, 189–192.

ELLIS, M. K., WAMPLER, S. N., and YAGER, R. N. 1966. Liquid scintillation method for determination of solvent-extracted phosphorus-32 in foods. Anal. Chim. Acta *34*, 169–174.

HAMILTON, E. I., MINSKI, M. J., and CLEARY, J. J. 1967. The loss of elements during the decomposition of biological materials with special reference to arsenic, sodium, strontium, and zinc. Analyst *92*, 257–259.

KIRCHMANN, R., and RODERBOURG, J. 1965. Determination of arsenic in vegetables by radioactivation and gamma spectrometry. Intern. J. Appl. Radiation Isotopes *16*, 457–460.

KRISTOFFERSEN, S. J., and HARPER, W. J. 1966. Determining volatile sulfur in cheddar cheese by use of ^{35}S. Intern. Dairy Congr., Proc. 17th, Munich *4*, 297.

NAGY, L. G., MESTER, Z., and TAKACS, G. 1967. Neutron activation analysis for metal contamination of ascorbic acid. Nagy. Kem. Foly. *73*, 137–138. Anal. Abstr. *15*, 3612, 1968.

TANAKA, G. 1968. Determination of radioactive iodine in natural water, milk, and thyroid gland. Japan Analyst *15*, 1068–1073. Anal. Abstr. *15*, 3001, 1968.

UNDERDAL, B. 1968. Mercury in foods determined by activation analysis. Nord. Veterinarmed Medlemsbl. *20*, 105–108.

YULE, H. P., 1966. Reactor neutron activation analysis. Anal. Chem. *38*, 818–882.

CHAPTER 20

Column Chromatography

No other separation process can match chromatography for simplicity, efficiency, and range of applications. Combined, these attributes make chromatography one of the great achievements of analytical chemistry.

Although the exact origin of the basic process may never be known, the work of Tswett who separated plant pigments by column techniques in 1906 is regarded as the first systematic study. Table 20.1 lists some of the types of chromatography currently used.

TABLE 20.1

TYPES OF CHROMATOGRAPHY

Stationary Phase	Mobile Phase	Technique	Physical Principle
Solid	Gas	Gas-solid Chromatography	Adsorption
Solid	Liquid	Column, Thin-layer, Paper	Adsorption Partition Ion-exchange Gel-permeation
Liquid	Liquid	Column, Thin-layer, Paper	Partition
Liquid	Gas	Gas-liquid	Partition

The basic techniques, their major subdivisions, and the physical principles will be discussed in the following chapters.

Displacement Chromatography

The basic procedure is to have a glass column a few centimeters in diameter and 10 to 20 cm in length, packed tightly with an inert filling. The solution, containing the compounds to be separated, is added to the top of the column in as thin a band as possible (Fig. 20.1A). Let us assume that both A and B are adsorbed strongly to the column packing, but B is adsorbed more strongly than A. A solvent, S, is added which is adsorbed more strongly to the column packing than either A or B.

When the solvent molecules, S, enter the top of the column, they displace both A and B leaving them momentarily free to migrate down the column. Since, in this case, B is more strongly adsorbed

257

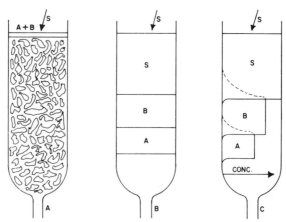

FIG. 20.1. DISPLACEMENT CHROMATOGRAPHY

than A, the molecules of B will on the average be readsorbed more readily on the next available site, thus forcing the A molecules to go further down the column before being adsorbed. If more S is added to the top of the column, a continuing displacement takes place. S displaces B, and B displaces A. The results are shown in Fig. 20.1B.

Figure 20.1C is a plot of the concentration of each component in 20.1B at its distance down the column. Notice that the bands actually overlap, and as a result this system is not used for quantitative analysis but is quite useful for preparative work.

Let us look at Fig. 20.1C a little closer. Notice that the concentrations per column length are different. The surface of the inert filling is really not inert but because of crystal imperfections, impurities, and surface geometry, has places on its surface that are more reactive than the rest of the surface. These are called *reactive sites*. As the A molecules proceed down the column they will be adsorbed on the reactive sites until all available sites are filled. The column is then *saturated* so we reach a concentration plateau.

There is a small region ahead of the main concentration. This is due in part to the fact that all of the A molecules do not have the same energy for adsorption. There are some A molecules that have a lower energy than most of the rest and when they arrive at an active site they do not have enough energy to react. Therefore they proceed down the column a bit further than the main body of A molecules. On the other hand there are some very active A molecules which can compete successfully with B molecules for the sites. These molecules provide the *tail*.

The B molecules are more energetic, as a group, than the A molecules, so reactive sites that would not interest an A molecule are quite appealing to a B molecule. Therefore the column saturation for B molecules is a bit higher than for the A molecules. It is even higher for the S molecules.

Partition Chromatography

The same equipment is used here as with displacement chromatography with one exception. The inert packing is coated with a film of a high boiling liquid which is strongly adsorbed to the particle surface. This is called the stationary phase. A second liquid, the mobile phase, is then passed through the column. This solution is selected so that (1) it will not dissolve or mix appreciably with the stationary phase, and (2) it will not adsorb to the inert phase as did the stationary phase.

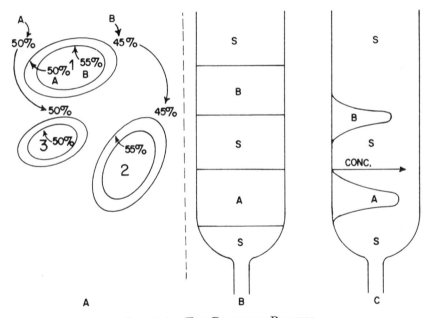

FIG. 20.2.   THE PARTITION PROCESS

When a solution of A and B is placed at the top of the column (Fig. 20.2B) and the mobile phase, S, is added, the compounds proceed slowly down the column. As an example let us assume that we have 100,000 molecules each of A and B. These molecules are crowded together in the mobile phase and as they pass the stationary phase some of each dissolve in it. Because of a difference in the molecular structures of A and B, 55% of the B molecules will dissolve in the stationary phase, whereas only 50% of the A molecules will dissolve.

The molecules remaining in the mobile phase are carried down to the next particle where the process is repeated again, 55 and 50% dissolving, and 45 and 50% staying in the mobile phase. What about the particles still retained in particle 1? Fresh solvent is now surrounding this particle and the molecules, crowded in the stationary phase, dissolve into the mobile phase in the 50-50 and 55-45 ratios. These molecules can then proceed to particle 2 or 3 where the entire process is repeated.

Notice that at no time is there a complete and immediate separation, but that every time the partitioning takes place a few more of the A molecules get ahead of the B molecules. If we continue the process enough times the two compounds will be completely separated from each other, although each component will be mixed with the eluting agent.

The concentration profile in Fig. 20.2C shows that the concentration of the band is not uniform as it was in displacement analysis, but that the complete separation of A from B is possible. Therefore partition analysis is preferred for quantitative analysis.

Tailing

There are times when the concentration profile looks like Fig. 20.3 rather than 20.2C. This phenomenon is called *tailing* and is very troublesome because it makes quantitative analysis difficult and sometimes impossible.

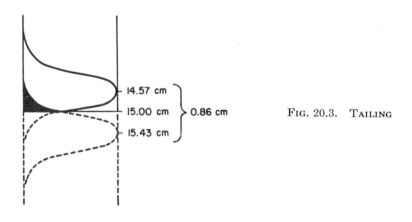

14.57 cm
15.00 cm } 0.86 cm
15.43 cm

Fig. 20.3. Tailing

In order to explain the shape of the curve we need to know the results of an entirely different experiment, the Langmuir adsorption isotherm.   To illustrate this, one adds some dilute acetic acid to a few grams of charcoal in a flask.   After thoroughly shaking the mixture, the charcoal is filtered off, and the acid is titrated to determine how much acid was adsorbed on the charcoal.   When this is done for several concentrations of acid and the data are plotted, the result looks like Fig. 20.4.

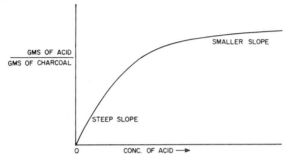

FIG. 20.4.   LANGMUIR ADSORPTION ISOTHERM

The significance of the isotherm is that at low concentrations of acid, the acid is adsorbed very strongly to the charcoal (steep slope), but as the acid concentration increases the acid does not adsorb as strongly (shallow slope).

Refer to Fig. 20.3.   We see that there is a slight leading edge. Earlier we said that this was due to the less energetic particles getting ahead of the main body.   This is true in part but notice that the molecules did not get very far ahead.   When those leading molecules get ahead of the main body, their concentration is low and now they become tightly adsorbed.   The main body coming along behind forms a normal curve.

What about those more energetic molecules?   They are strongly adsorbed, but as long as they are with the main body of the band where the concentration is high they move slowly along.   However, if the main body of the band passes over them, then they are in a low concentration region and again they are adsorbed strongly.   Finally there are no more molecules to come along and displace them, so a tail is formed.

Apparently these tailing molecules are being adsorbed on the "inert" surface.   What must be done is either to make the inert surface actually inert, or to use a more polar mobile phase to desorb

the adsorbed molecules.   The former can be done by acid washing
or silinizing the column to cover the active sites.   The latter can be
done by a process called gradient elution whereby you start out with
one solvent and then slowly mix into it another, more polar solvent.
The net effect is to slowly change the polarity of the mobile phase.

R_f Value

In later discussions the term R_f *value* will be mentioned.   This is
a parameter that measures how far a material moves with respect
to how far the solvent moves at the same time.   This ratio is defined as:

$$R = \frac{\text{distance the solute moves}}{\text{distance the solvent moves}} \qquad (20.1)$$

A problem arises when one has to determine how far the solute
has moved.   Where do you measure the distance from?   Since
most bands have some tailing it is almost impossible to find the center
of the band or the center of the concentration.   What is easy to find
is the front edge and this is the place to make the measurement.
To remind us that it is the front edge, an f subscript is used.

Solvents and Adsorbents

Unfortunately there are no formulas to calculate the proper solvent or adsorbent for the separation of a given compound.   Tables
20.2 and 20.3 show some of the materials that have been found to
work.

Example

It should be emphasized that you are not limited to a single column packing or to the chemical characteristics of a given inert phase.
Columns may be packed in segments and they may be made acidic or
basic, as desired.   As an example of this, the separation of noscapine

TABLE 20.2

ADSORBENTS IN DECREASING ORDER OF ADSORPTIVITY

Fullers earth (aluminum silicate)	Calcium phosphate
Charcoal	Potassium carbonate
Activated alumina	Sodium carbonate
Magnesium silicate (Florisil)	Talc
Silica gel	Inulin
Calcium oxide	Starch
Magnesium oxide	Powdered sugar
Calcium carbonate	

TABLE 20.3

SOLVENTS IN INCREASING ORDER OF POLARITY

Petroleum ether (30°–50°C)	Toluene
Petroleum ether (50°–70°C)	Esters of organic acids
Petroleum ether (50°–100°C)	1,2-dichloroethane, dichloromethane, chloro-
Carbon tetrachloride	form
Cyclohexane	Alcohols
Carbon disulfide	Water (varies with pH and salt concentration)
Ether	Pyridine
Acetone	Organic acids
Benzene	Mixtures of acids and bases with water, alcohol
	or pyridine

and pheniramine in cough syrup will be used. Figure 20.5 shows how the columns are packed.

Column A is placed over Column B, and 100 ml of ether is passed through the columns. This removes the flavor components from the syrup which would interfere later with the spectrophotometric determination. The middle tosic acid layer is a safety trap to ensure that the basic noscapine and pheniramine do not leave the top column during the ether wash. The NaHCO$_3$ is used to keep tosic acid from washing off the column, since tosic acid will also interfere with the spectrophotometric determination.

Then 150 ml of CHCl$_3$ is passed through both columns. Noscapine is completely removed from both columns and is in the eluate. Pheniramine is washed off of the weak tosic acid column but is retained on the stronger acid, H$_2$SO$_4$, column. Column A is discarded. In order to remove the pheniramine, column B must be made basic so a NH$_4$OH + Celite plug is added. When CHCl$_3$ is added, the NH$_4$OH is washed over the H$_2$SO$_4$ making this part of the column basic, and the pheniramine is now washed off and can be determined.

Generally, the sample is added in as narrow a band as possible at the top of the column. In this case, this was not necessary since the separation was so easy.

HIGH PRESSURE LIQUID CHROMATOGRAPHY

We have learned that the first chromatographic technique was column chromatography. Since then, paper, thin layer, and finally gas chromatography were developed. Now, 70 years later, we are going back to column chromatography! This seems like a strange sequence of events but there are good reasons for it. The original column chromatographic technique employed glass columns and either gravity flow or a slight vacuum to move the mobile phase through the column. This was also "slow" chromatography and "hard to reproduce" chromatography. It was, however, "extremely flexible" chromatography in that an almost unlimited variety of

Column A

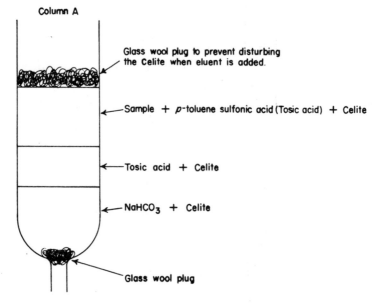

Glass wool plug to prevent disturbing
the Celite when eluent is added.

Sample + p-toluene sulfonic acid (Tosic acid) + Celite

Tosic acid + Celite

NaHCO$_3$ + Celite

Glass wool plug

Column B initially   Column B finally

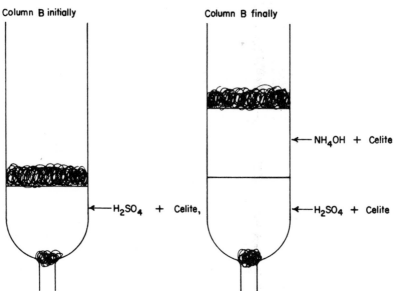

NH$_4$OH + Celite

H$_2$SO$_4$ + Celite,   H$_2$SO$_4$ + Celite

FIG. 20.5.   ILLUSTRATION OF COLUMN PACKING

solvents and column packings could be used, neither of these completely available to paper, thin layer or gas chromatography. It was because of this recognized flexibility that scientists re-examined column chromatography.

By going to steel columns and high pressures it was found that column chromatography could be "fast" and "reproducible" as well as flexible. The result is now usually called either "high pressure liquid chromatography" (HPLC) or "high performance liquid chromatography."

A basic instrument, shown in diagrammatic form in Fig. 20.6, consists of a solvent reservoir, a pump, a gradient chamber, injection port, column, detector, fraction collector, and a recorder. Depending upon the quality of the individual components the cost of such a combination may vary from $4,000 to $25,000.

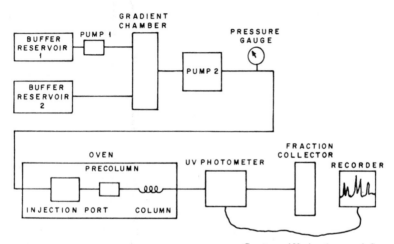

Courtesy of Varian Aerograph Co.

FIG. 20.6. COMPONENT DIAGRAM OF A VARIAN LCS 100 HIGH PRESSURE LIQUID CHROMATOGRAPH

Figure 20.7 is a photograph of a high pressure liquid chromatograph.

Of what value is this instrument to the food chemist? This instrument permits nutrition studies to be made that could only be dreamed about a few years ago. It permits both qualitative and quantitative examination of a much wider variety of nutrients so we can now determine their biological inter-relationships, and just what the specific nutrient requirements are to prevent disease or to promote growth and general good health. With a capillary column filled with a high performance ion exchange packing it is now possible to examine a wide variety of nutrients in the 1–10 nanomole range!

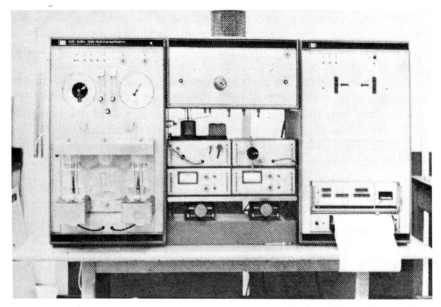

Courtesy of Hewlett-Packard Co.

FIG. 20.7.   A HEWLETT-PACKARD HIGH PRESSURE LIQUID CHROMATOGRAPH

Ever since man realized that he had to eat to live, the more curious of his breed have wondered how the process functioned. Much progress has been made in the past and the advent of the high pressure liquid chromatograph should take a large step toward further understanding these vital processes and how to control them for our benefit.

In this section we will briefly discuss the equipment, then the various types of liquid chromatography, and finish with a few practical examples.

EQUIPMENT

Solvents

The range of solvents that can be used in high pressure liquid chromatography is quite wide. Water and aqueous buffer solutions are common as are any nonaqueous solvents of low viscosity. High viscosity solvents are avoided because they require longer times to pass through the column which results in peak broadening and poorer resolution. This also requires a higher pressure to force the solvent through the column.

The solvents used must be free of suspended particles as well as chemical impurities. Suspended particles tend to plug up the column and chemical impurities produce spurious peaks at the detector.   If you are serious about performing a good HPLC separation you will purchase high purity solvents, carefully redistill them, then filter them through a millipore filter and store them where only you can find them.

The dissolved sample is also a problem.   You must be sure that there are no suspended or undissolved particles in the sample before it is injected onto the column.   To ensure that the sample is free of particles it is usually passed through a millipore type filter such as the one shown in Fig. 20.8.

A Teflon filter of about 0.45 μ is recommended when organic solvents are used and a cellulose acetate filter is used with aqueous systems.

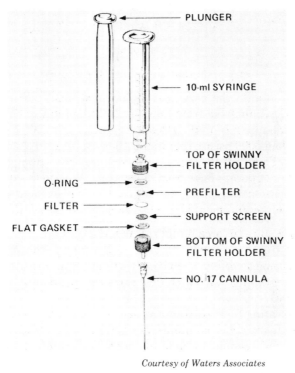

Courtesy of Waters Associates

FIG. 20.8.   SAMPLE CLARIFICATION KIT

Degassing may also be necessary if a reciprocating pump is used.   This type of pump (discussed later) has a tendency to produce gas bubbles from any dissolved gases during the intake stroke.   These bubbles do not always redissolve and can partially block a column or create noise at the detector.

Pumps

The pump is one of the major components of a high pressure liquid chromatograph. Equation 20.2 relates several of the variables producing a pressure drop across a column.

$$P = \frac{L\eta\nu}{\theta d^2}$$ (20.2)

where:

$$L = \text{column length}$$

$$\eta = \text{solvent viscosity}$$

$$\nu = \text{flow rate}$$

$$\theta = \text{a constant}$$

$$d = \text{particle diameter}$$

In order to obtain the very best separations small particles must be packed into the column but from Equation 20.2 you can see that this greatly affects the pressure required. Pressures greater than 500 psi and sometimes more than 5000 psi are required with usual pressures being from 700 to 1500 psi. To attain these pressures requires good pumps; and many different types of pumps have been developed. We will discuss two major types, the reciprocating pump and a positive displacement pump. The following seven requirements have been suggested for a HPLC pump.

1. High pressure to force liquid through the column.
2. Ability to obtain reproducible, stable flow rates in the 30 to 200 ml/hr range, using a variety of solvents with easy flow measurement, accurate resettability, and with flow rates independent of column back pressure.
3. Nonpulsating flow to attain full sensitivity of the detectors.
4. Ability to change even immiscible solvents easily and quickly.
5. Maximum freedom in choosing solvents including those that are volatile or corrosive.
6. Gradient elution capability and ability to blend solvents in desired proportions.
7. High reliability and minimum maintenance.

Reciprocating Pumps.—These pumps have the advantage of being able to pump solvent in unlimited quantities and are of use when long separations or preparative situations exist. Figure 20.9 shows a diagram of how one type functions.

The disadvantages are that they produce a pulsating pressure which affects most detectors; they present difficulties in providing reproducible flow control over a range of solvent types; pump volumes vary as the solvents change due to the compressibility of the solvents; and they cavitate—that is, on the inlet stroke when the pressure is released, gas bubbles can form either from dissolved gases or from high vapor-pressure solvents. This means that lower vapor pressure (usually also higher viscosity) solvents must be used and the solvents should be degassed. Bubbles can cause detector noise, variable volume delivery and "vapor lock" requiring the pump to be reprimed. The pulsating effect can be greatly reduced by using pulse suppressors.

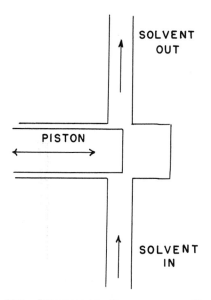

FIG. 20.9.   DIAGRAM OF A RECIPROCATING PUMP

Positive Displacement Pumps.—Figure 20.10 shows a pump of this type, a syringe pump.

A syringe pump has the disadvantage that only a single charge of solvents can be delivered, hopefully enough to complete the analysis. The advantages are: no cavitation, easy to change solvents, flow rates are reproducible and resettable at both high and low flow rates, and no check valves are required thus permitting more corrosive solvents to be used for longer periods of time.

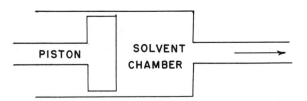

FIG. 20.10.   DIAGRAM OF A SYRINGE PUMP

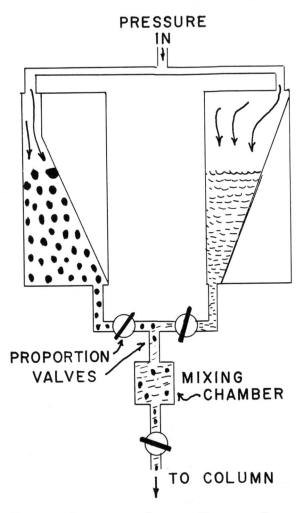

FIG. 20.11.   DIAGRAM OF A GRADIENT-PRODUCING DEVICE

Gradient Elution

If the solvent composition is kept constant throughout the entire separation the process is called "isocratic elution." If, however, the solvent composition is changed in any manner (pH, buffer strength, different solvent mixtures) then the process is called "gradient elution." Gradient elution is particularly useful when separating mixtures having widely varying characteristics. Column packings generally perform well over a narrow range of sample characteristics. However, the solvent characteristics can be progressively changed which in turn alters the behavior of the sample compounds. If done properly, a slow separation can be speeded up or groups of compounds that won't separate can be made to separate. One type of device used to produce a gradient (Fig. 20.11) consists of two wedge-shaped chambers, a mixing region, and proportioning valves.

More elaborate gradient systems can produce either linear, exponential or step gradients to meet the operator's specifications.

Sample Injection

Sample volumes of 5 to 50 μl are normal but they must be injected onto a column that is usually under several hundreds pounds of pressure. One way of doing this is shown in Fig. 20.12.

This injector allows sample introduction at full column pressure without interrupting the solvent flow. All of the parts are either stainless steel or teflon. A syringe is injected to an adjustable stop and a knurled knob is tightened, causing the teflon cylinder to form a pressure seal around the needle. The transverse seal is then opened, the syringe pushed forward and the injection made.

More elaborate chromatographs contain several columns of different characteristics. A sample injection device with 12 ports which can inject 2 samples at the same time onto 2 similar columns to obtain a differential comparison of 2 samples is shown in Fig. 20.13. One such use is to inject a food sample extract into one column and simultaneously inject a food sample extract of a second food which you suspect contains an unlisted additive. Only the differences are detected which greatly simplifies a very complex analysis.

Columns and Column Packing

Columns are commonly 30 to 100 cm in length and from $\frac{1}{8}$ to $\frac{3}{8}$ in. in diameter, the larger columns being used for preparative work. They are usually made out of stainless steel.

The high pressures employed require a very hard packing material. Furthermore, if high efficiency is to be achieved, then uniform packing must

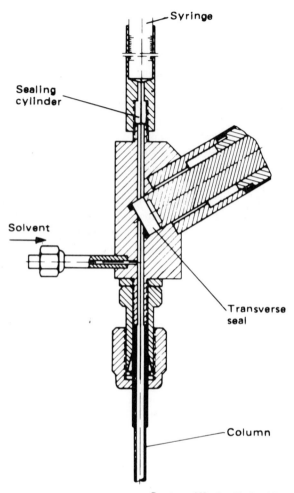

Courtesy of Hewlett-Packard Co.

FIG. 20.12.   A SEPTUMLESS SYRINGE INJECTION BLOCK

be achieved. These requirements usually result in a packing being made from silica or alumina and consisting of round particles.

There are currently three general types of particles: fully porous, pellicular, and microporous. These are shown diagrammatically in Fig. 20.14.

Fully Porous.—These come in various sizes but a popular size is about 50 μ in diameter. They have very large surface areas—300–500 m^2/g. This large area means high column capacity so these materials are used for

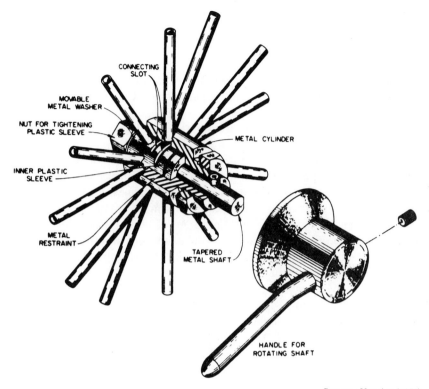

CONNECTING SLOT

MOVABLE METAL WASHER

NUT FOR TIGHTENING PLASTIC SLEEVE

INNER PLASTIC SLEEVE

METAL CYLINDER

METAL RESTRAINT

TAPERED METAL SHAFT

HANDLE FOR ROTATING SHAFT

Source: Veening (1973)

FIG. 20.13. HIGH PRESSURE 12 PORT INJECTION SYSTEM

preparative separations. The retention times are longer because the diffusion distances (25 μ in and 25 μ out) can be long. Examples: Porasil, Styragel, and the Durapak and Bondapak Porasils.

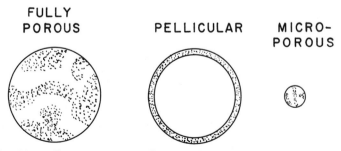

FULLY POROUS   PELLICULAR   MICRO-POROUS

FIG. 20.14. DIAGRAMS OF THE GENERAL TYPES OF COLUMN PACKINGS

Pellicular.—The pellicular particles consist of a solid core with a 2–3 μ crust etched onto the surface. They are also about 50 μ in diameter but their surface area is much less than the fully porous—10–30 m^2/g. Their column capacity is much lower than the fully porous but because of the smaller diffusion distances (2 μ in and 2 μ out) they are highly efficient and excellent for analytical separations. Examples are Corasil and the Durapak and Bondapak Corasils.

Microporous.—These are approximately 10 μ in diameter and are fully porous particles. They provide a highly efficient and high speed packing and are used for the most efficient separations. Their high porosity means heavier loading is possible than with the pellicular particles. In addition, since the diffusional distances are small you can have very good efficiencies. An excellent analogy of the advantages of a micro particle over a regular size particle is that given by the Waters Associates: "Consider for a moment a sewer pipe which is packed with basketballs and filled with water. Now, between each one of these basketballs, if you stop the flow of water, there is a certain volume of liquid; if there is anything dissolved in the water, each one of these volumes of liquid acts like a large mixing chamber. Consider now the same sewer pipe, only this time packed with baseballs. The volumes between the balls are much smaller; therefore the mixing chambers are much smaller. The total volume of the water in both cases is going to be nearly the same. But because the mixing chambers are smaller and the time between mixing and interaction between the liquid and the particle is smaller, the net result is that you end up with a lot less mixing with the smaller particles and therefore higher efficiencies." Examples are the micro-Porasils, Bondapaks, and Styragels.

Column Cleanups.—Liquid chromatograph columns are expensive, $150–500 per column. As a result, special care must be taken to protect them. They easily become inactive due to surface contamination and must be regenerated. A good policy is to flush the column at a slow rate (0.1 ml/min) overnight. If a quicker cleanup is required then the following can be tried: A silica column can be regenerated with 100 ml of iso-propanol followed by about 100 ml each of successively less polar materials such as acetone, chloroform and finally hexane, at a rate of 2–4 ml/min.

The reversed phase packings, such as the Bondapaks, require a water flush if buffers have been used which is then followed with 50 ml of methanol or acetonitrile.

Proteins are removed with 8M urea followed by water. The important concept is that columns do go bad but they can be regenerated and should not be thrown away too hastily.

Detectors

The detector is another of the critical components of a high pressure liquid chromatograph and, in fact, the practical application of liquid chromatography had to await a good detector system. There are now many types of detectors on the market but we will discuss only the two that are the most common, the ultraviolet absorption detector and the refractive index (RI) detector.

Ultraviolet Absorption Detector.—The ultraviolet absorption detector uses a low pressure mercury lamp as a source; the cells are about 1 cm in path length and have volumes of 8–30 μl. Most are double beam. Normally, they have absorbance ranges from 0.001 to 3 absorbance units which corresponds to a requirement of about 5×10^{-10} g/ml for a favorable compound.

The UV region is particularly useful for many food components in addition to additives, pesticides and drugs because these compounds contain —C=C—, —C=O, —N=O, and —N=N— functional groups which readily absorb UV radiation. Aromatic rings absorb very strongly at the 254 nm wavelength of the mercury lamp emission. By using a filter, the wavelength at 280 nm is also available although it is not as sensitive for most compounds.

Multiple wavelength detectors are available but they usually have somewhat higher detection limits at each wavelength because it is difficult to focus sufficient intensity of radiation through the cell.

Refractive Index Detectors.—The refractive index detector is applicable to all compounds although it is not as sensitive as the UV detector. An RI detector can detect differences of about 1×10^{-5} RI units which means that about 5×10^{-7} g/ml must pass through the detector for a favorable sample. As a general rule the sensitivity in milligrams of sample is almost equal to the reciprocal of the difference in refractive index between the solvent and the sample. Figure 20.15 shows the two basic methods of refractive index measurement.

Differential refractometers operate by utilizing one of the following principles. In the first type, the measurement is based on an optical displacement of the beam (Fig. 20.15A). A second method utilizes the Fresnel principle (Fig. 20.15B) which relates the transmittance of a dielectric interface to the refractive indices of the interface materials. Such an interface may be formed between a glass prism of selected optical properties and the liquid whose refractive index is to be measured.

These detectors are difficult to use with gradient elution systems and temperature control of the solvents is critical.

Angle of Deviation Method

A

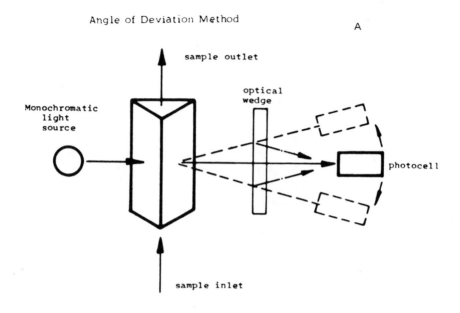

Fresnel Method

B

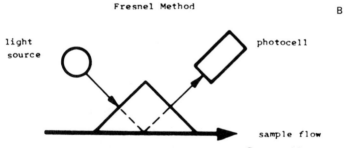

Courtesy of Journal of Chemical Education

FIG. 20.15.   THE TWO BASIC PRINCIPLES OF REFRACTIVE INDEX DETECTORS

TYPES OF LIQUID CHROMATOGRAPHY

There are five major types of liquid chromatography: liquid-solid, ion-exchange, liquid-liquid, paired-ion and gel permeation.   The first four of these are based on the differences in chemical properties of the materials to be separated while the latter is based on size differences.

Liquid-solid Chromatography

The title liquid-solid aptly describes the column conditions.   True, the packing is a solid, but of more importance, the outer layer of the packing

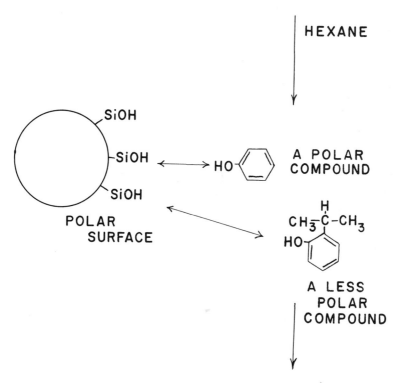

FIG. 20.16.   A LIQUID-SOLID SYSTEM

material that comes in contact with the mobile phase and sample compounds is a solid.   This is shown in diagrammatic form in Fig. 20.16.

Most solid column packings are clays or clay-type materials which means their surfaces are aluminates or silicates and consist of large numbers of terminal —OH groups that are highly polar.   Usually a nonpolar solvent such as hexane is used as the mobile phase.   When moderately polar compounds are dissolved in the mobile phase and passed over the column packing the more polar compounds are retained more strongly than the less polar compounds and a separation results.

Reversed Phase Liquid-solid Chromatography.—Later on when we discuss liquid-liquid systems you will find that one of the disadvantages of that system is that the liquid surface coating often strips off.   In addition, since the solid surfaces of the particles are polar, they are not very useful for separating nonpolar compounds since the nonpolar compounds have little affinity.   To solve this problem we now have what is known as "reverse phase bonded packing."   In this situation, shown in Fig. 20.17, the reactive

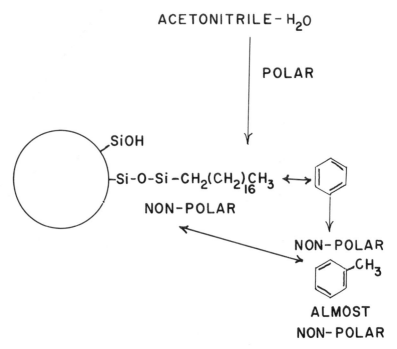

FIG. 20.17.   A REVERSE PHASE BONDED PACKING

surface of the particle is changed to a nonpolar compound and it is chemically bonded to the —OH group so it can't be stripped off.   A commercial system is Bondapak C_{18}.

Ion-exchange Liquid Chromatography

Ion-exchange is not a new process and you probably are already quite familiar with it.   While the normal ion-exchange resins which are made from styrene and divinyl benzene can be used in high pressure liquid chromatography, the beads are too soft, compress, and may plug up the column.   What is more commonly used is a bonded packing of typical ion-exchange functional groups on a hard silica particle.   This is shown in Fig. 20.18.

Generally, the compounds to be separated are placed in an aqueous system buffered about 1.5 pKa units above the highest sample pKa.   This ensures that the compounds are completely ionized and retained.   They are then eluted by:

(1)   Passing the solvent buffer through the column, the order of elution being the compound with the highest pKa eluting first.

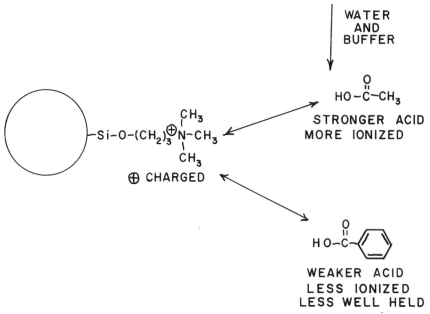

WATER
AND
BUFFER

$$HO-\overset{O}{\underset{\|}{C}}-CH_3$$

STRONGER ACID
MORE IONIZED

WEAKER ACID
LESS IONIZED
LESS WELL HELD

FIG. 20.18. A BONDED ION-EXCHANGE PACKING (CORASIL BONDAPAK-AX)

(2) Use an ionic strength gradient with a low ionic strength being used first.

(3) Change the pH. For anions, go more acidic as this changes the sample compounds back to neutral molecules and they will then elute. The reverse is done with cation exchangers.

Note: *The pH must be kept below 8 or the solid silica supports will begin to dissolve.*

Liquid-liquid Chromatography

This system consists of an inert phase solid support upon which is coated a liquid stationary phase. This is shown in Fig. 20.19.

The separation is based upon the relative solubilities of the sample compounds in the stationary phase. One difficulty is that under high pressure the stationary phases are often readily stripped from the solid support. One way to reduce this is to saturate the mobile phase with the stationary phase before the separation begins. See the precolumn arrangement in Fig. 20.6.

The amount of stationary phase used is about 1% by weight for each 15 m^2/g of solid phase surface area up to a maximum of about 15%.

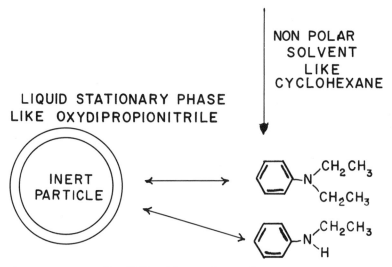

FIG. 20.19.   A LIQUID-LIQUID SYSTEM

Paired-ion Chromatography (PIC)

This is a technique, which in its most popular application, is a modification of reversed phase liquid-solid chromatography.   It is based entirely on concentration equilibrium and can be used to separate highly polar materials with a nonpolar surface.   A diagram of how this is done is shown in Fig. 20.20.

If the right system can be found, PIC usually provides better separation efficiencies than ion exchange.   The ion pair reagents for cations are organic sulfonic acids like $CH_3(CH_2)_6—SO_3^-$, H^+.

Gel Permeation (GPC) and Gel Filtration Liquid Chromatography

This is a technique based on size separation.   Probably the easiest way to describe this is to first take a look at Fig. 20.21.   Molecules of various size are drawn on the photo and it can be seen that the large molecules cannot fit at all and would be washed out of the column immediately.   The smallest compound that does not fit is called the "exclusion limit" and the volume required to elute the large molecules is called the "void volume," V_o.   The smaller compounds can permeate the gel; and, the smaller compounds permeating further, they therefore require more solvent to wash them out of the gel.   If the compounds are extremely small, then their molecular size compared to the pores is so small that they behave the same and require essentially the same amount of solvent to elute them.   This

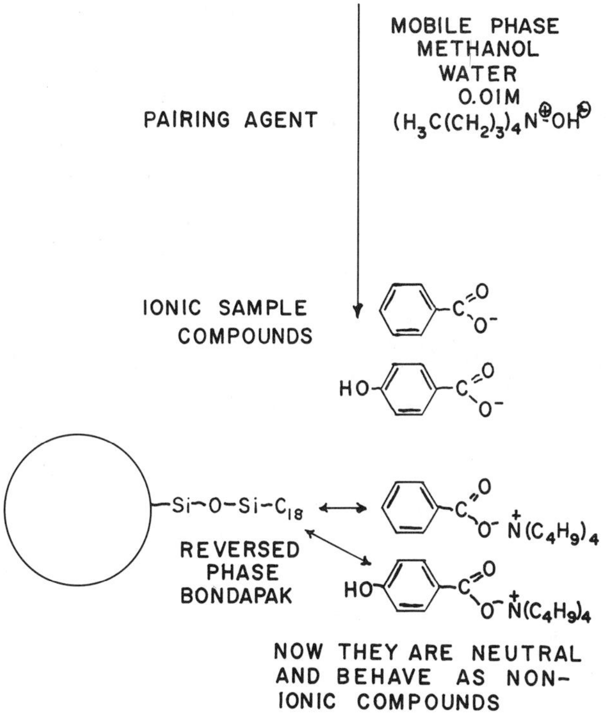

PAIRING AGENT

MOBILE PHASE
METHANOL
WATER
0.01M
$(H_3C(CH_2)_3)_4N^{\oplus}OH^{\ominus}$

IONIC SAMPLE
COMPOUNDS

-Si-O-Si-C$_{18}$

REVERSED
PHASE
BONDAPAK

NOW THEY ARE NEUTRAL
AND BEHAVE AS NON-
IONIC COMPOUNDS

Fig. 20.20.   The Paired-Ion Chromatography Concept

volume is called the "total permeation volume," V_t.   Figure 20.22 illustrates these terms. V_i is the "interstitial volume."

The size of the pores determines the molecular weight range of the compounds that can be separated.   GPC is often used to clean up an extracted solution.   For example, GPC is used to remove the fats and waxes from the pesticide residues that are extracted from a food composite.

A series of columns, each fitted with a different exclusion limit gel, can be used to separate a multiple component mixture.   A general rule is that compounds that differ by 10% can be separated in the same column.

This technique started with what is now called "gel filtration."   Polymers of dextran or starch were first used.   These use aqueous salt solutions for the mobile phase, they swell considerably and collapse rather easily.   In fact, the large pore gels must be flushed from the bottom to top in order to prevent the gel from collapsing and plugging the column.   These materials are produced by the Pharmacia Co. and they prefer that the term gel filtration be used with their product.

"Gel permeation" is the same concept except that the materials are highly

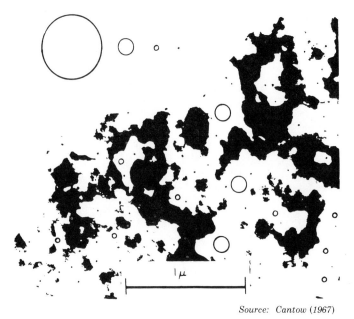

Source: Cantow (1967)

FIG. 20.21.   AN ELECTRON MICROGRAPH OF A GEL PARTICLE
MATRIX

crosslinked styrene-divinyl benzene (styragel, Poragel) or something
similar. Bio-gel-P for example is a polyacrylamide polymer.

These beads can be used with higher pressures, some like hydrogel can
withstand pressures up to 3000 psi. Organic solvents can be used; therefore,
a wide variety of compounds can be separated.

Some commonly-used solvents are cyclohexane, chloroform, methylene
chloride, dimethylformamide, and xylene.

Table 20.4 shows a few characteristics of three gels.

DEFINITION OF TERMS

We have discussed a few terms already: V_t, the retention volume; and
V_o, the void volume or the volume required to move an unreactive compound through the column. These are shown in Fig. 20.23 in a different
manner along with W, the peak width.

These measureable quantities are used to determine four parameters
useful in evaluating the quality of a liquid chromatographic system. These
parameters are k', the *capacity factor;* α, the *separation factor;* $N,$ the

SELECTED CHARACTERISTICS OF BIOGEL-P, PORASIL AND STYRAGEL

Material	Range	Exclusion Limit
Biogel P2	10^2–1.8×10^3	4×10^3
P4	8×10^2–4×10^3	8×10^3
P6	1×10^3–6×10^3	
P30	2.5×10^3–4.0×10^4	
P100	5×10^3–1×10^5	
P200	3×10^4–2×10^5	
P300	6×10^4–4×10^5	
Porasil A		4×10^4
B		2×10^5
C		4×10^5
D		1×10^6
E		1.5×10^6
F		$>4 \times 10^6$
Styragel 10^7A	5×10^5–5×10^8	5×10^8
10^6A	10^5–5×10^7	5×10^7
10^5A	5×10^4–2×10^6	2×10^6
10^4A	10^3–7×10^5	7×10^5
10^3A	500–5×10^4	5×10^4
10^2A	100–8×10^3	8×10^3

Source: Waters Associates, Milford, Mass.

number of theoretical plates; and *R, resolution.* Mathematically these are:

$$k' = \frac{V_i - V_o}{V_o} \tag{20.3}$$

$$\alpha = \frac{V_2 - V_o}{V_i - V_o} = \frac{k'_2}{k'_1} \tag{20.4}$$

$$N = 16 \left(\frac{V}{W}\right)^2 \tag{20.5}$$

$$R = \frac{2(V_2 - V_1)}{W_1 + W_2} \tag{20.6}$$

The k' is a measure of a compound's retention in terms of the column volume. A k' of 1 means that the sample comes through unretained, while a k' of 8 to 10 means that the sample takes a long time to come through the column. For rapid analysis a low k' is desired; while for good separation a high k' is needed. The compromise is a k' value of 2–6. The k' is usually controlled by changing the solvents. Table 20.5 lists several solvents and indicates their relative physical and chemical properties. We will see how this can be used later on.

The separation factor, α, is a ratio of the net retention time (equilibrium distribution coefficient) for any two components. If α is 1 then the sepa-

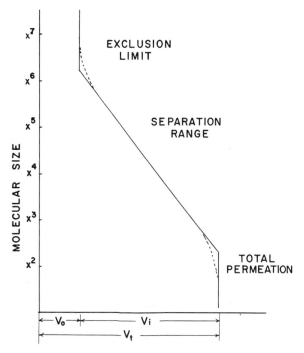

FIG. 20.22.   THE VOLUME TERMS IN GEL PERMEATION

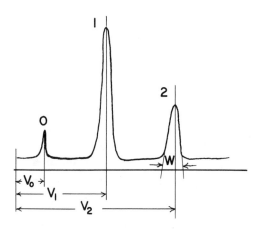

FIG. 20.23.   AN HPLC CHROMATOGRAM

ration between 2 components is zero. α is controlled by the solvent also and again Table 20.5, column 3, gives a relative measure of α.

The theoretical plate concept will be discussed in Chap. 22; and, in fact, the equation presented here is also presented in the gas chromatography chapter. N is the number of theoretical plates in a column. One way to increase N is to increase the column length. The separation time will be increased as well, but by increasing the pressure, the flow rate can be maintained. Another means of increasing N is to reduce the solvent flow rate but the separation then takes longer. If these two methods are not

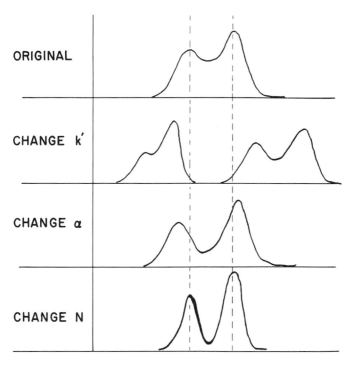

FIG. 20.24.   THE EFFECTS OF VARYING k', α, AND N

sufficiently effective then the column materials must be changed. Within the column, N depends upon the porosity of the particles, the size of the particles, and the homogeniety of the particle sizes. Usually the most porous, smallest and most uniform size particles provide the highest N, all other factors being equal.

High resolution is the ultimate goal of any separation process. Equation 20.6 showed how resolution can be measured. Equation 20.7 shows the theoretical factors involved in achieving good resolution.

TABLE 20.5

SOLVENT USE INDEX II

k', Polarity Index	Solvent	Dipole Moment	α Solvent Group	Viscosity
−0.3	n-Decane	0.00	0	0.92
−0.4	Isooctane	0.00	0	
−0.8	Squalane	0.00	0	
0.0	n-Hexane	0.00	0	
0.0	Cyclohexane	0.00	0	1.00
1.0	Carbon Disulfide	0.00	0	0.37
1.7	Butyl Ether	0.39	1	
1.7	Carbon Tetrachloride	0.32	7	0.97
1.8	Triethyl Amine	0.32	1	
2.2	Isopropyl Ether	0.35	1	
2.3	Toluene	0.44	7	0.50
2.4	p-Xylene	0.44	7	0.70
2.7	Chlorobenzene	0.42	8	0.80
2.7	Bromobenzene	0.42	8	
2.7	Iodobenzene	0.40	8	
2.8	Phenyl Ether	0.42	8	
2.9	Ethyl Ether	0.34	1	0.23
2.9	Ethoxy Benzene	0.44	7	
3.0	Benzene	0.43	7	0.65
3.1	Ethyl Bromide	0.40	6	
3.1	Tricresyl Phosphate	0.47	5	
3.2	1-Octanol	0.25	2	
3.3	Fluorobenzene	0.43	8	
3.3	Benzyl Ether	0.46	7	
3.4	Methylene Chloride	0.49	5	0.44
3.5	Methoxybenzene	0.41	7	
3.6	1-Pentanol	0.25	2	
3.7	Ethylene Chloride	0.45	5	0.79
3.9	Bis-2-Ethoxyethyl Ether	0.46	5	
3.9	n-Butanol	0.26	2	
3.9	1-Butanol	0.22	2	
4.1	1-Propanol	0.27	2	2.30
4.2	Tetrahydrofuran	0.40	3	0.51
4.3	Ethyl Acetate	0.42	6	0.47
4.3	2,6-Lutidine	0.35	3	
4.3	Isopropanol	0.26	2	2.30
4.3	Chloroform	0.33	9	0.57
4.4	Acetophenone	0.40	6	
4.5	Methyl Ethyl Ketone	0.47	6	0.43
4.5	Cyclohexanone	0.42	6	
4.5	Nitrobenzene	0.43	7	
4.6	Benzonitrile	0.39	6	
4.8	Dioxane	0.41	6	1.54
4.8	2-Picoline	0.30	3	
5.0	Tetramethyl Urea	0.40	3	

TABLE 20.5 *(Continued)*

k', Polarity Index	Solvent	Dipole Moment	α Solvent Group	Viscosity
5.0	Diethylene Glycol	0.33	4	
5.1	Triethyleneglycol	0.33	4	
5.2	Quinoline	0.33	3	
5.2	Ethanol	0.28	2	1.20
5.3	Pyridine	0.36	3	0.94
5.3	Nitroethane	0.42	7	
5.4	Ethylene Glycol	0.30	4	19.90
5.4	Acetone	0.40	6	0.32
5.5	Benzyl Alcohol	0.30	4	
5.5	Tetramethyl-Guanidine	0.37	1	
5.7	Methoxyethanol	0.36	4	
5.8	Triscyanoethoxypropane	0.41	6	
6.0	Propylene Carbonate	0.41	7	
6.2	Aniline	0.36	6	4.40
6.2	Acetonitrile	0.41	6	0.37
6.2	Methyl Formamide	0.36	3	
6.2	Acetic Acid	0.30	4	1.26
6.2	Oxydipropionitrile	0.39	6	
6.3	N,N-Dimethylacetamide	0.37	3	
6.4	Dimethyl Foramide	0.38	3	0.90
6.5	N-Methyl-2-Pyrolidone	0.28	3	1.65
6.5	Dimethyl Sulfoxide	0.38	6	2.24
6.5	Tetrahydrothiophene-1,1-Dioxide	0.38	6	
6.6	Methanol	0.30	2	0.60
6.6	Hexamethyl Phosphoric Acid Triamide	0.36	3	
6.8	Nitromethane	0.42	7	0.67
7.0	m-Cresol	0.25	9	20.80
7.3	Formamide	0.32	4	
7.9	Dodecafluoroheptanol	0.25	9	
9.0	Water	0.26	9	1.00
9.3	Tetrafluoropropanol	0.30	9	

Source: Waters Associates, Milford, Mass., and adapted from Table published in J. Chromatog. *92*, 223 (1974).

$$R = \frac{\sqrt{N}}{4} \left(\frac{k'}{k'+1}\right) \left(\frac{\alpha-1}{\alpha}\right) \qquad (20.7)$$

In practice, resolution is defined as the distance between the peak centers of two peaks divided by the average width of their respective bases. Figure 20.24 provides some idea of how k', α, and N are used to control resolution.

The following example will be used to illustrate how these various parameters are calculated.

Example calculation:

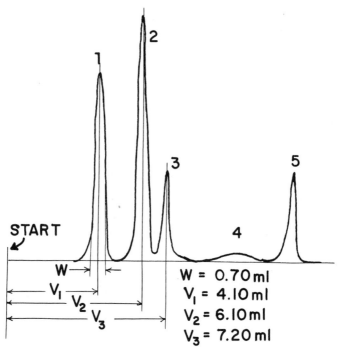

FIG. 20.25. CHROMATOGRAM OBTAINED USING A 92-CM COLUMN

The chromatogram shown in Fig. 20.25 was obtained using a 92-cm column. Calculate (A) k' for compound 1, (B) α for compounds 2 and 3, (C) N for compound 1, (D) R for compounds 2 and 3, and (E) HETP for compound 1. $V_o = 0.90$ ml, $W_2 = 0.80$ ml.

Answer:

(A) $k' = \dfrac{4.10 - 0.90}{0.90} = 3.6$

(B) $\alpha_{2,3} = \dfrac{7.20 - 0.90}{6.10 - 0.90} = 1.2$

(C) $N = 16\left(\dfrac{4.10}{0.70}\right)^2 = 549$

(D) $R_{2,3} = 2 \dfrac{(7.20 - 6.10)}{0.70 + 0.80} = 1.47$

(E) $HETP = \dfrac{920 \text{ mm}}{549 \text{ plates}} = 1.67 \text{ mm/plate}$

SOME RECENT APPLICATIONS

Stahl *et al.* (1974) used a Dowex (NX) column to separate sorbic, lactic, malic, portonic and citric acids in wine and fruit juices. Reversed phase chromatography (C_{18} Bondapak) was used by Wittmer *et al.* (1975) to investigate the food color tartrazine and its intermediates.

Cavins and Inglett (1974) separated the vitamin E isomers that he found in corn oil and wheat bran. A Corasil II column was used.

A novel gradient generator was used by Chilcote *et al.* (1974) to examine carbohydrates while Palmer and Brondes (1974) used a pressure of 400 psi and the cation exchanger, Aminex Q-150-S, to separate sucrose, glucose, and fructose in fruit juice, cereals, grains, and banana tissue. Goulding (1975) carried the use of cation exchangers (Aminex A-5) a lot further to separate sugars and the polyhydric alcohols, glycerol, and galactitol.

Phthalate esters in packaging materials and their transfer to food was studied by Persiani and Cukor (1975) using a porasil liquid-solid system.

The liquid chromatography of lipids in foods is discussed by Aitzetmuller (1975). Pesticides in foods have always been a problem. Moye (1975) reviews how HPLC can be used to separate pesticides in such samples as milks, citrus fruits, and cucumbers.

Vitamins in foods and as feed additives are examined by Conrad (1975) while Lee and Chang (1975) show how liquid-solid, liquid-liquid, and reverse phase columns can be used to separate the less volatile food flavors such as undecane, butyl acetate, monyl alcohol, 3-heptanone, 3-decanone, butyl benzene, and phenyl ether.

Wetzel (1975) has a complete literature survey of HPLC applications.

ION EXCHANGE

The equipment used in ion exchange separations is similar to that used in column chromatography and for most separations can be identical. Figure 20.26 is a diagram of one of the more elaborate columns.

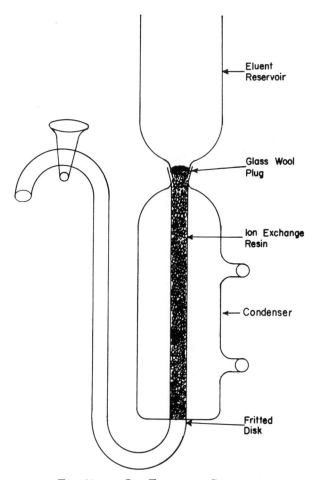

FIG. 20.26.   ION EXCHANGE COLUMN

The apparatus contains a water jacket and a means to prevent the column from going dry.

It has been said that ion exchange goes back to the Biblical time when Moses was leading the Israelites safely through the wilderness and made bitter water potable by stirring it with a tree. The oxidized cellulose may have exchanged with the bitter electrolytes of the water.

Be that as it may, Thompson in 1850 and Way in 1850 are credited with the first systematic study. Thompson wanted to know why $(NH_4)_2SO_4$ and KCl did not wash out of soils. When $(NH_4)_2SO_4$ was added at the top of the column, $CaSO_4$ was eluted. When KCl was added, $CaCl_2$ was eluted. In every case the cation was eluted.

Way was Thompson's student and continued the work. He arrived at the following conclusions which remain essentially unchanged over 100 yr later.

(1) The exchange of Ca^{2+} and NH_4^+ in soils noted by Thompson was verified. (2) Exchange of ions in soils involved the exchange of equivalent quantities. (3) Certain ions were more readily exchanged than others. (4) The extent of exchange increased with concentration, reaching a leveling-off value. (5) The temperature coefficient for the rate of exchange was lower than that of a true chemical reaction. (6) The aluminum silicates present in soils were responsible for the exchange. (7) Heat treatment destroyed the exchange capacity. (8) Exchange materials could be synthesized from soluble silicates and alums. (9) Exchange of ions differed from true physical adsorption.

Courtesy Rohm and Haas

FIG. 20.27. CATION EXCHANGE RESIN BEADS

Inorganic Exchangers

$Na_2Al_2Si_3O_{10}$ is a natural Zeolite in which 2 out of 10 atoms carry a negative charge. These crystals resemble open sponges with charges sticking out all over. The following equation involving another Zeolite shows what atoms exchange.

$$Na_2O \cdot Al_2O_3 \cdot 4SiO_2 \cdot 2H_2O + 2KCl =$$

$$K_2O \cdot Al_2O_3 \cdot 4SiO_2 \cdot 2H_2O + 2NaCl \quad (20.5)$$

Inorganic exchangers are limited in their capacity and they are destroyed by strong acids and bases.

Organic Exchangers

In 1935, Adams and Holmes prepared the first organic resins. These are more stable, have a larger capacity, and can be tailor-made to increase selectivity. The structural equation shows how the resins are made. The reactive group containing the exchangeable ion determines the resins characteristics. The amount of divinyl benzene determines how porous the bead will be.

TABLE 20.6

TYPICAL EXCHANGE GROUPS

Cation Exchangers		Anion Exchangers
—SO$_2$H		—N(CH$_3$)$_3$$^+Cl^-$
—COOH	Strong	$\underset{+}{-\text{N}}\!\!<\!\!\begin{array}{l}(\text{CH}_3\text{Cl})_2{}^-\\ \text{CH}_2\text{OH}\end{array}$
—CH$_2$SO$_3$H	↓	
—OH		—NR$_2$
—SH	Weak	—NHR
—PO$_2$H$_2$		—NH$_2$

Ion Exchange Affinity

(1)   At low concentrations and ordinary temperatures, the extent of exchange increases with increasing valency of the exchanging ion

$$Na^+ < Ca^{2+} < Al^{3+} < Th^{4+}$$

(2)   At low concentrations and ordinary temperatures and constant valence, the extent of exchange increases with increasing atomic number of the exchanging ion.

$$Li^+ < Na^+ < K^+ =$$
$$NH_4{}^+ < Rb^+ < Cs^+ < Ag^+ < Be^{2+} < Mn^{2+} < Mg^{2+} =$$
$$Zn^{2+} < Cu^{2+} = Ni^{2+} < Co^{2+} < Ca^{2+} < Sr^{2+} < Ba^{2+}$$

(3)   At high concentrations the differences in the exchange "potential" of ions of different valences (Na$^+$, Ca^{2+}) diminishes, and in some cases reverses.

(4)   At high temperatures, in nonaqueous media, or high concentrations, the exchange potentials of ions of similar charge become quite similar and even reverse.

(5)   The relative exchange potentials of various ions may be approximated from their activity coefficients. The higher the activity coefficient the greater the exchange.

(6)   The exchange potential of the H$^+$ and OH$^-$ ions varies considerably with the nature of the functional group, and depends upon the strength of the acid or base formed between the functional group

and either the hydrogen or hydroxyl ion.   The stronger the acid or
base the lower the exchange potential.
(7)   For a weak base exchanger the exchange follows:

$$OH^- > SO_4^{2-} > CrO_4^{2-} > \text{citrate} > \text{tartrate} > NO_3^- >$$
$$\text{arsenate} > PO_4^{3-} > MoO_4^{2-} \text{ acetate} = I^- = Br^- > Cl^- > F^-$$

Complex and Chelate Resins

In an effort to improve the selectivity of ion exchange resins,
various complexing and chelating agents have been built into the
resin structure.   Below are examples of some of these.

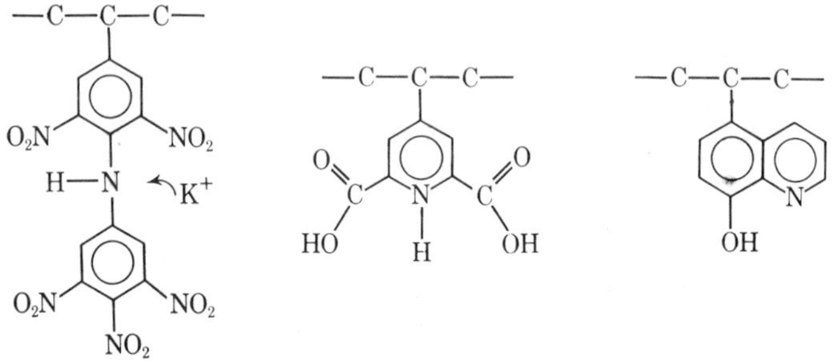

Ion Retardation Resins

An example of this is Bio-Rad AG 11A8 which is described below.
This resin is used primarily for the desalting of biological fluids.
It is made by polymerizing acrylic acid inside Dowex 1.   The result
is a spherical resin bead containing paired anion and cation exchange
sites.

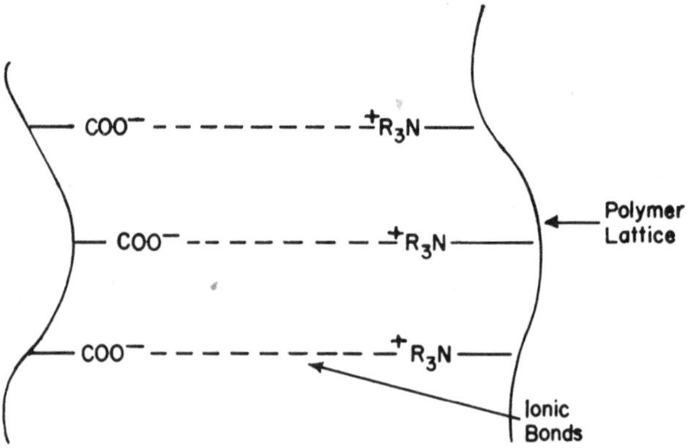

FIG. 20.28.   STRUCTURE OF AG 11A8

Retardation is due to the resin's paired anion and cation exchange sites, which attract mobile ions and associate weakly with them. Anions and cations are absorbed in equivalent amounts, but can be eluted with water. Organic molecules, even ionic species such as acidic and basic amino acids, are usually not absorbed by the resin.

The resin removes both acids and salts from solutions of amino acids, polypeptides, proteins, enzymes, nucleic acids, and sugars.

Ion Exclusion Resins

The following discussion is based on the book by Berg (1963) Ion exclusion is a process of separating ionic materials from nonionic materials. It is based on the inherent differences in the distribution of the two types of solutes between an ion exchange resin and a true aqueous solution. When an exchange resin is placed in a dilute electrolyte solution, the equilibrium electrolyte concentration is lower in the resin phase than in the surrounding solution. However, when the same resin is placed in a solution of a nonelectrolyte, the nonelectrolyte tends to establish equivalent concentrations in the two phases.

Ion exchange is not involved and both solutes can be extracted from the resin phase simply by diluting the external solution with water. Thus if a solution containing both ionic and nonionic materials is introduced to the top of a resin bed and washed through the column with water, the ionic material flows around the resin particles, whereas the nonionic material diffuses into the resin particles, as well as flows into the void places. By alternately passing water and sample solution through the column, alternate fractions of electrolytes and nonelectrolytes can be collected in the effluent. Since no exchange takes place, the resin is never exhausted and does not require regeneration.

Generally, the upper concentration limit for ionic materials is about 8%, and for the nonionic materials is about 40%.

The main disadvantages are (1) the resin must be of the same form as the electrolyte in solution, and (2) the volume of sample solution is limited by the volume of solution that can be absorbed by the resin.

Ligand Exchange

Helferich (1961) showed how to modify an ion exchange resin to separate nonionic compounds. The process was called *ligand exchange* (see Fig. 20.29). A complex such as $Cu(NH_3)_4{}^{2+}$ is passed through a weak acid resin, and the copper complex exchanges with

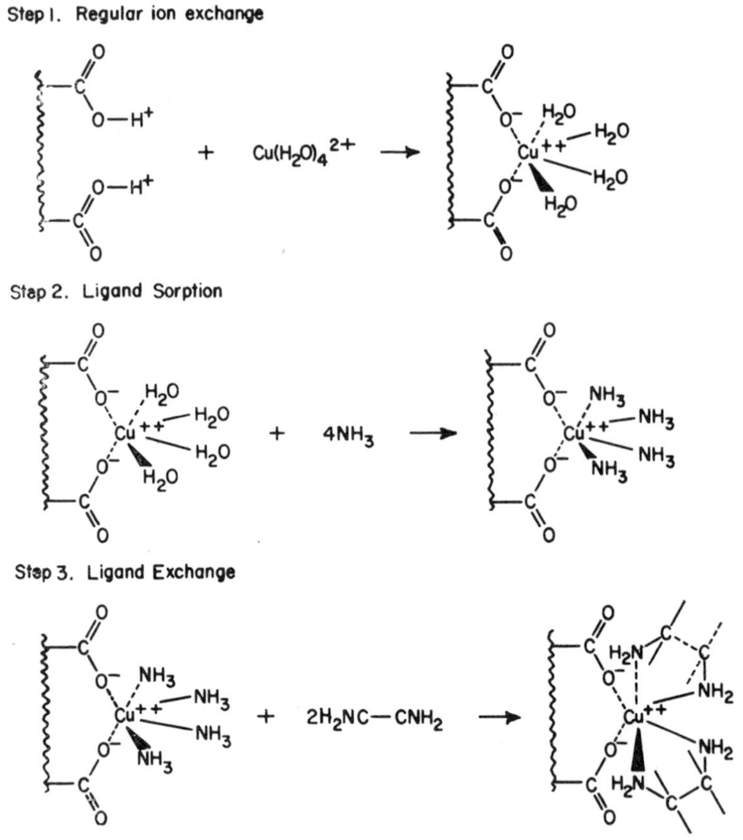

Step 1. Regular ion exchange

Step 2. Ligand Sorption

Step 3. Ligand Exchange

FIG. 20.29. LIGAND EXCHANGE

the H^+ on the resin. Now, if another amine is passed down the column it may exchange with the NH_3 of the copper complex, in effect performing a ligand exchange. If a mixture of ligands is added to the top of the column, these can then be separated because each has a different affinity for the copper complex.

The following metals have been used with success: Cu^{2+}, Ni^{2+}, Ag^+, Zn^{2+}, Co^{3+}, Fe^{3+}. Ligands that have been exchanged include NH_3, amines, amino acids, olefins, and polyhydric alcohols. The major problem is to prevent ion exchange of the cation.

Ion Exchange Kinetics

It is believed that ion exchange occurs within the bead because of the small outer surface and rather high capacity (2–10 meq/gm).

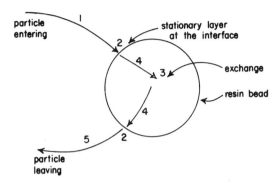

FIG. 20.30.   ION EXCHANGE KINETICS

The ion exchange process has been broken into five possible steps (Fig. 20.30) in an attempt to determine which step controls the rate of the exchange.

Steps 1 and 5 are not rate determining if the solution is agitated, such as happens when the solvent moves through the column. Step 3, the actual exchange is believed to be instantaneous and therefore is not a limiting factor.

Step 4, the migration through the bead to the exchange site depends on the degree of cross-linking and the concentration of the solution.   If the solution concentration is greater than 0.1 M, then this step controls the rate of exchange.

Step 2, the passage from the outer solution into the bead matrix is believed to control the rate if the solution concentration is less than 0.001 M.

For concentration ranges between 0.001 and 0.1 M, a combination of steps 2 and 4 controls the rate.

Liquid Ion Exchangers

A resin bead places some severe restrictions on the versatility of the ion exchange process.   One, the rate is slowed down in traversing the bead to get to the site, and another is that the necessity to polymerize a bead restricts the materials that can be introduced into the functional groups of the bead.   A solution to this problem can be found by using a liquid ion exchange resin rather than a bead. Thus, we can use not only the regular ion exchange properties, but also those of liquid-liquid extraction for which a great body of fundamental knowledge is already available.

The advantages are:

(1) there is a better control of the exchange by varying the con-

centration of the liquid resin; (2) a wider range of separations is possible; (3) more exchange structures are possible to build in selectivity; and (4) the system is more readily stripped of the extracted metal ions, and the separation can be easier and with smaller volumes. Some examples of liquid ion exchangers are:

(1) Amines   1-heptyloctylamine

$$H_2N\text{-}CH\text{-}(CH_2)_6\text{-}CH_3$$
$$|$$
$$H_2C\text{-}(CH_2)_5\text{-}CH_3$$

di-n-decylamine

$$HN[(CH_2)_9CH_3]_2$$

tri-iso-octylamine

$$N[\text{-}CH_2CH_2CH\text{-}CH_2\text{-}CH\text{-}CH_3]_3$$
$$\quad\quad\quad | \quad\quad | $$
$$\quad\quad CH_3 \quad CH_3$$

tetra-n-heptyl ammonium chloride

$$[N\text{-}(\text{-}(CH_2)_6CH_3)_4]^4Cl^-$$

(2) Acids   mono-n-butyl phosphoric acid

$$O$$
$$\|$$
$$(HO)_2\text{-}P\text{-}O\text{-}(CH_2)_3\text{-}CH_3$$

Fractionation According to Classes of Compounds (Heftman 1967)

Adsorption chromatography primarily separates nonpolar aliphatic and aromatic compounds according to the type and number of functional groups. The functional groups can be arranged in order of increasing adsorption affinity (and decreasing elution ease): $-Cl$ (chlorides); $-H$ (hydrocarbons);

$-OCH_3$ (ethers); $-NO_2$ (nitro compounds): $-N\overset{\diagup CH_3}{\diagdown CH_3}$ (tertiary amines);

$-C\overset{\diagup\!\!O}{\diagdown OCH_3}$ (esters); $-O-C\overset{\diagup\!\!O}{\diagdown CH_3}$ (O-acetyl compounds);

$-NH_2$ (primary amines);

$$-NH-\underset{\underset{CH_3}{\diagdown}}{\overset{\overset{O}{\diagup}}{C}}\quad(N\text{-acetyl compounds);}\quad -OH\;(\text{alcohols});$$

$$-\underset{\underset{NH_2}{\diagdown}}{\overset{\overset{O}{\diagup}}{C}}\quad(\text{amides); and}\quad -\underset{\underset{OH}{\diagdown}}{\overset{\overset{O}{\diagup}}{C}}\quad(\text{acids}).$$

If two or more constituents are present in a molecule, their influence on the rate of migration is roughly additive; but steric hindrances may be of greater importance.

Ion-exchange chromatography is particularly suitable for fractionation of compounds having ionized groups, i.e. sugar phosphates from sugars. Great differences in the net ionic charges of various classes lead, in turn, to class separations.

Partition chromatography is useful in fractionating hydrophilic components of widely different polarities into classes.

Reversed phase partition chromatography is one of the liquid-liquid systems in which the stationary liquid phase is nonpolar and the mobile phase is polar. If the substances being separated are only sparingly soluble in water, they merely move with the solvent system and no separation results. In such a case it may be advantageous to impregnate paper with a nonaqueous medium to act as the stationary phase. In normal chromatography the stationary phase, being aqueous, is more polar than the mobile phase. The technique where the mobile phase is more polar is called "reversed phase chromatography." The mobile phase is not necessarily water, though mixtures used normally contain some water. A number of substances have been employed as supports for the stationary phase. The greatest problem is to get evenly impregnated papers. Silicone-impregnated papers are available commercially.

Reversed phase partition chromatography is useful in separating lipophilic compounds into classes. Polar substances migrate ahead of less polar constituents of a mixture; i.e., the separation pattern is the reverse of that observed in adsorption or partition chromatography.

Fractionation Within Homologous Series

Adsorption chromatography is best for separating the first 5 to 7 members of a homologous series, or their simple derivatives, according to differences in molecular weight. Mixtures of homologues with more than eight carbon atoms migrate as classes, ahead of the shorter chain compounds.

Ion-exchange is suitable for fractionating mixtures of the lowest members of homologous series of ionic compounds.

Partition chromatography is used in separating hydrophilic homologues. As in adsorption chromatography, the higher members migrate ahead of the homologues of shorter chain lengths.

Reverse partition is best for efficient separation of homologous compounds with more than ten carbon atoms.

Fractionation According to Degree of Unsaturation

Adsorption chromatography separations are based primarily on the capacity to form complexes with $AgNO_3$.

Ion exchange does not resolve compounds differing only in the number of double bonds.

Partition and reverse phase partition can be used for separations of compounds varying in degree of unsaturation, but the separations are not very sharp. To overcome that limitation, derivatives of the compounds that are to be separated can be prepared and fractionated. Thus, treatment with Hg-acetate yields acetoxymercurimethoxy derivatives that can be more easily separated. After chromatography, the derivatives are treated with hydrochloric acid to obtain groups of original compounds.

AFFINITY CHROMATOGRAPHY (INCLUDING IMMOBILIZED ENZYMES)

The term "affinity chromatography" was first coined by Cuatrecasas *et al.* in 1968. (*See also* Cuatrecasas and Anfinsen 1971.) The potential applications of the technique are practically unlimited. They include purification, analysis, and characterization of biological substances; elucidation of mechanisms which govern biochemical systems; immobilization of enzymes, etc. Until the advent of affinity chromatography, practically all methods for purification of biological substances depended on gross chemical and/or physical differences between the substances that were to be separated. In gel filtration, separation is based on differences in molecular size and shape; in ion exchange chromatography separations are based on differences in electrical properties. In affinity chromatography, separations are based on specific interactions between interacting pairs of substances, i.e., a macromolecule and its substrate, cofactor, allosteric effector, or inhibitor (Guilford 1973). In principle, a ligand is attached covalently to a water-insoluble matrix which has been tailor-cut to adsorb from a mixture only the component(s) with a specific affinity for the ligand. All other components pass through the adsorbent unrestrained. The adsorbed molecules can then be eluted after some change in conditions, i.e., a solution of free ligand or by changes in pH and/or ionic strength (Fig. 20.31). The term affinity chromatography is somewhat imprecise as it is more characteristic of an extraction (purification) using an adsorbent, rather than chromatography. The technique parallels, in principle, the use of insolubilized antigens as immunosorbents in the purification of antibodies. Affinity chromatography has several advantages over other methods for separation: (a) specificity; (b) as a small proportion of the total protein is adsorbed from a crude mixture, only a small amount of adsorbent is required; (c) the adsorbed material is rapidly separated from proteolytic enzymes and can be stabilized by ligand binding at the "active" site; and (d) the adsorbent can be regenerated many times.

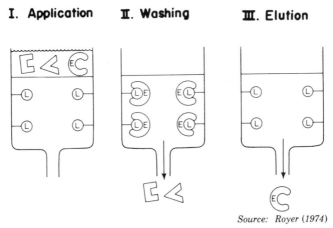

Source: Royer (1974)

FIG. 20.31. SCHEMATIC REPRESENTATION OF AFFINITY CHROMATOGRAPHY

The mixture of proteins containing the desired enzyme (EC is applied to the column that supports the ligand (L). Proteins (E <) which have no affinity for the ligand are washed through with the buffer used for application. The desired enzyme is re-covered by washing the column with a buffer of a different acidity.

The Matrix

A suitable system for affinity chromatography must meet several requirements. The matrix to which the ligands are attached must be capable of mild chemical modification without undergoing major physical changes, be practically free of residues capable of nonspecific interactions, have a loose lattice structure of sufficient porosity to allow unimpeded access of macromolecules to bound ligands, and be hydrophilic enough to permit interaction between two phases.

According to Royer (1974), criteria for the ideal matrix material are: adequate chemical functionality for reaction with various groups on proteins and ligands, low cost, resistance to microbial attack, dimensional stability, durability, hydrophilicity (some proteins denature at hydrocarbon-water interfaces), capacity to regenerate, high capacity for enzyme or ligand, and accessibility to solvent.

Many materials have been used as matrices. The organic materials include carbohydrates (cellulose, agarose, starch), vinyl polymers (polyacrylamide and others), polymers of amino acids and their derivatives, amine-containing resins, Nylon, Dacron, etc. (Royer 1974). The inorganic materials include glass (porous and solid beads), metals (nickel screens), colloidal silica, and aluminas. Conjugates of silicas or porous glass with organic polymers are an example of an inorganic-organic combination.

Beaded agarose, polyacrylamide and glass are used most widely because they meet the above requirements, can be used with certain nonaqueous systems, and are available commercially. The disadvantages of agarose are high cost and susceptibility to microbial attack. The fibrous, heterogeneous cellulose has a low porosity, poor flow rate, and has a significant proportion of carboxy-residues. Polystyrene and related polymers are unsuitable because they are highly lipophilic and because they adsorb strongly and nonspecifically many proteins (Guilford 1973).

A comparison of properties for several affinity matrices is given in Table 20.7 (according to Gelb 1973).

The ligand must interact specifically and reversibly with the molecule to be purified. The ligand must be suitable for coupling to a matrix with minimum of modification of the ligand structure which is essential for binding. Interactions which involve dissociation constants above 10^{-3} mol 1^{-1} are likely to be too weak. Almost all affinity chromatography systems involve a covalent bond between ligand and matrix. Before use, the ligand-carrier complex must be washed with several cycles of buffers of as wide a range of pH and ionic strength as the stability of the covalent bonds will allow. Detergents are sometimes used. Subsequent adequate washing of gels is essential to remove noncovalently bound ligand. (For details, see Anon. 1974.)

The amount of ligand on the carrier represents the maximum theoretical binding capacity. Since the matrix does not have an ideal porosity, only a fraction of the coupled molecules is accessible for binding. In addition, an adsorbed macromolecule may mask adjacent ligands and make them inaccessible. Three variations for fixing ligands to matrices are available:

TABLE 20.7

AFFINITY CHROMATOGRAPHY MATRICES

Matrix	Nonspecific Adsorption	Binding Capacity	Chemical Stability	Mechanical & Hydro-Dynamic Properties
Sephadex, PAA	Excellent	Low	Excellent	Excellent
Cellulose	Excellent	Low	Excellent	Fair
Silica	Poor	Moderate	Excellent	Excellent
Sepharose	Excellent	High	Excellent	Excellent

Source: Gelb (1973).

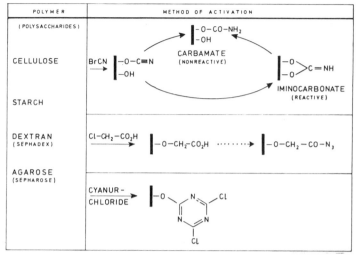

POLYMER	METHOD OF ACTIVATION

Source: Nelboeck and Jaworek (1975)

FIG. 20.32. COVALENT IMMOBILIZING OF ENZYMES THROUGH FUNCTIONALIZATION OF A POLYMER MATRIX (SCHEME I)

(a) direct linkage of ligand to carrier, (b) extension of ligand by an "arm" which is subsequently attached to the matrix, and (c) derivatization of the carrier with a spacer, the free end of which is functionally suitable for attaching the ligand (Guilford 1973).

Methods of covalent linking involving activation of polysaccharide matrices are shown in Fig. 20.32 and 20.33 (Nelboeck and Jaworek 1975). Activation of polysaccharides with cyanogen bromide at pH 11 is one of the most widely used methods. The active form, presumably an imino-carbonate, bound to Sepharose is available commercially.

Although the activation of Sepharose with CNBr is easy to perform, CNBr is at best an unpleasant material with which to work, and if handled improperly, may even be dangerous. In many instances, it may be preferable to use CNBr-activated Sepharose 4B, which has been activated, stabilized with lactose and dextran, and lyophilized. The following general method (from Gelb 1973) may be used as a guide for the coupling of ligands containing primary amino groups to CNBr-activated Sepharose 4B:

One gram of gel containing lactose and dextran is swollen for 10 min in 100 ml of 1 mM HCl, placed on a glass filter and aspirated almost to dryness. An additional 100 ml of 1 mM HCl is used to wash the gel and it is again aspirated to a semidry cake. Approximately 10 mg of the substance to be coupled is dissolved in 5 ml of 0.1 M NaHCO$_3$ containing 0.5 M NaCl. The coupling buffer containing the substance to be immobilized is added to the previously prepared gel and allowed to react for 2 hr at room temperature or overnight at 4°C. During this incubation period, the gel slurry should be agitated periodically. The use of magnetic stirring devices should be avoided, as this may cause frag-

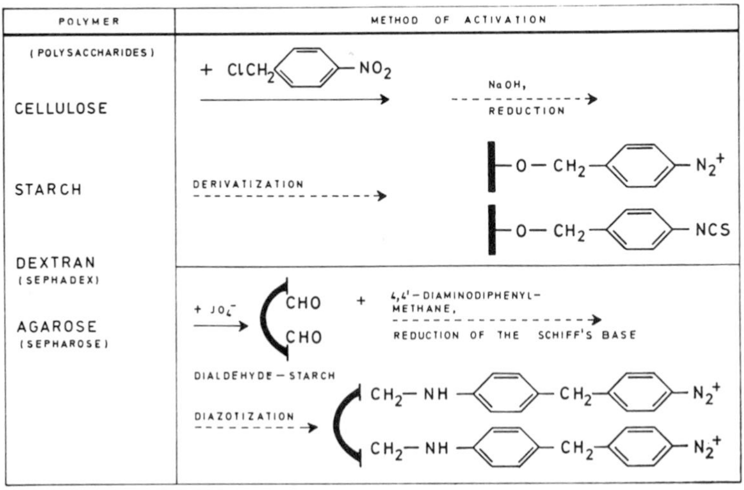

POLYMER	METHOD OF ACTIVATION
(POLYSACCHARIDES)	$+ \; ClCH_2$⟨⟩NO_2 $\xrightarrow{\quad}$ $\xrightarrow[\text{REDUCTION}]{NaOH,}$
CELLULOSE	
STARCH	DERIVATIZATION $\dashrightarrow$ $\vdash O-CH_2$⟨⟩N_2^+ $\vdash O-CH_2$⟨⟩NCS
DEXTRAN (SEPHADEX)	$+ \; IO_4^-$ ⟨CHO CHO⟩ $+$ $4,4'-$DIAMINODIPHENYL-METHANE, $\dashrightarrow$ REDUCTION OF THE SCHIFF'S BASE
AGAROSE (SEPHAROSE)	DIALDEHYDE – STARCH CH_2-NH⟨⟩CH_2⟨⟩N_2^+ DIAZOTIZATION $\dashrightarrow$ CH_2-NH⟨⟩CH_2⟨⟩N_2^+

Source: Nelboeck and Jaworek (1975)

FIG. 20.33. COVALENT IMMOBILIZING OF ENZYMES THROUGH FUNCTIONALIZATION OF A POLYMER MATRIX (SCHEME II)

mentation of the gel beads. Unbound material is removed by washing with 0.1 M $NaHCO_3$ containing 0.5 M NaCl and any remaining active groups are reacted with 1 M ethanolamine at pH 8 for 1–2 hr. Alternate washing with 0.1 M acetate buffer, 1 M NaCl, pH 4 and 0.1 M borate buffer, 1 M NaCl, pH 8 serves to remove any noncovalently adsorbed protein. Usually, three cycles with these washing buffers are sufficient.

Sepharose reacts primarily with primary amino groups of proteins. Another common matrix activation method involves azide-activated carboxymethyl cellulose. This results in binding of NH_2—, —SH, —OH, and histidyl residues. Activation of polysaccharides via triazine derivatives yields alkylating links. Replacing halogen atoms by other functional links significantly enhances variability of the method (Fig. 20.32).

Aldehyde-starch is a commercially available substrate for derivatization. Transformation with methylenedianilin into a Schiff's base and its reduction yields a product which can undergo diaziotization. The diazonium groups are highly reactive to amino, tyrosine, and histidine (and less reactive to —SH and tryptophan) residues in proteins (Fig. 20.33).

Polyacrylamide matrices are based on a cross-linked copolymer of acrylamide N,N'-methylenebisacrylamide, which comprises a hydrocarbon framework carrying carboxamide side chains which are resistant to hydrolysis at pH 1–10 (Guilford 1973). It is available commercially in spherical beads of various sizes and porosities. Several methods are available for modifying the structure with groups suitable for covalent bonding and preparation of biospecific adsorbents. Methods involving azide formation result in gel shrinkage and reduced porosity. This limits

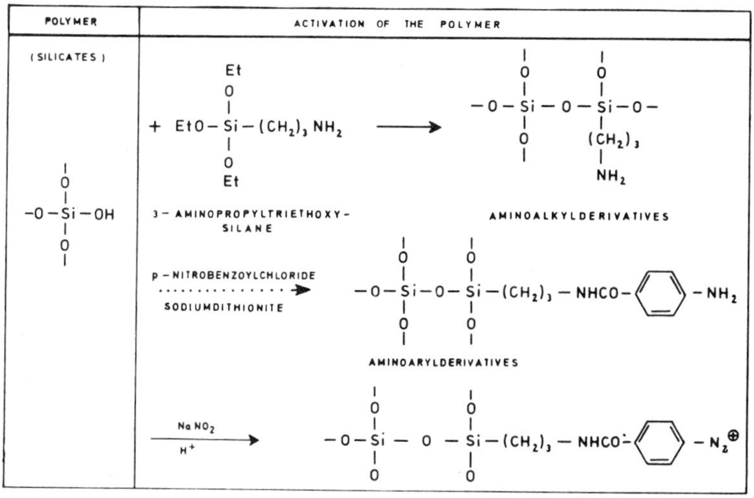

POLYMER	ACTIVATION OF THE POLYMER
(SILICATES) $-O-\overset{\displaystyle |}{\underset{\displaystyle |}{Si}}-OH$	$+\ EtO-\overset{\displaystyle Et}{\underset{\displaystyle |}{\overset{\displaystyle |}{\underset{\displaystyle |}{O}}{Si}}}-(CH_2)_3\,NH_2\ \longrightarrow\ $ $-O-\overset{\displaystyle O}{\underset{\displaystyle |}{\overset{\displaystyle |}{Si}}}-O-\overset{\displaystyle O}{\underset{\displaystyle (CH_2)_3}{\overset{\displaystyle |}{Si}}}-O-$ NH_2 3-AMINOPROPYLTRIETHOXY-SILANE AMINOALKYLDERIVATIVES p-NITROBENZOYLCHLORIDE ············▸ SODIUMDITHIONITE $-O-\overset{O}{\underset{O}{Si}}-O-\overset{O}{\underset{O}{Si}}-(CH_2)_3-NHCO-\!\!\bigcirc\!\!-NH_2$ AMINOARYLDERIVATIVES $\xrightarrow[H^+]{Na\,NO_2}$ $-O-\overset{O}{\underset{O}{Si}}-O-\overset{O}{\underset{O}{Si}}-(CH_2)_3-NHCO-\!\!\bigcirc\!\!-N_2^{\oplus}$

Source: Nelboeck and Jaworek (1975)

FIG. 20.34. COVALENT IMMOBILIZING OF ENZYMES THROUGH FUNCTIONAL-IZATION OF A POLYMER MATRIX (SCHEME III)

usefulness of polyacrylamide for affinity chromatography of large proteins. Polyacrylamide can also be made functional by treatment with glutaraldehyde to which ligands are attached as Schiff's bases.

Porous glass is an attractive support material for several reasons. Glass is resistant to changes in pH and solvent, mechanical damage, and microbial attack; it is rigid, regenerable, and provides a hydrophilic environment. On the other hand, porous glass is expensive, has less binding capacity than agarose, and erodes at basic pH. Coating with zirconium oxides improves stability but is costly. Activation of the porous glass surface starts with silanization with γ-amino-propyltriethoxysilane (Fig. 20.34). The amino groups of the resulting alkylamine glass may be reacted directly with the carboxyl groups of proteins (by using a carbodiimide) or ligands (by using dicyclohexyl carbodiimide). An arylamine derivative may be prepared from the amino alkyl derivative by p-nitrobenzyolation followed by reduction. The derivative can be diazotized and reacted with proteins or ligands through phenolic, imidazole, or amino groups.

Use of nylon involves controlled acid hydrolysis of the polycaprolactam chain to yield an "etched" product (Fig. 20.35, from Nelboeck and Jaworek 1975). Both the free carboxyl and amino groups can be derivatized. Adsorption involves formation of a Schiff base between the amino groups of the matrix and primary amino groups of the protein.

Another group of matrices involves copolymerization of an inactive monomer with a functionalized monomer capable of copolymerization. This approach is the basis of numerous matrices, including some which involve covalent binding of enzymes to copolymers with anhydride groups.

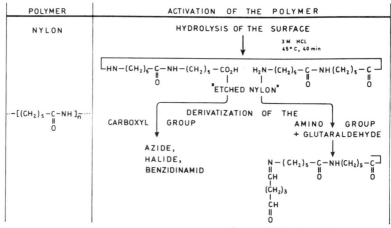

Source: Nelboeck and Jaworek (1975)

FIG. 20.35.   COVALENT IMMOBILIZING OF ENZYMES THROUGH FUNCTIONALIZA-
TION OF A POLYMER MATRIX (SCHEME IV)

The functional monomers can be maleic acid; the inactive comonomer,
ethylene. Functionalized maleic acid can be in form of anhydride, azide,
acid chloride, hydrazide, etc. Ethylene can be replaced by acrylamide or
vinyl ether; the principle of covalent immobilization of enzymes through
functionalization of monomers is summarized in Fig. 20.36.

Matrix Modification and Spacers

Because of steric considerations, it may be necessary to separate a small
ligand from the backbone of the matrix by a "spacer" to enhance binding
capacity. Effect of spacer groups on the accessibility of binding sites is
illustrated in Fig. 20.37 (from Gelb 1973). In addition, some ligands have
no NH_2 group for direct coupling. Activated agarose and polyacrylamide
can be modified with bifunctional reagents of the general structure NH_2
—R—X, where X is a functional group and R, chemically inert, determines
rigidity, hydrophilicity, and maximum length. The disadvantage of some
of the derivatives is lack of selectivity of the end product (Guilford
1973).

Immobilized Enzymes

The development of techniques of preparing immobilized enzymes which
retain catalytic activity along with greatly improved stability has resulted
in extensive use of bound enzymes (in place of free enzymes) in assay
methods (Anon. 1974).

According to recommendations of the "Meeting on Enzyme Engineering"
(Henniker, N. H., U.S.A., August 1971—from Nelboeck and Jaworek 1975)
immobilized enzymes can be classified as follows:

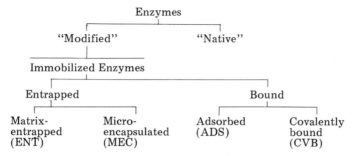

Whereas entrapped enzymes are isolated from large molecules which cannot penetrate into the matrix, bound enzymes may be exposed to molecules of all sizes. Immobilization of enzymes by affinity chromatography involves covalent binding. "Enzymes which are immobilized by covalent linkage to an insoluble carrier matrix have a number of unique advantages over the free enzymes. These are due to increased stability of the enzyme and to the ease with which it can be removed from the reaction mixture. Stability is increased both to thermal denaturation and, in the case of proteases, to autodigestion; the bound enzyme can thus be stored for long periods without loss of activity. As the enzyme can be removed from the

COMONOMER (INACTIVE)	COMONOMER (FUNCTIONALIZED)	CROSSLINK	ACTIVATED POLYMER
$CH_2 = CH_2$	$HC - CO_2H$ $HC - CO_2H$	$\begin{array}{l}COCH = CH_2 \\ NH \\ CH_2 \\ NH \\ COCH = CH_2\end{array}$	$CH - CON_3$ $CH - CON_3$
$CH_2 = CHCONH_2$			
$CH_2 = C-CONH_2$ CH_3	$HC - CO$ $HC - CO$ $\searrow O$	$\begin{array}{l}CH = CH_2 \\ O \\ CH_2 \\ CH_2 \\ O \\ CH = CH_2\end{array}$	$CH - CO$ $CH - CO$ $\searrow O$
$CH = CH_2$ O CH_3			CH_2
$CH = CH_2$ O CH_2CH_3	H_2C $H_3C-C-CO$ $\searrow O$ $H_3C-C-CO$ H_2C	$\begin{array}{l}COC = CH_2 \\ CH_3 \\ O \\ CH_2 \\ CH_2 \\ O \;\; CH_3 \\ COC = CH_2\end{array}$	$H_3C-C-CO$ $\searrow O$ $H_3C-C-CO$ CH_2

Source: Nelboeck and Jaworek (1975)

FIG. 20.36. COVALENT IMMOBILIZING ENZYMES THROUGH FUNCTIONALIZATION OF MONOMERS

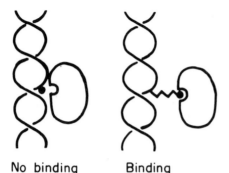

FIG. 20.37. EFFECT OF SPACER GROUPS
ON THE ACCESSIBILITY OF BINDING
SITES

(Left) Matrix with no spacer, i.e., ligand
directly coupled to CNBr-activated Sepharose 4B. (Right) Matrix with spacer, i.e.,
ligand coupled to either AH- or CH-Seph-
arose 4B.

No binding   Binding

reaction mixture quickly by filtration or centrifugation, the reaction can
be controlled without the addition of inhibitors. The reaction product is
obtained without contamination, and the recovered enzyme can be reused
many times. Since the immobilized enzyme can be packed in a column and
the reaction can be controlled without the loss of enzyme's activity, continuous flow reactions, either degrading or synthetic, can be carried out.
Enzyme reactions of this type are not possible with conventional soluble
enzymes" (Anon. 1974).

When an enzyme is immobilized, several changes in the enzyme's apparent behavior may occur. The optimal pH may shift, depending on the
carrier. There is, generally, an increase in K_m; the increase is related to
the change in the substrate and/or carrier, diffusion effects, and (possibly)
tertiary changes in enzyme configuration. In many cases, the rate of
thermal inactivation and denaturation of an immobilized enzyme is less
than that of the free enzyme. This does not necessarily imply superior
operational stability, as such stability depends also on carrier durability,
organic inhibitors, metal inhibitors, etc. (Weetal 1974).

Immobilized enzymes are generally less active than the corresponding
native enzymes (Orth and Brummer 1972). Enzymes, which act on low
molecular weight substrates, have after binding a residual activity of
40–80%; bound hydrolases retain only 5–40% of their activity for high molecular weight substrates. This partial loss of activity is offset by repeated
use. Some of the reasons for modified activity include steric hindrance
(resulting from attachment of proteases, amylases, and nucleases to several
points on the support), electrostatic interactions between the support and
similarly or oppositely charged enzyme substrates, and diffusion effects.

One of the novel and interesting uses of immobilized enzymes is for assays
in which the enzyme is the active element of an electro-chemical probe or
sensor. Updike and Hicks in 1967 described the "enzyme electrode" which
can be used to determine the glucose concentration in blood (even *in vivo).*

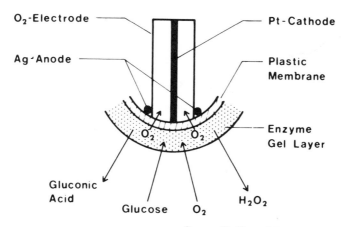

Source: *Updike and Hicks (1967)*

FIG. 20.38. PRINCIPLE OF THE ENZYME ELECTRODE

The electrode (Fig. 20.38) was prepared by polymerizing an enzyme-containing gelatinous membrane over a polarographic oxygen electrode. When the electrode is placed in contact with a biological solution or tissue, glucose and oxygen diffuse into the gel layer of the immobilized enzyme. The diffusion rate of oxygen through the membrane is lowered in the presence of glucose and glucose oxidase by the enzymatic oxidation of glucose to gluconic acid. The depletion of oxygen is measured by the oxygen electrode. At glucose concentrations below K_m for the immobilized enzyme, there is a linear relationship between glucose concentration and the measured oxygen depletion rate.

Since the pioneering studies of Updike and Hicks, many scientists have developed and employed numerous modifications of the enzyme electrode for intermittent and continuous assays of organic and inorganic compounds. For additional details, see Goldstein and Katchalski (1968), Katchalski *et al.* (1971), Guilbault (1972), Orth and Brummer (1972), Wingard (1972), Weetal (1972, 1974), Gough and Andrade (1973), Olson and Richardson (1974), Bergmeyer and Michal (1974), Moody and Thomas (1975), Nagy and Pungor (1975), and Falb and Grade (1976). Affinity chromatography is reviewed extensively in the book edited by Stark (1971); immobilized enzymes are the subject of the book by Zaborsky (1973).

USEFUL CHROMATOGRAPHIC DIAGRAMS

The following schematic diagrams shown in Fig. 20.39 through Fig. 20.47 demonstrate the principles of preparative separation in biochemistry for ultrafiltration, column electrofocusing, flat-bed electrofocusing, ion exchange chromatography, adsorption chromatography, affinity chromatography, gel filtration, electrophoresis and isotachophoresis. These

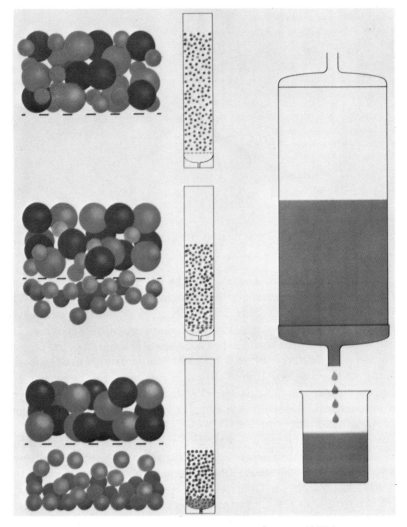

Courtesy of LKB Instruments

FIG. 20.39.   ULTRAFILTRATION

A membrane acts as a molecular filter in ultrafiltration.   Under an applied pressure larger molecules cannot penetrate the membrane and are thus retarded while smaller molecules and solvent pass through.   In this way molecules can be separated and concentrated.

schematics were developed by LKB Instruments, Rockville, Maryland and are reprinted with their permission.   They are included here in the belief that they constitute a useful summary for the reader.

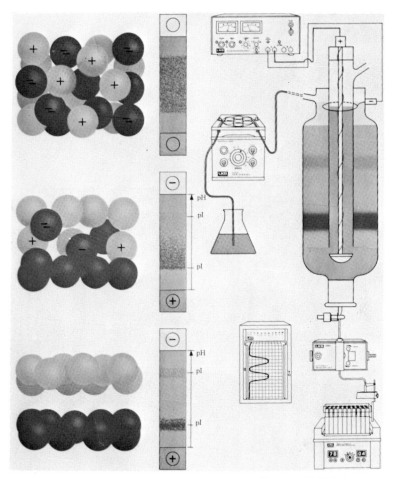

Courtesy of LKB Instruments

FIG. 20.40. COLUMN ELECTROFOCUSING

In electrofocusing the sample components are separated in a pH-gradient according to differences in isoelectric point (pI). Ampholine® carrier ampholytes are used to create the pH-gradient and when an electric field is applied the pH-gradient is rapidly formed. The sample is applied either as a zone or throughout the whole column. Charged sample components migrate toward the electrode of opposite charge. The net charge is continuously reduced during migration in the pH-gradient. At a pH-value equal to the pI of the sample, the net charge will be zero and migration stops. The sample components are focused at their respective pI-values forming concentrated and narrow zones.

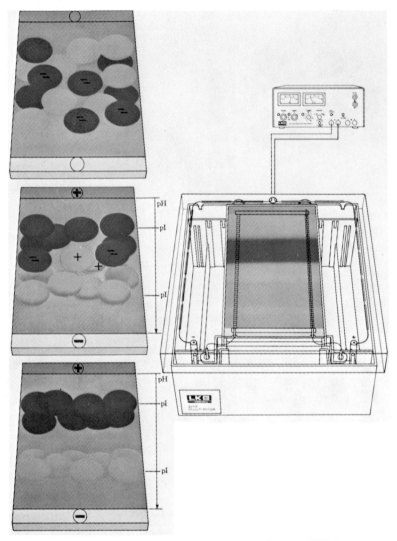

FIG. 20.41.   FLAT-BED ELECTROFOCUSING

Preparative electrofocusing can also be performed in a flat-bed of granulated gel. In this case the sample can also be applied as a zone or throughout the whole gel bed.   After the separation is completed, the gel bed is divided into sections by a fractionating grid and the sample components are eluted from the granulated gel.

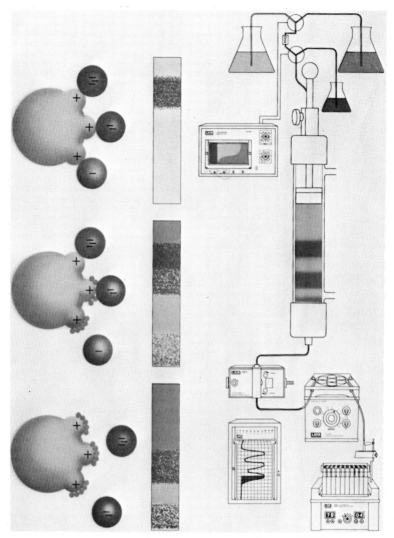

Courtesy of LKB Instruments

FIG. 20.42.   ION EXCHANGE CHROMATOGRAPHY

Ion exchange chromatography is a separation method in which components with different net charges are separated when a gradient of increasing ionic strength is used as the eluant.   The gel matrix can carry either positive or negative groups. In the case of a positively charged gel matrix sample components with negative net charges will be adsorbed.   During desorption the negatively charged sample components are exchanged by the negative ions from the salt gradient.   Each sample component is then desorbed at a specific ionic strength and continuously eluted from the column.

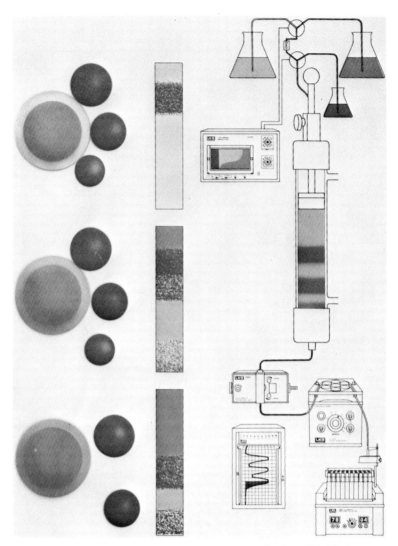

Courtesy of LKB Instruments

FIG. 20.43. ADSORPTION CHROMATOGRAPHY

In adsorption chromatography the sample components are separated based on their different polarities. The sample components are adsorbed to the granulated gel matrix by noncovalent bonds (such as hydrogen bonding), nonpolar interaction and Van der Waals' forces. In the desorption step a gradient, containing a nonpolar solvent, will exchange and release the sample components one at a time.

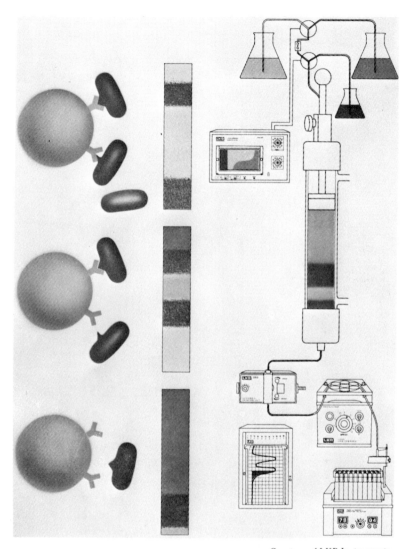

Courtesy of LKB Instruments

FIG. 20.44. AFFINITY CHROMATOGRAPHY

In affinity chromatography sample components are isolated and separated based on their different biological specificity, as in antigen-antibody or enzyme-inhibitor systems. A ligand with specific affinity for the sample components of interest is covalently coupled to the granulated gel matrix. When the sample is applied, only the components with an affinity for the ligand are adsorbed to the gel matrix while the other components are washed away. Desorption is obtained by a controlled altering of the pH and/or salt concentration of the gradient.

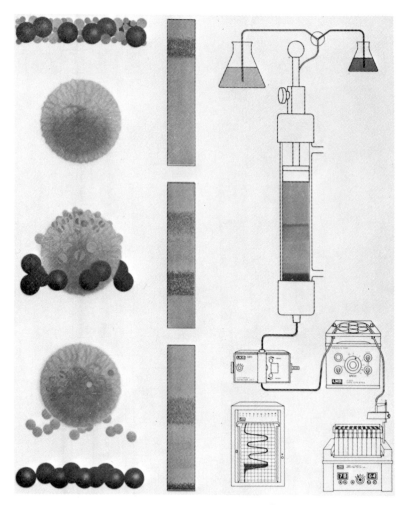

Courtesy of LKB Instruments

FIG. 20.45.   GEL FILTRATION

In gel filtration the gel acts as a molecular sieve separating molecules with differences in molecular size and weight.   The Ultrogel® matrix contains numerous porous beads with an eluant in between.   If sample mixture is applied at the top of
the column, the large molecules in the sample will not be able to enter the pores
in the beads but will pass between them and thus be eluted first.   Smaller molecules that have access to the pores are retarded in the gel to a certain extent and
will therefore be eluted after the large molecules in order of decreasing molecular
weight and size.

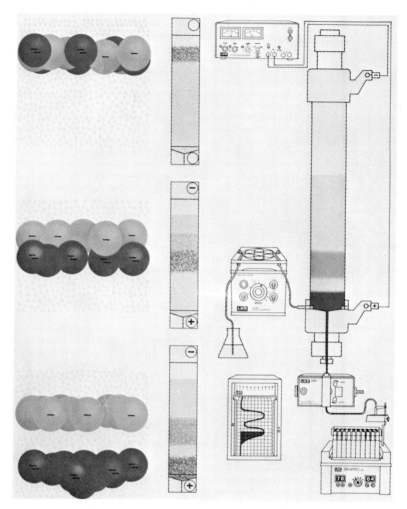

FIG. 20.46.   ELECTROPHORESIS

In conventional electrophoresis the sample components are separated based on their differences in net charge, size and shape.   The separation takes place at a constant pH and ionic strength.   Polyacrylamide gel, cellulose or granulated gels are used as stabilizing media.   The sample is applied as a narrow zone on top of the stabilizing gel and when the electric field is applied, the sample components will migrate into the gel.   Separation in the stabilizing media takes place because of different mobilities of the sample components.   The separated zones migrate one after the other out of the gel into the funnel-shaped elution chamber where they are then flushed out by a continuous stream of elution buffer.

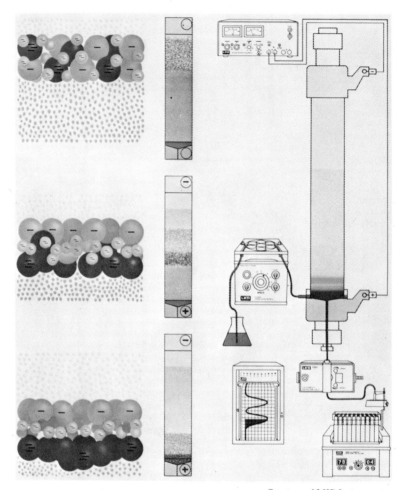

FIG. 20.47.   ISOTACHOPHORESIS

Isotachophoresis is an electrophoretic technique where sample components are separated based on differences in their net mobility. The separation is performed in a polyacrylamide gel utilizing a discontinuous buffer system. The sample components and the "spacer ions" are dissolved in the terminating electrolyte and leading electrolyte is initially applied throughout the polyacrylamide gel. When the electric field is applied separation of the samples takes place between the leading and terminating electrolytes during migration. When separation is completed, the separated zones and the spacer ions migrate with the same velocity in immediate contact with each other into the funnel-shaped elution chamber where they are then flushed out by a continuous stream of elution buffer. Isotachophoresis gives a high resolving power because of its concentrating effect. The zone boundaries are actively sharpened during the experiment, which prevents diffusional broadening.

BIBLIOGRAPHY

ADAMS, B. A., and HOLMES, E. L. 1935. Adsorptive properties of synthetic resins. J. Soc. Chem. Ind. *54*, 1–65.

AITZETMULLER, K. 1975. The liquid chromatography of lipids. J. Chromatogr. *113*, 231–266.

ANON. 1974. Affinity Chromatography, Principles and Methods. Pharmacia Fine Chemicals. A.B. Uppsala, Sweden.

BERG, E. W. 1963. Physical and Chemical Methods of Separation. McGraw-Hill Book Co., New York.

BERGMEYER, H. V., and MICHAL, G. 1974. Applications of enzymes in clinical biochemistry. *In* Industrial Aspects of Biochemistry. B. Spences (Editor). FEBS, 187–211.

CANTOW, M. J. R. 1967. Polymer Fractionation. Academic Press, New York.

CAVINS, J. F., and INGLETT, G. E. 1974. High resolution liquid chromatography of Vitamin E isomers. Cereal Chem. *51*, No. 5, 605–609.

CHILCOTE, D. D., MROCHECK, J. G., and KATZ, S. 1974. Recent advances in the use of liquid chromatography for carbohydrate analyses. Cereal Sci. Today *19* No. 9, 416–420.

CONRAD, E. C. 1975. Applying high pressure liquid chromatography to the rapid analysis of food nutrients. Food Prod. Rev. Sept., 97–102.

CUATRECASAS, P., and ANFINSEN, C. B. 1971. Affinity chromatography. Ann. Rev. Biochem. *40*, 259–278.

CUATRECASAS, P., WILCHEK, M., and ANFINSEN, C. B. 1968. Selective enzyme purification by affinity chromatography. Proc. Nat. Acad. Sci. (U.S.A.) *61*, 636–643.

FALB, R. D., and GRADE, G. A. 1976. Immobilized proteins and peptides. J. Macromol. Sci. Chem. *A10*, 197–221.

GELB, W. G. 1973. Affinity chromatography for separation of biological materials. Amer. Lab. Oct. 61–67.

GOLDSTEIN, L., and KATCHALSKI, E. 1968. Use of water-insoluble enzyme derivatives in biochemical analysis and separation. Z. Anal. Chem. *243*, 375–396.

GOUGH, D. A., and ANDRADE, J. D. 1973. Enzyme electrodes. Science *180*, 380–384.

GOULDING, R. W. 1975. Liquid chromatography of sugars and related polyhydric alcohols on cation exchangers. J. Chromatogr. *103*, 229–240.

GUILBAULT, G. G. 1972. Analytical uses of immobilized enzymes. Biotechnol. Bioeng. Symp. *3*, 361–376.

GUILFORD, H. 1973. Chemical aspects of affinity chromatography. Chem. Soc. Rev. (London) *2*, 249–270.

HEFTMAN, E. 1967. Chromatography. Reinhold Publishing Co., New York.

HELFERICH, F. 1961. Ligand exchange-a novel separation technique. Nature 198, 1001–1002.

KATCHALSKI, E., SILMAN, I., and GOLDMAN, R. 1971. Effect of the microenvironment on the mode of action of immobilized enzymes. Advan. Enzymol. *34*, 445–536.

LEE, S. C., and CHANG, S. S. 1975. Use of high pressure liquid chromatography for the fractionation of less volatile flavor compounds. J. Agri. Food Chem. *23*, No. 2, 337–339.

MOODY, G. J., and THOMAS, J. D. R. 1975. The analytical role of ion-selective and gas-sensing electrodes in enzymology. A review. Analyst *100*, 609–619.

MOORE, J. C. 1964. Gel permeation chromatography. I. A new method for molecular weight determination of high polymers. J. Polymer Sci. *A2*, 833–844.

MOYE, H. A. 1975. High speed liquid chromatography of pesticides. J. Chromatogr. Sci. *13*, 268–279.

NAGY, G., and PUNGOR, E. 1975. Enzyme electrodes, application of a voltammetric L-amino acid enzyme electrode to analysis in flowing solutions. Hung. Sci. Instr. *32*, 1–10.

NELBOECK, M., and JAWOREK, D. 1975. Immobilized enzymes and principles for their use in analytical and preparative processes. Chimia 29, 109–123. (German)

OLSON, N. F., and RICHARDSON, T. 1974. Symposium: Immobilized enzymes in food systems. Immobilized enzymes in food processing and analysis. J. Food Science *39*, 653–659.

ORTH, H. D., and BRUMMER, W. 1972. Carrier-bound biologically active substances and their applications. Angew. Chemie *11*, 249–260.
PALMER, J. K., and BRONDES, W. B. 1974. Determination of sucrose, glucose, and fructose by liquid chromatography. J. Agr. Food Chem. *22*, No. 4, 709–712.
PERSIANI, C., and CUKOR, P. 1975. Liquid chromatographic method for the determination of phthalate esters. J. Chromatogr. *109*, 413–417.
ROYER, G. P. 1974. Supports for immobilized enzymes and affinity chromatography. Chemtech. 694–670.
STAHL, G., LAUB, G., and WOLLER, R. 1974. Liquid chromatography of wines and fruit juices. 2. Preparation of samples, quantitative determination of acids and further practical applications. Z. Lebensm. Unters. Forsch *156*, 321–329.
STARK, G. R. (Editor). 1971. Biochemical Aspects of Reactions on Solid Supports. Academic Press, New York.
UPDIKE, S. J., and HICKS, G. P. 1967. The enzyme electrode. Nature *214*, 986–987.
VEENING, H. 1973. Recent developments in instrumentation for liquid chromatography. J. Chem. Ed. *50*, A529–A538.
WEETAL, H. H. 1972. Preparation, characterization and applications of enzymes immobilized on inorganic supports. *In* Immobilized Biochemicals and Affinity Chromatography. R. B. Dunlap, (Editor). Plenum Publishing Co., New York.
WEETAL, H. H. 1974. Immobilized enzymes: analytical applications. Anal. Chem. *46*, 602A–610A.
WETZEL, D. 1975. Guide to the literature of high performance liquid chromatography. Dep. Grain Sci., Kansas State Univ.
WINGARD, L. (Editor). 1972. Enzyme Engineering. Wiley-Interscience, New York.
WITTMER, D. P., NUESSLE, N. O., and HONEY, W. G., JR. 1975. Simultaneous analysis of tartrazine and its intermediates by reversed phase liquid chromatography. Anal. Chem. *47*, 1422–1423.
ZABORSKY, O. R. 1973. Immobilized Enzymes. CRC Press, Cleveland.

CHAPTER 21

Paper and Thin-Layer Chromatography

PAPER CHROMATOGRAPHY

Column chromatography was used only occasionally until 1931 when it was found that *cis-trans* isomers of organic compounds could be separated. In fact, Tswett, who died in 1920, never lived to see the results of his efforts. From 1931 until 1944 there was a rapid development in theory and applications, but no one really challenged the basic idea of using a column.

If you have ever done a column chromatographic separation, you know that sometimes it can take a long time. If you pack the column loosely so the eluent will come through fast, then the packing is subject to cracking and the nonuniform packing causes erratic results, but the packing can be removed rather easily from the column. If you pack the column tightly to get a better separation, then you have to apply a vacuum to pull the eluent through the column, and it is difficult to remove the packing from the column.

Why remove the packing? Suppose you wanted to separate 10 components. Some would come out fast, but many would still be in the column. Now you could keep eluting them from the column, but this dilutes the sample which then requires an additional concentration step. The other alternative is to push the packing out of the column after all of the components are separated, and cut up the packing into sections, extracting the sample from each section.

Most of these problems could be solved if the column could be disposed of. What about paper? Paper is almost pure cellulose so the inert phase is present. Paper normally has 2-5% moisture adsorbed to its surface (can be as high as 20%), so a stationary phase is present. A sample is placed as a spot on the paper and a mobile phase passed over it. Now separations that took hours and days could be done in minutes and hours because the paper was porous enough to allow the eluent to pass through freely, yet uniform enough to provide a good separation. Furthermore, once the sample is separated, the spots can be cut out or washed off of the paper which is considerably easier than removing the packing in column chromatography.

The principles of separation in paper chromatography are similar to column chromatography discussed previously (Consden *et al.* 1944). The main difference is that a sheet of paper is used for the inert phase.

The exact mechanism of the separation is not known. One theory says it is entirely an adsorption process, and that the water in the paper has no effect. Another theory is that the water in the paper acts as a stationary phase and that a partition process takes place. In view of the structure of the paper, the amount of water actually present, the shape, and concentration gradient of the spots, probably a combination of both mechanisms is actually taking place.

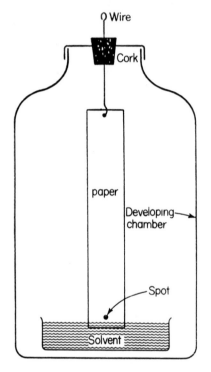

FIG. 21.1. ASCENDING CHROMATOGRAPHY

Ascending Chromatography

The sample is spotted $1/2$ in. from the end of a paper strip. The paper is suspended vertically, in a chromatographic chamber, with the spotted end down, and allowed to dip into about $1/4$ in. of a solvent. The solvent rises up the paper by capillary action, effecting a separation of the components as it ascends (see Fig. 21.1). The disadvantage is that the solvent will rise only 8–9 in.   The main advantage is the very simple equipment required.

Remember that paper normally has 2–5% water in it, but it can adsorb as high as 20% water.   What happens if a 50% water solution passes over the paper?   The paper picks up water until it reaches equilibrium.   The net result is that the first solvent to pass over the sample spot will be considerably different in composition than the solvent passing over it later on.   The R_f value changes radically and separation may not be possible.   However, if the paper is brought to equilibrium with the developing solvent at the start, a good separation is possible.

Descending Chromatography

The sample is spotted 2–3 in. from the end of a paper strip.   The strip is suspended vertically in a chromatographic chamber with the spotted end up.   The 2–3 in. wick is then bent over and placed in a solvent tank.   After the solvent rises up the wick, it descends across the spot and down the strip.   An antisiphon bar is placed at the top of the strip to prevent the solvent from simply siphoning out of the

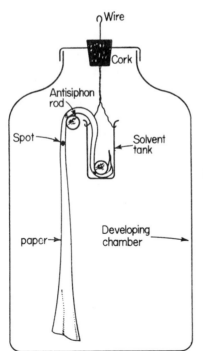

FIG. 21.2. DESCENDING
CHROMATOGRAPHY

tank (see Fig. 21.2). This technique permits a longer strip to be used than with ascending chromatography, and allows a corresponding increase in resolution.

Reversed Phase Chromatography

In this technique, the stationary phase becomes the nonpolar phase, and the mobile phase the polar. Dry the paper thoroughly and impregnate it with a nonpolar organic material. Use water or another polar solvent for the mobile phase. This technique is particularly suitable for separating substances like steroids, long chain fatty acids, or chlorinated insecticides. (See also Chapter 20.)

Two-Dimensional Chromatography

Two solvents are employed, one after the other. The solution is spotted in the corner of a square or rectangular sheet of paper. One solvent is used to develop the chromatogram in one direction. The paper is dried, turned 90°, and a second solvent is used. An example of the high degree of separation that can be obtained is shown in Fig. 21.3. By changing the characteristics of the solvent pair, a variety of compounds otherwise unseparable can be resolved.

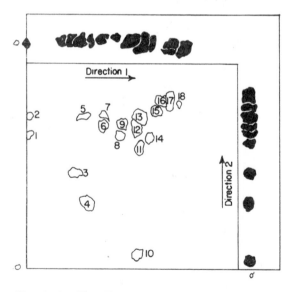

FIG. 21.3.   TWO-DIMENSIONAL CHROMATOGRAPHY

Horizontal Chromatography

This is also known as the _Rutter Technique._ The chamber may be a petri dish or a desiccator. The spot is placed in the center of a filter paper disc. A strip about $^3/_8$ in. wide is torn from the perimeter to the center of the paper disc, and this is bent down into the solvent. By means of this wick, the solvent is fed to the paper, and the resolved components are in the form of circles rather than spots.

General Methods in Paper Chromatography

Paper.—The nature of the paper used for chromatography is most important. Ideally, it contains pure cellulose and no lignins, copper, or other impurities. Examination of cellulose under the electron microscope reveals the crystallinity of the macromolecules forming closely attached fiber bundles, held together by hydrogen bonding. A small percentage of water is tightly bound. Other water molecules are adsorbed by the cellulose and fill interspaces of the fibers. It is not certain that this water participates in the partition process. The fact that the interspaces can be filled readily with a nonpolar solvent may exclude the adsorbed water as a factor in the partition process.

Chromatographic paper is usually available in sheets 18 × 22 in. or in strips 1 in. × several hundred feet. The long direction is the machine direction and should be used for solvent development or first direction for two dimensional work. It is important that the sample spot be as small, in diameter, as possible to get good fast separations. When paper is made and pulled through the mill, the fibers tend to line up in the machine direction. A solvent moving with these fibers meets little resistance with the result that the first direction separations are fast and there is little time for spreading due to diffusion. This means that the spots are still rather small when the second direction development starts. However, if you go against the machine direction with the first solvent, greater flow resistance is encountered and the spots get large. This makes it difficult to get complete separation in the second direction. Chromatographic paper contains organic impurities (lignin), and caution should be used when interpreting infrared or ultraviolet absorption of eluted spots of "unknown" compounds.

Preparation of the sample.—Conventional extraction procedures and evaporation techniques are applied. Depending on the nature of the compounds, different cleanup procedures are recommended: ion-exchange for amino acids and organic acids, pyridine solution for sugars. Since the final volume of the solution must be small, evaporations must be carried out *in vacuo*, or by lyophilization. Sometimes inorganic ions (Na^+, Mg^{2+}) interfere in chromatography and the solutions must be desalted. Usually commercial (Ion Exchange, Bio-Gel) desalters are used for this purpose.

TABLE 21.1

CHARACTERISTICS OF WHATMAN PAPERS

Fast	Medium	Slow
No. 4	No. 1	No. 2
No. 54	No. 7	No. 20
No. 540		

Spotting.—The principle is to spot a small amount of the solution by successive applications of small volumes of about 1–2 microliters. During the solvent development some diffusion occurs, and the original spot becomes larger during the process. Spots can be kept small by applying heat or a draft of warm air, as from a hair drier, while spotting. Micropipets are generally available and will empty when touched to the paper. Other pipets require a syringe control. The simplest qualitative method is to use a very small loop of platinum wire sealed into a glass rod.

Chromatographic Chamber.—The simplest arrangement can be made for the ascending technique. Here the upper edge of the paper may be fastened with a paper clip to a support and the bottom edge just immersed in the developing solvent. Test tubes, aquaria, biological specimen jars, pickle or large reagent jars are excellent chambers. Chrome-plated, insulated, explosion-proof commercial equipment is available for the more advanced researcher.

Solvents.—By knowing the partition coefficient of the solute in two partially miscible solvents (e.g. phenol-water), one can predict the movement of the solute or suggest a better pair of solvents. Here are some general rules.

(1)   If the substance moves too slowly, increase the solvent constituent favoring solubility of the solute. If the solute moves near the solvent front, increase the other solvent.

(2)   When using a two-phase system, the chromatographic chamber should be saturated with the stationary phase.

(3)   Some useful solvents are: Water—saturated phenol, Butanol-NH_4OH, Acetone-water, A dependable general solvent is n-butanol (40%)-acetic acid (10%)-water (50%) (shake and use the upper phase).

Ion Exchange Chromatography

This variation combines the specificity of ion exchange with the simplicity of paper chromatography. There are two basic areas. One involves impregnating the paper with ion-exchange resin, and the second involves chemical modification of the $-OH$ groups of the cellulose to produce acidic or basic groups. Both types of papers are commercially available. Table 21.2 shows some of the common types.

TABLE 21.2

SOME COMMERCIALLY AVAILABLE ION-EXCHANGE PAPERS

Type	Name	Ion Exchange Group
	Modified Cellulose	
Strong acid	Cellulose phosphate	$-OPO_3H_2$
Weak acid	Carboxymethyl cellulose	$-COOH$
	Cellulose citrate	$-COOH$
Strong base	Diethylaminoethyl cellulose (DEAE)	$-C_2H_4 \cdot NEt_2$
Weak base	Aminoethyl cellulose	$-C_2H_4 \cdot NH_2$
Weak base	Ecteola cellulose	Tert. amino
	Resin Loaded Papers	
		Resin Incorporated
Strong acid	Amberlite SA-2	IR-120
Weak acid	Amberlite WA-2	IRC-50
Strong base	Amberlite SB-2	IRA-500
Weak base	Amberlite WB-2	IR-4B

The advantages appear to be increased resolution and speed. Two-way separations can now be done in one day.

THIN-LAYER CHROMATOGRAPHY

Paper chromatography was a major advance over column chromatography, because it removed the restriction of the column. However, the paper itself served as a restriction in that cellulose was the only inert phase. Experiments with column chromatography had shown that such packings as silica gel, alumina, diatomaceous earth, and many other materials were good chromatographic materials. The problem was, how do you make a "paper" out of these? This problem was solved in 1956 by Stahl in Germany who added 2–5% of plaster of Paris ($CaSO_4$) to the silica gel, and "plastered" the silica gel to a glass plate. Now the plate could be held vertical and the normal paper chromatography techniques applied. However, the thin layer is much more uniform than paper, and sharp separations can be made much faster. Separations that require hours with paper chromatography take 15–20 min with thin-layer chromatography, hereafter referred to as TLC.

Thin-layer chromatography applications have increased at a very rapid rate since 1958. This rapid growth has been prompted by the many advantages of TLC.

(1) The method is quick; generally 20–40 min are sufficient for separations. (2) Inorganic layers are used as the sorbent; and more reactive reagents can be used to visualize spots than are used in paper chromatography. (3) The method is considerably more sensitive than paper chromatography. A lower limit of detection of 10^{-9} gm is possible in some cases. (4) A wide range of sample sizes can be handled. (5) Thin-layer chromatography can be scaled up and adapted to column preparative separations. (6) The equipment is simple and readily available. (7) No special manipulative skills are required. (8) Experimental parameters are easily varied to effect separations.

Thin-layer adsorption chromatography involves spreading a thin layer (about 250 μ thick) of a sorbent-water slurry on a glass plate. The sorbent generally contains a binder, ($CaSO_4$) or starch, to improve adhesion to the plate. After spreading, the layer is activated by drying. The activity is adjusted by the time and, more importantly, the temperature of drying.

Following activation, a drop of solution containing the mixture to be separated is applied to the sorbent near one end of the plate, and the carrier solvent is allowed to evaporate. The spotted plate is placed in a closed chamber in an upright position with the lower edge (nearest the applied spot) immersed in solvent at the bottom of the chamber. When the solvent rises through the sorbent layer by capillary action, the components of the applied spot separate into individual spots in a line perpendicular to the edge of the plate. After allowing the solvent to rise to the desired distance (10–15 cm), the plate is removed from the solvent and the solvent is allowed to evaporate from the plate.

Some separated spots are visible to the naked eye. Colorless spots can be made visible by a variety of methods. The spots may be charred by spraying the plate with a mixture of sulfuric acid and an oxidizing agent ($KMnO_4$ or $K_2Cr_2O_7$), and then heating the plate. Another method of visualizing spots involves incorporation of a phosphor with the adsorbent layer. Upon irradiation with an ultraviolet lamp, the phosphor glows and substances which absorb UV appear as black or colored spots against the fluorescing background. Finally, the plate may be sprayed with specific reagents. For each separation, pure compounds should be cochromatographed for identification of components in an analyzed mixture.

Principles

Adsorbants.—Most TLC applications depend upon the adsorptive properties of the thin-layer materials. This is true even though partition chromatography is well-established in various thin-layer applications, and ion exchange media are becoming more important. It is probably safe to say that cases of either pure adsorption or pure partitioning are rare. Rather, conditions are adjusted so that one predominates at the expense of the other. Although many materials have been tried as TLC adsorbents, the most frequently used is silica gel. Alumina follows but it is much less used than silica. Adsorption requires a high surface field strength, and activity increases with increasing lattice energy of the sorbent. Therefore, activity generally increases with hardness and melting point of the sorbent.

TABLE 21.3

ADSORBENTS FOR THIN LAYER CHROMATOGRAPHY

Solid	Used To Separate
Silica gel	Amino acids, alkaloids, sugars, fatty acids, lipids, essential oils, inorganic anions and cations, steroids, terpenoids
Alumina	Alkaloids, food dyes, phenols, steroids, vitamins, carotenes, amino acids
Kieselguhr	Sugars, oliogosaccharides, dibasic acids, fatty acids, triglycerides, amino acids, steroids
Celite	Steroids, inorganic cations
Cellulose powder	Amino acids, food dyes, alkaloids, nucleotides
Ion exchange cellulose	Nucleotides, halide ions
Starch	Amino acids
Polyamide powder	Anthocyanins, aromatic acids, antioxidants, flavonoids, proteins
Sephadex	Amino acids, proteins

Source: Stock and Rice (1967).

Adsorptive Processes.—Valence electrons at the surface of a solid are not saturated by adjacent atoms, and to a certain degree are available for bonding. Substances which are polar or polarizable undergo an electrostatic attraction to these electrons. Surface field strengths of refractory materials, as indicated by cold cathode emission of electrons, are of the order of 1,000,000 volts/cm. Therefore, polarization (field induced charge separation) is not unexpected. The surface forces are of short range, and greatly diminish upon building of 1 to 5 layers of molecules on the surface. First monolayers adsorbed are bonded 3–5 times more strongly than succeeding layers.

It is obvious from the above considerations that adsorption is a concentration dependent process and the adsorption coefficient, k, is not a constant. This phenomenon is in contrast to partition chromatography where the distribution coefficient is constant over a wide range. These relationships are illustrated in Fig. 21.4. Curve A represents an adsorption isotherm, and curve B represents partitioning behavior.

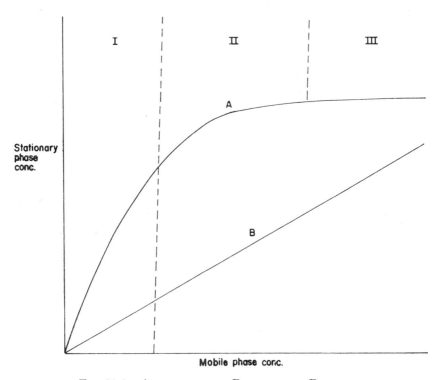

FIG. 21.4. ADSORPTION AND PARTITIONING BEHAVIOR

In region I, the adsorption coefficient is approximately linear, similar to partitioning behavior. In region II, however, a rapid change occurs, and in region III, saturation is reached. A sample size exceeding the adsorptive capacity therefore would be expected to result in relatively poor separation (tailing or smearing).

In general, substrates desirable for partition chromatography are porous, rather soft materials capable of holding tenaciously a liquid phase, whereas desirable substrates for adsorption are refractory, finely divided, strongly adsorbent powders.

Sorbent-Solvent Interactions.—Although adsorption is the major controlling factor in most TLC applications, simultaneous partitioning may also occur to a considerable extent. It is probably not well-recognized that even in cases where active material has been prepared by long drying at elevated temperatures, the sorbent still contains appreciable quantities of water. This water may behave as the stationary liquid phase. This is shown in Fig. 21.5 where it can be seen that as much as 9% water is still retained on a silica sorbent after 3 hr activation at 150°C.

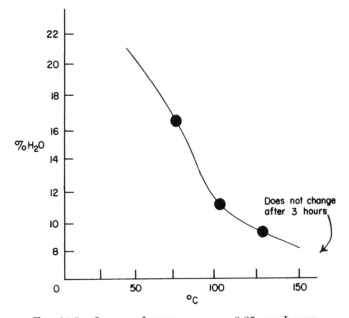

FIG. 21.5.   LOSS ON IGNITION FROM A 0.25-mm LAYER

When activated at lower temperatures, the opportunity for partition of hydrophilic substances is even greater. The properties of adsorbents commercially supplied may be modified considerably by incorporating either acids, bases, or buffer systems when the plate is prepared. In those systems, the reliance is primarily on partition rather than on adsorption processes.

Solvents.—As control of sorbent activity is somewhat restricted, adjustments needed to effect TLC separation are usually more readily accomplished by altering solvent composition. It is not surprising that, generally, solvent eluting power correlates with its dielectric constant as adsorption chromatography relies on electrostatic attraction. Assuming other things are equal, the attractive force varies inversely with the dielectric constant of the medium. On the other hand R_f values are inversely proportional to the attractive force between the sorbent and the material adsorbed; the weaker the force the higher the compound can climb on the plate and, therefore, the higher the R_f value. Table 21.4 shows a few solvents and their dielectric constants.

TABLE 21.4

DIELECTRIC CONSTANTS OF SELECTED SOLVENTS

Solvent	Dielectric Constant (25°C)	Solvent	Dielectric Constant (25°C)
n-Pentane	1.84 (20°)	Ethylene chloride	10.35
Cyclohexane	2.01	Pyridine	12.3
Carbon tetrachloride	2.23	Ammonia	16.9
Benzene	2.28	Acetone	20.70
Trichloroethylene	3.4	Ethanol	24.30
Chloroform	4.75	Methanol	32.63
Diethyl ether	5.02	Acetonitrile	37.5 (20°)
Ethyl acetate	6.02	Water	78.54
Tetrahydrofuran	8.20 (20°)	Formamide	110 (20°)

Blending solvents yields a solution with a dielectric constant approximately proportional to the quantities of the individual components. Thus a single solvent system of binary mixtures can be systematically devised to effect separation. Solvent dielectric constants are strongly dependent on purity, and for reproducible results only high purity solvents should be used.

Chemical Constitution and Adsorption Chromatography.— Strongly adsorbed substances require strong eluents, and weakly adsorbed materials weaker eluents. It is helpful to know which features of chemical constitution influence the strength of the adsorption bond. General rules have been summarized by Randerath (1963) as follows

(1)   Saturated hydrocarbons are either adsorbed weakly or not at all. Introduction of double bonds raises adsorption affinity in proportion to the number and degree of conjugation, because the polarizability of the molecule increases and, consequently, the strength with which it is bound to the surface of the adsorbent also increases.

(2)   The introduction of functional groups into a hydrocarbon

alters the adsorption affinity in the following order: COOH >
CONH₂ > OH > NHCOCH₃ > NH₂ > OCOCH₃ > COCH₃ >
N(CH₃)₂ > NO₂ > OCH₃ > H > Cl.   These rules are derived from
aromatic compounds and variations may occur for saturated compounds.

(3)   If there are several substituents in the same molecule, their
separate influences on the adsorption affinity are only approximately
additive.   Steric effects are important and can greatly vary the relative activity.

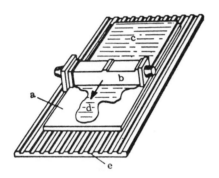

FIG. 21.6.   PREPARATION OF TLC
PLATES

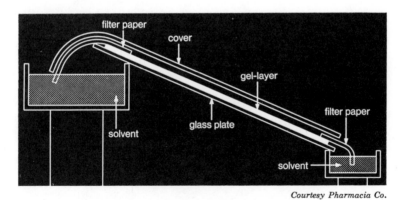

Courtesy Pharmacia Co.

FIG. 21.7.   THIN-LAYER GEL FILTRATION

Preparation of Plates.—Commercially prepared sheets of thin-
layer material already impregnated with phosphors are readily
available at reasonable cost.   Figure 21.6 shows how to prepare your
own, if desired.

Figure 21.7 shows the arrangement for a Sephadex TLC Plate.

BIBLIOGRAPHY

CONSDEN, R., GORDON, A. H., and MARTIN, A. J. P. 1944. Quantitative analysis of proteins, a partition chromatographic method using paper. Biochem. K. *5*, 224–232.

RANDERATH, K. 1963. Thin-layer Chromatography. Academic Press, New York.

STAHL, E., SCHROTER, G., KRAFT, G., and RENTZ, R. 1956. Thin layer chromatography (the method, affecting factors, and a few examples of application). Pharmazie *11*, 633–637.

STOCK, R., and RICE, C. B. F. 1967. Chromatographic Methods. Chapman & Hall, London, England.

CHAPTER 22

Gas-Liquid Chromatography

It was pointed out by Martin and Synge in 1941 that the use of a gas as a mobile phase might have certain advantages. The big problem in achieving fast and sharp separations lies in how fast the molecules can be moved around.

For fast separations we have to move our sample very rapidly to the stationary phase (and hopefully very rapidly out of it), and the more rapidly it moves the less time it has to diffuse in other directions, so we can get narrower bands and a sharper separation.

In 1952 Martin and James published the first paper on gas chromatography. In 1956 the first commercial instruments appeared and by 1959 over 600 articles had appeared in the literature. Figure 22.1 shows the necessary components.

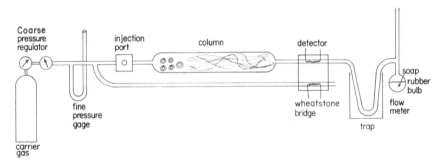

FIG. 22.1. SCHEMATIC OF A GAS CHROMATOGRAPH

Carrier Gas

The most common carrier gases (mobile phase) are helium, nitrogen, and argon. The type of carrier gas to be employed presents some choice, but is usually dictated by the type of detector used.

Pressure Differential

The pressure differential over the column plays an important role in the gas chromatograph. The gas velocity down the column is obtained by applying a pressure differential. Commonly the outlet pressure is atmospheric, and the inlet is maintained at a pressure somewhat above atmospheric. Generally a pressure ratio (p_1/p_2) of 2–3 is used. If the pressure ratio is too low, molecular diffusion remixes the separated components and efficiency decreases. If the ratio is too high, the resistance to mass transfer increases and again the efficiency of separation decreases. See the section on the van Deemter equation. For routine work the coarse pressure regulator gages on the gas cylinder can be used but for fundamental research an auxiliary precision pressure gage is necessary.

Injection Port

The injection port is the place where the sample is entered into the chromatograph. The injection port is usually heated so that liquids can be vaporized immediately upon injection. The better instruments have a separate heater for the injection port.

Sample addition is generally done by means of a syringe, a volume of 0.5–1 ml being required for gases, and 1–100 μl for liquids. However, gas sampling valves, backflushing valves, pyrolysis systems, inlet splitters, and solid samplers are among other sampling devices.

A gas sampling valve is shown in Fig. 22.2.

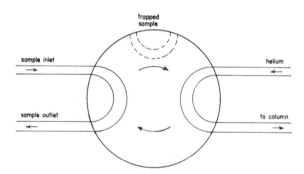

FIG. 22.2.   GAS SAMPLING VALVE

Backflushing allows flushing out of the column the high boiling components as a single composite peak after obtaining separation of the low boiling components, thereby shortening the time required before the next sample can be added.

Pyrolysis systems are used for solids and very high boiling liquids (paint film, alkaloids, plastics). The object is to partially combust the sample by heating the inlet port to a high temperature. These materials are degraded to volatile components characteristic of the starting material.

Splitters are used to divide a sample into two equal parts for dual column chromatography, or to take only a fraction of the initial sample.

A *solid sampler* is essentially a syringe within a syringe as shown in Fig. 22.3.

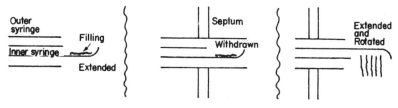

FIG. 22.3.   A SOLID SAMPLER

Column

Columns are usually made out of copper, stainless steel, aluminum, nickel, or glass with diameters of $1/8$ and $1/4$ in. and lengths of 3 to 10 ft being the normal sizes.   Columns $3/8$ in. and larger are considered to be *preparative columns*, and those smaller than $1/16$ in. are called *capillary columns*.

The column must be heated to about $2/3$ of the boiling point of the highest boiling material in the mixture to be separated.   If one part of the column is heated to a different temperature than another, separations may be made but they will not be reproducible and good physical constant measurements cannot be made.   Therefore in order to be able to heat the column more uniformly and at the same time save space, the column is usually coiled (3-in. diam for $1/8$ in. tubing, and about 6-in. diam for $1/4$ in. tubing) or bent into a U shape.   The better instruments have a separate heater for the column.

There will be times, when separating mixtures of several components, that the more volatile components come off in a short time and have nicely shaped peaks, whereas the higher boiling materials may take an hour or more to be eluted.   When a material takes this long to come off of the column it does so as a very broad band because there has been more time for diffusion.   This means that quantitative analysis is hard to obtain (see Fig. 22.4A).

However, once the low boiling materials have been separated, the column temperature can be raised, thus speeding up the emergence of the rest of the components, and as they come out of the column the temperature can be raised further to speed up the remaining materials. This can be done manually but usually a timing device can be installed which will increase the temperature at a set rate. The result is shown in Fig. 22.4B. Note that the analysis time has been shortened and the peak shape has been improved, so that good quantitative data can be obtained; this process is called temperature programming.

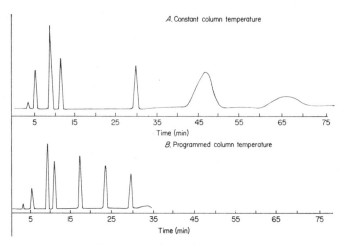

FIG. 22.4.   EFFECT OF TEMPERATURE PROGRAMMING

The disadvantages are (1) that the column must now be cooled to the original temperature before the next sample can be added, and (2) the temperature in the column lags behind the temperature in the surrounding air chamber so reproducible results are difficult to obtain.

Capillary Columns.—These are in the range of 0.005–0.02 in. diam, 100 ft to 500 ft long, and made from copper, stainless steel, glass, or nylon. The column itself acts as the inert phase. The stationary phase is added by filling the first foot with a 10–15% solution of the stationary phase, and pushing it through the column (with carrier gas) at 2–5 mm/sec. After the column is coated, the gas flow is continued at a faster rate for an hour or so to evaporate the excess solvent.

Sample sizes of a few tenths of a microliter are required which means that a sample splitter is necessary. Detectors that have a small volume must be used.

Because the thickness of the stationary phase layer is so small (0.3–2 μ), the resistance to mass transfer decreases and highly efficient, rapid analysis can be obtained. Several hundred thousand theoretical plates are common and there is one report of 7 components being separated in 25 sec.

Solid Supports

The function of the solid support is to act as an inert platform for the liquid phase in the column. Solid supports most often used are described below.

Chromosorb "P" is a pink diatomaceous earth material which has been carefully size graded and calcined. The surface area is 4–6 m^2/gm. This material is the least inert of the chromosorb supports but offers the highest efficiency.

Chromosorb "W" is a white diatomaceous earth material which has been flux-calcined with about 3% sodium carbonate and has a surface area of 1 to 3.5 m^2/gm. This material is more inert but less efficient than Chromosorb "P."

Chromosorb "G" to a very substantial degree combines the high column efficiency and good handling characteristics of Chromosorb "P" with the following advantages over present white diatomite supports: greater column efficiency, less surface adsorption, harder particles, and less breakdown in handling. Since Chromosorb "G" is about 2.4 times as heavy as Chromosorb "W," the 5% liquid loading on Chromosorb "G" is equal to 12% on Chromosorb "W." No more than 5% liquid phase should be used on Chromosorb "G."

Silanized supports are chromosorb supports that have been coated with dimethyldichlorosilane to reduce surface active sites of the diatomaceous earth material. Chromosorb supports deactivated with DMCS provide: less tailing, minimized catalytic effects, and improved results with low liquid loadings. Hexamethyldisilane, HMDS, has also been used. The combination of acid-washing and DMCS treating is particularly effective in reducing adsorption.

Fluoropak 80 and Chromosorb "T" are made from fluorocarbon polymer particles. They are chemically inert and can be used, when coated, to separate very polar compounds such as water, acetic acid, ammonia, and amines, without tailing.

Chromosorb "T" is made from 40/60 mesh Teflon 6 and has a surface area much greater than Fluoropak 80. This larger surface area generally results in columns with higher efficiencies.

Glass beads provide a low surface area, inert, solid support which is used with 0.25% or less stationary phase. It is recommended for use in separating highly polar compounds of high molecular weight.

Solid Adsorbents

Several important adsorbents are used in gas-solid chromatography. Charcoal, alumina, silica gel, and molecular sieves are used generally to separate gaseous mixtures. Molecular sieves are used to separate O_2 and N_2, but do not elute CO_2 under normal conditions. Silica gel on the other hand elutes CO_2, but does not separate O_2 and N_2. Liquid modified solid adsorbents are used for the analysis of higher molecular weight materials.

Poropak is finding many applications in gas chromatography. Poropak is a porous bead formed by the polymerization of monomers such as styrene with divinylbenzene as a cross linker. These beads serve both the function of solid support and liquid phase, although they may be modified with liquid phases. Elimination of the conventional solid support removes the adsorption sites which normally cause tailing. In addition, elimination of the liquid phase reduces bleeding. Poropak columns appear to be particularly useful for the analysis of gases and polar compounds.

Poropak "P" and "Q" are made with ethylvinylbenzenes or styrene monomers. Types "N," "R," "S," and "T" are similar, but have been modified with polar monomers. This changes the elution order of polar solutes, especially water.

Porous polymers also can be modified with conventional liquid phases. This reduces the absolute retention of all solutes and shifts the order of elution. The principal effect is to reduce tailing of polar materials such as amines and organic acids. A liquid coating of about 5% by weight is recommended.

Stationary Phases

Stationary phases may be classified as follows.

Nonpolar.—This class consists of hydrocarbon-type liquid phases including silicone greases (e.g. SE-30, squalene), but not aromatic materials. Although many exceptions are reported, generally nonpolar liquid phases separate solutes in order of increasing boiling points.

Polar.—Liquid phases containing a large proportion of polar groups are in this class (e.g. Carbowax, dimethyl sulfolane). These materials differentiate between polar and nonpolar solutes retaining only the polar materials.

Intermediate.—Polar or polarizable groups on a long nonpolar skeleton are typical (e.g. SE-52, diisodecyl phthalate, benzyl diphenyl). Members of this group dissolve both polar and nonpolar solutes with relative ease.

Hydrogen Bonding.—This special class of polar liquid phases contains compounds with a large number of hydrogen atoms available for hydrogen bonding (e.g. the hydroxyl groups of glycerols).

Specific.—Special purpose phases which rely on a specific chemical interaction between solute and solvent to perform separation (e.g. $AgNO_3$ for olefins).

Column Conditioning.—Columns should be preconditioned for at least 10 hr at about 20° above the maximum operating temperature you plan to operate the column, but below the maximum temperature limit for the stationary phase. This is to remove from the column, solvent and other volatile materials which will cause interference later by altering the column conditions and making reproducible results impossible. Very small amounts of carrier gas (5–10 ml/min) should be flowing through the column during the conditioning period. The columns can be conditioned in the chromatograph if the exit end is disconnected from the detector.

Common Abbreviations.—Table 22.1 lists some of the common stationary phases that have been used so often that they have abbreviations.

TABLE 22.1

ABBREVIATIONS OF SOME STATIONARY PHASES

Abbreviation	Chemical Name
Aroclor	Chlorinated biphenyls
DEGA	Diethylene glycol adipate
DEGSE	Diethylene glycol sebacate
DEGS	Diethylene glycol succinate
EGA	Ethylene glycol adipate
EGIP	Ethylene glycol isophthalate
EGSE	Ethylene glycol sebacate
EGS	Ethylene glycol succinate
HMPA	Hexamethylphosphoramide
IGEPAL	Nonyl phenoxy polyoxyethylene ethanol
NPGS	Neopentyl glycol succinate
SE-30	Methyl silicone gum rubber
THEED	Tetrahydroxyethyl ethylene diamine
TCP	Tri cresyl phosphate
TWEEN	Polyoxyethylene sorbitan monooleate

Detectors

A detector is a device that measures the change of composition of the effluent. There are dozens of different kinds of detectors. The ones that will be described here are thermal conductivity, hydrogen flame ionization, cross section, argon ionization, and electron capture.

Thermal Conductivity Detectors

If you blow air over a hot wire it will be cooled.   This idea is the basis of the thermal conductivity detector.   A thin filament of wire is placed at the end of the column and heated by passing a current through it.   The carrier gas molecules strike the hot wire, and as each molecule hits the wire it takes away some heat from the wire. As the wire is cooled, its resistance changes.   When a sample passes over the hot wire the sample molecules, because of different velocities and masses, will take a different amount of heat away from the hot wire than the carrier gas.   This means that the resistance of the wire will change from what it was and a signal will be produced.

The greater the difference in thermal conductivities of the carrier gas and the components of the mixtures, the greater the sensitivity of the method.   In Table 22.2 are listed the thermal conductivities of several representative materials.   Thus the two gases which present the largest differences from the other compounds are H_2 and He. Hydrogen would appear superior but presents safety hazards in the laboratory.   Helium is nearly as satisfactory and is noncombustible.

TABLE 22.2

THE THERMAL CONDUCTIVITY, K, OF SOME GASES AND VAPORS[1]

Substance	$K(\times 10^5)$	Substance	$K(\times 10^5)$
Hydrogen	41.6	i-Butane	3.32
Nitrogen	5.81	n-Pentane	3.12
Oxygen	5.89	i-Pentane	3.00
Air	5.83	n-Hexane	2.96
Ammonia	5.22	Ethylene	4.19
Water	3.5	Acetylene	4.53
Helium	34.8	Methyl chloride	2.20
Neon	11.1	Methylene chloride	1.61
Argon	3.98	Chloroform	1.58
Krypton	2.12	Methyl bromide	1.50
Xenon	1.24	Methyl iodide	1.13
CO	5.63	Freon-12	1.96
CO_2	3.52	Methanol	3.45
CS_2	3.70	Ethanol	3.5
Methane	7.21	Diethyl ether	3.6
Ethane	4.36	Acetone	3.27
Propane	3.58	Methyl acetate	1.61
n-Butane	3.22	Nitrous oxide	4.5
		Nitric oxide	5.71

[1] Taken largely from H. A. Daynes, *Gas Analysis by Measurement of Thermal Conductivity*, Cambridge University Press, London.

The wire is heated by an electric current to a temperature T_2, which is hotter than the wall of the detector, T_1.   The sensitivity of detection with these cells generally increases with a reduction in the wall temperatures (greater difference between T_2 and T_1), but one must not lower the wall temperature too greatly, or condensation of the components will occur in the cell.   The better instruments have a separate heater for the detector.

Figure 22.5 shows three different arrangements of thermal conductivity detectors.

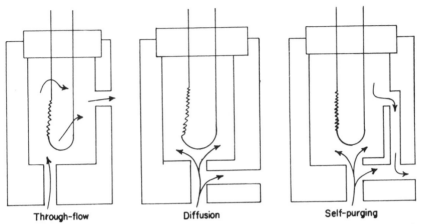

| Through-flow | Diffusion | Self-purging |

Courtesy Juvet and Dal Nogare, Gas-Liquid Chromatography

FIG. 22.5.   DETECTOR GEOMETRIES

The through-flow type has the advantage of a fast response time constant (1 sec), but is very susceptible to small changes in flow rate.   The diffusion type was designed to smooth out rapid flow rate changes, producing a more stable detector, but the response time constant is slow (20 sec).   The purging type is an attempt to combine the advantages of the two.

If one hot wire is used, the detector is unstable because it is sensitive to temperature changes, flow rate changes, and filament current changes.   To remedy this, the Wheatstone bridge circuit is used.

Characteristics: (Courtesy Varian-Aerograph Co.)
  (1)   Sensitive to all organic compounds (10^{-7} gm/sec)
  (2)   Linear dynamic range: 10,000
  (3)   Sensitive to flow rate and temperature changes
  (4)   Detector temperature limit: 500°C
  (5)   Carrier gases: He, H_2, N_2, Ar, CO_2

Hydrogen Flame Ionization Detector (FID)

The hydrogen flame ionization detector works as follows: When carrier gas (CG) and sample emerge from the column, hydrogen

(1H$_2$/1CG) and air (10 air/1CG) are added to the carrier gas to produce a flame of about 2100°C.   This flame produces + and − ions whose concentrations have been established at 10^{10} to 10^{12} ions/milliliter.   This concentration of ions is too large to come only from the known combustion products of the flame.   It has been suggested that carbon forms aggregates, since free carbon is known to polymerize, and the low work function, 4.3 v, of these carbon aggregates could explain the large ion concentrations observed.   This is further substantiated by the fact that this detector's sensitivity is roughly proportional to the carbon content of the sample.

The response per gram-mole of a given compound is directly proportional to the number of C atoms bound directly to H or other C atoms.   Carbons bonded to OH, amines, and X$_2$ provide a fractional contribution as do alkali and alkaline earth elements, but no signal is obtained from fully oxidized C atoms (COOH, C≡O) and inert gas molecules.   The net result is that this detector is sensitive to all organic compounds except HCOOH and insensitive to (1) all inorganic compounds except those containing alkali and alkaline earth elements, and (2) Ar, He, Ne, O$_2$, H$_2$S, CO, CO$_2$, NO$_2$, and H$_2$O.

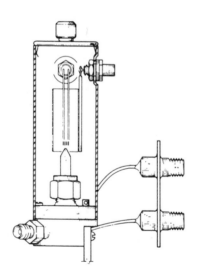

Flame Ionization

Courtesy Varian-Aerograph Co.

FIG. 22.6. FLAME IONIZATION
DETECTOR

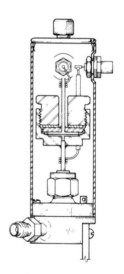

Cross-section

Courtesy Varian-Aerograph Co

FIG. 22.7. MICRO CROSS SEC-
TION DECTOR

The ionization detector is about 1000 times more sensitive than a thermal conductivity cell and can easily detect 1 nanogram samples. As a generalization, a TC detector is not as good as your nose, a FID detector is as good as your nose, and the electron capture detector to be discussed later is better than your nose.

Caution: (1)   A hydrogen flame is colorless so use a piece of paper and not your finger to check if the flame is on.   (2)   Dust and dirt contain alkali metals so be sure your air supply is clean or erratic results will be obtained.

Characteristics: (courtesy of Varian-Aerograph Co.)
(1)   Sensitive to all organic compounds (10^{-12} gm/sec)
(2)   Linear dynamic range: 1,000,000
(3)   Insensitive to temperature change (ideal for programming)
(4)   Detector temperature limit, 400°C
(5)   Carrier gas: N_2, He, or Ar

Figure 22.6 shows the physical arrangement of a FID detector.

Cross-Section Detectors

Rather than use a flame to produce ions, high speed electrons (β particles) can be used. Sr^{90} and H^3 (tritium) are used as the source of β particles.   When the β particles strike the carrier gas they can produce $+$ ions and electrons.   These electrons can be captured by other gas molecules and become $-$ ions, or the electrons may ionize other gas molecules.   The net effect is an increase in the charged particles in the gas stream with the result that the conductivity of the gas increases.   If 300–500 v are applied to the collector plate the charged species can be collected, producing a signal.   When the sample comes along it will also produce ions but because of the usually larger size of the sample molecules, more electrons are utilized and the gas conductivity is different than the carrier gases.

The adsorption cross section is roughly proportional to the mass of the molecule so to get maximum sensitivity, H_2 and He are usually used as the carrier gases. β particles do not have a high penetrating power and in order to be efficient, the detector volume should be small.   The β sources are quite susceptible to high temperature and these detectors are limited to 200°–225°C.

The cross-section detector is sensitive to all organic and inorganic compounds, and is the only detector capable of measuring gas sample concentrations up to 100% within the detector.   It is about as sensitive as a thermal conductivity cell.

Characteristics: (courtesy of Varian-Aerograph Co.)
  (1) Sensitive to all vapors and gases, comparable to a good TC detector. About 10^{-7} gm/sec
  (2) Linear dynamic range: 5×10^5 to 100% concentration
  (3) Insensitive to temperature change
  (4) Insensitive to carrier gas molecules
  (5) Detector temperature limit $220\,^\circ$C
  (6) Carrier gas H_2 to $100\,^\circ$C or He + 3 CH_4 (1:3)

Argon Ionization Detectors

Lovelock found that the cross-section process could be made much more efficient if argon was used as the carrier gas. The β particle would produce an excited Ar* atom which would then transfer its energy very efficiently to the sample molecules producing ions and electrons and a very strong signal, 10–100 times more sensitive than a hydrogen flame ionization detector. The sequence of reactions is suggested as:

$$\text{Background}\begin{cases} \text{Ar} + \beta^- & = \text{Ar}^+ + \text{e}^- \\ \quad \text{high} \\ \quad \text{energy} \\ \text{Ar} + \beta^- & = \text{Ar*} \ (10^{-4} \text{ sec half life}) \\ \quad \text{low} & \quad (11.6 \text{ ev}) \\ \quad \text{energy} \\ \text{Ar*} + \text{impurities} = \text{Ar} + \text{M}^+ + \text{e}^- \end{cases}$$

Signal $\qquad$ Ar* + sample $= $ Ar + M^+ + e^-

Materials that can be ionized by 11.6 ev can be detected, which means that most materials (except the noble gases, some inorganic gases, and fluorocarbons) can be detected. ^{90}Sr, ^{85}Kr, RaD, and T are usually used to provide the β particles. The argon carrier gas must be very dry.

The output stability of the argon detector is such that no reference cell is required. Moderate variations in carrier flow rate and temperature has negligible effect on the detector base current.

Electron Capture Detector

With the cross-section detector there was an increase in the number of ions in the gas, so there was an increase in signal and all types of compounds would respond. To make a more selective detector, use of electron capture was investigated. Suppose that the electrons striking the sample molecule had just enough energy to penetrate the electrical field of the molecule and be captured but not enough to break up the molecule into ions.   If this were to happen, the original electrical signal, based on the free electrons, would now decrease because of the electrons lost due to being captured.

Halogenated compounds, conjugated carbonyls, nitriles, nitrates, and some organometallic compounds are quite sensitive to the electron capture process.

Characteristics: (courtesy of Varian-Aerograph Co.)
  (1)   Variable sensitivity–0.1 pg (10^{-13} gm) for CCl_4 or 5×10^{-14} gm/sec
  (2)   Linear dynamic range: 500
  (3)   Sensitive to temperature change on column, programming not recommended
  (4)   Insensitive to most organic compounds except those containing halogens, sulfur, nitrogen (nitriles, nitrates), and conjugated carbonyls.
  (5)   Detector temperature limit: 220°C
  (6)   Carrier gas: N_2, must be pure and dry
  (7)   Ionization source—250 mC tritium

Flow Rate

If retention times are to be duplicated from day to day from laboratory to laboratory, one of the parameters that must be controlled is the carrier gas flow rate.   Many precise flow meters are commercially available.   However, in their absence, a buret, fitted with a side arm and a rubber squeeze bulb full of liquid soap, will work quite well (see Fig. 22.1).   All you do is squeeze the bulb until the soap raises so that the carrier gas will form a film out of it.   This film will be pushed to the top of the buret and by measuring how many milliliters it rises in a given time, you can determine the flow rate.   A spot of grease at the top of the buret breaks the soap film and it runs back into the squeeze bulb.

Recently, flow programming has been suggested as an alternative to temperature programming for heat sensitive materials.   In this case the flow rate is increased as the separation progresses.

To see the full effect of flow rate changes, see the section on the van Deemter equation which follows later.

Height Equivalent to a Theoretical Plate (HETP)

The determination of HETP for a column gives the operator some idea how efficient his column is and whether or not it would be worthwhile to spend time trying to improve the separation. Two equations for obtaining the number of theoretical plates are:

(a)   the Glueckauf equation:

$$N = 8(t/\beta)^2 \tag{22.1}$$

(b)   the van Deemter equation:

$$N = (4t/W)^2 \tag{22.2}$$

where N = the number of theoretical plates of the column, β = peak width at H/e of the peak height, and W = the distance between the intersections of tangents to inflection points on the base line.   These are shown schematically in Fig. 22.8.   Note that t is shown as the time lapse between the injection point O and the middle of the peak. t in units of distance could also be used in determining N.

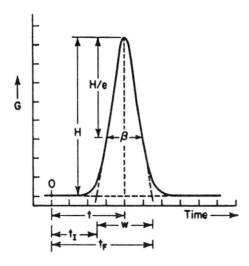

FIG. 22.8.   DEFINITION OF GAS
CHROMATOGRAPHIC TERMS

Expressions very close to that of equation 22.2 are

$$N = \left(\frac{4t_I}{W} + 2\right)^2 \tag{22.3}$$

and

$$N = \left(\frac{4t_F}{W} - 2\right)^2 \tag{22.4}$$

and may also be used to determine N; in particular, the latter is quite useful when unsymmetrical peaks are obtained.

Example

Let us use Fig. 22.8 to calculate the number of theoretical plates, N, by means of equations 22.1 to 22.4. From Fig. 22.8, $H = 42$, $t = 40$, the peak area $= 540$, β is 15, $t_I = 29$, and $t_F = 51$. Thus:

(a) From equation 22.1

$$N = 8(40/15)^2 = 8 \times 7.11 = 56.9$$

(b) From equation 22.2

$$N = (4 \times 40/22)^2 = (7.273)^2 = 52.9$$

(c) From equation 22.3

$$N = \left(\frac{4 \times 29}{22} + 2\right)^2 = (5.273 + 2)^2 = 52.9$$

(d) From equation 22.4

$$N = \left(\frac{4 \times 51}{22} - 2\right)^2 = (9.273 - 2)^2 = 52.9$$

The height equivalent to one theoretical plate is defined by

$$\text{HETP} = L/N \tag{22.5}$$

where L is the length of the column. Thus using equation 22.2 we obtain

$$\text{HETP} = L(W/4t)^2 \tag{22.6}$$

Of course the units of W and t must be the same.

Example

Let us assume that the column used to obtain the separation shown in Fig. 22.8 was 4 ft in length. That is, the length, $L = 4$ ft $\times$ 12 in./ft $\times$ 2.54 cm/in. $= 122$ cm. From equation 22.5 and using an average value of $N = 54$ obtained in example 1, we see that

$$\text{HETP} = 122/54 = 2.26 \text{ cm}$$

Van Deemter Equation

The theoretical plate concept is useful in determining the efficiency of any given column, but it does not indicate the effects of various operational parameters. Here, however, the Van Deemter rate theory proves valuable. The Van Deemter equation is

$$\text{HETP} = A + \frac{B}{V} + CV \tag{22.7}$$

where $A = 2\lambda \bar{d}$ (22.8)

$B = 2\gamma D_g$ (22.9)

$C = 8kd_f/\pi^2 D_l(1 + k)^2$ (22.10)

V = velocity

λ = a quantity characteristic of the column packing

$\bar{d}$ = average particle diameter

D_g = the diffusion coefficient in the gas phase

k = the fraction of the sample in the liquid phase divided by the fraction in the vapor phase

d_f = the average liquid film thickness

D_l = the diffusion coefficient in the liquid phase

The first term in equation 22.7 is due to Eddy diffusion (turbulence and the column packing), the second term to molecular diffusion, and the third term to the resistance to mass transfer. These component parts of the van Deemter equation are illustrated graphically in Fig. 22.9.

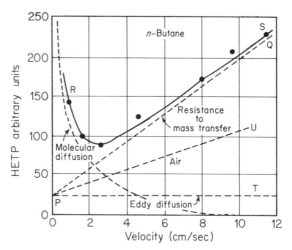

FIG. 22.9. VARIATION OF HETP AS A FUNCTION OF CAR-
RIER GAS VELOCITY

In Fig. 22.9 data are shown for the variation of the HETP of a column prepared by coating Celite with tri-cresyl phosphate, as a function of the linear velocity of the carrier gas, in cm/sec. The curve R-S is the actual variation of the HETP with velocity. The tangent to this curve, P-Q allows the extrapolation to zero velocity. Thus the Eddy diffusion is shown as the straight line, P-T. The resistance to mass transfer is represented by the contributions between lines P-T and P-Q for n-butane, and P-T and P-U for air. The molecular diffusion contribution is represented by the difference between P-Q and R-S. Thus at very low velocities of carrier gas the molecular diffusion is more important, whereas at high gas velocities the mass transfer resistance becomes more important.

The optimum velocity of carrier gas to be employed is at that point where the HETP-velocity curve is at a minimum.

Identification of Compounds

Under a given set of conditions and with a given chromatographic column, the retention time or retention volume is characteristic of a particular component. However, in certain cases, two or more materials may have the same retention volume. They seldom have the same retention time on stationary phases varying in polarity, so if in doubt, change columns and see if other peaks appear. The result is that you can never be completely sure of a positive identification of an unknown from retention data alone. Nevertheless, the characteristic retention volume (or retention time) is extremely valuable in identifications and particularly in separations of components in mixtures. If some knowledge is already available concerning the possible constituents of the mixture, the gas chromatographic results may be sufficient. In cases of doubt, small quantities of the separated material may be trapped in cold traps after they emerge from the column. Mass spectrometric identification or fast scan infrared identification can be employed for positive identification.

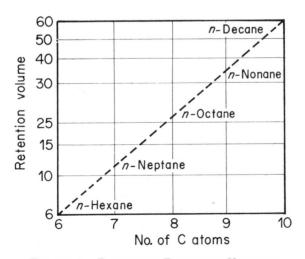

FIG. 22.10. PREDICTING RETENTION VOLUMES

A useful relation which exists in gas chromatography and which can be very useful for predicting retention volumes of new compounds, is that the logarithm of the retention volume is often a linear function of the number of carbon groups, CH_2 groups, or the like. This is shown in Fig. 22.10 for the straight chain hydrocarbons.

Example

From the data for straight chain hydrocarbons in Fig. 22.10, we can calculate the retention volumes for the higher homologues in the series. We may write the equations:

$$\log 6 = 6x + y$$
$$\log 10.5 = 7x + y$$
$$\log 18.6 = 8x + y$$
$$\log 32.7 = 9x + y$$

Taking any 2 of these equations, we may solve 2 simultaneous equations to find x and y. Thus,

$$\log 18.6 = 8x + y$$
$$-\log 6.0 = 6x + y$$
$$\overline{\log 18.6 - \log 6.0 = 2x}$$

$$1.26951 - 0.77815 = 2x = 0.49136$$

$$x = 0.24568$$

and therefore

$$y = 0.77815 - (6 \times 0.24568)$$
$$= 0.77815 - 1.47408 = -0.69593$$

As a check, we will calculate the retention volume for n-decane:

$$\log V_R = 10(0.24568) - 0.6593$$

$V_R = 57.7$ and is in good agreement with the experimental value of 57.5.

Dense Gas Chromatography

Analysts have long thought of combining the high solubility and efficient separating ability of liquid chromatography with the speed of gas chromatography.

A step in this direction is the recent advent of dense gas chromatography (high-pressure chromatography or critical temperature chromatography).

It has been known for some time that at about 1000 atm pressure a gas would have the same solubility characteristics as if it were a liquid.

In order for a gas to remain a gas at 1000 atm pressure, it must be heated above its critical temperature, which for most potential solvents is not very high, less than 150°C.

Klesper *et al.* (1962) first demonstrated that this would work by separating 2 porphyrins at 140 atm using Freon as the carrier gas.

Perhaps the best summary of the status of the technique at this time is the statement by Myers and Giddings (1966) "working at pressures of up to 2000 atm, more than 10 times higher than previous gas chromatography, we used the solvent power of dense gases to enable migration of chromatographic substances of molecular weights as high as 400,000. Carotenoids, cortical steroids, sterols, nucleosides, amino acids, carbohydrates, and several polymers have been caused to migrate, separated and detected in NH_3 and CO_2 carrier gases at temperatures of 140° and 40°C, just above the respective critical points."

The main barrier to making this a routine analytical technique is the detection system. When the materials emerge from the column they tend to form a fog which produces a noisy signal.

BIBLIOGRAPHY

KLESPER, G., CORWIN, A. A., and TURNER, D. A. 1962. High pressure gas chromatography above critical temperatures. J. Org. Chem. 27, 700–701.

MARTIN, A. J. P., and JAMES, A. T. 1952. Gas liquid partition chromatography. The separation and micro-estimation of volatile fatty acids from formic acid to dodecanoic acid. Biochem. J. 50, 679–690.

MARTIN, A. J. P., and SYNGE, R. L. M. 1941. A new form of chromatography employing two liquid phases. Biochem. J. 35, 1358–1368.

MYERS, M., and GIDDINGS, J. C. 1966. Ultrahigh pressure gas chromatography in microcolumns to 2000 atmospheres. Separ. Sci. 1, 761–767.

CHAPTER 23

Extraction

A large number of analytical methods for foods require an extraction as a cleanup procedure, as a concentration step, to remove a slightly soluble material, or to aid in the identification of a component.

This involves liquid-liquid and liquid-solid extractions using batch, continuous, and discontinuous counter-current techniques. A *liquid-liquid* extraction is based on a partition of a material between 2 immiscible liquids, usually 1 is organic and 1 is water. The usual way to express the extent of this partition is the *distribution ratio, D.*

$$D = \frac{\text{total grams solute in the organic phase}}{\text{total grams solute in the aqueous phase}} \quad (23.1)$$

We cannot quantitatively predict what a D will be for any given system, but it is known that for a substance to be extracted it must be neutral. Ions are made neutral by chelation or ion association. For example, lead in food coloring dyes can be chelated and neutralized by diethyl dithio carbamate, whereas iron in ferbam, a plant insecticide, can form an ion association system with HCl.

$$\text{H}_5\text{C}_2 \quad \text{S}^- \quad \text{S} \quad \text{C}_2\text{H}_5$$
$$\text{N-C} \quad \text{Pb}^{2+} \quad \text{C-N}$$
$$\text{H}_5\text{C}_2 \quad \text{S}^- \quad \text{S} \quad \text{C}_2\text{H}_5$$

$$\text{FeCl}_4^-\text{H}^+$$

chelate
extracts into $CHCl_3$

ion association
extracts into ether

Percent Extraction

Equation 23.2 is used to calculate the percent of a compound that can be extracted if the D and the volume of each phase is known.

$$\% \text{ Extraction} = \frac{100D}{D + V_w/V_o} \quad (23.2)$$

where V_w and V_o = volume of water and organic layers, respectively.

In order to separate two components they must not only have different distribution ratios, they must have the correct range.

Example

System 1:   Cpd A has a D of 1 and cpd B has a D of 10.
System 2:   Cpd X has a D of 100 and cpd Y has a D of 1000.
Both systems have a difference in D's of 10 yet system 1 is far easier to separate than system 2, why?

$$A\% = \frac{100 \times 1}{1 + 1/1} = 50\% \qquad X\% = \frac{100 \times 100}{100 + 1/1} = 99.0\%$$

$$B\% = \frac{100 \times 10}{10 + 1/1} = 90.9\% \qquad Y\% = \frac{100 \times 1000}{1000 + 1/1} = 99.9\%$$

B is not completely extracted but it is separating from A.

Both are almost completely extracted and no separation takes place.

Multiple Extractions

A common mistake is made when an analyst is in a hurry.   The directions call for three, 10-ml extractions, but time is short so one 30-ml extraction is made and it is assumed it is the same; it is not. Equation 23.3 provides the answer.

$$\checkmark \qquad W_m = W \left(\frac{V_w}{DV_o + V_w} \right)^n \qquad (23.3)$$

W_m = weight of solute remaining in the water after n extractions
W = original weight of solute in the water layer

Example

A fat is to be removed from a meat sample by an ether extraction. Three 30-ml extractions are recommended to remove 0.1 gm of fat from 1 gm of meat dispersed in 30 ml of H_2O.   Assume $D = 2$. Which is best, one 90-ml extraction, or three 30-ml extractions?
For a single 90-ml extraction,

$$W_m = 0.1 \left(\frac{30}{(2 \times 90) + 30} \right) = 0.014 \text{ gm left}$$

For the 3 extractions,

$$W_m = 0.1 \left(\frac{30}{(2 \times 30) + 30} \right)^3 = 0.0036 \text{ gm left} \quad \text{(almost a four fold. improvement)}$$

The forementioned equations refer to liquid-liquid extractions in a batch process. This type of extraction is generally carried out in a separatory funnel.

Continuous Extraction

Liquid–Solid Systems.—Extracting a material from a solid generally requires more time because it is difficult to get the extracting solvent into direct contact with the solute. Because of this, the effective distribution ratios are low and large volumes of solvent are sometimes necessary.

An efficient method for the liquid-solid extraction which eliminates the use of large solvent volumes is the *Soxhlet extractor* shown in Fig. 23.1.

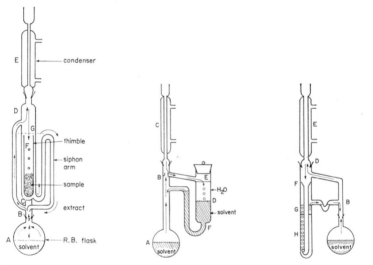

FIG. 23.1. SOXHLET EX-   FIG. 23.2. SOLVENT   FIG. 23.3. SOLVENT
TRACTOR   HEAVIER EXTRACTOR   LIGHTER EXTRACTOR

The extracting solvent is placed in the flask at the bottom (A). When the solvent is heated it goes up the side arm at B. C is closed off and the siphon return (right side of B) fills with liquid almost immediately and is also closed off. The solvent vapor is condensed in the condenser and drips onto the sample. When the extractor fills up to G, the solvent will siphon back into the reservoir flask and the process will start again. The solvent is used over and over again so while a large volume of fresh solvent attacks the sample, actually only a small total volume is used.

An example of where this apparatus is used is the determination of diethylstilbestrol, a growth enhancer, in poultry meat.

Continuous Extractors, Solvent Heavier.—The Soxhlet extractor works well with solids but not with liquids. Continuous extractors are excellent for use with materials having a low distribution ratio.

The extractor shown in Fig. 23.2 is a continuous extractor for use with solvents heavier than water.

The solvent is placed in the flask, A, and heated. The vapor rises to B and then up to C where it is condensed. The liquid cannot get back in the flask because of the seal at B. It runs into the extractor, D, and drips through the water layer, E, and collects, at F. The excess solvent containing the extract runs out the bottom at F and back into the flask. This is a continuous process with the extract collecting in the flask.

By using chloroform, caffeine can be extracted easily from cola beverages in about 45 min.

Continuous Extractors, Solvent Lighter.—The solvent is placed in the flask, A, and heated. The vapor rises past B and on to D. It cannot go to F past C because after an initial few minutes, C is filled with liquid. The vapor condenses in E and drips down into the flared tube, F. The liquid will stay in tube F until it builds sufficient pressure to force its way out of the bottom opening and then it bubbles up through the sample solution H and collects at the top, G. The excess runs back into the flask by tube C.

The ipecac alkaloids can be extracted from expectorants with ether by this technique in about 3 hrs.

Discontinuous Countercurrent Extractions

Craig and Post (1949) developed this concept. It is a method whereby several extractions are performed simultaneously with the top phase shifting one tube after each extraction. Figure 23.4 shows the steps required for 1 extraction and 1 transfer.

Consider a set of perhaps 100 tubes like that shown above, all placed side by side. All of the tubes are filled with the lower phase in the volume indicated. The upper phase is then added to tube 0 and an extraction performed. The sample is in the lower phase of tube 0. The tubes are rocked gently a few times and then stopped to allow the layers to separate. The bank of tubes is then turned to an upright position and the top layer, which contains some of the extracted sample, transfers into the transfer tube. When the tube is then turned upright again, the original top layer drains out of the

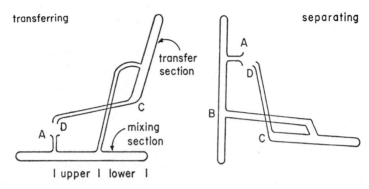

transferring

separating

transfer
section

mixing
section

| upper | lower |

FIG. 23.4. CRAIG EXTRACTION PROCESS

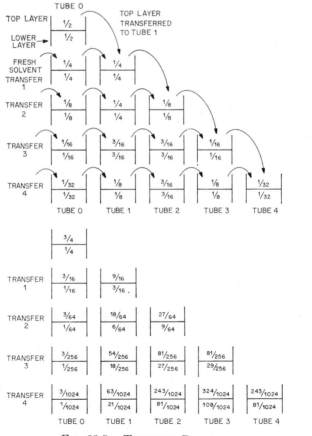

FIG. 23.5. TRANSFER PROCESS

transfer tube at C and goes to the next tube in the line. D of one tube fills the second tube at A and so forth. Fresh solvent is added to the original tube at A the same time the other tubes transfer.

The Figure that follows shows the process for 5 transfers with 2 different materials. Notice how the compounds begin to separate. If we neglect diffusion processes, this is basically the same step by step procedure that takes place in partition chromatography. Notice that at no time is a complete separation accomplished, but that at each step the separation becomes a little more complete.

Assume that this compound has a D of 1. If this is so, then $1/2$ goes into the top layer and $1/2$ goes into the lower layer. The top layer is transferred to the next tube and this $1/2$ is again divided in half, so we have $1/4$ in the top layer and $1/4$ in the lower layer, the total being $1/2$. Fresh solvent is then added to the first tube and the lower layer equilibrated with it. This process is repeated until the desired separations occurs.

If the compound has a D of 3, this means that $3/4$ will go into the top layer and $1/4$ into the lower layer. In order to get an idea of how the extraction is proceeding, compare the amount of compound 1 in tube 3 with the amount of compound 2 in the same tube.

By calling the first tube 0, the mathematics of the process can be made to fit existing binomial expansion equations and the fraction in any tube after any given number of transfers has been made can be calculated.

Extraction P Values

This technique was developed primarily to help identify pesticides in foods. There is no reason why it could not be extended to other compounds found in foods.

There are times when the gas chromatographic retention times of two or more materials are so close that an identification is difficult even if two different columns have been used. A thin-layer chromatography separation can then be used, but this generally requires a lengthy cleanup and concentration because of the small (nanogram) samples involved. A much quicker method is the use of P values. This method is based on the distribution of the pesticide between two immiscible phases.

The procedure of Beroza and Bowman (1965) which follows shows how it is done. Assume a hexane-acetonitrile system.

A 5-ml aliquot of the upper phase containing a given pesticide is analyzed by gas chromatography. To a second 5-ml aliquot in a graduated glass stoppered 10-ml centrifuge tube is added an equal volume of lower phase, the tube shaken for about 1 min, and upper phase is analyzed exactly like the first 5-ml aliquot.   The ratio of the second analysis to the first is the P value—the amount of the pesticide on the upper phase (2nd analysis) divided by the total amount of pesticide (1st analysis).

Table 23.1 (Beroza and Bowman 1965) shows how valuable this can be. o,o'-DDT and TDE have identical retention times.

Aliquots of the equilibrated phases are measured with volumetric pipets with the phase volume being noted before and after equilibration.   Emulsions are separated by centrifugation.   After the distribution, the upper phase is passed through a 1-in. layer of anhydrous Na_2SO_4 to dry the solvent.

The P value is quite reproducible (± 0.02), provided the 2 phases are saturated with each other before distribution.

Table 23.2 shows P values of several insecticides (Beroza and Bowman 1965), and Table 23.3 (Manske and Frasch, 1966) shows

TABLE 23.1

P VALUES OF TWO INSECTICIDES HAVING INDISTINGUISHABLE RETENTION TIMES

	Cyclohexane Methanol	Hexane 90% Dimethyl Sulfoxide	Iso Octane 80% Dimethyl Formamide
o,o-DDT	0.61	0.55	0.44
TDE	0.37	0.08	0.16

TABLE 23.2

P VALUES OF INSECTICIDES

Pesticide	Hexane Acetonitrile	Hexane 90% Aq. Dimethyl Sulfoxide	Isooctane 85% Aq. Dimethyl Formamide	Isooctane Dimethyl Formamide
Aldrin	0.73	0.89	0.86	0.38
Carbophenothion	0.21	0.35	0.27	0.04
Gamma chlordane	0.40	0.45	0.48	0.14
p,p'-DDE	0.56	0.73	0.65	0.16
o,o'-DDT	0.45	0.53	0.42	0.10
p,p'-DDT	0.38	0.40	0.36	0.08
Dieldrin	0.33	0.45	0.46	0.12
Endosulfan I	0.39	0.55	0.52	0.16
Endosulfan II	0.13	0.09	0.14	0.06
Endrin	0.35	0.52	0.51	0.15
Heptachlor	0.55	0.77	0.73	0.21
Heptachlor epoxide	0.29	0.35	0.39	0.10
1-Hydroxychlordene	0.07	0.03	0.06	0.03
Lindane	0.02	0.09	0.14	0.05
TDE	0.17	0.08	0.15	0.04
Telodrin	0.48	0.65	0.63	0.17

TABLE 23.3[1]

P-VALUES OBTAINED BY SPIKING THE 6% AND 15% ELUATES OF VARIOUS CROPS

Pesticide →	Aldrin Eluate		Heptachlor Eluate		DDE Eluate		Ronnel Eluate		Kelthane Eluate		Dieldrin Eluate	
Sample	6%	15%	6%	15%	6%	15%	6%	15%	6%	15%	6%	15%
Raspberries	0.72	0.86	0.50	0.62	0.56	0.65	0.22	0.30	0.15	0.20	0.37	0.42
Blueberries	0.72	0.79	0.52	0.58	0.56	0.58	0.24	0.31	0.15	0.18	0.38	0.43
Beet pulp	0.75	0.81	0.57	0.59	0.59	0.50	0.23	0.25	···	0.15	0.41	0.34
Wheat	0.78	0.80	0.53	0.57	0.59	0.60	0.26	0.29	0.16	0.17	0.39	0.39
Potatoes	0.81	0.83	0.54	0.61	0.62	0.62	0.23	0.29	0.14	0.16	0.37	0.41
Lettuce	···	···	0.56	···	0.60	···	···	···	···	···	···	···
Strawberries	0.83	0.70	0.56	0.50	0.62	0.57	0.27	0.29	0.16	0.17	0.38	0.41
Pea silage	0.70	0.76	0.50	0.52	0.56	0.58	0.27	0.30	0.17	0.20	0.38	0.41
Cream	0.73	0.69	0.51	0.48	0.56	0.52	0.24	0.28	0.15	0.18	0.35	0.39
Pea silage	···	···	···	···	···	···	0.21	···	0.13	···	···	···
Beets	···	···	···	···	···	···	···	···	···	···	···	···
Avg	0.75	0.78	0.53	0.56	0.58	0.58	0.25	0.29	0.15	0.18	0.38	0.40
Avg Dev.	0.04	0.05	0.02	0.04	0.02	0.04	0.02	0.01	0.01	0.01	0.01	0.02

[1] Further information can be obtained from Burke and Guifrida (1964); Johnson (1965); and Mills et al. (1963).

the results for several foods. The 6 and 15% eluates are from a Florisil cleanup.

Some Recent Applications

Thaler and Arneth (1968) developed a scheme for the separation of the polysaccharides of green coffee beans, and Thomas *et al.* (1966) used extraction to separate hesperidin and citroflavonoids in fruit pigments.

Investigations of the effectiveness of various solvent systems for the extraction of lipids from beef were done by Hagen *et al.* (1967), who found a chloroform-methanol mixture particularly good for extracting phospholipids. Novak (1965) extracted free fatty acids from serum by extracting the cobalt soap with heptane and then forming a cobalt complex with 1-nitroso-2-naphthol. The structure of unsaturated glycerides of vegetable oils was determined by the use of a countercurrent distribution apparatus, 200 tubes, and lipase hydrolysis by Evans *et al.* (1966).

PROBLEMS

(1)   The aldrin in a sample of strawberries was extracted with hexane and concentrated to 5 ml. 5 ml of 90% dimethyl sulfoxide was added, and it was found that 83% of the aldrin was in the hexane layer. What is the distribution ratio for this compound between these solvents?   Ans. 4.8

(2)   Assume you wanted to have 98% of the aldrin in problem 1 extract into the hexane. How many total milliliters of hexane must be used?   Ans. 51

(3)   Suppose you know that cholesterol in egg noodles can be extracted with ethyl acetate. The D is 0.3 (assumed). How many 50-ml extractions will it take to remove 95% of the cholesterol in a batch of egg noodles (20 ml H_2O) if they contain 2% cholesterol? Assume a 10-gm sample.   Ans. 5.3 or 6 extractions

(4)   Referring to Fig. 23.5, what is the amount of the second compound $(D = 3)$ in the upper and lower phases in the 3rd tube after 5 transfers?   Ans. upper 729/4096, lower 351/4096

BIBLIOGRAPHY

BEROZA, M., and BOWMAN, M. C.   1965.   Identification of pesticides at nanogram levels by extraction p-values.   Anal. Chem. *37*, 291–292.

BURKE, J., and GUIFRIDA, L.   1964.   Investigations of electron capture G.C. for the analysis of multiple chlorinated pesticide residues in vegetables.   J. Assoc. Offic. Agr. Chemists *47*, 326–342.

CRAIG, L. C., and POST, O.   1949.   Apparatus for countercurrent distribution.   Anal. Chem. *21*, 500–504.

EVANS, C. D., McCONNELL, D. G., SCHOLFIELD, C. R., and DUTTON, J. H. 1966. The structure of unsaturated glycerides. J. Am. Oil Chemists' Soc. *43*, 345–349.

HAGEN, S. N., MURPHY, G. W., and SKELLY, L. M. 1967. Extraction of lipids from raw beef. J. Assoc. Offic. Anal. Chemists *50*, 250–255.

JOHNSON, L. Y. 1965. Collaborative study of a method for multiple chlorinated pesticide residues in fatty foods. J. Assoc. Offic. Agr. Chemists *48*, 668–675.

MANSKE, D., and FRASCH, D. L. 1966. Identification of pesticides by extraction p-values. Food Drug Admin. Interbureau Bylines *4*, 171–174.

MILLS, P. A., ONLEY, J. H., and GAITHER, R. A. 1963. Rapid method for chlorinated pesticide residues in non-fatty foods. J. Assoc. Offic. Agr. Chemists *46*, 186–191.

NOVAK, M. 1965. Colorimetric ultra-micro method for the determination o free fatty acids. J. Lipid Res. *6*, 431–433.

THALER, H., and ARNETH, W. 1968. Coffee and coffee substitutes. XI. Polysaccharides of green coffee arabica beans. Z. Lebensm. Untersuch. Forsch. *138*, 26–35.

THOMAS, F., CARPENA, O., and ABRISQUETA, C. 1966. Spectrophotometric determination of hesperidin and citroflavonoids. Ann. Bromat. *18*, 393–406.

CHAPTER 24

Centrifugation

Centrifugation is used for separation of solids from liquids and from immiscible solvents, and for resolution of emulsions that are formed during extraction. Centrifugation at high speeds (ultracentrifugation) is useful for the concentration of high molecular weight materials, and for estimating their molecular weights.

Basic Considerations

In the equation

$$F = ma = m\omega^2 r \qquad (24.1)$$

F = force on a particle in dynes, m = mass of a particle in grams, ω = angular velocity of rotation in radians per second, and r = radial distance of a particle from the axis of rotation in cm.

We can express the centrifugal force F' (in grams):

$$F' = m\omega^2 r/g \qquad (24.2)$$

where g = gravitational constant 980.7 cm/sec^2 or

$$F' = 1.118 \times 10^{-5} \times m \times r \times N^2$$

where N = speed of rotation in rpm

Another useful parameter is:

$$RCF = \omega^2 r/g \qquad (24.3)$$

where RCF is relative centrifugal force defined as the force acting on a given particle in a centrifugal field in terms of multiples of its weight in the earth's gravitational field, and

$$RCF = 1.118 \times 10^{-5} \times r \times N^2 \qquad (24.4)$$

If the axis of rotation is horizontal, the difference between RCF at the tip and at the surface of the liquid is generally twice, or more. As the radial distance from the axis of rotation and the depth of the liquid in the centrifuged tube vary widely with different types of centrifuges, it is inadequate to report results in terms of the speed of rotation in rpm. Reporting data in terms of centrifugal force ($\times g$) calculated from the latter formula is much more meaningful.

Generally, the centrifugal force at the middle of the liquid depth is reported; reporting the centrifugal force at both the tip end and the free surface of the liquid is preferable. If the axis of rotation is vertical, the downward pull of gravity on the rotating parts, is insignificant if RCF is $25g$ or above.

The effective sedimenting or centrifugal force must be corrected by the bouyancy factor, from the Archimedes law,

$$m_{eff} = m - V\rho \tag{24.5}$$

in which the particle volume is V in a solution that has a density of ρ and

$$F_{eff} = (m - V\rho)\omega^2 r \tag{24.6}$$

If the particle is a sphere

$$F_{eff} = \left(\Delta\rho \frac{\pi}{6} D^3\right) \omega^2 r \tag{24.7}$$

where D = diameter of the particle, and $\Delta\rho$ = difference between the density of the particle and that of the fluid in which it is suspended. This is the force that is available to move the particle through the liquid medium. The sedimenting molecules are subjected to a frictional force $F\rho$ exerted by the medium. For a slowly moving small particle, the frictional force is given by Stokes' law

$$F = 3\pi\eta D V_s \tag{24.8}$$

where η = viscosity of the liquid, and
V_s = velocity of particle moving through the liquid phase

Combining the latter two equations we have

$$V_s = \Delta\rho D^2 \omega^2 r/18\eta \tag{24.9}$$

from which we can calculate the velocity of a particle at a distance r from the axis of rotation.

Equipment

There are two basic types of centrifuges, and a third that may be considered a combination of the two. (1) Solid wall centrifuges in which separation or concentration is by subsidence or flotation. (2) Perforated wall centrifuges (centrifugal filters) in which the solid phase is supported on a permeable surface which the fluid phase is free to pass. (3) Combinations of the two in which the primary concentration is effected by subsidence followed by drainage of the liquid phase away from the solid phase (Amber and Keith 1956).

Bottle centrifuges have a motor-driven vertical spindle on which various heads or rotors can be mounted The rotors carry metal containers into which fit glass tubes or bottles (from 2 to 16 or more) of a total capacity of up to about 1 gal. The bottle centrifuges are particularly useful for analytical and small-scale preparative separations. The time, speed, and (in some), the temperature can be closely controlled. They are generally enclosed for safety in a metal guard bowl. Most bottle centrifuges have a horizontal swinging type rotor. The glass tubes (bottles) are placed into metal shields on a rubber cushion, and the shields or containers are supported in trunnion rings which are set in slots in the rotor head. Since the center of gravity of the assembly is below the trunnions, the tubes hang vertically at rest. As the rotor starts to turn they gradually swing out to a horizontal position where they remain as long as the head is rotating. The advantage of the method is fractional sedimentation across the tube length; the disadvantages are the long path of travel of some particles and hindered settling near the bottom. To overcome the disadvantages, prolonged centrifugation at high speeds is required.

In the angle or conical type centrifuges, the tubes (in metal shields) are held at a fixed angle of about 45° in holes in the rotor, both at rest and during rotation. The rotor can be used for relatively high centrifugation speeds. The particles travel in free sedimentation only a distance equal up to the diameter of the tube times the secant of the angle of inclination from the vertical. The particles that strike the glass wall of the tube aggregate and slide to the bottom. The sedimenting particles deposit at an angle.

The bottle centrifuge can accomodate a large variety of tubes and bottles of various sizes: plain and graduated cylinders, plain tubes with round or conical bottoms, plain and graduated pear-shaped tubes, and separatory funnels. Special refrigerated bottle centrifuges can be used for work with low-melting point compounds or heat-labile biochemical materials. Heated centrifuges are used in fat determination in dairy products.

The hydrostatic pressure exerted on a filled glass container in a laboratory bottle centrifuge may reach high enough values to rupture the container. This may be largely offset by filling the space between the glass tube and the metal cup with a liquid (water, glycerin, ethylene glycol) or by using semielastic, plastic containers that deform sufficiently (but do not rupture) under pressure to carry the hydrostatic pressure directly on the wall of the metal shield. All rotating parts of the centrifuge are subject to stresses created by centrifugal forces; those stresses impose limitations on the maximum permissible speed of the centrifuge. The effects of those stresses can be minimized in properly designed centrifuges by careful balancing before loading, increasing slowly the speed to that desired, and proper maintenance and following manufacturer's instructions. To prevent remixing after centrifugation, deceleration should be slow.

Ultracentrifugation

Separation and fractionation of macromolecules by subjecting them to a strong centrifugal force originated with Svedberg and co-workers, who in the early twenties invented and developed the instrument called the ultracentrifuge. In 1924, Svedberg and Rinde described the use of ultracentrifugation in studies of colloidal particles. Two years later, Svedberg and Fahraens (1926) described the determination of the molecular weight of proteins from sedimentation in an ultracentrifuge by measuring the diffusion coefficient. The centrifugal forces that can be attained in ultracentrifugation are in the order of 500,000 times gravity. The high forces can be used for the determination of molecular weights of macromolecules or for preparative fractionations.

The theory and practice of ultracentrifuge measurements have been described in a number of monographs and reviews, preeminent is the classic work of Svedberg and Pederson (1940). More recent books include those by Williams (1963), Schachman (1959), and by Fujita (1962); the last two are devoted exclusively to theory.

Analytical Ultracentrifuge

The commercially available analytical ultracentrifuge provides a photographic record of the migration of high molecular weight substances in a strong gravitational field. The instrument has automatic temperature and speed controls for the rotor, a high vacuum chamber to reduce friction, an optical system for measuring the rate at which individual peaks (representing different proteins) move towards the bottom of the cell, an automatic photographic system for recording changes in concentration at specified intervals, and special cells. Three optical systems are available: Schlieren, interference, and absorption. The most commonly used system is the astigmatic Schlieren optics. The photographic record of the sedimentation pattern using Schlieren optics gives the concentration gradient in the cell in terms of the refractive index gradient.

In using the ultracentrifuge in determining the molecular weight of proteins, two lines of approach are possible: the sedimentation equilibrium, and the sedimentation velocity methods.

In the sedimentation equilibrium a relatively low centrifugal force is applied to a solution till the distribution of protein throughout the column of liquid in the centrifuge tube reaches a steady state. What one measures is the stage of equilibrium between outward sedimentation of proteins and backward diffusion. At this stage, the molecular weight (M) can be determined from the formula:

$$M = [2RT \ln (c_1/c_2)]/[\omega^2(1 - \rho/\sigma) (x_1{}^2 - x_2{}^2)] \quad (24.10)$$

where R = the gas constant
      T = absolute temperature
      c_1, c_2 = concentrations at distances x_1 and x_2 from the center of rotation
      ρ = the density of the solvent
      σ = density of the protein
      ω = angular velocity.

The term

$$\rho/\sigma = \rho v \quad (24.11)$$

where v is the partial specific volume of the protein (increase in volume when 1 gm of dry protein is added to a large volume of liquid) and can be assumed to be 0.75. The angular velocity

$$\omega = V/x \quad (24.12)$$

where V is the velocity of the centrifuged solution, and x the distance from the center of the rotor. If the number of revolutions per second is z, then

$$V = 2r\pi z \quad (24.13)$$

and

$$\omega = 2\pi z \quad (24.14)$$

It is customary to give the velocity in rpm, and $z = \text{rpm}/60$, or

$$\omega = \frac{2 \times 3.14 \text{ rpm}}{60} = 0.105 \text{ rpm}$$

As mentioned earlier, in the sedimention equilibrium method relatively low centrifugal velocities are used. If very high velocities were used, there would be packing of the protein at the bottom with little difference in c_1 and c_2 or x_1 and x_2 values. The main disadvantage of the sedimentation equilibrium method is that it requires several days for a determination. In the approach to equilibrium techniques (Trautman 1964), concentrations of protein are measured at various times at the meniscus of the solution and at the bottom of the tube. From those data, c_1 and c_2 are extrapolated. The procedure is especially useful with small protein molecules which sediment slowly. If special small cells are used, ultracentrifugation may take as little as 45 to 70 min.

In the sedimentation velocity method, high speeds at which the protein particles sediment at a fast rate are used. If the molecular weight of the particles is high, their rate of diffusion can be neglected. The sedimentation velocity depends on the shape and the hydration of the protein molecules, whereas the sedimentation equilibrium is independent of those factors and depends only on the molecular weight. Nevertheless, many of the molecular weight determinations are made by the sedimentation velocity method, combined with diffusion measurements. The principal reason for this is the shorter time of centrifugation which eliminates the danger of bacterial growth and decomposition of the proteins.

The rate of sedimentation is usually expressed in terms of the sedimentation constant, s, which is the velocity per unit centrifugal field force, and which has the dimensions of time. For most proteins, s ranges from 1 to 200×10^{-13} sec. A sedimentation constant of 1×10^{-13} is called a Svedberg unit, and sedimentation constants generally are given in Svedberg units.

Preparative Ultracentrifugation.—Moving boundary ultracentrifugation is of limited use in preparative work. While the lightest protein in a mixture can be obtained in pure form, the heavier fractions contain some of the lighter ones. For preparative work several methods are available, which eliminate diffusion of particles and make complete separations possible. The methods are called *density gradient ultracentrifugation.* They depend on the formation of a gradient, the density of which increases with distance from the axis of rotation. Actually, usefulness of density gradient ultracentrifugation is threefold: it provides information on the size of the separated molecules in a mixture, permits separation on a preparative scale, and can be carried out with low concentrations of solute.

Density gradient separations can be classified into three categories according to the way in which the gradient is used (Anon.

1960). In *stabilized moving boundary centrifugation*, one uses a shallow gradient formed during the centrifugation. Its function is to stabilize against convection. Material to be fractionated is distributed throughout the solution (Pickels 1943). The sample in the preparative centrifuge is separated into fractions that are analyzed to determine the position of the boundary. If several components are present, several boundaries form and the sedimentation coefficient of each can be calculated.

In *zone centrifugation*, a solution containing particles of varied characteristics is layered on top of a steep gradient (Brakke 1951). The steep gradient keeps convective stirring at a minimum. Each substance sedimenting at its own rate forms a band or zone in the fluid column. Those zones of solutes will be separated from one another by distances related to their sedimentation rates. After centrifugation, each substance can be drawn off separately for the determination of its sedimentation rate and for further analysis. This fractionation technique is probably the most widely used in biochemical ultracentrifugation. It has two limitations: only small amounts of material can be separated at high speeds, and the separation is incomplete due to the wall effect (from some particles being reflected from the tube walls back into the solution, sticking to the walls, or clumping). In the zonal ultracentrifuge developed by Anderson (1962), high-speed rotors with increased capacity are used for sharp sample zone separations under conditions that minimize wall effects. Sedimentation takes place in sector-shaped compartments in hollow rotors of capacities up to 120 times that of a high speed swinging bucket rotor of comparable separating capabilities. The gradient solutions are introduced and recovered while stabilized in a centrifugal field. Continuous flow zonal centrifugation is also available.

In *isopycnic gradient centrifugation* (Messelson *et al.* 1957), separation is based on differences in density of the macromolecules in a sample solution that is usually distributed evenly throughout the gradient column before ultracentrifugation. The gradient column must cover the entire density range of the particles, and centrifugation is continued until the particles reach positions at which the density of the surrounding liquid is equal to their own.

One should distinguish clearly between zone (or band) and isopycnic (or equilibrium) density centrifugation. Zone sedimentation is a kinetic method that determines the sedimentation coefficient of the macromolecules (Szybalski 1968). The gradient is preformed and is usually substantially less dense than that of the macromolecules.   The purpose of the gradient is to stabilize the sedimenting band of macromolecules and prevent convection.   The macromolecules move continuously through the gradient and settle at the bottom of the tube if centrifugation is allowed to proceed for a very long period.   Isopycnic density gradient ultracentrifugation is an equilibrium (static) method which determines the buoyant density of the macromolecules.   The gradient is self-generating in the centrifugal field, and its density range is so adjusted as to be denser at the bottom of the tube and less dense near the meniscus, than the macromolecules in the column of solution.   Thus, the macromolecules, most frequently nucleic acid or viruses, form a definite band at a definite level in the column and remain there irrespective of the length of centrifugation.

Materials used to form gradients should be chemically inert to the studied system, nontoxic, soluble in water and salt solutions, have a high density and molecular weight, and have a low viscosity.   For separation and analysis of proteins, the gradient material should contain no nitrogen.   High-density materials are required to form a steep gradient; if the molecular weight is high, the osmotic pressure gradient in the tube is relatively small.   Low viscosity permits easy handling during gradient formation, and rapid sedimentation and fractionation.   Sucrose is the most widely used material for the gradient.   Its main disadvantage is high viscosity.   Ficoll is a commercially available water-soluble neutral colloid with properties similar to that of a polysaccharide.   Its average molecular weight is about 50,000 and it is stable in nonoxidizing neutral or alkaline solutions.   Its viscosity and density are lower than those of sucrose. For the separation of nucleic acids, inorganic salts (mainly cesium chloride or rubidium chloride) are used.

Gradients can be produced manually, by layering sucrose (or other) solutions of decreasing density into a centrifuge tube.   After 24 hr or more, the gradient becomes semicontinuous as a result of diffusion.   Commercially available or mechanical devices constructed in the laboratory to produce various stable gradients (linear, exponential, concave, convex, or S-shaped) are increasingly popular. The use of a specific gradient depends on the distribution of sedimentation rates or densities.   Thus, a linear gradient is most useful for a solution containing a relatively high range of densities.   An S-shaped gradient would give best results for a mixture containing two components with similar densities and one with a considerably higher density.

The position of particles in the centrifuged tube can be determined visually (under ordinary or ultraviolet light) or after the contents of the tube has been separated into small fractions. This is done most commonly by carefully puncturing a hole at the bottom and removing dropwise the tube content.

BIBLIOGRAPHY

AMBLER, C. M., and KEITH, F. W. 1956. Centrifuging. *In* Technique of Organic Chemistry, Vol. III, A. Weissberger (Editor). Interscience Publishers, New York.

ANDERSON, N. G. 1962. The zonal ultracentrifuge. A new instrument for fractionating mixtures of particles. J. Phys. Chem. *66*, 1984–1989.

ANON. 1960. An Introduction to Density Gradient Centrifugation, Tech. Rev. 1. Beckman Instruments Inc., Spinco Division, Palo Alto, Calif.

BRAKKE, M. K. 1951. Density gradient centrifugation: a new separation technique. J. Am. Chem. Soc. *73*, 1847–1848.

FUJITA, H. 1962. Mathematic Theory of Sedimentation Analysis. Academic Press. New York.

MESSELSON, M., STAHL, F. W., and VINOGRAD, J. 1957. Equilibrium sedimentation of macromolecules in density gradients. Proc. Natl. Acad. Sci. *43*, 581–594.

PICKELS, E. G. 1943. Sedimentation in the angle centrifuge. J. Gen. Physiol. *26*, 341–365.

SCHACHMAN, H. K. 1959. Ultracentrifugation in Biochemistry. Academic Press, New York.

SVEDBERG, T., and FAHRAENS, R. 1926. A new method for the determination of the molecular weight of the proteins. J. Am. Chem. Soc. *48*, 430–438.

SVEDBERG, T., and PEDERSEN, K. O. 1940. The Ultracentrifuge. Clarendon Press, Oxford, England.

SVEDBERG, T., and RINDE, H. 1924. The ultracentrifuge, a new instrument for the determination of size and distribution of size of particle in amicroscopic colloids. J. Am. Chem. Soc. *46*, 2677–2693.

SZYBALSKI, W. 1968. Equilibrium sedimentation of viruses, nucleic acids, and other macromolecules in density gradients. Fractions *1*, 1–15.

TRAUTMAN, R. 1964. Ultracentrifugation. *In* Instrumental Methods in Experimental Biology, D. W. Newman (Editor). MacMillan Publishing Co., New York.

WILLIAMS, J. W. 1963. Ultracentrifugal Analysis in Theory and Experiment. Academic Press, New York.

CHAPTER 25

Densimetry

INTRODUCTION

Determination of density is one of the most common, simple measurements in food analyses. It is made primarily on liquids, though it can also be used in analyses of solids.

The determination in liquids provides a useful parameter whenever the density of a mixture of two compounds is a function of its composition, and when the composition can be read off calibration graphs or tables. This procedure is used to determine the sugar or alcohol content of aqueous solutions.

The smaller the change in volume on mixing the components of a solution, the more closely is the concentration related to density. If in addition, changes in density per unit change in concentration are large, densimetric methods yield an accurate as well as rapid method of analysis.

Densimetry also can be used in analyses of more complex systems, such as tomato juice or milk. In those instances, density is an index of total solids contents.

The density of many substances is a characteristic physical property and serves for identification purposes. Density is a function of both the length of the carbon chain of the glyceride fatty acids and the degree of unsaturation of glyceride fatty acids. The determination is used in evaluating commercial fats and oils. In addition, several rapid methods are available to determine oil content from measuring density of extracts of lipids obtained under specified conditions.

Density is an important criterion of seed purity; texture and softness of fruits; maturity of such products as peas, sweet corn, and lima beans; and has been proposed as an index of soundness of dried prunes or plumpness and dryness of raisins.

Definitions

Density, d is defined as

$$d = \text{mass/volume} = m/v \tag{25.1}$$

In accordance with the cgs system, *absolute density*, d^t, at a temperature t, is defined as

$$d^t = m \text{ gm}/V_{cc} \text{ cc} \tag{25.2}$$

In practice, density measurements are expressed in grams of weight per milliliter which under most conditions is equivalent to grams of mass per milliliter. The *relative density* at $t°C$ is defined as

$$d_4^t = m \text{ gm}/V_{ml} \text{ ml} \tag{25.3}$$

By definition, 1 ml $= 0.001$ part of the volume of 1 kg of pure water at $3.98°C$. Therefore, d_4^t gives the ratio of absolute density at $t°C$ to the absolute density of water at $3.98°$.

Originally, it was intended that 1 cc should equal 1 ml, but precise measurements have shown that

$$1 \text{ cc} = 0.999973 \text{ ml, and that} \tag{25.4}$$

$$V_{cc} = 1.000027 \, V_{ml} \tag{25.5}$$

For most food analyses, the difference between d^t (gm/cc) and d_4^t (gm/ml) is negligible.

The weight W is related to the mass, m, by the equation

$$W = mg \tag{25.6}$$

where $g =$ the acceleration due to gravity is 980.665 cm/sec^2 at a latitude of $45°$ and at sea level.

Based on the proportionality between mass and weight of calibrated weights and mass and weight of an analyzed sample, at a given geographical location, there is no change in the numerical value of density by using grams of weight instead of grams of mass.

In practice, d_4^t is used because it can be determined accurately by comparing the weights of equal volumes of the substance at $4°C$ and of water at $3.98°C$. Sometimes the term *specific gravity* d_t^t, is used. It is a dimensionless number defined as the mass, m, of a substance at $t°C$ relative to mass, m_o, of an equal volume of water at $t°C$.

$$d_t^t = mV/m_o V \tag{25.7}$$

For the calibration of instruments used in density determinations, pure water or mercury is used.

Density is affected by temperature. For water at room temperature, the density decreases by about 0.03% per °C rise in temperature, and the temperature coefficient of cubical expansion is

$$\beta_t = 3 \times 10^{-4} \text{ deg}^{-1} \tag{25.8}$$

Some organic solvents (aliphatic hydrocarbons) have β_t values up to five times greater than water; the density of organic solids usually changes somewhat less rapidly with temperature than that of liquids.

If an error of not more than ± 0.001 in density is permitted, temperature control should be within $1\,^\circ C$ (Bauer 1945). For higher accuracy, correspondingly better control is needed. However, the required temperature control depends also on the method used. When pycnometric determinations are made, the temperature effect is somewhat reduced by the expansion of the glass container.

Atmospheric pressure differences are important if they deviate considerably from the standard one. Ordinary fluctuations in pressure encountered in the laboratory have little effect on density.

During mixing of two or more liquids, the volume increases or decreases above the theoretical one. The decrease is as high as 9.5% in mixing equimolar amounts of water and sulfuric acid, and 2.56% with water and ethanol. In the case of acetone—disulfide there is an increase of 1.2%. The changes are concentration-dependent and the resultant effects on density must be established empirically (Mahling 1965).

Measurement of the Density of Liquids

Pycnometers.—The most common method of density determination consists of measuring the weight of a known volume of liquid in a vessel, the volume of which was calibrated in terms of weight of pure water which the vessel holds. This is best done by using a pycnometer. If the weighing is done with an error of not more than 0.1 mg, for a vessel up to 30 ml, an accuracy of $\pm 5 \times 10^{-6}$ can be attained (Bauer 1945). Increasing the size of the vessel, increases theoretically the precision, but is accompanied by compensating errors resulting from such factors as insensitivity of the balance and difficulty in maintaining a uniform temperature of a large volume of liquid.

Pycnometers should preferably be made from resistant glass with a low coefficient of thermal expansion such as Pyrex or Vycor, or fused quartz. Figure 25.1 shows five types of commercially available pycnometers.

Type A—is a Gay-Lussac bottle with ground-in perforated glass stopper and outer ground-on cap to reduce evaporation.

Type B—is similar to A, but has an evacuated jacket to stabilize the temperature of the material being weighed.

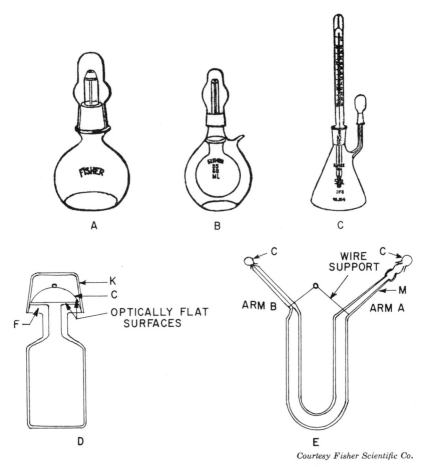

Courtesy Fisher Scientific Co.

FIG. 25.1. DIFFERENT TYPES OF PYCNOMETERS

Type C—has a thermometer connected to the flask by a ground glass joint.

Type D—has a wide neck opening and can be used for both solids and liquids. The outside cap minimizes evaporation and leakage.

Type E—is a standard Sprengel-Ostwald pycnometer. It has two openings and is easy to fill and clean even though the side arms are made of small-bore capillaries. It can be filled without danger of loss by evaporation. A simple, though less precise, Sprengel pycnometer can be made from an ordinary pipet.

Before use, the pycnometer must be thoroughly cleaned with a mixture of potassium dichromate and concentrated sulfuric acid, rinsed thoroughly with water, and dried. To avoid changes in volume, the pycnometer must not be subjected to excessive temperature or pressure changes.

Pycnometers can give reproducible results with an accuracy in the fourth decimal place. For fifth place accuracy, the weights must be checked against one another to obtain their relative values. Metal-plated weights are preferrable to lacquered weights as the latter are more susceptible to humidity effects. An error of about 2×10^{-5} in density can result from inequalities in the lengths of the balance arms. The error can be eliminated by determining the average of 2 weighings, 1 on each balance pan. A pycnometer dried at room temperature under vacuo, can adsorb several milligrams of water by standing in a humid atmosphere. The adsorbed moisture can be removed by carefully wiping the glass surface with a lintless cloth, and allowing the vessel to stand in the balance case for about 15 min before weighing (Bauer 1945).

Bubbles are the most common sources of relatively large errors. Whenever possible the liquid should be boiled and cooled shortly before filling. It is advisable to cool the liquid about 1°C below the thermostat temperature just before filling to ensure an excess of liquid when equilibrium is reached. The excess liquid is removed by fine capillaries or strips of filter paper. Special filling devices are available to accurately fill the pycnometer and remove all air.

Buoyancy Methods.—Several methods for measuring density are based on the Archimedes principle which states that the upward buoyant force exerted on a body immersed in a liquid is equal to the weight of the displaced liquid. This principle can be applied in several ways to measure density.

An ordinary gravimetric balance can be adapted to precise density measurements. This is done by removing one of the pans and attaching to the balance arm a fine, freshly platinized (black) platinum wire at the end of which is suspended a sinker (cylinder of glass or metal). The sinker is immersed in a column of liquid below the balance case. The combined volume (V) of the sinker and wire (up to the point of immersion) is determined by calibration with water. If the weight in air is W_o and in the liquid W_L, the apparent weight loss $W_o - W_L$ is equal to the mass of displaced liquid and the density d

$$d = (W_o - W_L)/V \qquad (25.9)$$

The Mohr balance, as improved by Westphal, is a commercially available direct reading instrument based on the above principle.

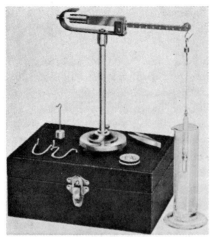

FIG. 25.2. THE WESTPHAL BALANCE

Courtesy A. H. Thomas Co.

This balance can be used to determine the specific gravity of liquids heavier or lighter than water to the fourth decimal (Fig. 25.2).

The glass plummet displaces 5 g of water at 15°C; the largest rider weighs 5 gm and equals 1 when placed on the same hook with the plummet. The other riders weigh 0.5, 0.05, and 0.005 gm, respectively. A 15-gm weight is supplied for initially balancing the beam in the air.

The riders are placed on the beam in succession until it balances. The sum of the settings shows the specific gravity. If the liquid is lighter than water, the position of the heaviest rider shows tenths of a unit; the smaller riders give respectively the second, third, and fourth decimal places. If the liquid is heavier than water, a second 5-gm rider is placed on the plummet hook, where it has the value of 1. The balance is widely used for rapid determination of density of nonvolatile solutions and nonhygroscopic liquids.

Wagner *et al.* (1942) determined density with a quartz float attached to an elastic quartz helix mounted in a temperature controlled tube containing the investigated liquid. The float is free to move and the helix is stretched by the buoyant force acting on the float. The density is determined from the change in length of the helix. The method is rapid, can be used to determine density of liquids in a closed container, and is very precise provided temperature control is adequate. However, calibration of the helix is quite time-consuming.

Hydrometry is based on the principle that the same body displaces equal weights of all liquids in which it floats. If V_1 and V_2

denote the volumes of two liquids displaced by the same floating body, and D_1 and D_2 their respective densities,

$$V_1 D_1 = V_2 D_2 \qquad (25.10)$$

and

$$D_1 / D_2 = V_2 / V_1 \qquad (25.11)$$

Thus, the volumes of different liquids displaced by the same floating body are inversely proportional to the densities of the liquids. If the floating body is an upright cylinder of uniform diameter, the volumes displaced are proportional to the depths to which the body sinks

$$D_1 / D_2 = H_2 / H_1 \qquad (25.12)$$

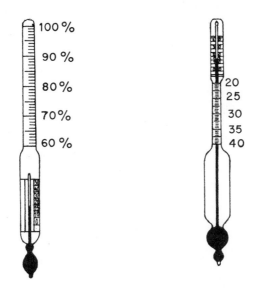

FIG. 25.3. HYDROMETERS: LEFT ALCOHOLOMETER; RIGHT, LACTODENSIMETER

Hydrometers (Fig. 25.3) are hollow glass bodies that have a broad and heavy bottom and a narrow upper stem. The lower part is loaded by a metallic insert of appropriate weight so that the whole hydrometer sinks in the tested solution to such a depth that the upper calibrated stem is in part above the liquid. The total weight of the hydrometer must be smaller than that of the liquid it displaces. The deeper the hydrometer sinks, the lower the density of

the solution.   The calibrated stem shows increasing density values
from top to the bottom of the stem.   For a wide range of densities,
a rather long glass stem that might break easily would be required.
Generally, for orientation purposes a wide-range and less precise
hydrometer is used.   This is followed by more precise determinations
with a set of overlapping narrow-range hydrometers.

The precision of hydrometers is generally limited to three decimal
places.   The container, generally a glass cylinder, should be at least
twice as wide as the diameter of the bottom part of the hydrometer.
A density determination with a hydrometer must be corrected for
deviation from the standard temperature.   Some hydrometers have
a built-in thermometer and correction scale.

The scale can be calibrated in density units or in percentage composition as related to density.

Alcoholometers are used for determining the percentage of alcohol
by volume (Tralle Scale) and deviations from "Proof" (50% on
Tralle Scale = 0 on Proof Scale).   The Tralle Scale is graduated
from 0 to 100% in 1% divisions, and the Proof Scale from 100 under
proof to 100 over proof in single proof divisions.

Baumé hydrometers are of two kinds: heavy Bé for liquids
heavier than water, and light Bé for liquids lighter than water.
In the first, 0° corresponds to a density of 1.000 and 66° to a density
of 1.842.   In the lighter than water scale, 0° Bé is equivalent to the
density of a 10% solution of NaCl, and 60° Bé corresponds to a
density of 0.745.   For Bé degrees on a scale of densities above
1.000,

$$\text{density} = \frac{145}{145 - \text{Bé reading}} \qquad (25.13)$$

Saccharometers can have the Balling scale (% of sugar) or the
Brix scale (% of sugar by weight at 20°C).

Salometers are used to determine the percentage of saturation of
salt brines.

Oleometers for vegetable and sperm oils have a scale of 50° to 0°
that corresponds to densities of 0.870–0.897.

Lactometers are used to determine dry milk solids (and adulteration
by water addition) of milk from density determinations.   The
Soxhlet lactometer has a scale of 25–35 (density 1.025–1.035) subdivided into suitable divisions.   The Quevenne scale has a range of
14° to 42° Quevenne in 1° divisions, and corresponding scales to
indicate percentage of water in whole or skim milk.   A thermometer

scale above the Quevenne scale indicates corrections that are to be made when temperatures are above or below 60°F.

The density of milk also can be estimated from the sinking of plastic colored beads of specified density (Erb and Manus 1963).

Measurement of the Density of Solids

Methods for measuring the density of solids are considerably less precise than those for liquids mainly due to nonhomogeneity of solid foods, partial solubility, and presence of occluded air bubbles. Most methods depend on immersing the solid in an inert liquid of known density. The volume of known weight of solid can be determined from the volume of a fluid that is displaced by a submerged solid. In the pycnometric method, the volume of the solid is determined from the change in weight when the vessel is successively filled with a liquid of known density, with solid in air, and with solid in liquid.

When a solid neither rises nor falls through a liquid in which it is submerged, the density of the solid and liquid are equal. The use of this principle was previously mentioned in connection with determining density of milk with calibrated beads. By preparing a series

Courtesy Henry Simon, Ltd.

FIG. 25.4. APPARATUS FOR DETERMINATION OF TEST
WEIGHT OF CEREAL GRAINS

of liquid mixtures of known density, the density of solid foods can be determined.

In many instances, apparent (rather than absolute) density is determined. Thus, volume of bread (determined by displacement of dwarf rapeseed) is an important criterion of the breadmaking potentialities of a wheat flour. For many years, plumpness has been considered an important characteristic of good grain. This quality is indicated in a general way by the test weight of a grain. Minimum test weight is specified by standards for cereal grains and some oil seeds (flax and soybeans) (U. S. Dept. Agr. 1947).

The equipment used in test weight determinations is shown in Fig. 25.4. In the United States and Canada, test weight is determined in pounds per bushel; in most other countries using the metric system in kilograms per hectoliter.

Test weight is influenced mainly by kernel shape and size as they govern the packing of grain in a container. Kernel size for a specific grain is of less significance. The other important factor is density of grain that depends on the structure of the grain and its chemical composition including moisture content. Also wetting, subsequent drying, and even mechanical handling in grain elevators influence the test weight. Determinations of test weight must be made under rigidly controlled conditions to obtain reliable results. Theoretically, conversion factors can be applied to convert test weight values from one system to another, provided tests are made in containers of comparable size.

BIBLIOGRAPHY

BAUER, N. 1945. Determination of density. In Physical Methods of Organic Chemistry, Vol. I, A. Weissberger (Editor). Interscience Publishers, New York.

ERB, R. E., and MANUS, L. J. 1963. Estimating solids non-fat in herd milk using the plastic bead method of Golding. J. Dairy Sci. 46, 1373–1379.

MAHLING, A. 1965. Density. In Handbook of Food Chemistry, Analysis of Foods, Physical and Physico-Chemical Assay Methods, Vol. II, Part I, J. Schormuller (Editor). Springer-Verlag, Berlin (German).

U.S. DEPT. AGR. 1947. Grain Grading Primer, Misc. Publ. 740, U.S. Dept. Agr., Washington, D.C.

WAGNER, G. H., BAILEY, G. C., and EVERSOLE, W. G. 1942. Determining liquid and vapor densities in closed systems. A precise method. Ind. Eng. Chem. Anal. Ed. 14, 129–131.

CHAPTER 26

Refractometry

When a ray of electromagnetic radiation strikes a flat surface at an angle, the ray may be bent upward (*reflected*) or bend downward (*refracted*) as in Fig. 26.1. Notice that the ray did not go straight through the material.

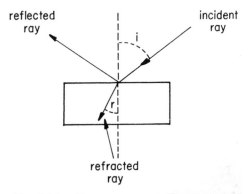

reflected ray

incident ray

refracted ray

FIG. 26.1. REFLECTION AND REFRACTION

The amount of refraction is a characteristic of every substance and is called the refractive index.

$$n = \sin i / \sin r \qquad (26.1)$$

Refractive index measurements have long been used for the qualitative identification of unknown compounds by comparing the refractive index of the compound with literature values. When density measurements are added, a certain amount of information about the structure of the compound can be obtained. Each atom, bond, or group will contribute to the overall refractive index. When these individual contributions are taken together the result is a *specific refractivity*. If the specific refractivity is multiplied by the molecular weight of the compound, then the *molecular refractivity* is obtained. The Lorentz-Lorenz equation (26.2) shows this relation.

$$R = \frac{(n^2 - 1)}{(n^2 + 2)} \times \frac{M}{d} \qquad (26.2)$$

383

The relation is particularly useful in the identification of unknown materials. Suppose you have isolated a new compound and suspect that it may be one of several possible structures. Using the data in Table 26.1 you can calculate the expected refractive index for each compound and then see which structure actually fits.

TABLE 26.1

ATOMIC REFRACTIONS

Element	Na_D	Element	Na_D
F	1.0	Nitro group in	
C	2.42	Aromatic nitro compounds	7.30
H	1.10	Nitroamines	7.51
O in OH	1.52	Nitroparaffins	6.72
O in ester OR	1.64	Alkyl nitrites	7.44
O⁻	2.21	Alkyl nitrates	7.59
Cl	5.97		
Br	8.86	Nitroso group in	
I	13.90	Nitrates	5.91
S in SH	7.69	Nitrosoamines	5.37
S in R_2S	7.97	Structural units	
S in RCNS	7.91		
S in R_2S_2	8.11	Double bond	
O in ether	1.64	no radicals	1.51
N in I° aliph. amines	2.32	$RCH{=}CH_2$	1.60
in II° aliph. amines	2.49	$RCH{=}CHR$	1.75
in III° aliph. amines	2.84	$R_2C{=}CHR$	1.88
in I° arom. amines	3.21	$R_2C{=}CR_2$	2.00
in II° arom. amines	3.59	Triple bond	2.40
in III° aromatic amines	4.36	3 membered ring	0.71
in hydroxylamines	2.48	4 membered ring	0.48
in hydrazines	2.47		
in aliphatic cyanides	3.05	Diazo group	
in aromatic cyanides	3.79	$-C{\Big\langle}\genfrac{}{}{0pt}{}{N}{N}$	8.43
in aliphatic oximes	3.93		
in amides	2.65	$-N{\Big\langle}\genfrac{}{}{0pt}{}{N}{N}$	7.47
in II° amides	2.27		
in III° amides	2.71		

Sources: Eisendohr (1910, 1912).

Example

Calculate the refractive index of dimethoxymethane (CH_3OCH_2-OCH_3, $M = 76.10, d = 0.8560$)
From the atomic refractions:

$$
\begin{array}{llll}
3C & = 3 \times 2.42 = & 7.26 \\
8H & = 8 \times 1.10 = & 8.80 \\
2 \text{ ether O} & = 2 \times 1.64 = & \underline{3.28} \\
& R = & 19.34
\end{array}
$$

$$19.34 = \left(\frac{n^2 - 1}{n^2 + 2}\right) \times \frac{76.10}{0.8560}$$

$$(n^2 - 1)/(n^2 + 2) = 0.21754$$
$$n^2 - 1 = 0.2175\ n^2 + 0.4351$$
$$0.7825\ n^2 = 1.4351$$
$$n^2 = 1.834$$

thus
$$n_D = 1.354$$

The value reported in the literature is 1.3534.

Refractive Index of Mixtures

This determination is very important because it can be used to determine the concentration of eluents from paper, column, and thin-layer chromatography where the sample is mixed with the solvent.

The refractive index of mixtures will vary linearly with the mole fraction of the components. Refractive index measurements provide one way of measuring mixtures of salts that would otherwise be difficult to determine.

The fact that specific and molecular refractivities are an additive property of substances, makes it possible to set up an equation for calculating the refractivity of a mixture from the refractivity of the solvent and the solute. The equation is:

$$Zr_m = Xr_a + Yr_b \qquad (26.4)$$

where X grams of compound A, and Y grams of compound B with specific refractivities r_a and r_b respectively, make up the mixtures of Z grams with specific refractivity r_m.

Example

Twenty milliliters of a mixture of xylene and carbon tetrachloride had a density of 1.2156 and $n_D{}^{25} = 1.4338$. Pure xylene has a density of 0.8570 and $n_D{}^{25} = 1.4915$, whereas pure carbon tetrachloride has a density of 1.5816 and $n_D{}^{25} = 1.4562$. Calculate the weight % of this mixture.

First calculate the specific refractivity of the pure compounds.

$$r = \left(\frac{n^2 - 1}{n^2 + 2}\right) \times \frac{1}{d}$$

for xylene:

$$r = \frac{(1.4915)^2 - 1}{(1.4915)^2 + 2} \times \frac{1}{0.8570} = 0.3382$$

for carbon tetrachloride:

$$r = \frac{(1.4562)^2 - 1}{(1.4562)^2 + 2} \times \frac{1}{1.5816} = 0.1719$$

for the mixture:

$$r = \frac{(1.4738)^2 - 1}{(1.4738)^2 + 2} \times \frac{1}{1.2156} = 0.2311$$

let X = gm of carbon tetrachloride:

Y = gm of xylene

gm of mixture = 20 ml $\times$ 1.2156 = 24.31 gm

Then set up two equations.

$$X + Y = 24.31$$

$$0.1719X + 0.3382Y = (24.31)(0.2311) = 5.618$$

Solving the two equations:

$$X = 15.66 \text{ gm and } Y = 8.65 \text{ gm}$$

Thus the mixture is 64.4% by weight carbon tetrachloride, and 35.6% by weight xylene.

Refractometers

Now that some of the applications of refractive index measurements have been discussed, the instruments for obtaining refractive indices will be examined. The symbol n_D^{20} is used to represent refractive index. The figure 20 means that the temperature at which the measurements were made was 20°C (1°C changes the refractive index 0.00045 so temperature control of ±0.2°C should be maintained.) The subscript $_D$ signifies that the refractive index is that of the sodium D line (the yellow doublet at 589 nm).

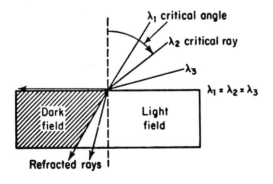

FIG. 26.2.   ILLUSTRATION OF THE CRITICAL RAY AND CRITICAL ANGLE

Several instruments use the critical ray approach. Refer to Fig. 26.2.

Three rays of monochromatic radiation strike a medium of different density. Two of these rays are refracted and would produce light on the other side of the medium. However, the third ray and all other rays having an angle of incidence equal to or greater than λ_2 are not refracted. Thus no light gets through the medium at this point and a dark field is produced. This *critical ray* is utilized in refractometers to measure the refractive index of various substances, since the critical angle is different for each substance. Each wavelength has a critical angle and if white light were used, no sharp division would occur between the light and dark fields, and there would be a light field followed by a rainbow of colors and the dark field. We can eliminate the rainbow of colors by the *Amici prism* discussed next.

An instrument is constructed so that it will measure the critical angle of the sodium D line. The rays of other wavelengths are then disposed of by means of an Amici prism (Fig. 26.3).

Notice that only the sodium D line comes through in the same direction as the incident light. This permits the use of white

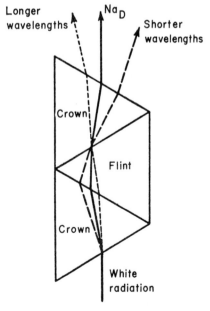

FIG. 26.3.   AMICI PRISM

light since all the rays, other than the one of interest, are scattered
out of the optical path.

Abbe Refractometer

The Abbe refractometer is probably the most common type employed today and a schematic diagram of it is shown in Fig. 26.4.
This refractometer covers the refractive index range from 1.3 to 1.7
with a precision of ±0.0003 units, it can be used for direct reading,
and requires only 1–2 drops of sample. The procedure involves
placing the sample between the two lower prisms, and then rotating
the connecting arm until the critical ray is centered in the eyepiece.
If the division line between the light and dark field is colored, the
Amici prism (the compensator) is turned until a sharp division
appears. Never use ether or acetone to clean off the sample
from the prisms; these solvents evaporate quickly and in that process change the temperature of the prism sufficiently to cause a
significant error in the measurement. Use ethanol, but if you must
use ether or acetone, wait at least 10 min for temperature equilibrium
to be reestablished.

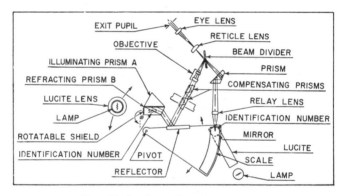

FIG. 26.4.   ABBE REFRACTOMETER

Refractive Index of Solids

Probably the easiest method is to use a microscope and solutions
of different refractive index. A crystal of the solid is placed on a
slide and a few drops of a liquid of known refractive index are placed
around the crystal. A slide cover is placed on top to prevent the
solution from evaporating.

A thin band of visible light outlines the crystal when a narrow
axial illuminating core and an objective of low aperature are em-

ployed. This line around the crystal is called the *Becke Line.* This line moves toward the medium of higher refractive index if the focus is raised, and toward the medium of lower refractive index if the focus is lowered. By changing the surrounding solutions, the refractive index can be determined to less than 0.1 of a RI unit.

Differential Refractometry

The Abbe refractometer can determine n-values to ± 0.0003 n, but for many cases this is not accurate enough, particularly when determining column eluates. The differential refractometer produces more accurate readings by using two hollow prisms set opposite to each other, so that the image displacement by one prism tends to offset the displacement by the other. This keeps the net shift small, and when this shift is compared to a fixed reference point (observed by a high power microscope) differences of the order of ± 0.000005 n can be obtained. Figure 26.5 is a schematic diagram of such an instrument. A monochromatic source and very close temperature control are necessary.

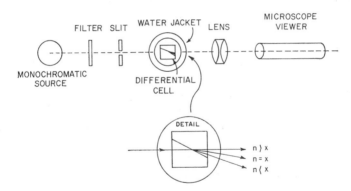

FIG. 26.5. DIFFERENTIAL REFRACTOMETER

Interferometer

The interferometer is a technique based upon the interference of light waves discussed previously in connection with the interference filter, the diffraction grating, and X-ray diffraction. Figure 26.6 shows the setup.

Let us take two rays of radiation coming from the same source and have them "race" to the detector plate. The lens is used to get the rays in the direction we want them to go, and the baffle is used to ensure that only the two rays of interest continue on with the

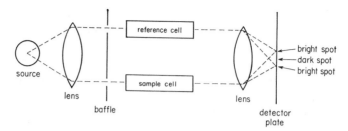

FIG. 26.6. SCHEMATIC DIAGRAM OF AN INTERFEROMETER

race. Assume for the moment that both the reference cell and the sample cell are empty. The two rays continue through the cells, are focused by the lens onto the detector plate, and (since both rays have covered the same distance) they are in phase when they hit the detector plate; consequently, each ray reinforces the other and a bright spot is formed.

If however, some sample is added to the sample cell, that ray may find the going a bit tougher and be slowed down. If it is slowed down by one-half a wavelength, when it hits the detector plate it will completely interfere with the other ray and be destroyed, thus producing a dark spot. If the sample ray is slowed down by one full wavelength, it will again be in phase with the reference ray and again a bright spot will be produced. This bright spot will be offset from the first bright spot because the slower ray is focused differently by the lens. The net result is a series of bright and dark spots (*fringes*) if monochromatic radiation is used, and a series of rainbows if white light is used. Equation 26.5 is used to calculate the results.

$$N = b(n - n_0)/\lambda \qquad (26.5)$$

where N = the number of fringes
   b = the cell thickness
   n = the refractive index of the reference
   n_o = the refractive index of the sample
   λ = the wavelength of radiation used

How good is an interferometer? Using a 1-m cell it is possible to determine differences in n of ± 0.0000001 for liquids and ± 0.00000001 for gases.

Associated Problems

(1) Calculate the molecular refractivity for diethylamine; $n_D = 1.3873$, $d = 0.7180$. Ans. 23.98

(2)   The molecular refractivity of 6,8-dimethyl quinoline is 41.73. What is its refractive index if its density is 1.0660?   Ans. 1.4776

(3)   Mixtures of acetic acid (n_D = 1.3718) and cetyl acetate (n_D = 1.4358) produce a linear relationship between refractive index and mole fraction. What would be the refractive index of a compound that would seem to disappear when placed in a solution containing 0.152 mole fraction of acetic acid?   Ans. 1.4261

BIBLIOGRAPHY

EISENDOHR, F.   1910.   A new calculation of atomic refractions. Part 1. Z. Physik. Chem. 75, 585–607.
EISENDOHR, F.   1912.   A new calculation of atomic refractions. Part 2. Z. Physik. Chem. 79, 129–146.

CHAPTER 27

Polarimetry

Let us examine in more detail the light coming from the lamp you are using to read this page. Recall from Chap. 5 that we said that this radiation behaved as if it had an electric component and a magnetic component, and that they acted as if they were at right angles to each other. One such ray might be represented as in Fig. 27.1.

However there are millions of rays coming from your lamp and the direction of the electric and magnetic components are purely random and may look like Fig. 27.2. This radiation is said to be unpolarized. If, however, by some means we can get all of the rays to have their electric and magnetic components all in the same direction, then we say that the radiation is *plane polarized*. Of what value is this to us? Consider the molecule below:

$$
\begin{array}{c}
\text{Cl} \\
| \\
\text{H—C—OH} \\
| \\
\text{Br}
\end{array}
$$

We see that it is unsymmetrical. No matter where we placed a mirror in the compound, we could not get an exact mirror image. If unpolarized radiation strikes this molecule, and the electric component of one ray interacts more with one side than the other we might notice an effect; however, since the radiation is unpolarized, there are always some rays oriented opposite to our first ray that react in just the opposite manner and cancel out any initial effect we might have seen.

If polarized radiation strikes the molecule there is no way for an opposite effect to take place, and as shown in Fig. 27.3 the radiation is "rotated."

Compounds which can do this are said to be *optically active*. If you are looking toward the light source and the rotation is clockwise, it is said to be *dextro* (+) and if counterclockwise, *levo* (−).

If a compound has a plane of symmetry this effect does not happen even with polarized radiation, because what effect occurs on one

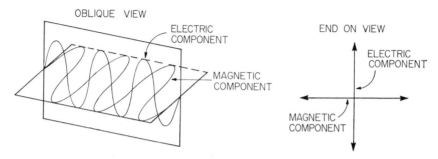

FIG. 27.1.   COMPONENTS OF RADIATION

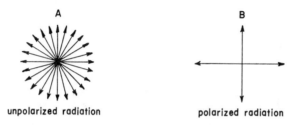

FIG. 27.2.   VECTOR DIAGRAM OF RADIATION

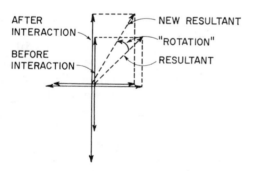

FIG. 27.3.   ROTATION OF RADIATION

side of the molecule is canceled out by an opposite effect on the other side of the molecule.

When we have an optically active atom in a molecule it is generally starred and is called an *asymmetric center*.

Examples

```
      H    O
       \  //
        C
        |
    H—C*—OH
        |
    H—C*—OH
        |
   HO—C*—H
        |
    H—C—OH
       /
      H
```

optically
active

inactive

Instruments

The instrument for measuring optical rotation is called a *polarimeter.* In order to see how it works, let us take this book and two picket fences with vertical pickets and have the fences placed about 10 ft apart. The book will represent a ray of unpolarized radiation initially. Suppose I ask you to throw the book through the picket fence. You know that it will only go through the fence when the length of the book lines up with the pickets. This first picket fence will then polarize our book, i.e., line it up in one direction as it comes through, and the picket fence is called a *polarizer.* If nothing happens to the book as it crosses the 10 ft to the second picket fence, it will be lined up properly and go through it also. But suppose that a gust of wind tipped the book slightly after it left the first picket fence. In other words, it has rotated just like a polarized ray would be rotated by an asymmetrical center of a molecule. Now the book would not get through the second picket fence. If we picked up one end of the fence (rotated it) until the book lined up with the openings, the book could then pass through. Suppose we put a big wheel on this second picket fence and marked degrees of rotation on it. It would now be a simple matter to see how many degrees we had to rotate the fence to allow the book (radiation) to get through. This second picket fence would then be an *analyzer.*

A polarimeter does just this. A monochromatic source is used; a polarizer polarizes the radiation; the sample rotates the radiation; and an analyzer tells us how much the rotation is.

Sources

Optical rotation depends upon the wavelength of radiation used. For this reason it is easier to interpret the data if monochromatic radiation is used. The sodium vapor lamp and the mercury vapor lamp are the two most common sources.

Polarizers and Analyzers

The two most common means of producing polarized radiation are the Nicol and the Glan-Thompson prisms. These are shown in Fig. 27.4. These prisms are made from quartz or calcite, and are cut in such a way that the ordinary ray is totally reflected, permitting the extraordinary ray to emerge polarized.

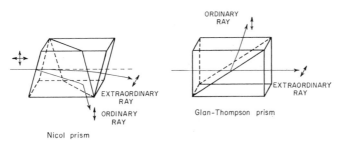

Nicol prism

Glan-Thompson prism

FIG. 27.4. POLARIZERS

The Glan-Thompson prism has the advantage that the emerging radiation is going in the same direction as the entering radiation, and the prism can be turned without distorting the beam intensity.

End-Point Devices

The analyzer can be turned in two directions to measure the optical rotation. If we turn it one way, we get a very bright field, and if we turn it in the other direction, we get a black field. The problem is that our eyes are not able to tell where the brightest and blackest field is since they have no reference.

The purpose of the end-point device is to provide a reference. The human eye is very good at comparing shades of colors, so the end-point device is designed to produce a gray shade over half of the viewing eyepiece which is then matched by rotating the analyzer. The most common type of end point device is the *Lippich prism* which is a small Glan-Thompson prism, tipped slightly, covering half of the field. This is shown in Fig. 27.5 along with the other components of a polarimeter.

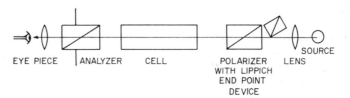

FIG. 27.5.   COMPONENTS OF A POLARIMETER

Basic Equation of Polarimetry

The basic equation used for analytical purpose is:

$$[\alpha]_\lambda^t = \alpha/lc \qquad (27.1)$$

where $[\alpha]_\lambda^t$ = specific rotation in degrees,
   t = the temperature in °C
   λ = the wavelength of radiation employed
   α = the observed angle of rotation in degrees
   l = length of the tube in DECIMETERS
   c = concentration in gm/ml (for a pure liquid, density replaces c)

Example

A solution of an organic compound is placed in a 20-cm tube and produces a rotation of 38.73°. A tube containing only solvent has a rotation of 1.46°. If the specific rotation is 62.12°, what is the concentration of the solute?

$[\alpha]_\lambda{}^t = 62.12°$
$\alpha = (38.73° - 1.46°) = 37.27°$ corrected
$l = 20$ cm or 2 dm

$$62.12 = \frac{37.27}{2 \times c}$$

$c = 0.30$ gm/ml

Effects of Temperature

The above calculation is valid providing a number of variables are controlled, one of which is the temperature. When the temperature of the solution in the polarimeter tube is different than the temperature at which the specific rotation was obtained, a correction is necessary. The equation for this is

$$[\alpha]_\lambda^{t_1} = [\alpha]^{t_2} + n(t_1 - t_2) \qquad (27.2)$$

where t = temperature of the solution in °C

n = constant

Example

A solution of sucrose was placed in a polarimeter and the specific rotation of sucrose was determined at three different temperatures. From the following data, determine the constant, n, and the value of $[\alpha]_D^{20}$

t_1 (°C)	$[\alpha]_D^t$
14.0	66.57
22.0	66.375
30.0	66.18

From this graph the slope, equal to n, is found to be -0.0244, over the temperature range studied.

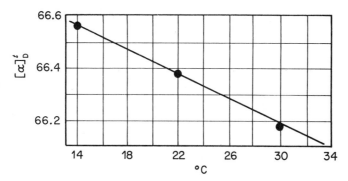

FIG. 27.6. EFFECT OF TEMPERATURE ON SPECIFIC ROTATION

To obtain $[\alpha]_D^{t=20}$, we calculate

$$[\alpha]_D^{20} = [\alpha]_D^{22} - 0.0244\ (20\text{--}22)$$
$$[\alpha]_D^{20} = 66.375 + 0.0488$$
$$[\alpha]_D^{20} = 66.42$$

Effect of Concentration

A second correction that must be applied in some cases is that optical rotation changes with concentration. Biot has developed three equations which may be used to make a suitable correction:

$[\alpha] = A + Bq$	(linear)	(27.3)
$[\alpha] = A + Bq + Cq^2$	(parabolic)	(27.4)
$[\alpha] = A\ [Bq/(C + q)]$	(hyperbolic)	(27.5)

A, B, and C are constants and q is the weight fraction of solvent in the solution. [α] is generally replaced with (α/lpd) where p is the weight fraction of solute, d is the density of the solution, and l is the tube length in decimeters. To determine which equation to use, plot [α] vs. q, and see if the resulting curve more closely approaches a parabola, hyperbola, or a straight line.

Example

Three solutions of a saccharide containing 20, 50, and 80% solute were prepared. The solutions were placed in a 20-cm tube and the following data obtained:

p	q	d	α	[α]
0.80	0.20	1.116	18.26	10.22
0.50	0.50	1.290	13.74	10.65
0.20	0.80	1.464	6.47	11.04

Determine which of the Biot's equations can be used and evaluate the constants.

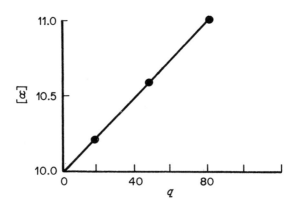

FIG. 27.7.   EFFECT OF CONCENTRATION ON SPECIFIC ROTATION

A plot of [α] vs. q resulted in a straight line. Therefore equation 27.3 applies.

$$10.23 = A + 0.20B \quad (1)$$
$$10.65 = A + 0.50B \quad (2)$$
$$11.04 = A + 0.80B \quad (3)$$

Solve any pair of these equations simultaneously; in this case $A = 9.955$ and $B = 1.37$.

Effect of Wavelength

Specific rotation also varies with wavelength. This relationship is expressed by the Drude equation

$$[\alpha] = \frac{k_1}{\lambda_0^2 - \lambda_1^2} + \frac{k_2}{\lambda_0^2 - \lambda_2^2} + \frac{k_3}{\lambda_0^2 - \lambda_3^2} \qquad (27.7)$$

k_1, k_2, k_3 along with $\lambda_1, \lambda_2, \lambda_3$ are constants. λ_0 is the wavelength employed, generally expressed in angstroms. For most purposes, when λ is far removed from an optically-active band, the simplified equation (27.8) can be used.

$$[\alpha] = k/(\lambda_0^2 - \lambda_1^2) \qquad (27.8)$$

If λ is very far removed, so that $\lambda_0 \gg \lambda_1$ then

$$[\alpha] = k/\lambda_0^2 \qquad (27.9)$$

Example

An organic compound had a $[\alpha]$ of 14.63° when radiation of 589 nm was employed, and 15.27° when radiation of 436 nm was used. Calculate the Drude equation constants and predict the $[\alpha]$ when a λ_0 of 707 nm is used.

$$14.63 = \frac{k}{(0.589)^2 - \lambda_1^2}$$

$$15.27 = \frac{k}{(0.436)^2 - \lambda_1^2}$$

These can be further reduced to

$$5.08 - 14.63 \lambda_1^2 = k$$
$$2.90 - 15.27 \lambda_1^2 = k$$
$$\overline{2.18 + 0.64 \lambda_1^2 = 0}$$

Thus,

$$\lambda_1^2 = -3.41, \text{ and } k = 54.97$$

Use these two values to obtain the specific rotation at 707 nm.

$$[\alpha] = 54.97/[(0.707)^2 + 3.41]$$
$$[\alpha] = 14.06$$

Saccharimeters

A modification of a polarimeter used extensively by the sugar industry is the quartz wedge saccharimeter. The *rotatory dispersion* of quartz and most sugars are about the same. This means that white light can be used rather than monochromatic radiation, because any wavelength effects can be canceled out. See Fig. 27.8. When a sugar sample is added to the cell a dextrorotation is usually obtained. The levo quartz wedge is then moved to exactly compensate this rotation. The smaller wedge of levo quartz is to eliminate refractive dispersion. The piece of dextro quartz is used for those sugars that produce levo readings.

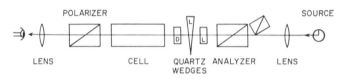

POLARIZER   SOURCE

LENS   CELL   QUARTZ   ANALYZER   LENS
                            WEDGES

FIG. 27.8. SACCHARIMETER

Associated Problems

(1) A solution of l(−) glutamic acid containing 1.085 gm/20 ml of 0.37 N HCl had a rotation of −1.63° when placed in a 100-mm tube. Calculate the specific rotation of the glutamic acid. Ans. −30.04°

(2) l(−) diiodotyrosine has a specific rotation of +2.890. What would be the expected optical rotation of a solution containing 4.41 gm/60 ml of 1.1 N HCl using a 5-cm tube? Ans. 1.062°

(3) The specific rotation of l(+) alanine is +14.7°. If 20-cm tubes are used and an optical rotation of +1.70° is observed, what is the concentration of alanine in the solution? Ans. 0.057 gm/ml

(4) l(−)-cystine has a specific rotation of −214.0° at 24.35°C. At 20.0°C 0.997 gm/200 ml gave a rotation of −2.0 using a 20-cm tube. What would be the specific rotation at 30°C? Ans. 224.4°

(5) Determine which of Biot's equations best express the following data.

p	q	$[\alpha]$
0.97	0.03	15.84
0.71	0.29	15.00
0.50	0.50	14.50
0.35	0.65	14.26
0.24	0.76	14.13
0.10	0.90	14.06

Determine the constants required.
Ans. A = 14.06, B = −3.78, C = 1.89

(6)   Nicotine has a specific rotation of $-162°$ using radiation of 5993 Å, and $-126°$ when radiation of 6563 Å is used. What is its specific rotation when radiation of 4861 Å is used?   Ans. $-318°$

(7)   A solution of santonin dissolved in alcohol produced a specific rotation of $+442°$ using radiation of 686.7 nm, and 991° using radiation of 526.9 nm.   What would be the wavelength required to have a specific rotation of $+1323°$?   Ans. 488.3-nm

Some Recent Applications

The measurement of glucose and maltose in potato starch syrup has been described by Rychlik and Fedorowska (1967) using optical rotation and an iodine titration.   Methods for starch in meat products have been described by Ojtozy (1966B) using polarimetry and gravimetry.   Ojtozy (1966A) also used polarimetry to determine starch in the presence of glycogen by separating the glycogen with alcoholic potassium hydroxide. Saccharase activity in honey has been determined by Hadorn and Zuercher (1966) using optical rotation measurements to follow the reaction.

BIBLIOGRAPHY

HADORN, H., and ZUERCHER, K.   1966.   An improved polarimetric method for the determination of the saccharase number in honey. Deut. Lebensm. Rundschau 62, 195–201.

OJTOZY, K. 1966A.   A polarimetric determination of starch and glycogen in sausage containing liver.   Elelmiszervizsgalati Kozlemen 12, 323–329.   Anal. Abstr. 14, 6409, 1967.

OJTOZY, K. 1966B.   Starch content in meat preparations.   Elelmiszervizsgalati Kozlemen 13, 76–81.   Chem. Abstr. 69, 26085t, 1968.

RYCHLIK, M., and FEDOROWSKA, Z.   1967.   Determination of glucose and maltose in potato starch syrup.   Roczniki Panstwowego Zakladu Hig. 18, 735–741. Chem. Abstr. 68, 94727w, 1968.

CHAPTER 28

Rheology

RHEOLOGICAL PARAMETERS

Rheology is concerned with stress-strain time relations of materials which show a behavior intermediate between those of solids and liquids. Stress can be compressive, tensile, or shear. The passage of time does not itself cause changes in materials. Time is, however, often introduced in measuring rates of changes of forces and deformations. Chemical changes in foodstuffs often occur in time and may be studied by rheological methods. Temperature is also important and often appears in rheological equations. Strain is measured by deformation. Figure 28.1 classifies deformations according to a procedure proposed by the British Rheologist Club, as modified by Bushuk. This is a useful classification, though the limits of each category are arbitrary. For convenience, all deformations are divided into elastic deformations and flow. By an elastic deformation, we mean one that can be recovered, irrespective of whether the recovery is spontaneous or not. Elastic deformations can be divided into ideal (no time effect), and nonideal which shows time effect. Ideal deformations may be Hookean or nonHookean, in which latter case strain may increase more or less rapidly than stress. Nonideal deformations may be completely recoverable, and the strain may vary more rapidly than, proportionally to, or less rapidly than stress.

The incompletely recoverable deformations are linked with flow, and also lead to plastoelastic and plastoinelastic groups. On the flow side, the figure divides into plastic and viscous groups. The plastic deformations may be Newtonian or nonNewtonian, and the latter is subdivided into viscoelastic and viscoinelastic groups. The Newtonian and Bingham systems are listed as separate subcategories on account of their exceptional importance. Figure 28.2 classifies fluids according to their behavior under stress. Viscosity is given as slope of curves at top and is plotted at bottom as a function of shear rate, which affects viscosity of nonNewtonian fluids (Hallikainen 1962). Viscosity terminology and equations are summarized in Table 28.1.

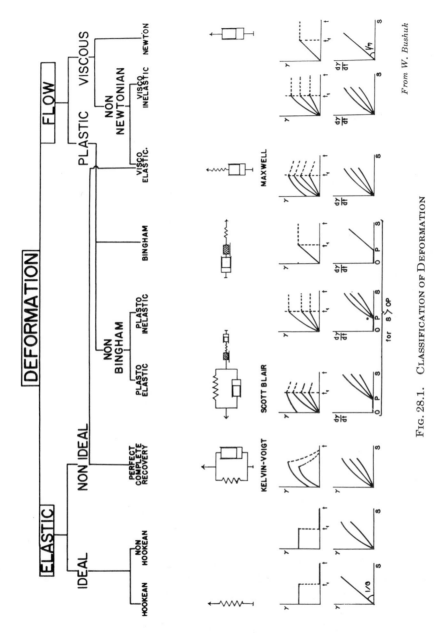

Fig. 28.1. Classification of Deformation

From W. Bushuk

Texture can be regarded as a manifestation of the rheological properties of a food.   It is an important attribute in that it affects processing and handling (Charm 1962), influences food habits, and

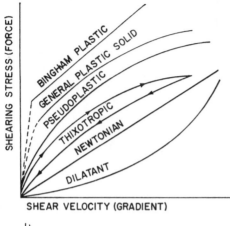

Fig. 28.2.   Classification of
Fluids

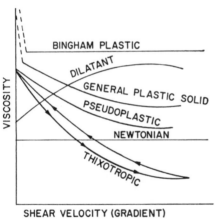

From Hallikainen (1962)

affects shelf-life, and consumer acceptance of foods (Matz 1962). Research indicates that the consumer is highly conscious of food texture, and that in certain foods texture may be even more important than flavor (Szczesniak and Kleyn 1963).

Analytical vs Integral Approach

Measurement of rheological properties of foods can be based on either the analytical or the integral approach. In the first, the properties of a material are related to such simple systems as Newtonian fluids or Hookean solids. When the material approximates any of the above systems, the appropriate equations are used with a suitable correction. For materials removed from perfect elastic and true viscous behavior, the hypothesis is made that the material consists of two or more parts, and that the effects are additive.

TABLE 28.1

VISCOSITY TERMINOLOGY AND EQUATIONS

Stress—Force/Area—F/A

Velocity Gradient (Shear)—Rate of change of liquid velocity across the stream—V/L for linear velocity profile, dV/dL for nonlinear velocity profile. Units are V/L = ft/sec/sec = sec^{-1}.

Absolute (dynamic) viscosity η—Constant of proportionally between applied stress and resulting shear velocity (Newton's Hypothesis):

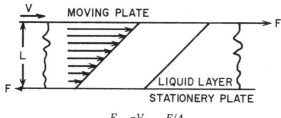

$$\frac{F}{A} = \frac{\eta V}{L} \quad \eta = \frac{F/A}{V/L}$$

Poise (η)—Unit of dynamic or absolute viscosity—1 dyne-sec/cm²

Kinematic Viscosity η—Dynamic viscosity/density = η/ρ

Stoke—Unit of kinematic viscosity (v)

Hagen-Poiseuille Law (flow through a capillary)

$$Q = \frac{\pi R^4}{8 \eta L} (P_1 - P_2)$$

Saybolt Viscometer (Universal, Furol)—Measures time for given volume of fluid to flow through standard orifice: units are seconds

Fluidity—Reciprocal of absolute viscosity: unit in the cgs system is the rhe, which equals 1/poise.

Specific Viscosity—Ratio of absolute viscosity of a fluid to that of a standard fluid, usually water, both at same temperature.

Relative Viscosity—Ratio of absolute viscosity of a fluid at any temperature to that of water at 20°C (68°F). As water at this temperature has an η of 1.002 cp, the relative viscosity of a fluid equals approximately its absolute viscosity in cp. As density of water is 1, kinematic viscosity of water equals 1.002 cs at 20°C.

Apparent Viscosity—Viscosity of a non-Newtonian fluid under given conditions. Same as consistency.

Consistency—Resistance of a substance to deformation. It is the same as viscosity for a Newtonian fluid, and the same as apparent viscosity for a non-Newtonian fluid.

Saybolt Universal Seconds (SUS)—Time units referring to the Saybolt Viscometer.

Saybolt Furol Seconds (SFS)—Time units referring to the Saybolt viscometer with a Furol capillary, which is larger than a Universal capillary.

Shear Viscometer—Viscometer that measures viscosity of a non-Newtonian fluid at several different shear rates. Viscosity is extrapolated to zero shear rate by connecting the measured points and extending curve to zero shear rate.

From Hallikainen (1962)

In the second, integral approach, a simple and initially empirical relation between stress, strain, and time is sought. This approach is often dictated by the fact that its results correlate better with sensory evaluation than descriptions based on rheological models or simple dimensional terms.   Foods seldom have simple rheological properties.   In addition, most rheological measurements refer to the arbitrary conditions imposed by a particular instrument.   What we are generally measuring is not a pure rheological parameter, but the way in which the properties vary under some standardized system of applied forces.

Yet, the relationship between stress and strain systems and their time derivatives are often expressed in terms of models—a pictoral presentation of the analytical approach.   The models may have sometimes no more than a symbolic significance; in other cases they throw light on existing configurations and are a very valuable rheological tool (Scott-Blair 1953).

Rheological Models

The rheological models are also useful as they present in an easily comprehensible way the mechanical behavior of materials, and facilitate mathematical description of their behavior.   The following is a description of various types of rheological models, their respective deformation-time curves, and the rheological shorthand frequently used:

Figure 28.3 (of elementary and composite rheological models) and and Fig. 28.4 (of a rheological wheat flour dough model) are from a recent review by Lerchenthal and Muller (1967).   The four basic components include: a dashpot (piston sliding in a cylinder filled with oil) depicting Newtonian viscosity (N); the spring depicting the Hookean elasticity element (H):   the spring clip or St. Venant body $(St.\ V)$; and the shear pin (SP).   The dotted lines above the deformation time plots represent the timing of loading and unloading.   Upon loading, the spring (H) lengthens immediately by an amount dependent on the load, whereas in the dashpot (N) it is the speed (or "rate") of movement which is proportional to the load. When the dashpot (N) is unloaded, the deformation remains; when the spring (H) is unloaded there is instantaneous and complete recovery.

The St. Venant element $(St.\ V)$ is shown by a friction weight.   It describes ideal plasticity, i.e. an irrecoverable deformation caused at a specific stress above the yield value.   Below this, no deformation takes place; above it, deformation continues at a constant rate as long as the yield stress persists.   The shear pin (SP) allows presentation of the rupture of an element under a specific stress.

A dashpot and a spring combined in a series $(H\text{-}N)$ give a model of a Maxwell body (Fig. 28.3).   On loading, there is at first an im-

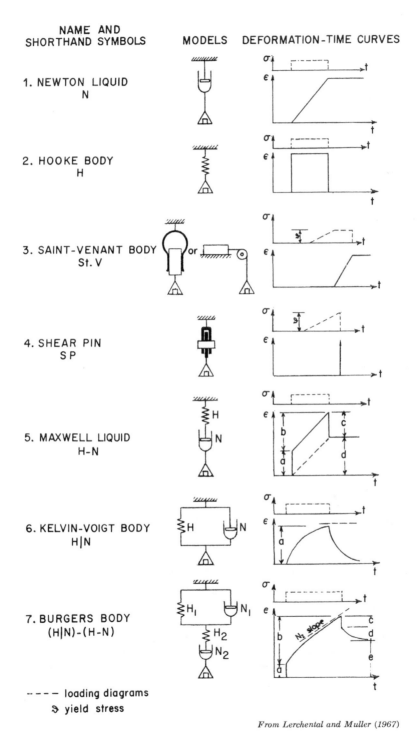

NAME AND SHORTHAND SYMBOLS	MODELS	DEFORMATION-TIME CURVES

1. NEWTON LIQUID
   N

2. HOOKE BODY
   H

3. SAINT-VENANT BODY
   St. V

4. SHEAR PIN
   S P

5. MAXWELL LIQUID
   H-N

6. KELVIN-VOIGT BODY
   H|N

7. BURGERS BODY
   (H|N)-(H-N)

- - - - loading diagrams
   ∿ yield stress

From Lerchental and Muller (1967)

FIG. 28.3. ELEMENTARY AND COMPOSITE RHEOLOGICAL MODELS

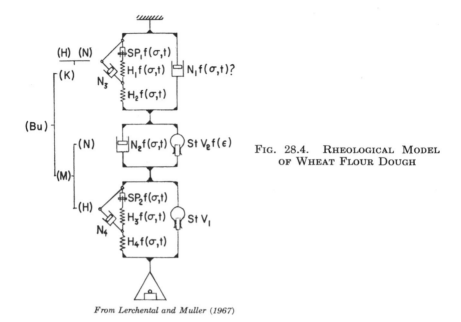

FIG. 28.4. RHEOLOGICAL MODEL OF WHEAT FLOUR DOUGH

From Lerchental and Muller (1967)

mediate deformation of (a) due to the spring followed by a slow deformation (b) due to the dashpot. On unloading, there is an immediate recovery of (c) equivalent to (a), and irrecoverable residual deformation of (d) equal to (b).

When the spring and dashpot are combined in parallel (H/N), the model is referred to as the Kelvin or Voigt type. When applied, the load is gradually transferred from the dashpot to the spring which eventually tends to hold the load entirely. This causes the deformation time curve to level out asymptotically. On unloading, there is slow but complete recovery as the spring recovers against the friction of the dashpot.

When a Maxwell and a Kelvin-Voigt body are placed in series (H_1/N_1)–(N_2-H_2), we have a Burger's body. The deformation-time curve shows at first an immediate response (a) due to the free spring H_2 of the Maxwell element. The curve then rises (b), and due to the free dashpot N_2 does not level out as it would in the Kelvin-Voigt model. On unloading, there is an immediate response (c) due to H_2, followed by a delayed elastic response (d) caused by H_1 working against the resistance of N_1. The residual deformation (e) is due to the free Newtonian element N_2 of the Maxwell component.

The rheological model of wheat flour dough (Fig. 28.4) illustrates the complexity of model systems in describing actual food materials.

RHEOLOGICAL METHODS

Empirical Methods

Numerous schemes have been proposed to classify the available rheological methods used in food evaluation. Scott-Blair (1958) classified rheological methods of texture measurement under the headings; fundamental, empirical, and imitative. According to Szczesniak (1966), fundamental tests measure basic rheological parameters and relate the nature of the tested product to basic rheological models. Because of the complexity of food products and the necessity to relate the obtained measurements to functional properties, the fundamental tests have only limited use. However, they can serve to define the system and provide a more scientific basis for empirical and imitative tests. Thus, fundamental tests are seldom used for quality control in either flour mills or bakeries, but the information gained from them has led to the development of a number of useful empirical tests. Similarly, fundamental research on the relation between concentration and viscosity of emulsions was useful in the manufacture of condensed milk. Basic parameters determined in working molten chocolate and studies on the rheology of hydrogenated fats are additional examples of fundamental studies which have led to applied empirical tests. However, most empirical tests measure poorly defined properties that have been shown by experience to be related to textural quality. The most common instruments used in indirect empirical tests are penetrometers. They are used to determine: (1) rigidity of gels in terms of strain produced by a load; (2) the force required just to penetrate the material; and (3) the consistency of the material as measured by resistance to further penetration either by the rate of sinking of a column or needle or by the total depth of penetration following impact. Penetrometers have been much used for measuring firmness of dairy products, and tenderness of fruit, vegetables, and fish. Much work has been published on the Bloom "gelometer" and its various modifications. Sinkers and line-spread consistometers are also useful tools in indirect tests. They include "plummets" and "bobs" of all kinds, and are used to determine the consistency of chocolate, creams, spreads, sauces, and fillings.

Imitative Methods

Imitative tests measure various properties under conditions similar to those to which the food is subjected in practice. Devices used measure the properties of the material during handling, and the properties of the food during consumption.   In the first group, are instruments such as butter spreaders which give a measure of the spreading properties and various mixing and load-extension meters which measure dough-handling properties.   In the second group, are various tenderometers which simulate the chewing action of teeth.

Scott-Blair (1958) pointed out that imitative tests do not necessarily correlate better with actual performance than tests somewhat removed from the practical situation.   Thus, in testing butter, an instrument has been devised in which a small cube of butter is sheared by a knife edge at standard loads and speeds; the amount of shear being measured to indicate "spreadability."   Yet, the correlation with spreadability scores of an experienced panel were higher with simple viscosity measurements, than with the test imitating the conditions to which the material is subjected in practice.   The reason is that the imitation is far from perfect, that a "standard" surface (differing substantially from that of bread crumb) is used, and that spreadability is not related only to "softness" assessed by determining shearing forces.

Classification

In addition to the classification of Scott-Blair (1958), instruments used to measure textural quality have been classified according to the foods they evaluate (Matz 1962; Amerine et al. 1965).   The weakness in this kind of classification is that many instruments are used on more than one of the group of foods.   Drake (1961) developed a classification system based on the geometry of the rheological apparatus.   This classification contains the following types: (1) rectilinear motion (parallel, divergent, and convergent); (2) circular motion (rotation, torsion); (3) axially symmetric motion (unlimited, limited); (4) defined other motions (bending, transversal); and (5) undefined motions (mechanical treatment, muscular treatment).   Each subheading in Drake's classification is subdivided further on the basis of the geometry of the apparatus.

The classification system proposed by Bourne (1966) is based upon the variable (or variables) that constitute the basis of the measurement.   This classification system which stresses the nature and dimensional units of the property being measured is given in Table 28.2.

Force-measuring instruments are the basis of the most common method used in measuring food texture.   The measured variable is force, usually maximum force, and distance and time are held either

TABLE 28.2

CLASSIFICATION OF OBJECTIVE METHODS
FOR MEASURING TEXTURE AND CONSISTENCY

Method	Measured Variable	Dimensional Units	Examples
(1) Force-measuring	Force F	$m\,l\,t^{-2}$	Tenderometer
(2) Distance-measuring	(a) Distance	l	Penetrometers
	(b) Area	l^2	Grawemeyer Consistometer
	(c) Volume	l^3	Seed Displacement (bread volume)
(3) Time-measuring	Time T	t	Ostwald viscometer
(4) Energy-measuring	Work $F \times D$	$m\,l^2t^{-2}$	Farinograph
(5) Ratio-measuring	F, or D, or T, or $F \times D$, measured twice	Dimensionless	Cohesiveness (G.F. Texture Profile)
(6) Multiple-measuring	F, and D, and T, and $F \times D$	$m\,l\,t^2$, l, t, $m\,l^2t^{-2}$	G. F. Texturometer
(7) Multiple-variable	F, or D, or T (all vary)	Unclear	Durometer
(8) Chemical analysis	Concentration	Dimensionless (% or ppm)	Alcohol insoluble solids
(9) Miscellaneous	Anything	Anything	Optical density of fish homogenate

constant or replicatable. The group includes instruments measuring compression or tensile strength. In distance-measuring instruments, force and time are held constant or replicatable, and distance, area, or volume is measured. The instruments measure depth of penetration, compression and flow, or volume of either end product (i.e. bread) or expressed liquid (i.e. juice).

Viscosimeters are the main time-measuring instruments. Energy-measuring instruments determine work or energy and include recording dough mixers or meat grinders. If a force-distance curve is drawn during a test, the area under the force-distance curve is a measure of work.

For ratio-measuring methods, at least two measurements of the same variable must be taken. Thus, cohesiveness can be computed from the ratio of the work done during the first and second bites of food.

Instruments in the multiple-measuring group can measure various forces, distances, areas, etc., and record the results. The instruments can be used for a large number of products. They include the Allo-Kramer Shear Press, the General Foods Texturometer, and the Instron (Bourne *et al.* 1966). Such instruments are highly versatile and useful. Generally, a special test cell for each type of product is necessary for meaningful evaluation (Kramer and Backinger 1959). Commercial universal testing machines are expensive;

construction of a simple mechanism for food research was described by Voisey *et al.* (1967). Multiple-variable instruments have more than one uncontrolled variable, but only one variable is measured. Although sometimes those instruments correlate well with sensory evaluation, it is difficult to evaluate the results in terms of fundamental rheological parameters or to relate the measurements with data from other instruments.

Chemical analyses do not measure texture directly, but are often highly correlated with physical texture measurements or subjective panel tests. A typical chemical determination of this kind is the content of alcohol-insoluble solids, an accepted index of green peas' maturity. In addition, there exist objective methods which are highly correlated with texture or consistency of a food, and do not fit any of the above groups. This group of miscellaneous methods includes various refractometric, polarimetric, electrical, and sound-testing devices.

Comprehensive reviews on physical characterization of foods were published by Mohsenin and co-workers (Mohsenin 1968; Morrow and Mohsenin 1968).

INSTRUMENTS

Elements

According to Szczesniak (1966), all food texture-measuring devices have five essential elements:

(1) Driving mechanism—varying from a simple weight and pulley arrangement to a more sophisticated variable-drive electric motor or hydraulic system.

(2) Probe element in contact with food—flat plunger, shearing jaws, tooth-shaped attachment, piercing rod, spindle, or cutting blade.

(3) Force—simple or composite—applied in a vertical, horizontal, or levered manner of the cutting, piercing, puncturing, compressing, grinding, shearing, or pulling type.

(4) Sensing element—ranging from a simple spring to a sophisticated strain gage. Proving ring dynamometers and transducers may also be used.

(5) Read-out system—maximum force dial, an oscilloscope, or a recorder tracing the force-distance relationship.

Texture Profiles

In selecting a rheological test, it is essential to establish the type of information sought and its potential usefulness. For intelligent interpretation of results, it is generally desirable also to know the expected precision and applicability of results to various conditions. For quality control, generally one textural parameter determined on a composite sample may be useful. For research purposes, a texture "profile" in terms of several parameters determined on a small homogenous sample may be desirable. The mechanical textural characteristics of foods which govern, to a large extent, the selection of a rheological procedure and instrument can be divided into primary parameters of hardness, cohesiveness, viscosity, elasticity, and adhesiveness; and into secondary (or derived) parameters of brittleness, chewiness, and gumminess (Szczesniak 1966).

The basic measurement in hardness determination involves the load-deformation relationship. Cohesiveness may be measured as the rate at which the material disintegrates under mechanical action. Tensile strength is a manifestation of cohesiveness. Cohesiveness is usually tested in terms of secondary parameters: brittleness, chewiness, and gumminess. The three characteristics: brittleness, crunchiness, and crumbliness which can be placed on a continuum can be measured as the ease with which the material yields under an increasing compression load; the smaller the deformation under a given load, the lower the cohesiveness and the greater the "snappability" of the product. Tenderness, chewiness, and toughness are characteristics measured in terms of energy required to masticate a solid food. They are the most difficult to measure precisely, because mastication involves compressing, shearing, piercing, grinding, tearing, and cutting, along with adequate lubrication by saliva at body temperatures.

Gumminess is characteristic of semisolid foods with a low degree of hardness and a high degree of cohesivenes. The rate at which a food returns to its original condition after removal of a deforming force is an index of the food's elasticity. Adhesiveness is measured in terms of the work required to overcome the attractive forces between the surface of a food and the surface of other materials with which the food comes into contact.

Sherman (1969) examined the texture profile of Szczesniak and proposed several modifications. The modified profile is shown in Fig. 28.5. The proposed scheme consists of primary, secondary, and tertiary categories. Primary attributes are analytical composition, particle size and size distribution, particle shape, etc. The three secondary attributes are elasticity (E), viscosity (η), and adhesion (N). The tertiary characteristics are the responses most often used in sensory analyses of texture. Panel responses associated with

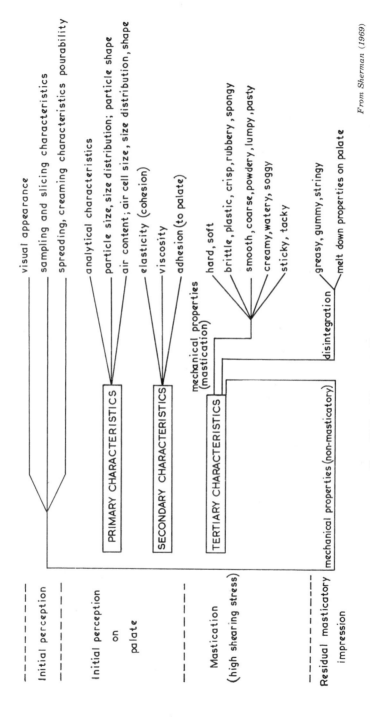

Fig. 28.5. Modified Texture Profiles

From Sherman (1969)

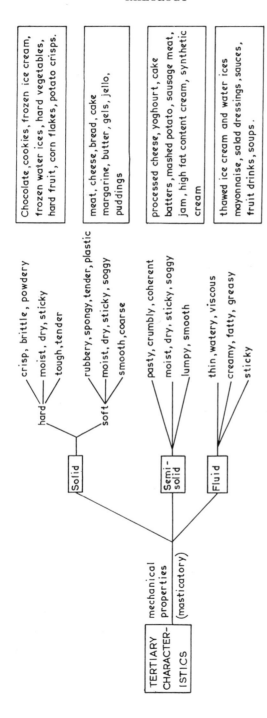

From Sherman (1969)

FIG. 28.6.   PANEL RESPONSES ASSOCIATED WITH MASTICATORY TERTIARY CHARACTERISTICS OF MODIFIED TEXTURAL PROFILE

masticatory tertiary characteristics are summarized in Fig. 28.6. Tertiary characteristics are derived from a complex blending of two or more sensory attributes, and can be regarded as falling within a three dimensional continuum which has the secondary attributes as coordinate axes. All tertiary attributes can be represented by rectangular coordinates of the form (αE, $\beta \eta$, γN) in which α, β, and γ represent the respective magnitudes of the secondary attributes. Sherman (1969) postulated that since solids, semisolids, and fluids have characteristic values of these attributes, it should be possible to predict panel responses from mechanical strain-time tests, which are carried out at approximately constant rate of shear operative during mastication and adhesion tests.

Classification

Kramer and Twigg (1959) used the following classification as basis for mechanical measurement of texture.

A. Finger feel
   (1) Firmness
   (2) Softness or yielding quality
   (3) Juiciness

B. Mouth feel
   (1) Chewiness
   (2) Fibrousness
   (3) Grittiness
   (4) Mealiness
   (5) Stickiness
   (6) Oiliness

The parameters measured to determine the above sensory characterists are: compression, shearing, cutting, tensile strength, and shear

TABLE 28.3

RELATION OF KINESTHETIC CHARACTERISTICS TO PHYSICAL METHODS OF APPLICATION OF FORCE

Sensory Reaction	Physical Test	Instruments or Procedures for Measurement
Firmness	Compression	Pressure tester; shear-press
Yielding quality	Compression	Pressure tester; shear-press; ball compressor
Juiciness	Compression (juice extraction)	Puncture tester; succulometer; shear-press; moisture tests
Chewiness	Shear-pressure	Tenderometer; texturemeter; shear-press; specific gravity; solids
Fibrousness	Cutting; comminuting	Fibrometer; shear-press; fiber analysis
Grittiness	—	Comminution; elution; sedimentation
Mealiness	—	Starch, and/or gum analysis
Stickiness	Tensile strength	Jelly strength, pectin, and/or gum analysis

pressure. Table 28.3 summarizes the physical tests and procedures used to measure the sensory reactions.

According to Scott-Blair (1945), there are eight basic rheological properties measured in textural examinations.

I. Viscosity
II. Elastic modulus
III. Elastic hysteresis (or fore- and after-effects)
IV. Structural viscosity; i.e., immediately recoverable fall in viscosity with rise in shearing stress
V. Sharpness of curvature of flow curve, i.e. extent to which material approximates yield—value behavior
VI. Extent of breakdown under shear not immediately recovered (includes irreversible breakdown, thixotrophy, etc.)
VII. Extent of work hardening or dilatancy produced by strain
VIII. A strength factor—generally tensile strength, but sometimes compressive strength

To measure the above properties, the following methods are available.

(1) Capillary tube viscosimeters or plastometers (Ostwald)
(2) Jet or orifice viscosimeters (Saybolt or Engler viscosimeters)
(3) Rotating bodies (spheres, discs, cylinders—MacMichael)
(4) Falling sphere viscosimeters
(5) Rising sphere viscosimeters
(6) Displaced sphere viscosimeters
(7) Hot wire anemometers
(8) Ergometer, extensimeter (Chopin)
(9) Mixographs measuring work input in stirring
(10) Falling solid cylinder in large hollow concentric cylinder (i.e. sinker viscosimeter)
(11) Extensimeters—load extension meters
(12) Stretching of spheres into bands
(13) Compression or extension of cylinders or spheres, extension of strips (plastometers and parallel plate consistometers)
(14) Compression of materials in bulk
(15) Cutting tests (curd-o-meters, sectilometers)
(16) Penetrometers (also used for compression without penetration)
(17) Bending bars and rods (torque and flexion)
(18) Tests by dropping weights on sample

(19)   Fatigue measurements of failure
(20)   Indentation hardness testers
(21)   Resilience (bounce) and pendulum methods
(22)   Vibration methods (without failure)
(23)   Tensile strength (direct tension)
(24)   Subjective testing by handling

The following is a partial list of the properties and methods used in textural evaluation of foods.

	Properties	Methods
Milk	I	1, 3, 10
Honey	I	1, 3, 4, 23
Butter	I, IV, VI, VII	2, 13, 15, 16
Hard cheese	I, II, III, IV, V, VII	13, 15, 16, 24
Soft cheese	I, II, VII	13, 15, 16, 24
Cream	I, VI	1, 2
Dough	I, II, III, IV, VII	2, 8, 9, 11, 12, 13, 16, 19 21, 23, 24

An extensive bibliography by Scott-Blair (1945) lists sources of information regarding the various instruments. A more recent and detailed description of instruments which can be used for measurement of viscosity and consistency of foods was given by Kramer and Twigg (1959). Some of the modern and sophisticated instruments included an ultrasonic viscosimeter which measures the viscosity or consistency electronically by means of a magnetostrictive sensing element which vibrates longitudinally, radioactive density gages which measure the absorption of gamma rays, and several instruments which give a continuous indication and record of the viscosity of a product under actual processing conditions.

After this general outline, we are ready to discuss some of the applications of rheological measurements in the evaluation of various foods. Those have been classified as cereals, dairy products, meat, fruits and vegetables, and miscellaneous foods.

APPLICATIONS

Cereals

Cereal technologists are faced with a variety of rheological problems. Starting with the milling process, the brittleness of the bran, and the pliability of the germ are of major importance in separating the endosperm from the outer layers. Viscosity, elastic modulus, and tensile strength are the outstanding factors in determining the behavior of wheat flour dough; compressibility, crispness, and breaking strength of the baked or fried product play a role in acceptability by the consumer.

Rheological properties of dough are of particular importance for several reasons (Bloksma and Hlynka 1964). Dough is one of the main stages in breadmaking. With increased mechanization and automation in the baking industry, dough properties are important from the mechanical viewpoint. A number of common laboratory tests for determining flour quality depend upon the measurement of physical dough properties. A study of the physical properties of dough contributes to our understanding of baking quality. Finally, in dough, the chemical, physical, and biological aspects interact in bringing out the full potential of the system.

The history of the rheological studies of dough was reviewed by Muller (1964) and of instrumentation by Brabender (1965). Basic considerations of dough properties were discussed by Greup and Hintzer (1953) and by Bloksma and Hlynka (1964). A review of major factors governing rheological properties and performance of wheat flour doughs was presented by Miller and Johnson (1954), Kent-Jones and Amos (1957), and by Hlynka (1967).

Kernel Hardness.—Objective hardness measurements are mainly useful in differentiating between soft or hard wheats in plant breeding programs (Miller and Johnson 1954). Kernel hardness (or at least results of determination of kernel hardness) appears to be affected by protein and moisture contents. Extremely hard milling characteristics are usually reflected in increased power requirements and reduced yields of flours of acceptable quality (i.e. ash content or color). Alternatively, when milled to a fixed extraction, the mineral content and color are impaired. Extensive softness interferes with efficient bolting and increases requirements for sieving space.

The methods available for measuring hardness of cereal grains include the following.

(1) The Smetar hardness test which utilizes the penetration of a diamond-shaped stylus into a section of a kernel with measurement of length of indentation at the surface as a measure of hardness (Smeets and Cleve 1956). (2) The laboratory pearler tester (Taylor et al. 1939). The test consists of grinding 20 gm of wheat in a Strong-Scott barley pearler for 1 min. The harder the wheat, the smaller the amount of wheat removed. Though widely used in soft wheat laboratories, the pearling test has not received wide acceptance for evaluating hard wheat. (3) Modification of a commercial portable

tester for metal, known as the Barcol Impressor, so that it can be used to test hardness of wheat sections (Katz *et al* 1961).   (4) Measurement of the amount of work required to mill a known weight of grain; the most commonly used is the Brabender hardness tester. The tester consists of a small burr mill fitted to the dynamometer coupling and recording of a farinograph.   The device measures the torque required to operate the mill as the grain is ground under standardized conditions, and the height on the graph paper indicates hardness (Paukner 1951).   Use of the method in evaluating physical properties of weathered wheat was described by Milner and Shellenberger (1953).   Anderson *et al.* (1966) modified the procedure and evaluated the results by determining the particle size distribution of wheat passed through the tester.   (5) The wheat is passed through a pin mill and particle size distribution is determined (Anderson *et al.* 1966).   (6) The wheat is ground and sifted under standardized conditions, and a granulation index is calculated for each wheat (Cutler and Brinson 1935).

Physical Dough Testing.—Flow occurs in dough only if the stress exceeds a certain limiting value or yield value.   In this respect dough behaves as a Bingham body.   At lower stresses the deformations disappear completely after the stress is released (Bloksma and Hlynka 1964).   Whether the elastic deformation is accompanied by permanent viscous deformation (at higher stresses), or not (at lower stresses), the attainment of the equilibrium shape after the release of stress requires some time.   Dough shows retarded elasticity (elastic aftereffect) like that of a Kelvin body.

The viscosity of dough and, generally, its resistance against deformation increase with increasing shear; its rate of shear decreases correspondingly.   The changes in properties are temporary, and the phenomenon was termed "structural activation."   Apart from its dependence on the amount of shear, the apparent viscosity decreases with increasing stress.   Gluten behaves essentially like dough.   It does not show an appreciable yield value, and its retardation time is shorter.   Satisfactory bread doughs are characterized by a high viscosity:modulus quotient and a small decrease in apparent viscosity with increasing stress.   The quotient corresponds to the relaxation time for a simple Maxwell body.   Bread doughs seem to differ from durum flour doughs in that deviations from linear behavior are more pronounced in elastic deformations, whereas durum flour doughs show those deviations most clearly in viscous deformations.

Hlynka and associates conducted systematic studies of doughs

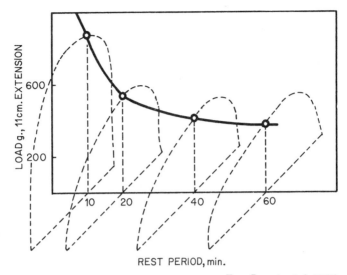

From Dempster et al. (1953)

FIG. 28.7. DERIVATION OF A STRUCTURAL RELAXATION CURVE

rested for various periods between rounding and shaping of a dough and its stretch. By plotting extensigram heights (see description of load-extension meters) at an arbitrarily selected fixed sample extension, against rest period, t, they obtained a structural relaxation curve (Fig. 28.7 from Dempster *et al.* 1953). The curve can be described by an equation of a hyperbola asymptotic to the load axis, and to a line parallel to the time axis and at a distance L_a above it (Fig. 28.8 from Hlynka and Matsuo 1959).

$$L = L_a + c/t \qquad (28.1)$$

where L = load
L_a = asymptotic load
c = structural relaxation constant
t = test period

The semiaxis constant a, that equals $\sqrt{2c}$, is useful in studies of the effects of oxidants in breadmaking. The parameters c and L_a can be found graphically by plotting the product of L and t vs t (Fig. 28.9 from Dempster *et al.* 1955). This results in a transformation to a straight line given by

$$(Lt) = L_a t + c \qquad (28.2)$$

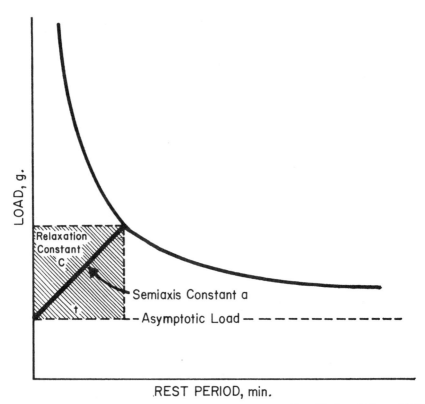

From Hlynka and Matsuo (1959)

FIG. 28.8.   HYPERBOLIC REPRESENTATION OF THE STRUCTURAL RELAXATION
CURVE

The intercept and slope of this line are c and L_a, respectively.

Muller *et al.* (1961, 1962, and 1963) have analyzed the stretching process in the Brabender extensigraph.   They converted the load-extension curve recorded into a stress-strain curve in which both stress and strain were expressed in cgs units.   By interrupting the stretching process and allowing the test piece to recover, they divided the total extension into a recoverable elastic part and an irrecoverable viscous one.

Assuming that the rheological behavior of dough can be described by a generalized Maxwell body, one can similarly divide the work performed on a test piece into an elastic and viscous part.   Bloksma (1967) analyzed the above "work technique" and developed a theory that leads to an alternative presentation of experimental results, and which permits under the conditions of some simplifying assump-

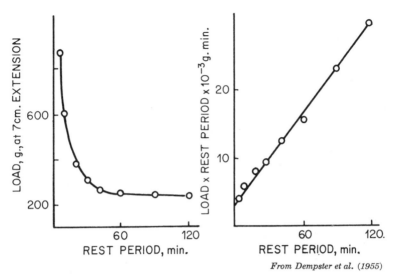

From Dempster et al. (1955)

FIG. 28.9. STRUCTURAL RELAXATION CURVE AND ITS LINEAR TRANS-
FORMATION

tions direct conclusions on changes in modulus and viscosity.

The main value of physical dough testing devices is in providing excellent supplementary information on changes occurring during a short period of the breadmaking process. This concept is illustrated in Fig. 28.10. Physical dough testing devices are most useful in evaluation and prediction of plant breeder samples, in quality control during flour milling and breadmaking, and in basic rheological studies. They provide information on specific properties which cannot be obtained by other means. Physical dough testing methods have many limitations which should be realized in the interpretation of results. Thus, rheological properties may differ in flours varying little in their breadmaking performance. Physical dough testing curves may be affected considerably by modification of flour constituents or by certain additives, which have little effect on actual breadmaking.

Most rheological measurements are made on unyeasted doughs; fermentation is known to modify considerably both rheological properties and breadmaking characteristics. And finally although physical dough tests are useful analytical tools because they measure changes at certain stages of the breadmaking process, they fail to measure the effects of interaction between various factors on bread quality.

	FARINOGRAPH	EXTENSOGRAPH	AMLYOGRAPH
Nature of Method	dynamic	static	dynamic
Type of Indication	Farinogram  — Resulting curve of above instrument	Extensogram 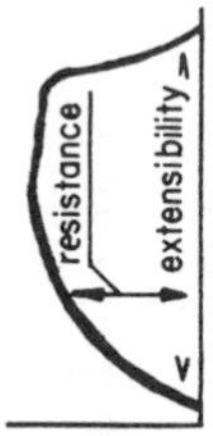 (resistance, extensibility) — Resulting curve of above instrument	Amylogram  — Resulting curve of above instrument
Information Obtained	Absorption, Mixing time, Mixing tolerance (same as general strength)	$\dfrac{Resistance}{Extensibility}$ = degree of maturing (indication of physical dough properties)	Crumb formation characteristics (gelatinization properties)
Correction Possibilities in a flour mill	changes in wheat blend	adjustment in "bleach" (chemical maturing agents)	mill stream switching and/or malt additions (diastasing)
in a bakery	changes in absorption and mixing	adjustment of yeast food	malt additions

Courtesy C. W. Brabender Instruments, Inc.

FIG. 28.10.   THE THREE-PHASE CONCEPT OF BREADMAKING

Recording Dough Mixers.—The two most widely used types of recording dough mixers are the Brabender farinograph and the mixograph. Both instruments record the power that is needed for mixing dough at a constant speed. The record consists of an initial rising

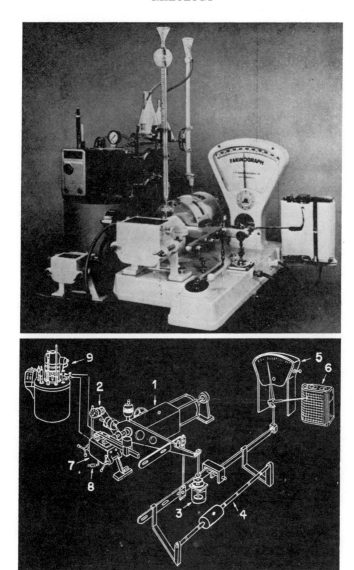

Courtesy C. W. Brabender Instruments Inc.

FIG. 28.11. BRABENDER FARINOGRAPH

part which shows an increase in resistance with mixing time, and is interpreted as dough development time. The point of maximum resistance (or minimum mobility) is generally identified with optimum dough development, and is followed by a second part of more or less rapid decrease in consistency and resistance to mixing.

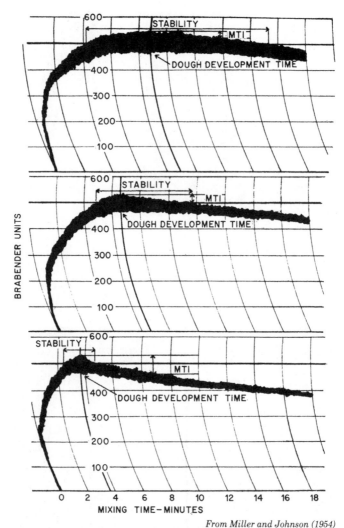

From Miller and Johnson (1954)

FIG. 28.12.   TYPICAL FARINOGRAMS OF FLOUR SAMPLES

Long, medium and short dough development times, showing
values for stability and mixing tolerance index (MTI).

A picture and diagram of the Brabender farinograph are given in
Fig. 28.11.   The instrument measures plasticity and mobility of
dough subjected to a prolonged, relatively gentle mixing action at
constant temperature.   Resistance offered by the dough to mixing
blades is transmitted through a dynamometer to a pen that traces a
curve on a kymograph chart.   The general practice has been to

determine in a preliminary "titration curve" the amount of water required to give a standard consistency (generally 500 BU). The farinograph provides also information on optimum mixing time and dough stability. Figure 28.12 compares farinograms of three types of flours and shows the main parameters calculated. Many investigators calculate the valorimeter value, which is determined by the length of time required to mix the flour to minimum mobility and the descending slope. The larger the valorimeter value, the stronger— in terms of breadmaking potential—the flour.

The mixograph is a miniature high-speed recording dough mixer (Swanson and Working 1933). The mixer (Fig. 28.13) has 4 vertical pins revolving about 3 stationary pins in the bottom of the bowl. As the gluten develops, a gradually increasing force is required to push the revolving pins through the dough. The increased force is measured by the tendency to rotate the bowl which is placed in the center of a lever system. A record of the torque produced on the lever system is made on a chart moving at a constant rate of speed. Mixograms of a weak and a strong flour are shown in the upper part of Fig. 28.13.

From K. F. Finney

FIG. 28.13. THE MIXOGRAPH

Load Extension Meters.—In a number of commercially available instruments, dough mixed and shaped under standardized conditions is stretched until rupture; and a curve of load versus elongation is recorded.   From the curve, resistance to deformation, extensibility, and energy needed to rupture the dough is computed. The instruments are particularly useful in studying the effects of oxidants on dough properties.

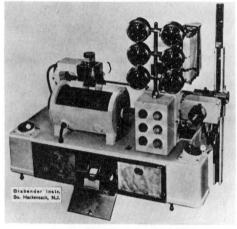

FIG.   28.14.   THE   BRABENDER
EXTENSIGRAPH

Courtesy C. W. Brabender Instruments, Inc.

The Brabender extensigraph (Fig. 28.14) was introduced in 1936 to supplement the information provided by the farinograph.   The tested flour is mixed in the farinograph into a dough reaching its maximum consistency at 500 BU.   It is taken out of the mixer and two pieces are shaped into balls by means of a rounder-homogenizer and then into cylinders by means of a roller. The dough cylinders are clamped in cradles and allowed to rest for 45 min at 30°C. After stretching, the dough is shaped, allowed to rest, and stretched again. Generally, 3 stretching curves are obtained at 45-min intervals. It is usual to make the following measurements: (1) extensibility— length of the curve in mm, (2) resistance to extension—height of the extensigram in BU measured 50 mm after the curve has started, (3) strength value—area of the curve, and (4) proportionality figure—ratio between resistance and extensibility.

Muller and Hlynka (1964) discussed the following factors affecting extensigraph determination and evaluation: choice of dough mixed and mixing procedure, consistency, the use of sodium chloride, and repeated extension of the same dough piece.

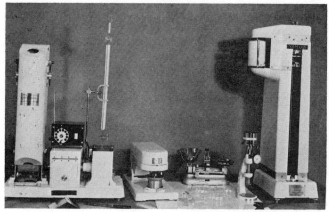

Courtesy Simon Engineering Ltd.

FIG. 28.15. THE EXTENSOMETER

The research extensometer designed by Halton and his associates (Halton 1938, 1949) from the Research Association of British Flour Millers is similar in many respects to the Brabender extensometer. The extensometer is part of a four-unit equipment (Fig. 28.15) which includes a water absorption meter, mixer, shaper, and stretching unit. The water absorption meter measures the extrusion times of yeasted doughs. For load-extension curves, the doughs are mixed at optimum absorption, shaped, and impalled on a two-part peg of the research extensometer. The upper half of the peg is fixed; the lower half moves at a fixed rate and stretches the impalled dough. The force that is exerted on the upper peg is transmitted and recorded on a graph paper.

The alveograph (Chopin extensometer) consists of three parts: the mixer, the bubble blowing portion, and a recording manometer. The procedure involves using air pressure to blow a bubble from a disc of a flour-water-salt dough. The bubble is expanded to the breaking point, and the recording manometer which is operated hydraulically records a curve from which three basic measurements are taken: distance (in mm) that the dough stretches before it ruptures; resistance to stretching at peak height; and curve area. The alveograph is quite popular in several European countries and is the most commonly used physical dough testing device in France.

Bloksma *et al.* (1962) reported results of an international collaborative test carried out in 13 laboratories from 8 countries within the framework of the International Association for Cereal Chemistry. Generally, the repeatability of the measurements within a laboratory was 10–15% for the alveograph and extensigraph; for the farinograph water absorption was 0.2% or better. As expected, variations between laboratories were greater than within laboratories. The differences were traced to individual factors and to differences in instruments and in their conditions.

Starch.—For the evaluation of starch products, three aspects of starch rheology are of paramount importance (Hofstee and de Willigen 1953): (1) the alterations of rheological properties during pasting; (2) hot-paste viscosity and its variations with time; and (3) change in rheological properties during and after cooling of the paste (cold paste viscosity; jelly strength). In all cases, the properties of the pastes are governed by the pasting or cooking method employed. On heating with water, a series of changes occurs, the most important being: loss in birefringence, change in translucence, change in X-ray pattern, swelling of the grain, and abrupt change in viscosity. The changes do not occur at the same time and not at a sharply defined temperature. They take place at a small interval of temperature, conveniently called the gelatinization point.

To study the complex rheological changes in starch pasting, it is essential to heat the paste under standardized conditions and record continuously the changes. A series of instruments have been designed to meet these requirements. They include the Brabender amylograph, the Corn Industries Recording Viscometer, and the VI Viscometer. The amylograph and its modifications (i.e. the Viscograph) are most commonly used and are described in detail later. All recording hot-paste recording viscometers work on a similar general principle. From the curve obtained the following parameters are generally recorded: (1) the time (temperature) of initial rise in viscosity; (2) temperature and height of maximum viscosity; and (3) viscosity after prolonged heating beyond the point of peak viscosity. There is a linear relation between the logarithm of maximum viscosity (η max) and logarithm of concentration (c).

$$\log \eta \, max = a + b \log c \qquad (28.3)$$

Although the reproducibility of the data from all the above recording viscosimeters is satisfactory, they differ in their performance and utility. The Corn Industries Recording Viscosimeter allows for varying the rate of stirring. Its disadvantage is the large amount of starch required (100 gm) and large volume of sample beaker (1000 ml). The VI Viscograph is cheap and simple. It has several disadvantages: determination of paste temperature is difficult, and no provision is made for the condensation of evaporated water or for keeping a fixed temperature. It is commonly used in textile and paper industries. The Brabender amylograph (Fig. 28.16) is a torsion viscosimeter that provides a continuous automatic record of changes in viscosity of starch, as the temperature is raised at a constant rate of approximately 1.5°C per min (Anker and Geddes 1944). The instrument consists of a cylindrical stainless bowl that holds a suspension of 100 gm flour in 460 ml of a buffered solution. The bowl is rotated at 75 rpm in an electrically heated air bath by a synchronous motor which also operates the recording and temperature control devices. A steel arm which dips into the bowl is connected through a shaft to a pen which records changes in viscosity of the heated flour suspension in the bowl. Depending on the change in viscosity of the heated suspension, a torque is exerted on the steel arm and is recorded on an arbitrary scale. Viscosity of the slurry is generally recorded as the temperature rises from 30° to 95°C. Modifications of the amylograph which record viscosity changes at a uniformly rising or falling temperature are useful in studies of gelatinization characteristics of various cereal starches. In Europe, the amylograph is used widely to predict the baking performance of rye flours and to detect excessive amounts of flours from sprouted grain. In the United States, the amylograph is used primarily to control malt supplementation. Use of the Brabender amylograph in quality control of potato flakes was described by Tape (1965).

Consumer Goods.—Many consumers frequently use softness as a criterion of bread freshness. Consequently, there is justification for using crumb compressibility in evaluating staleness. The methods used to measure loaf softness are based on three principles (Thomas and Thunger 1965): (1) The crumb is subjected to a

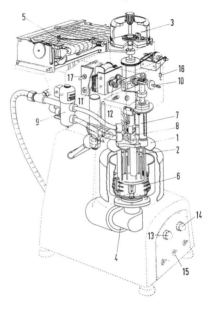

Courtesy C. W. Brabender Instruments, Inc

FIG. 28.16. BRABENDER AMYLOGRAPH

constant load for a fixed time, and the deformation is measured to find a softness index. (2) The force required to give a fixed deformation is measured and crumb firmness is determined. (3) The crumb firmness is subjected to shearing or squeezing forces.

The latest edition of *Cereal Laboratory Methods* (AACC 1962) gives two tests for measuring staleness of bread, one based on organoleptic evaluation by a trained panel, and one using a compressimeter which measures crumb firmness (Platt and Powers 1940). An instrument which measures quantitatively the effects of squeezing a loaf of bread was described by Hlynka and Van Eschen (1965).

The various physical tests are widely used as convenient means of obtaining measurements of changes in crumb properties as bread ages. It should be realized, however, that the various tests record different rates of change, and that they are not entirely consistent with one another. It has also been shown that such methods frequently do not agree with human judgment of freshness (Bechtel 1955).

Mechanical strength of prepacked cereal foods (cookies, macaroni, breakfast cereals, snack foods, cereal bars, and various fried or baked crisp or flaky products) is of great economic importance. Measurements of the texture of such food items are generally made by methods which measure the force necessary to break a piece of the material, and which determine changes in particle size resulting from subjecting the food to the impact of crushing forces.

Dairy Products

Most liquid foods are essentially Newtonian fluids and may be considered to have their texture or "mouth feel" adequately described by their viscosity.

Measurements of the viscosity of milk were made before the end of the last century, mainly with a view to finding a property which might prove useful as an index of quality; thus, avoiding lengthy analyses. This hope was never fulfilled (Scott-Blair 1953). Milk deviates but little from Newtonian behavior though its viscosity is not easily measured accurately because of creaming. Cream shows a variety of rheological anomalies, and it is difficult to define any suitable set of properties by which to describe what the consumer means by "body." The flow properties of condensed milk are important during manufacture but are commonly assessed subjectively. Rheological properties of frozen ice cream, mix, and melted ice cream by a coaxial cylinder viscosimeter were described by Shama and Sherman (1966), and by Sherman (1966).

The rheological properties of butter are important in relation to quality. The whole complex of those properties is so characteristic in butter that other substances are often said to have a butter-like consistency (Mulder 1953). Sectility, spreadability, and eating texture and consistency affect consumer evaluation of butter (and margarine). Butter should have a good spreadability accompanied by some elasticity. It should be firm enough not to collapse, but not too firm to break down into pieces; it should be moderately extensible, long, and tough; and should not be oily or sticky or have a mealy or sandy taste. The various methods used to estimate rheological properties of butter include penetrometers, plastometers (deformation under load), cutting methods, the sagging beam methods, and various spreadability procedures. In most cases, the properties which are determined by the various methods differ widely. It is difficult to expect that a single method would serve to evaluate all the important parameters or express consistency as it is judged by butter graders by organoleptic tests. Davis (1937) calculated the viscosity of butter by dividing shearing stress by the rate of deformation of a cylinder of butter subjected to various loads. The elastic modulus was computed from deformation changes in samples after the loads were removed. Kruisheer and den Herder (1938) described a mechanically driven plastometer devised for the determination of permanent deformation and "yield values"— important parameters of butter consistency. An apparatus constructed to measure the force required to cut with a wire a sample of butter was described by Dolby (1941), and a penetrometer by Mohr and Wellm (1948). Kapsalis *et al.* (1960) developed an instrument called the "consistometer" for measuring both the spreadability and the hardness of butter samples. The instrument consists of a constant speed motor driving a pendulum which carries the knife or wire used to measure, respectively, spreadability and hardness. The butter is held in a stainless steel frame mounted at the lowest point of an arc described by the pendulum. The sample is raised $1/16$ in. for the knife, and $1/8$ in. for the wire before each determination. The resistance offered by the sample is equivalent to the sum of forces exerted by fixed weights and the motor. The weights act through a pulley system and can be measured directly while the force exerted by the motor is registered on a torque meter.

The great variety of cheeses and their tremendous variations in consistency and composition stimulated development of a large number of compressimeters, penetrometers. and sectimeters. From the earliest times, the assessing of "body" of cheese and of curd (partly made cheese) by subjective means has been regarded as important. Faults in body at certain stages in manufacture affect considerably the quality of the finished cheese. Poor body in "green" (unripe cheese) foreshadows inferior keeping quality and even serious deterioration during ripening (Baron and Scott-Blair 1953).

Consistency of rennetted cheese is partly defined by the nature of the surface in rupture (clean or flocculated), and partly by the resistance of the curd to compression. The latter is best expressed by the shear modulus of the gel. For displacements under small loads applied for short times, curd gels behave approximately as Hookean elastic solids. If measurements are repeated on the same surface of the gel, there is an irreversible deformation due to some elastic fatigue. Thus readings should be taken at different spots on the surface. Numerous methods have been devised to measure the rigidity of rennetted curd by drawing a cutter through the coagulum, and recording the resistance to cutting as the tension exerted on an attached spring.

An instrument manufactured under the trade name of "American Curd-O-Meter" (and its modifications) combines features of an automatic compression type of spring balance recorder with an instrument that measures the curd tension of milk in terms of the pull in grams required to draw a 10-pronged star-shaped knife through 100 ml of coagulated milk at 35°C.

Viscosity of rennetted milk can be measured by various rotating and oscillating viscosimeters.

Meat

Tenderness is probably the most important factor affecting consumer evaluation of meat quality and acceptability. Although fibrousness is a very important textural factor in meats, the toughness of the fascia, the fat content, and the quality and amount of collagen are some additional factors that might be expected to influence tenderness (Matz 1962).

Mechanical methods of measuring tenderness were the subject of a Symposium (Pearson 1963) and were reviewed by Schultz (1957), Kaufmann (1959), Sale (1960), DeFelice et al. (1961), and by Matz (1962). Sale (1960) divided meat tenderness devices into four types depending on their mode of action: (1) shear, (2) penetration, (3) biting, and (4) mincing. Historical development of a wide range of shear, penetration, mastication, compression, tensile, and mincing devices was given by De Felice et al. (1961).

Despite limitations inherent in the difficulty of objectively measuring meat tenderness, the Warner-Bratzler shear testing device (Bratzler 1949) represents one of the most commonly used and accepted mechanical devices to evaluate consumer acceptance of meat (Schultz 1957). Sensory evaluation of meat tenderness was recently compared with mechanical shear measurements by the Warner-Bratzler and L.E.E.-Kramer (Bailey et al. 1962) instruments (Sharrah et al. 1965 A,B,; Voisey and Hansen 1967). Evaluation of frankfurter texture by a penetrometer was described and compared with taste-panel evaluation by Simon et al. (1965). Kulwich et al. (1963) developed and evaluated a device for estimating the tenderness of slices of cooked meat. The slice-tenderness evaluator consisted of a sample holder in which the slice of cooked meat was mounted, and a penetrator that first punctures and then shears a

$^3/_8$-in. diam circular portion of the sample. The evaluator was used in conjunction with a commercial materials-testing instrument that provided continuous recording of the force-penetration curve.

The tenderness press, developed by Sperring et al. (1959) uses a cam and gear-reduction box to operate the hydraulic pump of a laboratory press. In this technique, a meat sample is compressed between a plunger and cylinder base until a portion of sample extrudes through a small orifice in the cylinder base. The hydraulic pressure required to extrude the first portion through an orifice is used as a measure of meat tenderness. The usefulness of the Warner-Bratzler shear press, slice-tenderness evaluator, and improved tenderness press was evaluated and compared with panel tenderness scores (Alsmeyer et al. 1966).

Estimating meat tenderness by measuring the energy consumed in grinding a sample of food was the basis of a procedure described by Miyada and Tappel (1956). Proctor et al. (1955) devised a denture tenderometer simulating actual chewing conditions. Carpenter et al. (1965) evaluated pork samples by objective and subjective methods. The data from the Warner-Bratzler shear, grinder tenderometer, and denture tenderometer indicated that all three instruments possess a satisfactory potential for use in prediction of pork tenderness.

Fruits and Vegetables

Methods used for evaluating the texture of fruits and vegetables include: (1) compressimeters; (2) penetrometer; (3) shear-testing devices; and (4) to a smaller extent instruments measuring energy consumption during grinding and sonic techniques.

The Magness-Taylor apparatus (1925) has served as the basis for several pressure testers. The apparatus is composed of a plunger surrounded by a metal collar and enclosed in a tubular handle. The plunger is connected to the handle by a calibrated steel spring. The spring is extended to an amount proportional to the force applied to the handle when the plunger tip is in contact with the test object. As the plunger is forced into the sample, the fruit surface pushes the collar into contact with an electrical junction. The light bulb which is energized by this contact determines the end point. The pressure is indicated on a scale attached to the handle which is part of the plunger. Plunger-type pressure testers are simple to use, and give readings which in foods of homogenous consistency are highly correlated with overall texture as determined organoleptically.

The Kattan Firm-O-Meter (Kattan 1957) and a commercial device based on the same principle (Garrett *et al.* 1960) crush whole tomatoes; the decrease in fruit diameter is an indication of fruit firmness. The instrument measures attributes similar to those estimated by squeezing the fruit by hand. Results correlated at a highly significant level with panel rankings of sample firmness.

Virtually all fruits and vegetables contain some components which are perceived as fibers when the food is eaten (Matz 1962). Fibrousness is a particularly serious problem with flat beans, but may also be encountered with round varieties harvested at late maturity. Although they are sometimes used on products such as asparagus and snap beans, pressure testers incorporating a narrow pointed or rounded plunger probably cannot give an accurate measurement of fibrousness. Most investigators use instruments based on shear-type actions for measuring fibrousness. The fibrometer devised by Wilder (1948) tests individual asparagus spears by a standardized cutting device. A wire held rigidly in a frame is arranged in such a way that the wire may be lowered upon the tested spear and allowed to press down with a definite shearing force. The presence of excessive fibrousness is indicated by the failure to cut through the spear. Wilder's asparagus fibrometer was modified by Gould (1951) to evaluate maturation of snap beans.

The most important single factor determining the quality of commercially processed green peas is their harvest maturity. This has been recognized by the United States standards for grades of canned and frozen peas. The unsatisfactory nature of personal assessment of grade has led to the development of a number of objective tests which can be classified as chemical, physical, and mechanical (Lynch *et al.* 1959).

Most of the mechanical tests measure the force required to crush, puncture, or shear samples of peas.

Research instruments which test single peas or small pea samples include: (1) a denture tenderometer which uses artificial dentures operated mechanically to simulate frequency and motion of chewing; (2) various crushing instruments which measure the force required to penetrate completely a test piece by a standard cylindrical rod, or to crush a test material to a specified fraction of its diameter; and (3) an apparatus which measures the force required to puncture a single pea with a cylindrical pin.

Commercial instruments include: (1) a hand operated texturemeter which measures the force required to drive 25 steel pins through a sample of peas in a cylinder; (2) a tenderometer which measures the force required to shear a sample of peas; (3) a shear press designed as a multipurpose instrument for measuring the force required to cut, shear, and puncture different foods (the mode of action of the shear press is similar to that of the tenderometer, and results from both are highly correlated); (4) a maturometer which measures the resistance to puncture 143 peas by pins; (5) a hardness meter in which compression by a spring-loaded piston is recorded; and (6) a succulometer which presses juice from a sample by application of thrust developed hydraulically, and in which the volume of expressed liquid is measured.

Lynch *et al.* (1959) reported that correlations of most instrument readings with determinations of the AIS (alcohol insoluble solids) content—a widely accepted standard against which other methods are compared—were generally high and significant. Changes in AIS equivalent to the standard deviations or replicate measurements with a tenderometer and a maturometer were similar and almost identical with the standard deviation calculated from replicate determinations of AIS on the same peas. Angel *et al.* (1965) reported recently that the correlations with taste panel scores of tenderness were respectively 0.93 for the AIS method and 0.91 for a shear press equipped with an extrusion cell to measure the kinesthetic properties of canned peas. The application of the extrusion principle in texture measurement of fresh peas was described by Bourne and Moyer (1968).

Johnson *et al.* (1965) and Sistrunk and Moore (1967) found that texture measurements by shear press correlated with sensory analyses of fresh, frozen, and gamma-irradiated strawberries.

Estimating firmness of the entire fruit or vegetable is preferable to determining the texture of parts of the sample or of individual tissues. Estimating fruit firmness by the thumb test has probably been in existence ever since people began eating fruit (Bourne 1965). The thumb test is more useful in evaluating maturity of fruits that soften considerably as they ripen (i.e. tomatoes), and less useful in determining maturity of firm fruits (i.e. apples). Objective evaluations of fruit hardness have been performed for over 40 yr in the United States by the previously described Magness-Taylor pressure tester (Magness and Taylor 1925), and its modifications (Haller 1941). The Delaware jelly-strength tester was adapted to measure the resistance to crushing of cylinders cut from apples (Whittenberger 1951) or carrots (Powrie and Asselbergs 1957).

Bourne (1965) made punch tests on apples with Magness-Taylor pressure tips mounted in a commercial testing machine which draws a force-distance curve for each punch.   The force increased rapidly with little deformation until the yield point was reached when the pressure tip began to penetrate the fruit tissue.   The results indicated that depth of penetration of the tip should be greatly reduced to obtain a consistent measurement of yield point pressure tests. Skin on apples raised the pressure test by adding the shear strength of the skin to the pressure test.   The speed of travel of the tip affected the pressure test to only a small degree.

Since the softness of the whole potato is used as a measure of quality, there is a demand for an objective method of measuring the softness of potatoes and a need to determine the factors that affect potato firmness.   Sawyer and Collins (1960) used an instrument which was designed for measuring rubber and similar products. The instrument (Durometer) consists of a spring-loaded indentor which protrudes through a metal anvil.   A circular dial calibrated from 0 to 100 indicates the force exerted on the spring.   Finney et al. (1964) reported on the deformation of potatoes that were ruptured under the load of a small cylindrical punch.   Bourne and Mondy (1967) evaluated firmness of potatoes by measuring in a commercial testing machine the deformation of cylinders cut from potato tissues with a cork borer and of whole potatoes under a metal punch using a constant force.   Firmness measurements were highly correlated with sensory scores.   Somers (1965) evaluated the usefulness of a commercial instrument in providing fundamental rheological parameters and applied information on the viscoelastic properties of storage tissues from potato, apple, and pear.

Abbott et al. (1968) described two ways of using sonic energy to measure the inner texture of fruits and vegetables.   In one, sections cut from the fruit or vegetable were vibrated at their natural frequencies, and from the data obtained internal friction and the elastic modulus (ratio of stress to strain) were calculated.   In the second method, sonic energy was applied to the whole fruit or vegetable, and the frequency of a resulting series of resonances was measured. When the frequency of one of the resonances was squared and multiplied by the mass of the food, a parameter called "stiffness coefficient" that measured stiffness or firmness was obtained.   Finney et al. (1967) used a sonic technique to measure elasticity of cylindrical sections of fresh, firm bananas.   Softening of the banana during ripening was associated with a decrease in modulus of elasticity from 272 to 85 $\times$ 10^5 dynes/cm^2.   Modulus of elasticity was positively correlated with starch content, and negatively correlated with luminous reflectance and the log of reducing sugars.

Miscellaneous Food Products

The rheology of miscellaneous food products was reviewed by Harvey (1953). Sucrose, invert sugar, and confectioners' glucose are three raw materials basic to the sugar confectionery industry, and the rheological properties of their syrups and mixed syrups are of prime importance. The confectionery syrups are typical Newtonian liquids and solutions of sucrose are considered suitable for calibrations of viscometers.

Sucrose solutions show the characteristic viscous behavior of nonelectrolyte solutions in that their relative viscosity falls with rising temperature. For sucrose, invert sugar, and confectioners' glucose a simple linear relation exists between the log of viscosity and the concentration of soluble solids (expressed as the ratio of sugar dissolved per unit weight of water).

Honeys differ widely in composition and display a wide range of rheological properties. In general, honeys are Newtonian liquids, but heather honey, which is one of the most important European honeys, is thixotropic and various honeys derived from eucalypti are dilatant (Pryce-Jones 1953) (these terms are defined later in this chapter).

Addition of corn syrup to "Newtonian" honeys markedly increases the viscosity. The viscosity range for various honeys is, however, very wide and markedly affected by water contents. For most "Newtonian" honeys, there is a linear inverse relationship between the logarithm of viscosity (using a falling ball apparatus) and logarithm of water content between the limits of 12.8% water (475 poises) and 22.4% water (50 poises). For honey samples ranging in water content from 12.4 to 19.7%, the equation correlating moisture (w), viscosity (η_t, determined by a falling ball viscosimeter with a tube of 21.2 mm internal diameter and steel ball of 16 mm diam) and absolute temperature (T) is

$$w = (62,500 - 156.7T)\,[T(\log \eta_t + 1) - $$
$$2.287\,(313 - T)\,] \qquad (28.4)$$

A comparison of the results obtained by viscosity measurements and the official vacuum drying method showed that they did not differ on the average by more than 0.2%.

Good quality heather honey sets into a rigid gel when left to rest, but flows again on stirring. Thus, this honey shows properties of thixotropy according to the definition of Freundlich and Peterfi; namely, an isothermal reversible gel-sol-gel transformation induced by shearing and subsequent rest. Thixotropy of honey largely depends on the protein content of the sample, and its true thixotropic nature has been questioned.

The term "dilatancy" was used originally to describe the property of a granular system, which increases in volume when sheared laterally. The meaning has been recently expanded to include any system whose viscosity increases with increasing rate of shear beyond a critical minimum value. Several honeys from eucalypti display the property of "Spinnbarkeit." They can be drawn out into long strings and show dilatancy. Dilatant honeys contain 6–7% dextrans which can be removed by precipitation with acetone. After the honey has been restored to its original concentration, it behaves like a true Newtonian liquid. The peculiar rheological properties of of honeys from eucalypti can be simulated by adding dextran to clover or sainfoin honey.

For mixtures containing a constant proportion of gelatin by weight and constant proportion of nongelatin dissolved solids, there exists, at constant temperature, a simple linear relationship between the logarithm of viscosity and the concentration of total solids (including gelatin), when the latter is expressed in terms of weight of total solids per unit weight of water. At constant total solids concentration, there exists a simple relationship between the logarithm of viscosity and the percentage of gelatin by weight.

The ability of gelatin solutions to form a gel is the characteristic that accounts for the major uses of gelatin. The standard measure of gel strength, Bloom, is the main rheological property considered in grading gelatin for food uses. Bloom is measured with an instrument called the Bloom gelometer (Bloom 1925). This method measures the force required to depress 4 mm a $^1/_2$-in. diam plunger into the top center of a gel prepared under standardized conditions. The force is provided by lead shot flowing into a cup on top of the plunger. Determination of Bloom in gelatin solutions at nonstandard concentration was discussed by Kramer and Rosenthal (1965). Kramer and Hawbecker (1966) reviewed methods for measuring and recording rheological properties of gels. Most of the instruments measure the force required to break a gel. The instruments include: (1) the jelmeter (Baker 1934), a simplified version of an Ostwald viscosimeter used to determine correct proportions of sugar, pectin, and acid in making jellies, jams, and marmelades; (2) the Bloom Gelometer (Fellers and Griffith 1928); and (3) the penetrometer (Underwood and Keller 1948) which provides a measure of the force required to penetrate into a gel. In addition to the determination of gel strength, it may be desirable to measure uniformity, adhesiveness (stickiness), and deformation prior to breakdown of gel structure. The latter three parameters can be measured by using special extrusion-type cells in the Allo-Kramer shear press (Kramer and Hawbecker 1966). Charm and McComis (1965) reviewed critical physical measurements of gum solutions and gave examples of use of the information obtained.

Most of the studies in rheological properties of chocolate have involved measurements made on the melted substance. The most obvious usefulness of such data is in processing. For example, the physical behavior of molten chocolate is of particular importance when it is to be used for coating purposes. The instrument most frequently used in this country for measuring the viscosity of molten chocolate is the MacMichael viscosimeter. Capillary devices and viscosimeters of the falling sphere type have also been used (Matz 1962).

Heiss (1959) described an apparatus for measuring surface stickiness of candies. The instrument is intended to quantitate the evaluation obtained by pressing for a short time a "dry and clean finger" on to the candy. An upper plate cushioned with rubber is connected to a calibrated spring. A lower movable plate carries the candy piece which is made flat on the upper surface. The upper plate is initially pressed to the candy with a weight of 11.5 oz for 10 secs, and then the lower plate is gradually withdrawn at a constant rate of speed until the candy surface and the rubber surface are pulled apart. The distance traveled before the release occurs is a measure of the degree of stickiness of the candy surface.

BIBLIOGRAPHY

ABBOT, J. A. et al. 1968. Sonic techniques for measuring texture of fruits and vegetables. Food Technol. 22, No. 5, 101–112.
ALSMEYER, R. H., THORNTON, J. W., HINER, R. L., and BOLLINGER, N. C. 1966. Beef and pork tenderness measured by the press, Warner-Bratzler, and STE methods. Food Technol. 20, 115–117.
AM. ASSOC. CEREAL CHEMISTS. 1962. Cereal Laboratory methods, 7th Edition. Am. Assoc. Cereal Chemists. St. Paul, Minn.
AMERINE, M. A., PANGBORN, R. M., and ROESSLER, E. B. 1965. Principles of Sensory Evaluation of Food. Academic Press, New York.
ANDERSON, R. A., PFEIFER, F. F., and PEPLINSKI, A. J. 1966. Measuring wheat kernel hardness by standardized grinding procedures. Cereal Sci. Today 11, 204–209.
ANGEL, S., KRAMER, A., AND YEATMAN, J. N. 1965. Physical methods of measuring quality of canned peas. Food Technol. 19, 96–98.
ANKER, C. A., AND GEDDES, W. F. 1944. Gelatinization studies upon wheat and other starches with the amylograph. Cereal Chem. 21, 335–360.

BAILEY, M. E., HEDRICK, H. B., PARRISH, F. C., and NAUMANN, H. D. 1962. L. E. E.-Kramer shear force as a tenderness measure of beef stock. Food Technol. *16*, No. 12, 99–101.

BAKER, G. L. 1934. New methods for determining the jelly power of fruit juice extraction. Food Ind. *6*, 305–307.

BARON, M., and SCOTT-BLAIR, G. W. 1953. Rheology of cheese and curd. *In* Foodstuffs, Their Plasticity, Fluidity, and Consistency, G. W. Scott-Blair (Editor). Interscience Publishers, New York.

BECHTEL, W. G. 1955. A review of bread staling research. Trans. Am. Assoc. Cereal Chemists *13*, 108–121.

BLOKSMA, A. H. 1967. Detection of changes in modulus and viscosity of wheat flour doughs by the "work technique" of Muller *et al.* J. Sci. Food Agr. *18*, 49–51.

BLOKSMA, A. H., FRANCIS, B., and ZAAT, J. C. A. 1962. A comparison of data obtained in different laboratories from the alveograph, farinograph, and extensigraph. Cereal Sci. Today *9*, 308–310, 312–313, 330.

BLOKSMA, A. H., and HLYNKA, I. 1964. Basic considerations of dough properties. *In* Wheat Chemistry and Technology, I. Hlynka (Editor). Am. Assoc. Cereal Chemists. St. Paul, Minn.

BLOOM, O. T. 1925. Penetrometer for testing jelly strength of glues, gelatins, etc. U. S. Pat. 1,540,979.

BOURNE, M. C. 1965. Studies on punch testing of apples. Food Technol. *19*, 113–115.

BOURNE, M. C. 1966. A classification of objective methods for measuring texture and consistency of foods. J. Food Sci. *31*, 1011–1015.

BOURNE, M. C., and MONDY, N. 1967. Measurement of whole potato firmness with a universal testing machine. Food Technol. *21*, 97–100.

BOURNE, M. C., and MOYER, J. C. 1968. The extrusion principle in texture measurement of fresh peas. Food Technol. *22*, 1013–1018.

BOURNE, M. C., MOYER, J. C., and HAND, D. B. 1966. Measurement of food texture by a universal testing machine. Food Technol. *20*, No. 4, 170–174.

BRABENDER, C. W. 1965. Physical dough testing—past, present, and future. Cereal Sci. Today *10*, 291–304.

BRATZLER, L. J. 1949. Determining the tenderness of meat by use of the Warner-Bratzler method. Proc. 2nd Ann. Reciprocal Meat Conf. Nat. Livestock Meat Board.

CARPENTER, Z. L., KAUFMANN, R. G., Bray, R. W., and Weckel, K. G. 1965. Objective and subjective measures of pork quality. Food Technol. *19*, 118–120.

CHARM, S. E. 1962. The nature and role of fluid consistency in food engineering applications. Advan. Food Res. *11*, 356–435.

CHARM, S. E., and McCOMIS, W. 1965. Physical measurements of gums. Food Technol. *19*, 58–63.

CUTLER, G. H., and BRINSON, G. A. 1935. The granulation of whole wheat meal and a method of expressing it numerically. Cereal Chem. *12*, 120–129.

DAVIS, J. G. 1937. The rheology of cheese, butter, and other milk products. J. Dairy Res. *8*, 245–264.

DEFELICE, D. *et al.* 1961. Fundamental aspects of meat texture. Quartermaster Contract Res. Project Rept. 2, Project 7-84-13-002.

DEMPSTER, C. J., HLYNKA, I., and ANDERSON, J. A. 1953. Extensograph studies of structural relaxation in bromated and unbromated doughs mixed in nitrogen. Cereal Chem. *30*, 492–503.

DEMPSTER, C. J., HLYNKA, I., and ANDERSON, J. A. 1955. Influence of temperature on structural relaxation in bromated and unbromated doughs mixed in nitrogen. Cereal Chem. *32*, 241–254.

DOLBY, R. M. 1941. The rheology of butter. I. Methods of measuring the hardness of butter. J. Dairy Res. *12*, 329–336.

444 FOOD ANALYSIS

DRAKE, B. K.   1961.   An attempt at a geometrical classification of rheological apparatus.   Unpublished.   Cited by M. C. Bourne.   1966.   A classification of objective methods for measuring texture and consistency of foods.   J. Food Sci. *31*, 1011–1015.

FELLERS, C. R., and GRIFFITHS, F. P.   1928.   Jelly strength measurements of fruit jellies by Bloom gelometer.   Ind. Eng. Chem. *20*, 857–859.

FINNEY, E. E., BEN-GERA, I., and MASSIE, D. R.   1967.   An objective evaluation of changes in firmness of ripening bananas using a sonic technique.   Food Technol. *32*, 642–646.

FINNEY, E. E., HALL, C. H., and THOMPSON, N. R.   1964.   Influence of variety and time upon the resistance of potatoes to mechanical handling.   Am. Potato J. *41*, 178.

GARRETT, A. W., DESROSIER, N. W., KUHN, G. D., and FIELDS, M. L.   1960.   Evaluation of instruments to measure firmness of tomatoes.   Food Technol. *14*, 562–564.

GOULD, W. A.   1951.   Quality evaluation of fresh, frozen, and canned snap beans.   Ohio Agr. Expt. Sta. Res. Bull. *701*.

GREUP, D. H., and HINTZER, H. M. R.   1953.   Cereals.   *In* Foodstuffs, Their Plasticity, Fluidity, and Consistency, G. W. Scott-Blair (Editor).   Interscience Publishers, New York.

HALLER, M. H.   1941.   Fruit pressure testers and their practical applications.   U. S. Dept. Agr. Circ. 627.

HALLIKAINEN, K. E.   1962.   Viscosimetry.   Instr. Control Systems *35*, No. 11, 82–84.

HALTON, P.   1938.   Relation of water absorption to the physical properties and baking quality of flour doughs.   Cereal Chem. *15*, 282–294.

HALTON, P.   1949.   Significance of load extension tests in assessing the baking quality of wheat flour doughs.   Cereal Chem. *26*, 24–45.

HARVEY, J. G.   1953.   The rheology of certain miscellaneous food products.   *In* Foodstuffs, Their Plasticity, Fluidity, and Consistency, G. W. Scott-Blair (Editor).   Interscience Publishers, New York.

HEISS, R.   1959.   Prevention of stickiness and graining in stored hard candies.   Food Technol. *13*, 433–440.

HLYNKA, I.   1967.   Progress in the area of dough rheology.   Brot Geback *21*, 125–130.

HLYNKA, I., and MATSUO, R. R.   1959.   Quantitative relation between structural relaxation and bromate in dough.   Cereal Chem. *36*, 312–317.

HLYNKA, K., and VAN ESCHEN, E. L.   1965.   Studies with an improved load softness tester.   Cereal Sci. Today *10*, 84–87.

HOFSTEE, J., and DE WILLIGEN, A. H. A.   1953.   Starch.   *In* Foodstuffs, Their Plasticity, Fluidity, and Consistency, G. W. Scott-Blair (Editor).   Interscience Publishers, New York.

JOHNSON, C. F., MAXIE, E. C., and ELBERT, E. M.   1965.   Physical and sensory tests on fresh strawberries subjected to gamma radiation.   Food Technol. *19*, 119–123.

KAPSALIS, J. G., BETTSCHER, J. J., KRISTOFFERSEN, T., and GOULD, I. A.   1960.   Effect of chemical additives on the spreading quality of butter.   I.   The consistency of butter as determined by mechanical and consumer panel evaluation methods.   J. Dairy Sci. *43*, 1560–1569.

KATTAN, A. A.   1957.   Changes in color and firmness during ripening of detached tomatoes, and the use of a new instrument for measuring firmness.   Proc. Am. Soc. Hort. Sci. *70*, 379–386.

KATZ, R., COLLINS, N. D., and CARDWELL, A. B.   1961.   Hardness and moisture content of wheat kernels.   Cereal Chem. *38*, 364–368.

KAUFMANN, R. G.   1959.   Techniques of measuring some quality characteristics of pork.   Proc. 12th Ann. Reciprocal Meat Conf., Nat. Livestock Meat Board.

KENT-JONES, D. W., and AMOS, A. J. 1967. Modern Cereal Chemistry, 6th Edition. Food Trade Press, London.

KOVATS, L. T., and LASZTITY, R. 1965. New developments in dough rheology. Periodica Polytech. (Budapest) 9, No. 1, 57–67.

KRAMER, A., and BACKINGER, G. 1959. Textural measurement of foods. Food 28, 85–86, 95.

KRAMER, A., and HAWBECKER, J. V. 1966. Measuring and recording rheological properties of gels. Food Technol. 29, 111–115.

KRAMER, A., and ROSENTHAL, H. 1965. Determination of bloom of gelatin in solutions of nonstandard concentration. Food Technol. 19, 111–114.

KRAMER, A., and TWIGG, B. A. 1959. Principles and instrumentation for the physical measurement of food quality with special reference to fruit and vegetable products. Advan. Food Res. 9, 153–220.

KRUISHEER, C. I., and DEN HERDER, P. C. 1938. Investigations on the consistency of butter. Chem. Weekblad 35, 719–730.

KULWICH, R., DECKER, R. W., and ALSMEYER, R. H. 1963. Use of a slice-tenderness evaluation device with pork. Food Technol. 17, 83–85.

LERCHENTHAL, C. H., and MULLER, H. G. 1967. Research in dough rheology at the Israel Institute of Technology. Cereal Sci. Today 12, 185–187, 190–192.

LYNCH, L. J., MITCHEL, R. S., and CASIMIR, D. J. 1959. The chemistry and technology of the preservation of green peas. Advan. Food Res. 9, 61–152.

MAGNESS, J. R., and TAYLOR, G. F. 1925. An improved type of pressure tester for the determination of fruit maturity. U. S. Dept. Agr. Circ. 350.

MATZ, S. A. 1962. Food Texture. Avi Publishing Co., Westport, Conn.

MILLER, B. S., and JOHNSON, J. A. 1954. A review of methods for determining the quality of wheat and flour for breadmaking. Kansas Agr. Expt. Sta. Tech. Bull. 76.

MILNER, M., and SHELLENBERGER, J. A. 1953. Physical properties of weathered wheat in relation to internal fissuring detected radiographically. Cereal Chem. 30, 202–212.

MIYADA, D. S., and TAPPEL, A. L. 1956. Meat tenderization. I. Two mechanical devices for measuring texture. Food Technol. 10, 142–145.

MOHR, W., and WELLM, J. 1948. Viscosity measurement of butter. Milchwissenschaft 3, 181–185.

MOHSENIN, N. N. 1968. Physical properties of plant and animal materials. Part II. Dept. Agr. Eng. Pennsylvania State Univ. College Park, Penna.

MORROW, C. T., and MOHSENIN, N. N. 1968. Dynamic viscoelastic characterization of solid food materials. Food Technol. 23, 646–651.

MULDER, H. 1953. The consistency of butter. In Foodstuffs, Their Plasticity, Fluidity, and Consistency, G. W. Scott-Blair (Editor). Interscience Publishers, New York.

MULLER, H. G. 1964. Dough rheology, I. Early developments before 1900. Brot Geback 18, 117–121.

MULLER, H. G., and HLYNKA, I. 1964. Brabender extensigraph techniques. Cereal Sci. Today 9, 422–424, 426, 430.

MULLER, H. G., WILLIAMS, M. V., RUSSELL-EGGITT, P. W., and COPPOCK, J. B. M. 1961. Fundamental studies on dough with the Brabender Extensograph. I. Determination of stress-strain curves. J. Sci. Food Agr. 12, 513–523.

MULLER, H. G., WILLIAMS, M. V., RUSSELL-EGGITT, P. W., and Coppock, J. B. M. 1962. Fundamental studies on dough with the Brabender Extensograph. II. Determination of the apparent elastic modulus and coefficient of viscosity of wheat flour dough. J. Sci. Food Agr. 13, 572–580.

MULLER, H. G., WILLIAMS, M. V., RUSSELL-EGGITT, P. W., and COPPOCK, J. B. M. 1963. Fundamental studies on dough with the Brabender Extensograph. III. The work technique. J. Sci. Food Agr. 14, 663–672.

PAUKNER, E. 1951. Objective measurement of softness for determination of degree of malt solubility. Brauenwissenschaft *11*, 187–190.

PEARSON, A. M. 1963. Objective and subjective measurement of meat tenderness. Proc. Meat Tenderness Symp., Campbell Soup Co., Camden, N. J.

PLATT, W., and POWERS, R. 1940. Compressibility of bread crumb. Cereal Chem. *17*, 601–621.

POWRIE, W. D., and ASSELBERGS, E. A. 1957. A study of canned syrup-pack whole carrots. Food Technol. *11*, 257–277.

PROCTOR, B. E., DAVISON, S., MALECKI, G. J., and WELCH, M. 1955. A recording strain gage denture tenderometer for foods. I. Instrument evaluation and initial tests. Food Technol. *9*, 471.

PRYCE-JONES, J. 1953. The rheology of honey. *In* Foodstuffs, Their Plasticity, Fluidity, and Consistency, G. W. Scott-Blair (Editor). Interscience Publishers, New York.

SALE, A. J. H. 1960. Measurement of meat tenderness. *In* Texture in Foods, Monograph 7. Soc. Chem. Ind., London.

SAWYER, R. L., and COLLINS, G. H. 1960. Black spot of potatoes. Am. Potato J. *37*, 115.

SCHULTZ, H. W. 1957. An evaluation of the methods of measuring tenderness. Proc. 10th Ann. Reciprocal Meat Conf., Natl. Livestock Meat Board.

SCOTT-BLAIR, G. W. 1945. A Survey of General and Applied Rheology. Sir Isaac Pitman and Sons, London.

SCOTT-BLAIR, G. W. 1953. Foodstuffs, Their Plasticity, Fluidity, and Consistency. Interscience Publishers, New York.

SCOTT-BLAIR, G. W. 1958. Rheology in food research. Advan. Food Res *8.*, 1–61.

SHAMA, F., and SHERMAN, P. 1966. The texture of ice cream. II. Rheological properties of frozen ice cream. J. Food Sci. *31*, 699–706.

SHARRAH, N. KUNZE, M. S., and PANGBORN, R. M. 1965A. Beef tenderness: sensory and mechanical evaluation of animals of different breeds. Food Technol. *19*, 131–136.

SHARRAH, N., KUNZE, M. S., and PANGBORN, R. M. 1965B. Beef tenderness: comparison of sensory methods with the Warner-Bratzler and L. E. E. Kramer shear press. Food Technol. *19*, 136–143.

SHERMAN, P. 1966. The texture of ice cream. III. Rheological properties of mix and melted ice cream. J. Food Sci. *31*, 707–716.

SHERMAN, P. 1969. A texture profile of foodstuffs based upon well defined rheological properties. J. Food Sci. *34*, 458–462.

SIMON, S., KRAMLICH, W. E., and TAUBER, F. W. 1965. Factors affecting frankfurter texture and a method of measurement. Food Technol. *19*, 110–113.

SISTRUNK, W. A., and MOORE, J. N. 1967. Assessment of strawberry quality—fresh and frozen. Food Technol. *21*, 131A–135A.

SMEETS, H. S., and CLEVE, H. 1956. Determination of conditioning by measuring softness. Milling Production *21*, No. 4, 5, 12–14, 16.

SOMERS, G. F. 1965. Viscoelastic properties of storage tissues from potato, apple and pear. J. Food Sci. *30*, 922–929.

SPERRING, D. D., PLATT, W. T., and HINER, R. L. 1959. Tenderness of beef muscle as measured by pressure. Food Technol. *8*, 155–158.

SWANSON, C. O., and WORKING, E. B. 1933. Testing the quality of flour by the recording dough mixer. Cereal Chem. *10*, 1–29.

SZCZESNIAK, A. S. 1966. Texture measurements. Food Technol. *20*, 52, 55–58.

SZCZESNIAK, A. S., and KLEYN, D. H. 1963. Consumer awareness of texture and other food attributes. Food Technol. *17*, 74–77.

TAPE, N. W. 1965. Viscosity of potato flake slurries. Food Technol. *19*, 180–192.

TAYLOR, J. W., BAYLES, B. B., and FIFIELD, C. C. 1939. A simple measure of kernel hardness in wheat. J. Am. Soc. Agron. *31*, 775–784.

THOMAS, B., AND THUNGER, L. 1965. Physical measurements of bread crumb. Brot Geback *19*, 65–74.

UNDERWOOD, J. C., and KELLER, G. J. 1948. A method of measuring the consistency of tomato paste. Fruit Prod. J. *28*, 103.

VOISEY, P. W., and HANSEN, H. 1967. A shear apparatus for meat tenderness evaluation. Food Technol. *21*, 37A–42A.

VOISEY, P. W., MACDONALD, D. C., and FOSTER, W. 1967. An apparatus for measuring the mechanical properties of foods. Food Technol. *21*, 43A–47A.

WHITTENBERGER, R. T. 1951. Measuring the firmness of cooked apple tissues. Food Technol. *5*, 17–20.

WILDER, H. K. 1948. Instructions for use of the fibrometer in the measurement of fiber content in canned asparagus. Natl. Canners Assoc. Res. Lab. Rept. *12313-C*, San Francisco.

CHAPTER 29

Serology, Immunochemistry and Immunoelectrophoresis

INTRODUCTION

Immune phenomena play a vital role in the health and disease of higher animals. Among these phenomena are: (1) the protection which is conferred on higher animals against certain viral- and bacterial-induced diseases, either by their survival of an initial attack of the disease or by the deliberate injection of a modified form of the disease-producing agent (immunization); (2) the rejection of tissue or organ grafts from one individual to another of the same species; and (3) the occurrence of various diseases such as certain types of food allergy (Perlman 1964; Singer 1965).

The importance of an immunological approach in preventive medicine, in diagnosis, and in therapy is well-established. Less appreciated, however, is the fact that similar procedures can be used to yield valuable information concerning the nature of chemical components of biological materials (Treffers 1944). In the study of proteins, these contributions have been of several kinds. Immunological methods have helped to discover a number of proteins with unique properties. As microanalytical tools, serological methods have been used to detect small amounts of proteins in the presence of large amounts of other substances. Finally, serological methods have contributed materially to our knowledge of the specificity of proteins, complementing physical methods, and in several cases serving to differentiate proteins where other methods have failed.

In this chapter, the authors' primary aim has been to make the reader cognizant of the possibilities that immunochemical methods offer to the food analyst. No attempt has been made to survey the extensive and rapidly growing literature. The review is limited to a short outline of basic considerations followed by a description of some of the useful procedures of particular interest to food analysts.

Basic Considerations and Definitions

When an animal receives one or more parenteral injections of certain foreign materials, there generally appear in the serum within a few days substances which have the property of reacting with the material injected.   These substances are called *antibodies*, and the materials which have stimulated their production are called *antigens* (Carpenter 1956).   Most known antigens are proteins; some are polysaccharides or lipid-carbohydrate-protein complexes.   Although antibodies generally react only with the antigen used (the homologous antigen), certain exceptions, termed cross reactions, have been noted.   Those reactions occur with substances other than the homologous antigens (Kabat 1961).   The reactions are due to structural similarities between the antigens concerned.   Antigens are sometimes referred to as foreign proteins, which signifies that they are proteins not normally found in the body they invade.   The antibodies that form in the organism in response to the introduction of an antigen are proteins.   The antibodies are globulins, have molecular weights appropriate to globulins, and are called *immunoglobulins*. The nomenclature of immunoglobulins and their genetic factors have been recommended by committees of the World Health Organization (1964).   There are three main classes of immunoglobulins present in all mammalian sera that have been studied, though their relative amounts vary considerably from one species to the other.   Those classes have been named *IgG* (gamma), *IgM* (macro), and *IgA*, and in human serum their relative amounts are 71, 7, and 22%, respectively.   Their common characteristics are that they all may have the same four-peptide chain structure (Porter 1967).   Antigenically, they are related but not identical.   The molecular weight of IgG is 150,000; of IgM 900,000, and of IgA about 400,000, but the latter has a tendency to polymerize and dissociate.   Immunoglobulins of all types and of all species tested contain carbohydrates (Boyd 1966).   The carbohydrate content of human IgG is 3%, of IgM 12%, and of IgA 10%.

The Antigens

The antigens must be administered parenterally to organisms of another species.   Oral administration is generally not satisfactory, since the ingested proteins are digested by proteolytic enzymes of the gastrointestinal tract.   The breakdown products are generally not antigenic (Haurowitz 1960).

Certain proteins of various species exist in more than one antigenic form, called *allotypes*.   Thus rabbit gamma globulin exists in at least six different forms, and the differences seem to be hereditary. There is also evidence for the existence of such allotypes in man. Although we do not have the complete explanation of what makes a substance antigenic, we know (Boyd 1966) that:

(1)	Antigens must be foreign to the circulation of the experimental animal, and the more foreign, i.e. the more remote the source taxonomically, the more antigenic a protein will be.

(2)	Antigens must have more than a certain minimal degree of complexity and a certain molecular size. That is suggested by the nonantigenicity of the protamines and of the relatively simple molecule of gelatin, and by the high antigenicity of the large molecule of the hemocyanins.

It seems that a minimum molecular weight of approximately 5,000 to 10,000 is a prerequisite for antigenicity (Haurowitz 1960, 1961). Formation of antibodies is sometimes observed after the administration of small molecules of highly reactive substances, such as iodine. It is generally assumed that the substances combine with proteins at the site of injection and that the conjugated proteins are the true antigens. Denatured proteins may show reduced reactivity or fail to react with antibodies to native proteins, depending on the extent of denaturation.

As mentioned before, specificity of antibodies formed in response to antigens, varies. Although antibodies generally react only with the antigen used, cross reactions have been noted. The reactions are due to structural similarities between the antigens concerned. Thus, antibodies to horse serum protein can also react with donkey serum protein, and ovalbumin from ducks' eggs reacts with antibodies to albumin from hens' eggs. Serum albumins of different mammalian species are evidently antigenically similar, though not absolutely identical. The more closely related any two species are, the greater the serological likeness of their corresponding proteins. Thus, hemoglobin and myoglobin are serologically different; there is only a slight relation, serologically, between hemoglobin of beef and man. The closer the phylogenetic relation between two protein antigens, the more extensive the cross reaction. Thus if antibodies for bovine serum albumin are reacted with a variety of other albumins, the maximum amount of precipitation is obtained in the homologous interaction. There is, however, considerable cross reaction with albumins from sheep and several other animals.

Proteins from the same species which have different functions in the body, usually differ widely in specificity. Thus blood hemoglobin differs serologically from proteins of the kidney tissue, and the serum globulins are different from the serum albumins. There is evidence that most organs possess special proteins or carbohydrates peculiar to them, and organ-specific antisera have been obtained (Boyd 1966).

The specificity of antigens resides in the structural pecularities of their molecules (Marrack 1938; Landsteiner 1945; Pappenheimer 1953; Karush 1962). The dependence of specificity on chemical structure is shown by several lines of evidence: (1) purified proteins that exhibit chemical differences can generally be differentiated serologically; (2) simple chemical substances and carbohydrates, when chemically similar, give serological cross reactions; (3) chemical alteration of antigens generally alters their specificity; and (4) corresponding proteins of different species which are functionally, and thus probably structurally, related generally cross-react. Available evidence indicates that the reactivity of an antigen resides, primarily in a small prosthetic group called *hapten*. The word "hapten" was first used by Landsteiner (1921) to describe simple organic residues that react specifically with antibodies. The word is derived from a Greek verb, the meaning of which is to touch, grasp or fasten. The term has been used to describe substances which, in themselves, are incapable of inducing antibody formation, but which when attached to ordinary immunogens, such as proteins and polysaccharides, can induce antibodies against themselves. Day (1966) defined hapten as the specific chemical grouping to which a single antibody site conforms and with which it reacts. As long as the hapten occupies an antibody site, no other hapten can occupy the same site. A hapten, whether free or attached to a carrier in an exposed position, will react with an antibody site.

Immunochemical Methods

Types.—There are several ways, *in vivo* and *in vitro*, of observing the result of the combination of an antigen with its specific antibodies (Munoz 1959). If the antigen is a macromolecule, combination with its antibodies under appropriate conditions results in precipitation. If the antigen is part of a cellular surface, agglutination of the cells is observed. The formation of specific antigen-antibody aggregate usually results in coprecipitation of certain components of the serum called *complement*. This complement fixation is directly related to the extent of the antigen-antibody combination. If the antigen is on a cellular surface, reaction with antibody in the presence of complement is often cytotoxic, resulting in cell lysis (Singer 1965).

When a crude extract containing a mixture of antigens is injected, several antibodies specific for each antigen are formed. The various antibodies formed can be removed separately from the serum by adding each antigen singly. Quantitative studies have indicated that even antibodies to a single antigen, such as crystalline egg albumin, exhibit microheterogeneity. They may still be considered to behave as a single substance when comparisons are made between a number of totally unrelated antigens, since the antibody molecule to each antigen can react specifically with their own antigen(Kabat 1961).

Precipitin Reactions.—Available immunological methods vary in sensitivity and usefulness. Thus it is possible to observe a positive precipitin reaction with a few micrograms of antibody per milliliter, and a passive hemagglutination reaction can be obtained even with concentrations a 1000 times smaller. Yet, the precipitin reaction is the most commonly used because it is directly visible and can be quantitated. It has the disadvantage of being limited to antigens in aqueous solutions.

The *in vitro* methods for measuring antibody potency can be divded into several types.

(1) Dilution methods. The classical methods use as the end point the highest dilution of serum added to a constant amount of antigen at which a detectable reaction such as precipitation, agglutination, or lysis occurs. The serum dilution giving this endpoint is known as the *titer*.

(2) Optimal proportions method. The procedure determines the ratio of antigen and serum at which flocculation is most rapid if the relative proportions are optimal. Various dilutions of antigens are mixed with a constant volume of serum in a series of tubes, the total volume in each tube being constant. The tube in which flocculation first occurs is noted, and the ratio of antigen dilution to serum dilution is calculated.

(3) Quantitative chemical methods. They are generally based on determining the protein-N content in a precipitate or agglutination product.

The dilution methods are semiquantitative and are useful in estimating the order of magnitude of an antigen in a mixture. Greater precision may be obtained by the optimal proportions method. The most precise method of estimating amounts of antigens in unknown solutions involves using a calibration curve prepared by analyzing a series of washed specific precipitates formed by adding known volumes of antigen to a measured volume of antiserum.

The quantitative precipitin method is capable of widespread application in the analysis of mixtures containing immunologically reactive substances. It offers several advantages over the usual analytical chemical methods. It is highly specific and permits estimation of a given constituent of a mixture without chemical fractionation or purification.   It requires small amounts of material for analysis, since the amount of specific precipitate analyzed is several times the amount of antigen in the sample, except for antigens of very high molecular weight. With the use of sensitive methods of protein determination (i.e. the Folin-Ciocalteau colorimetric method), samples containing as little as 1 to 5 μg of antigens are adequate.   The method is capable of considerable precision, ± 2 to 5% under suitable conditions.

If small amounts of a chemically pure and serologically active product suitable for the preparation of a calibration curve with antiserum are obtainable, it is possible to estimate the amount of the substance in crude starting materials.   This procedure can be used in estimating specific polysaccharides and proteins.   Immunochemical precipitation methods can be used to supplement or to replace less precise bioassays for estimation of proteins with biological activity.   In the estimation of such substances (enzymes, protein hormones, toxins in mixtures), estimations of biological activity are compared with quantitative immunochemical analysis.   In many instances, inactivated material or the precursor of the active substances may still precipitate with antiserum.   In such cases, use of both methods of estimation may provide evidence for the existence of altered products.   Immunochemical criteria are useful in determining the homogeneity and purity of proteins and carbohydrates. Such determinations are generally used in conjunction with physicochemical methods and, whenever applicable, biological assays.

One of the main hazards in carrying out the quantitative precipitin reaction is the possibility of nonspecific precipitation (Cohn 1952). This is particularly marked with crude extracts.   Nonspecific precipitation can sometimes be avoided by choice of conditions of precipitation or by fractionation of antisera.   In certain cases, it occurs only at protein concentrations much higher than those used in the the specific reaction.   If control tests indicate that nonspecific precipitation cannot be avoided, a control can be set up for each mixture and this value subtracted from that obtained with the sample.   Results under those conditions are at best rough approximations.

Gelified Media.—The precipitin reaction is the only immunochemical method which can be used in gelified media (Grabar 1958). The use of gelified media for the precipitin reaction was introduced at the turn of the century.   Better knowledge of the principles governing the precipitin reaction allowed a development, in recent years, of several methods of specific precipitation in gels (Crowle 1960, 1961).

Oudin (1946, 1952) devised a method by which complex antigen-antibody systems could be analyzed by allowing them to react in a capillary tube filled with agar. The antibody solution is mixed with warm agar, which is then allowed to harden in the tube. When the antigen solution is added, the reaction of antigen and antibody forms a precipitation zone. The number of such zones is less than or equal to the number of independent precipitation systems (i.e. antigen-antibody reactions) present in the mixture examined (Fig. 29.1).

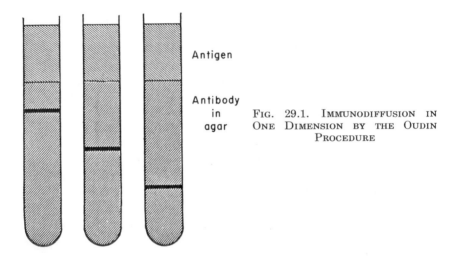

Antigen

Antibody in agar

Fig. 29.1. Immunodiffusion in One Dimension by the Oudin Procedure

Ouchterlony's technique (called the *double diffusion method*) (1958, 1962, 1967) is similar, but differs from Oudin's in that it permits both antigen and antibody to diffuse into an agar-filled glass dish that initially contains neither reagent. A few drops each of antigen and antibody solution are placed separately in small wells cut into the agar. Antigen and antibody diffuse outward toward each other at a rate related to their concentration and their diffusion coefficients. A line of precipitate forms where an antigen interacts with its antibody. Because of differences in diffusion rates, the lines are distinctly separated (Fig. 29.2). The clean separation of lines on the Ouchterlony plates makes it possible to distinguish more reactions with them than with the Oudin tubes. Consequently, the plates are more useful in studies of complex systems (Williams 1960). The line of precipitation formed by an individual reaction can be identified if either the antigen or the antibody is available in relatively pure form.   Experience has shown that immunochemical reactions carried out in agar gel allow the identification of a single component in far smaller microquantities than would be possible with chemical or physicochemical methods.

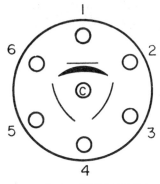

FIG.   29.2.   PRECIPITATION   BANDS   OF UNABSORBED   BEEF   ANTISERUM   ON   GEL DIFFUSION   PLATES

C BEEF ANTISERUM

I BEEF SERUM

2 PORK EXTRACT

3 BEEF EXTRACT

4 HORSE EXTRACT

5 LAMB EXTRACT

6 SALINE

From Warnecke and Saffle (1968)

Immunoelectrophoresis.—To improve the resolution and interpretation of the double diffusion method, the *immunoelectrophoretic* procedure was developed by Grabar and co-workers (Grabar and Williams 1953;   Grabar 1950, 1957, 1964;   Wilson 1964;   Grabar *et al.* 1965;   Wunderly 1961).   The principle of the method (Fig. 29.3) is the following.   The mixture to be studied is subjected to electrophoresis in a transparent gel.   At the end of electrophoresis, the components of the analyzed mixture have been separated by the electrical current in the gel and occupy different positions depending on their relative mobilities.   An immune serum, rich in precipitating antibodies is then poured into elongated troughs made in the gel and parallel to the axis of migration.   The antibodies diffuse perpendicularly to this axis, and as in the double diffusion

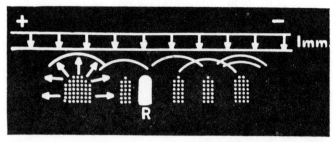

From Grabar and Nummi (1967)

FIG. 29.3. SCHEMATIC REPRESENTATION OF IMMUNOELECTRO-
PHORETIC ANALYSIS

R indicates the hole for the antigen in which the protein mixture
is poured before the run. The white, spotted areas represent
the location of the different antigen constituents after their
electrophoretic separation. Imm indicates the trough for the
immune serum (to be filled after the run). The arrows indicate the diffusion of proteins.

method of Ouchterlony, when antibodies and antigens meet in suitable proportions there is specific precipitation in the form of arcs.

Immunoelectrophoresis permits one to separate and determine
antigenic (or heptagenic) constituents of a liquid. In addition to
characterization by precipitation, colorimetric methods and enzymic
reactions may be employed. The most commonly used gel is from
agar. For certain purposes, pectin or starch can be used. However, starch is opaque and the concentration needed to form a gel
is, sometimes, excessively high.

The advantages and disadvantages of immunoelectrophoresis, the
most sophisticated analytical tool of immunochemistry, were summarized by Grabar (1959). The method can be used in studies of
natural mixtures without any previous treatment. Conditions can
be chosen (pH, temperature, etc.) to minimize loss of a particular
biological activity of the substance to be studied. Very small
amounts are necessary to perform a complete analysis. As the gel is
rich in liquid (98–99%), the electrophoresis resembles free electrophoresis in liquid and there is less danger of encountering effects of
liquid-solid interfaces as in the case of electrophoresis on paper,
starch, or cellulose. At the same time, electrophoresis in a gel has
the advantage over electrophoresis in a liquid medium of slowing
down the free diffusion of the macromolecular substances at the end
of the electrophoresis, thus facilitating their detection.

The use of the specific precipitation reaction allows fine and specific identification of components of a liquid. Substances possessing

even identical mobilities can be distinguished and impurities of less than 0.1% can be detected.   Enumeration of the constituents of a mixture by the precipitation reaction is greatly facilitated by their preliminary dispersion during the electrophoresis.   The components can be characterized and defined by 2 or 3 methods: electrophoretic mobility, immunological specificity, and other characteristics (color, enzymatic activity).

Common to all immunochemical procedures, immune sera are biological products and are difficult to standardize (Maaloe and Jerne 1952).   The methods are time-consuming and require the facilities of a serological laboratory.   The methods are qualitative and, generally, can give only rough quantitative information.   Yang *et al.* (1966) described conditions required for improved quantitation of immunoelecotrphoretic analyses.

Applications

Precipitin formation in food extracts is affected by all factors which modify proteins.   Lipids may interact with proteins and form lipoproteins.   Under certain conditions, tannins can prevent precipitin formation.   High concentrations of sucrose and low concentrations of glucose may reduce precipitin formation.   Similar interference was observed in the presence of urea and saponins.

Intensity of precipitation depends, in addition to ratios of the various components, on the presence and concentration of electrolytes.   Precipitation normally occurs in the neutral range; it is somewhat enhanced in slightly acidic media, but prevented in the presence of excess of inorganic acids and bases.   Once formed, the precipitins are soluble in dilute alkali or acid.   Antigen-antibody combination increases with temperature increase over the range 0° to 30°C, though the optimum range varies with the interacting components.   Precipitation itself may decrease at elevated temperatures.

Serological differentiation of plant proteins is relatively simple, provided interfering materials (lipids, tannins, etc.) are removed. Quantitative serological methods have been applied for many years to taxonomic investigations (De Talozzo and Jaffe 1969; Leone 1964).   Liuzzi and Angeletti (1969) described application of immunodiffusion in detecting 5% barley flour in wheat flour.   One cannot distinguish, however, serologically among proteins of several of the commonly eaten legumes.   Serological examinations are useful in distinguishing between proteins from meat, milk, or eggs.   The most commonly used procedure to determine the source of meat and possible adulteration by meat from other sources or other protein rich materials, is the precipitation test and its modifications (Warnecke and Saffle 1968). It can be used in studying fresh, frozen, processed (dried, smoked, and even heated at low temperatures), and damaged meat. Heating at elevated temperatures may reduce or eliminate value of a precipitation test. It is not completely clear whether the effect of such heat treatment is actually a modification and destruction of serologically reactive groups or a decreased solubility and availability of the reactive groups. This reactivity can be restored, in part at least, by defatting, homogenization, and lyophilization.

Heating of milk impairs little its capacity to give reactive antigens. Similarly, manufacture of yoghurt or cultured milk and processing into cheese or dried milk do not affect adversely the results. The precipitation reaction can distinguish clearly between proteins of the white and of the yolk of an egg. It is rather difficult to establish, serologically, the presence of egg yolks in baked products, but not in margarine or alcoholic beverages. Admixture of fish eggs to caviar, determination of honey purity (depending on the presence of specific protein), and source of nuts in sweetened goods or of proteins in bread are some examples of the use of serological tests in food evaluation (Kotter 1967; Lietz, 1969).

BIBLIOGRAPHY

BOYD, W. C. 1966. Fundamentals of Immunology, 4th Edition. Interscience Publishers, New York.

CARPENTER, P. L. 1956. Immunology and Serology. Saunders Publishing Co., Philadelphia.

COHN, M. 1952. Techniques and analysis of the quantitative precipitin reaction. A reaction in liquid media. Methods Med. Res. 5, 301–335.

CROWLE, A. J. 1960. Interpretation of immunodiffusion tests. Ann. Rev. Microbiol. 14, 161–176.

CROWLE, A. J. 1961. Immunodiffusion. Academic Press, New York.

DAY, E. D. 1966. Foundations of Immunochemistry. The Williams and Wilkins Co., Baltimore.

DE TALOZZO, A., and JAFFE, W. F. 1969. Immunoelectrophoretic studies with bean proteins. Phytochemistry 8, 1255–1258.

GRABAR, P. 1950. Immunochemistry. Ann. Rev. Biochem. 19, 453–486.

GRABAR, P. 1957. Agar-gel diffusion and immunoelectrophoretic analysis. Ann. N.Y. Acad. Sci. 69, 591–607.

GRABAR, P. 1958. The use of immunochemical methods in studies of proteins. Advan. Protein Chem. 13, 1–33.

GRABAR, P. 1959. Immunoelectrophoretic analysis. Methods Biochem. Analy. 7, 1–38.

GRABAR, P. 1964. Immunoelectrophoretic analyses of cell and tissue components. Colloq. Ges. Physiol. Chem. 15, 36–46.

GRABAR, P., ESCRIBANO, M. J., BENHAMOU, N., and DAUSSANT, J. 1965. Immunochemical study of wheat, barley and malt proteins. Agr. Food Chem. 13, 392–398.

GRABAR, P., and NUMMI, M.   1967.   Recent immunoelectrophoretic studies on soluble proteins in their transformation from barley to beer.   Brewers' Dig. *42*, No. 3, 68–74.
GRABAR, P., and WILLIAMS, C. A.   1953.   Methode permettant l'etude conjugee des prorpietes electroproretiques et immunochimiques dun melange de proteines.   Application au serum sanguin.   Biochim. Biophys. Acta *10*, 193–194.   (French)
HAUROWITZ, F.   1960.   Immunochemistry.   Ann. Rev. Biochem. *29*, 609–634.
HAUROWITZ, F.   1961.   Use of radioisotopes in immunochemical research. Ergeb. Mikrobiol., Immunitatsforsch. Exp. Therapie. *34*, 1–26.
KABAT, E. A.   1961.   Kabat and Meyer's Experimental Immunochemistry, 2nd Edition.   Thomas Publishing Co., Springfield, Ill.
KARUSH, F.   1962.   Immunologic specificity and molecular structure. Immunology *2*, 1–40.
KOTTER, L.   1967.   Serological methods in protein investigations.   *In*, Handbook of Food Chemistry, Analysis of Foods, Detection and Determination of Food Components, J. Schormuller (Editor).   Springer-Verlag, Berlin.
LANDSTEINER, K.   1921.   Heterogenic antigens and haptens.   XV, Antigens. Biochem. Z. *119*, 294–306.
LANDSTEINER, K.   1945.   The Specificity of Serological Reactions.   Harvard Univ. Press, Cambridge.
LAWRENCE, M.   1964.   The techniques of immunoelectrophoresis.   Am. J. Med. Meth. *30*, 209–221.
LEONE, C. A.   1964.   Taxonomic Biochemistry and Serology.   The Ronald Press.   New York.
LIETZ, A.   1969.   Quantitation of food adulterants by multiple radial immunodiffusion.   II.   Wheat in bread sold as "wheat-free" for the use of allergy patients.   J. Assoc. Offic. Anal. Chemists *12*, 995–998.
LIUZZI, A., and ANGELETTI, P. A.   1969.   Application of immunodiffusion in detecting the presence of barley in wheat flour.   J. Sci. Food Agr. *20*, 207–209.
MAALOE, O., and JERNE, N. K.   1952.   The standardization of immunological substances.   Ann. Rev. Microbiol. *6*, 349–366.
MARRACK, J. R.   1938.   The Chemistry of Antigens and Antibodies.   His Majesty's Stationers Office, London, England.
MUNOZ, J. J.   1959.   Some newer immunological techniques.   Anal. Chem. *31*, 981–985.
OUCHTERLONY, O.   1958.   Diffusion-in-gel methods for immunological analysis. Progr. Allergy *5*, 1–78.
OUCHTERLONY, O.   1962.   Quantitative immunoelectrophoresis.   Acta. Pathol. Microbiol. Scand. Suppl. *154*, 252–254.
OUCHTERLONY, O.   1967.   Handook of Immunodiffusion and Immunoelectrophoresis.   Ann Arbor Science Publishers, Ann Arbor, Mich.
OUDIN, J.   1946.   Immunochemical methods of analysis involving selective precipitation in solidified media.   Compt. Rend. Acad. Sci. *222*, 115–116.
OUDIN, M.   1952.   Specific precipitation in gels and its application to immunochemical analysis.   Methods Med. Res. *5*, 335–378.
PAPPENHEIMER, A. M. (Editor).   1953.   The Nature and Significance of the Antibody Response.   Columbia Univ. Press, New York.
PERLMAN, F.   1964.   Immunochemistry.   *In* Proteins and Their Reactions, H. W. Schultz, and A. F. Anglemier (Editors).   The Avi Publishing Co., Westport, Conn.
PORTER, R. R.   1967.   The structure of immunoglobulins.   *In* Essays in Biochemistry, P. N. Campbell, and G. D. Greville (Editors).   The Biochemical Society, England *3*, 1–24.
SINGER, S. J.   1965.   Structure and function of antigen and antibody proteins. *In* The Proteins-Composition, Structure and Function, H. Neurath (Editor). Academic Press, New York.

TREFFERS, H. P. 1944. Some contributions of immunology to the study of proteins. Advan. Protein Chem. *1*, 69–119.

WARNECKE, M. O., and SAFFLE, R. L. 1968. Serological identification of animal proteins. I. Mode of injection and protein extracts for antibody production. J. Food Sci. *33*, 131–135.

WILLIAMS, C. A. 1960. Immunoelectrophoresis. Sci. Am. *202*, No. 3, 130–134, 136, 138–140.

WILSON, A. T. 1964. Direct immunoelectrophoresis. J. Immunol. *92*, 431–434.

WORLD HEALTH ORGANIZATION. 1964. Nomenclature for human immunoglobulins. Bull. World Health Organ. *30*, 447–450.

WUNDERLY, C. 1961. Immunoelectrophoresis: Methods, interpretation, results. Advan. Clin. Chem. *4*, 207–273.

YANG, W. K., YAEGER, R. G., and MILLER, O. N. 1966. Quantitative considerations on immunoelectrophoretic analysis. Bull. Tulane Univ. Med. Fac. *25*, 73–84.

CHAPTER 30

Enzymatic Methods

INTRODUCTION

Enzymology, although a relatively new science, is a rapidly growing one. In recent years, there has been a great increase in the number of enzymes newly described; the total number now known amounting to more than 700. Classification and nomenclature of the enzymes and units of enzyme activity were published and standardized by the International Union of Biochemistry (1961). Current reviews on various aspects of enzymology are published yearly in the form of Advances in Enzymology and other review books covering various areas of biochemistry. Several journals are devoted entirely to reports of enzyme studies. In addition, articles on enzyme-related subjects are covered in most scientific journals in biochemistry, physiology, microbiology, and general biology. The importance of enzymes in the preservation of agricultural products and their processing into foods, and in the acceptance of foods by the consumer cannot be overemphasized. Enzymes have been reviewed in books (Anderson 1946; Soc. Chem. Ind. 1961; Schultz 1960; Reed 1966), selected chapters in books (Cruess 1943; Geddes 1946; Hildebrandt 1947; Ford and Vitucci 1948; Bergstrom and Holman 1948; Joslyn 1949, 1964; Joslyn and Ponting 1951; Bernfeld 1951; Anderson and Alcock 1954; Deuel and Stutz 1958; Acker 1962; Cook 1962; Schultz 1962; Forsyth and Quesnel 1963; Schultz and Anglemier 1964; Hlynka 1964; de Becze 1965; Pomeranz 1968), and in review articles (Geddes 1950; Johnson and Miller 1953; Mohler 1958; Pomeranz 1964A, B, 1966; Beckhorn et al. 1965; Taufel 1965). General textbooks and reference books on enzymes have been published. Some of the more comprehensive books in this field include those by Dixon and Webb (1958); a series on methods in enzymology (Colowick and Kaplan since 1955), and several volumes of detailed description of various enzymes (Boyer et al. 1959–1963).

Methods of Enzymatic Analysis

An analytical method is of value when its specificity, reproducibility, and sensitivity are high and when the expenditure of labor, time, and material are low (Bergmeyer 1963). Theoretically, most of these requirements can be met admirably by enzymatic analysis. The term enzymatic analysis is generally understood to mean analysis with the aid of enzymes. The major advantages of enzymes in analysis lie in their ability to react specifically with individual components of a mixture. This avoids lengthy separations of the components and reduces the time needed for an analysis. The amount of substrate (sample) required for analysis is small; owing to the mild conditions employed, enzymes often allow the detection and the determination of labile substances which can only be estimated rather inaccurately by other methods. Enzymatic techniques in analytical chemistry have been limited to biochemical and clinical laboratories (Amador and Wacker 1965), because of the difficulty of preparing the enzymes. Today enzymes are used much more widely as analytical tools in analysis of agricultural and food products than in the past. New methods have been developed for the use of commercially available pure preparations. Use of enzymes in analytical chemistry has admittedly certain limitations. Reliability depends on meticulous adherence to, and satisfactory control of, standardized assay conditions. In all enzymatic analyses, the presence of contaminants in the sample which can partially or totally inhibit the enzyme-catalyzed reactions is a serious problem. Heavy metals, oxidants, and SH-blocking reagents may affect results. Compounds structurally similar to the enzyme but without biological activity are competitive inhibitors. And finally, the purity of a preparation is crucial. Some enzymatic impurities may enhance, others reverse a reaction. Instability of enzymes may limit their usefulness; it can, however, be minimized by using proper conditions.

The relatively high cost of using purified preparations can be reduced by preparing immobilized (insolubilized) enzymes without loss of activity (Guilbault 1966A; Kay, 1968; Goldstein and Katchalski 1968). Preparations immobilized on polytyrosyl polypeptides or on a collodion matrix could be used in the form of columns for continuous or automated assay procedures.

Methods of enzymatic analysis were comprehensively reviewed in a book edited by Bergmeyer (1963), and in three excellent general reviews by Schormuller (1957A, 1968), Devilin (1959), and Guilbault (1966A). Detailed information on the methods employed in enzymatic analysis is described in the series of books on *Methods in Enzymology* edited by Colowick and Kaplan (since 1955). Use of enzymes in evaluation of certain food products is given in detail in the standard AOAC methods (Assoc. Offic. Agr. Chem. 1965), AACC methods (Am. Assoc. Cereal Chem. 1962), Methods of the Corn Industries Research Foundation (Anon. 1967A), and of the American Society of Brewing Chemists (1958).

Determination of Substrates

According to the Michaelis Menten equation:

$$E + S \underset{K_{-1}}{\overset{K_1}{\rightleftarrows}} ES \overset{K_2}{\longrightarrow} E + P \tag{30.1}$$

the substrate, S, combines with the enzyme, E, to form an intermediate complex, ES, which subsequently breaks down to product P and free enzyme. The equilibrium constant for the formation of the complex is called the Michaelis constant, and is denoted K_m. It is defined as $(K_2 + K_{-1})/K_1$. The rate of reaction, v, is a function of the enzyme and substrate concentration, and is affected by the presence of activators or inhibitors. At a fixed enzyme concentration, the initial velocity is:

$$v_0 = (V_{max}[S]_0)/(K_m + [S]_0) \tag{30.2}$$

The initial rate increases with substrate until a nonlimiting excess of substrate is reached, after which additional substrate causes no increase in rate.

Methods.—The concentration of a substance modified by an enzymatic reaction can be determined in two ways. The most useful enzymatic method is the determination of a product of an enzyme-catalyzed reaction which is formed in a definite stoichiometric relation from the compound to be assayed. The ideal condition for this type of assay is a reaction which goes virtually to completion. If the substrate is consumed only partly, quantitative utilization of the substrate can be achieved by increasing the concentration of some reactants or by removing one of the products of enzymatic action. If it is not feasible to alter the reaction so that all of the substrate is consumed, the equilibrium mixture of reactants and products can be determined and compared to a standard curve prepared under defined conditions.

For determination of total change either direct measurements or measurements with the aid of coupled reactions are made. If the substrate has a characteristic absorption spectrum, the decrease in absorbance at a selected wavelength can be measured, i.e. measurement of absorbance at 293 nm in the determination of uric acid with uricase (Plesner and Kalckar 1956). In the determination of ethanol with alcohol dehydrogenase, NAD (nicotinamide-adenine dinucleotide) is reduced to NADH which has an absorbance peak at 340 nm. The transfer of the hydrogen of the substrate to the pyridine ring of NAD or NADP (nicotinamide-adenine dinucleotide phosphate) and measurement of the absorbance peak at 340 nm are the basis of many enzymatic assays. The basic procedure can be used in the oxidation of ethanol in which NAD accepts the substrate hydrogen or in the reduction of pyruvate to lactate in which NADH donates hydrogen. Use of the methods in the determination of organic acids, carbonyls, alcohols, and sugars in wine was reviewed by Peynaud *et al.* (1966). Bruns and Werners (1962) discussed distribution, function, and assay of dehydrogenases in biological fluids and tissues. Lundquist (1959) compared physical and chemical methods of determining ethanol in blood and tissues with the enzymatic assay, and Talalay (1960) reviewed use of enzymatic methods that utilize purified hydroxysteroid dehydrogenases for the specific sensitive microdetermination of steroid hormones. Another example of direct measurement of the enzymatically modified substrate is the titration (or manometric measurement of CO_2 released from a bicarbonate buffer) of glucose with gluconic acid formed in the determination of glucose with glucose oxidase (Free 1963).

In measurement with the aid of a coupled reaction, an auxiliary and an indicator step are involved. Dyes can serve as indicators in coupled reactions by reacting with a product of the auxiliary reaction. An example is the determination of glucose with glucose oxidase and peroxidase using o-dianisidine as the indicator:

$$\text{Glucose} + H_2O + O_2 \xrightarrow[\text{oxidase}]{\text{glucose}} \text{gluconic acid} + H_2O_2$$

$$H_2O_2 + \text{leuco-dye} \xrightarrow{\text{peroxidase}} H_2O + \text{dye}$$

The second general method involving enzymes as analytical reagents is based on measuring the rate of the enzyme-catalyzed reaction. With limiting quantities of substrate and under closely controlled conditions, the initial rate of an enzyme-catalyzed reaction may be used to determine the concentration of the substrate. The region in which linearity is achieved, and in which an analytical determination of substrate concentration can be based on the rate of reaction, lies below 0.1 Km. The rate of an enzyme-catalyzed reaction depends on pH, temperature, and ionic composition of the assay medium. Small variations in the assay conditions cause serious errors in the determinations.

The temperature coefficient of the enzyme reaction rate is roughly 10% per degree, and a 10°C rise in temperature may result in a twofold increase in the reaction rate. Hence, constant temperature is essential in the assay of enzyme activity. In enzyme reaction rate studies, standards are usually run with each test. If normal precautions are taken, the error in measurement of enzyme activity is usually less than ±5% (Guilbault 1966A). In automated systems, better accuracy and replicability can be attained.

Polysaccharides.—The list of substrates which can be assayed enzymatically is very large, and only a few of those which are of particular significance in food analysis will be reviewed here. Lee and Whelan (1966) determined starch and glycogen concentration after hydrolysis with amyloglucosidase. The enzyme hydrolyzes both α-1 → 4 and α-1 → 6 linkages of the polysaccharide and gives glucose. The latter can be assayed with glucose oxidase.

All forms of cellulose can be described as β-1,4 polyglucosides. As a result of the stability of native undegraded cellulose, there is no standard method for the enzymatic determination of this kind of substrate. However, partly degraded forms of cellulose are rapidly and completely hydrolyzed by enzymes (Halliwell 1959). If the cellulose is treated with phosphoric acid, it swells and is transformed into a more reactive form without any great change in the degree of polymerization. Swollen cellulose or swollen cotton fibers are rendered 97% soluble in 22 hr by the action of cell-free culture filtrates from *Myrothecium verrucaria*. The extent of the enzymatic hydrolysis is determined by the loss of weight of the insoluble substrate. The loss can be determined directly or indirectly. In the direct procedure (Halliwell 1958), the cellulase-rich filtrate is allowed to act on defatted, insoluble cellulose. The reaction mixture is filtered through sintered glass; the residue is oxidized with a dichromate-sulfuric acid mixture, and the color is measured at 430 nm. The difference between the color of the control and of the digested residue can be correlated with the amount of cellulose hydrolyzed by the enzyme. The amount of cellulose is computed from a calibration curve with glucose, and by multiplying the glucose equivalent by 0.9 to obtain cellulose content.

In the indirect procedure (Halliwell 1960), the soluble carbohydrates formed on enzymatic hydrolysis are determined instead of the insoluble residue. The procedure is more rapid, but the tested solution must be protein-free or low in soluble proteins.

Hemicelluloses are plant cell-wall polysaccharides that are insoluble in water and ammonium oxalate solution, but are soluble in dilute alkali.   They are bound with cellulose and lignin, and must be isolated prior to assay.   A further complication in the determination of hemicelluloses is that they exist in a variety of chemical structures. Because of this structural complexity, no standard methods for the determination of hemicelluloses are available.   Certain simple soluble hemicelluloses of the arabino-xylan or xylan type are completely hydrolyzed to their constituent pentoses, i.e. to arabinose and xylose by washed suspensions of toluene-treated bacteria from rumen of sheep, or by a cell-free hemicellulase preparation obtained after treatment of bacteria from sheep rumen with aqueous butanol and centrifugation.   After deproteinization of the hydrolysate, the reducing value of the carbohydrates is determined (Halliwell 1957 A, B).

Mono and Oligosaccharides.—Raffinose is a trisaccharide of glucose, galactose, and fructose.   It is present in sugar beet, beet juice, crude beet sugar, and less pure syrups containing sugar beet molasses.   Raffinose can be hydrolyzed by invertase to fructose and the disaccharide, melibiose.   Melibiase hydrolyses melibiose to glucose and galactose.   The optical rotation and the reducing value change during this hydrolysis.   Both changes can be used in raffinose determination.

The optical rotation method in sugar assays is recognized in the United States by AOAC (1965), and internationally by the International Commission for Uniform Methods of Sugar Analysis. The rotation is measured: (1) before the hydrolysis; (2) after the action of invertase; and (3) after the action of melibiase.

Analogous to the optical rotation method, the reducing value is determined (i.e., titrimetrically with Fehling's solution) before and after the action of invertase, and invertase plus melibiase.   In the reducing value method, it is not necessary to carry out the preliminary lead precipitation for the removal of color substances which interfere with the measurement of optical rotation.

The most common method of sucrose determination is by measuring optical rotation.   The method is precise and rapid in samples free of interfering optically active compounds, and relatively low in sugars other than sucrose.   The reducing sugar formed on inversion can be determined chemically if the sample contains virtually no other reducing substances.   Sucrose is hydrolyzed by invertase to glucose and fructose.   The two hexoses are phosphorylated by ATP to the corresponding hexose-6-phosphates in a reaction catalyzed by hexokinase.   Fructose-6-phosphate is isomerized to glucose-6-phosphate by phosphoglucoisomerase, and glucose-6-phosphate is oxidized by NADP and glucose-6-phosphate dehydrogenase to 6-phosphogluconate. The increase in optical density at 340 nm due to the formation of NADPH is a measure of the overall reaction. The reaction proceeds rapidly and quantitatively if the measurements are made at the pH optima of invertase (4.6) and hexose determination (7.6). The reaction can be carried out on any type of sample and requires 10 γ sucrose or less.

A lactose determination involves using β-D-galactosidase which catalyzes the hydrolysis of lactose into glucose and galactose. In the presence of ATP and hexokinase, the glucose is phosphorylated. Oxidation of the formed glucose-6-phosphate with glucose-6-phosphate dehydrogenase in the presence of NADP yields NADPH. The NADPH formed is measured by an increase of absorbance at 340 nm.

Glucose can also be assayed with glucose oxidase. At 20 °C, the enzyme oxidizes β-D-glucose 150 times faster than the α-isomer. Yet, the enzyme can be used to determine glucose in solutions containing an equilibrium mixture of the isomers, because even highly purified preparations of glucose oxidase contain mutarotase and the reaction time selected for analytical assays is long enough to oxidize all of the α-glucose. The determination is specific for D-glucose. Among the commonly occurring sugars, only mannose and galactose react and give about 1% of the value with glucose. The measurement of glucose with glucose oxidase can be made manometrically, but is generally made today colorimetrically. For the analysis of mixtures of sugars, it is necessary to use highly purified enzyme preparations. The procedure has been used to assay glucose in biological fluids and in various foods (corn syrup, hydrolysates of polysaccharides, fermentation liquors, etc).

As indicated previously, glucose oxidase catalyzes the reaction:

$$\text{Glucose} + H_2O + O_2 \rightarrow \text{gluconic acid} + H_2O_2 \qquad (30.3)$$

The hydrogen peroxide is decomposed in the presence of peroxidase and the oxygen liberated oxidizes a hydrogen donor DH (i.e., o-dianisidine) to a colored derivative D

$$H_2O_2 + DH_2 \rightarrow 2H_2O + D \qquad (30.4)$$

The amount of the dye D formed from DH_2 is a measure of the glucose oxidized. The dye has maximum absorbance around 460 nm. The extinction coefficient depends on the experimental conditions, and the measured absorbance is related to a glucose standard. In addition to o-dianisidine, o-toluidine, 2,6-dinitrophenolindophenol, and other dyes have been suggested.

Many modifications of the basic method have been suggested.

The reaction can be accelerated by using higher temperatures (i.e. 37°C for 20 min) provided the rapid color fading is prevented. The method has been adapted for automated assays. The measurements are made after 1 min, and the error is only 2%. Filter paper strips soaked with the enzymes and dye (o-toluidine) can be used for semi-quantitative determination of glucose in biological fluids.

Alcohols and Acids.—The determination of glycerol in biological materials, foods, and industrial products usually requires an extensive purification of the samples in order to remove contaminants. Enzymic determination of glycerol is specific, consequently, there is no need to purify the sample. Glycerol is phosphorylated by glycerokinase and ATP to give L-(−)-glycerol-1-phosphate; the latter is oxidized with glycerol-1-phosphate dehydrogenase in the presence of NAD to give dihydroxyacetone phosphate and NADH. The amount of NADH formed is equivalent to the amount of glycerol present. For quantitative oxidation in the second step, the reaction is carried out at pH 9.8 and the dihydroxyacetone phosphate formed is trapped with hydrazine.

Muscle lactate dehydrogenase catalyzes the oxidation of L-lactate by NAD to pyruvate and NADH. To obtain quantitative oxidation, the reaction products must be removed from the reaction mixture. This is done by making the medium alkaline (pH 9.5) and trapping the pyruvate as hydrazone. The course of the reaction is followed spectrophotometrically by the increase in absorbance due to the formation of NADH. Determination of lactate with yeast lactic dehydrogenase—unlike the assay with the enzyme from muscles—is not NAD-linked. It can be used to determine lactic acid in the presence of a large excess of pyruvate and is a useful routine biological assay. Yeast lactic dehydrogenase, a flavocytochrome, transfers hydrogen from lactate to potassium ferricyanide. The decrease in color on reduction of the ferricyanide ion can be followed at 405 nm.

The isolation from yeast and mammalian liver in pure form of enzymes which specifically catalyze the dehydrogenation of primary alcohols to aldehydes (alcohol dehydrogenase) opened the possibility to determine alcohol colorimetrically, since the enzymes from both sources require NAD as coenzyme. NADH has an absorbance maximum at 340 nm, and NAD has virtually no absorbance at that wavelength. At neutral reaction conditions and with the amount of NAD ordinarily employed, less than 1% of the alcohol is oxidized to acetaldehyde. It is, however, possible to drive the reaction nearly to completion when a high pH is chosen and the aldehyde formed is removed by reaction with semicarbazide. The use of the 3-acetylpyridine analog of NAD may also be used to change the equilibrium. With this substance as the reactant, the equilibrium constant increases about 200 times. Generally, only butanol might affect the enzymatic assay of ethanol. NADH may be also measured by fluorescence, and the sensitivity of the assay is substantially improved.

In addition to the compounds discussed, methods are available to determine acetate, formate, citrate, malate, fumarate, and other organic acids present in foods. Several amino acids can be determined after dehydrogenation by measuring disappearance of NADH. In some cases, the first step involves transamination. Thus, L-alanine is converted to pyruvate by glutamate-pyruvate transaminase and α-oxoglutarate. Similar assay procedures are available for the determination of aspartate and glutamate. With threonine, the amino acid is first oxidized by periodate at neutral pH to acetaldehyde and glyoxylate, the latter undergoing a relatively rapid oxidation to formate and carbon dioxide. After destruction of excess periodate with a mercaptan, the acetaldehyde is determined with alcohol dehydrogenase and reduced NADH. Enzymatic methods of determining α-keto acids were reviewed by Neish (1957). The assays include decarboxylation of pyruvate by yeast carboxylase, oxidation of pyruvate to acetic acid, inhibition of anaerobic fermentation of certain microorganisms by pyruvate, and formation of γ-lactone of α-keto-γ-hydroxybutyric acid from formaldehyde and pyruvate, and measuring the disappearance of formaldehyde.

Decarboxylases.—Certain bacteria, grown under suitable conditions, produce specific L-amino acid decarboxylases. In most cases, the pH optima for the decarboxylases are in the acid range so that the carbon dioxide produced can be measured manometrically. The decarboxylases catalyze the reactions of the type

$$\overset{\displaystyle NH_2}{\underset{\displaystyle |}{R-CH}}-COOH \rightarrow R-CH_2-NH_2 + CO_2$$

The carbon dioxide produced is a measure of the amino acid content of the sample. Specific decarboxylases are available for the determination of lysine, ornithine, tyrosine, histidine, aspartic acid, glutamic acid, glutamine, β-hydroxyglutamic acid, phenylalanine, diaminopimelic acid, valine, and leucine (Gale 1945, 1946, 1957). In the determination of amino acids which have a pH of about 5.8 at the end of decarboxylation (lysine, ornithine, and aspartic acid), manometer vessels with double side arms are used with a Warburg apparatus (Umbreit *et al.* 1957). The second side arm contains dilute acid, which at the end of the decarboxylation is tipped into the main compartment to release the retained carbon dioxide. Spectrophotometric assays were described by Dickerman and Carter (1962).

Amino acid decarboxylases attack free amino acids only. If an excess of decarboxylase and limiting amounts of peptidase are added to a peptide preparation, the carbon dioxide formed is an index of peptidase activity. The decarboxylases can be used to fractionate D and L amino acids by destroying the L isomer with decarboxylase which does not attack the D isomer. Free amino acids can be determined in the presence of proteins. Many transamination reactions involve glutamic acid as one of the components. Estimation of glutamic acid by the decarboxylases can be used to follow transaminases. The decarboxylases possess pyridoxal phosphate as a prosthetic group and in some cases, e.g. tyrosine decarboxylase, it is possible to remove the prosthetic group leaving an inactive apo-decarboxylase preparation. The apo-enzyme is activated by pyridoxal phosphate and the rate of decarboxylation is proportional over a limited range to the concentration of pyridoxal phosphate.

Miscellaneous.—Urease catalyses the reaction

$$O=C \begin{matrix} NH_2 \\ \\ NH_2 \end{matrix} + H_2O \rightarrow CO_2 + 2NH_3$$

Both reaction products can be used to determine the urea content of a sample. The carbon dioxide can be determined gasometrically or colorimetrically; the ammonia can be determined directly after distillation into a receiver, or by titration after diffusion. The ammonia can also be determined colorimetrically with Nesslers' reagent after isolation by distillation or by ion exchange chromatography.

Several enzymatic methods are available for the determination of lipids. The ultraviolet absorbance of conjugated diene hydroperoxides, arising from the lipoxidase catalyzed oxidation of *cis*-polyunsaturated fatty acids can be used to measure the concentration of the enzyme in the presence of excess substrate (Bergstrom and Holman 1948; Holman 1955). Alternatively, substrate concentration can be determined by oxidation in the presence of excess enzyme. All polyunsaturated fatty acids, whose double bonds are separated by *cis*-methylene groups (linoleic, linolenic, and arachidonic), react quantitatively. The extinction coefficients of the diene hydroperoxides at 234 nm are the same for all the polyunsaturated fatty acids thus far examined.   The method cannot differentiate the individual polyunsaturated fatty acids as only one mole of diene hydroperoxide is formed for each mole of fatty acid, irrespective of the number of double bonds.   The method is very sensitive and allows the determination of as little as 5γ of linoleic acid.   It has been used in foods, agricultural products, and biological materials (MacGee 1959).   Fatty acid esters must be saponified, before determination of the conjugated double bonds.

The determination of lecithin (i.e. as a measure of the egg content of foods) can be made by catalyzing the following reaction with lecithinase (phospholipase) D:

$$\text{Lecithin} + H_2O \rightarrow \text{phosphatidic acid} + \text{choline}$$

Lecithin and phosphatidic acid are soluble in ether, choline in aqueous solutions.   After the enzymatic reaction, phosphatidic acid can be separated from the choline by extracting with ether, and the choline can be precipitated from the aqueous phase as reineckate. The red-violet reineckate is soluble in acetone and the amount is measured colorimetrically at 520 nm (Casson and Griffin 1959; Griffin and Casson 1961).

Several methods are available for the determination of inorganic compounds.   Hydrogen peroxide is decomposed by peroxidase. The oxygen liberated in this process oxidizes a colorless hydrogen donor to a colored compound.   o-Dianisidine—or other aromatic amines—can be used as a hydrogen donor, and yields a red brown dye with a broad absorption maximum around 460 nm.   Nitrate reductase from certain strains of *Escherichia coli* catalyzes the reduction of nitrate to nitrite in the presence of a hydrogen donor (i.e., formic acid) and a hydrogen acceptor (i.e. methylene blue). The nitrite formed can be measured colorimetrically.

Magnesium ions activate isocitric dehydrogenase.   The enzyme catalyzes the reaction:

$$\text{isocitrate} + \text{NADP} \xrightarrow{Mg^{2+}} \text{NADPH} + H^+ + \alpha\text{-oxoglutarate} + CO_2$$

With constant amounts of enzyme, the rate is dependent on the magnesium ion concentration.

All inorganic pyrophosphatases so far described require magnesium ion for their action.   The enzyme catalyzes the formation of orthophosphate, the concentration of which can be determined colorimetrically.

TABLE 30.1

CLASSIFICATION OF ENZYMES

Type of Enzyme	Examples	Remarks
Hydrolyzing		
Proteases, peptidases	Pepsin, rennin, trypsin, bromelin, papain, ficin, etc.	Hydrolyze —CO—NH— links in proteins, peptides, etc.
Carbohydrases	α- and β-amylases, amyloglucosidase, oligoglucosidase, invertase, maltase, cellulase etc.	Hydrolyze glycosidic links
Esterases	Lipases, cholinesterases, pectinesterases, phytase, phosphatases, tannase, etc.	Hydrolyze ester links in fats, alcohol esters, phosphoric esters, sulfuric esters, thioesters and phenolic esters
Other	Arginase, urease, deaminase, etc.	
Adding		
Add or remove water	Aconitase, enolase, fumarase, glyoxalase, serine diaminase, etc.	
Add or remove CO_2	Amino acid decarboxylase, carboxylase (in yeast), malic decarboxylase, oxaloacetic decarboxylase, pyruvic oxidase, etc.	Thiamine pyrophosphokinase,
Other	Aldolase, aspartase, etc.	
Transferring		
Oxidoreductases	D- and L-amino acid oxidases, 1-amino acid dehydrogenase, L-glutamate dehydrogenases, proline oxidase, Co II cytochrome reductase, alcohol dehydrogenase, glycerol dehydrogenase, glucose oxidase, aldehyde dehydrogenase, aldehyde oxidase, succinate dehydrogenase, cystine reductase, oxalate oxidase, cytochrome oxidases, peroxidases, catalases, etc.	Transfer hydrogen
Transfer nitrogenous groups	Transaminases, glutamate transaminase, glycine transamidinase, γ-glutamyl transpeptidase, γ-glutamyl transferase, etc.	
Transfer phosphate groups	Hexokinase, glucokinase, ketohexokinase, fructokinase, ribokinase, thiamine pyrophosphokinase phosphoglucomutase (intramolecular transfer) etc.	
Transacylases	Amino acid transacetylase, glucosamine transacetylase, choline transacylase, glyoxylate transacetase, etc.	Transfer acyl groups
Transglycosylases	Dextrin transglucosylase, maltose transglucosylase, sucrose transglucosylase, hyaluronidase (animal) etc.	Transfer glycosyl groups
Transfer coenzyme A	CoA transferase	

TABLE 30.1

CLASSIFICATION OF ENZYMES (*Continued*)

Type of Enzyme	Examples	Remarks
Transmethylases	Nicotinamide transmethylase, guanidinoacetate transmethylase, betaine-methionine transmethylase, etc.	Transfer methyl groups
Other	Transaldolase, transketolase, thiaminase, transoxidase, etc.	
Isomerizing		
Isomerases (stereoisomerases)	Alanine racemase, glutamate racemase, lactate racemase, mutarotase, etc.	
Miscellaneous		
Synthetases	Glutamine synthetase, asparagine synthetase, tryptophan hydroxamate synthetase, etc.	Catalyze synthetic reactions
Add groups to double bonds	Fumarase, aconitase, D- and L-serine dehydrase, dihydroxy acid dehydrase, phosphopyruvate carboxylase, aldolase, citrase, isocitrase, etc.	

Source: de Becze (1965).

Determination of Enzymes

Assays of enzymatic activity in foods are conducted to determine the soundness of foods, to follow changes in heat processing, and establish damage. The common techniques of quantitatively following enzyme reactions include spectrophotometric, manometric, electrometric, polarimetric, chromatographic, and chemical procedures. The details of such assay methods have been described in reference books, textbooks, and manufacturers' catalogs, and will not be discussed here. The review is limited to a general outline of methods used in food analysis and a review of the principles involved in the individual assays.

Table 30.1 (deBecze 1965) summarizes types of enzymes and Table 30.2 (Pulley 1969) the commercial applications of enzymes in food production.

According to the recommendations of the International Union of Biochemistry (1961), an *enzyme unit* is defined as the amount of enzyme required to modify under standardized conditions 1 μ mole of substrate (or in the case of polysaccharides and proteins 1 μ equivalent) per minute. The assay should be carried out—if possible—at 25°C, at optimum and defined pH, and the substrate concentration should provide for enzyme saturation Measurements should be made in the range of initial velocity, so that the kinetics follow a zero order reaction and substrate modifications are related in a linear manner to enzyme concentration. This re-

TABLE 30.2

COMMERCIAL APPLICATIONS OF ENZYMES IN FOOD PRODUCTION

Processing Difficulty or Requirement	Enzyme Function	Enzyme Used	Enzyme Source
Milling and Baking			
High dough viscosity	Catalyzes hydrolysis of starch to smaller carbohydrates by liquefaction, thus reducing dough viscosity	Amylase	Fungal
Slow rate of fermentation	Accelerates process	Amylase	Fungal
Low level of sugars resulting in poor taste, poor crusts, and poor toasting characteristics	Converts starch to simple sugars such as dextrose, glucose and maltose by saccharification, thus increasing sugar levels	Amylase	Bacterial
Staling of bread	Enables bread to retain freshness and softness longer	Amylase	Bacterial
Mixing time too long for optimum gas retention of doughs	Reduces mixing time and makes doughs more pliable by hydrolyzing gluten	Protease	Fungal
Curling of sheeted dough for soda crackers as dough enters continuous cracker ovens	Prevents curling of sheeted dough	Protease	Fungal
Poor bread flavor	Aids in flavor development	Lipoxidase Protease	Soy Flour Fungal
Offcolor flour	Bleaches natural flour pigments and lightens white bread crumbs	Lipoxidase	Soy Flour
Low loaf volume and coarse texture	Hydrolyzes pentosans	Pentosanase	Fungal
Meats			
High fat content	Aids in removal or reduction of fat content	Lipase	Fungal
Upgrade meat	Hydrolyzes muscle protein and collagen to give more tender meat	Protease	Ficin (figs) Papain (papaya)
Serum separation of fat in meat and poultry products	Produces liquid meat products and prevents serum separation of fat in meat products and animal foods	Protease	Fungal
High viscosity of condensed fish solubles	Reduces viscosity while permitting solids levels over 50% without gel formation of the condensed fish solubles	Protease	Fungal
Protein shortage	Prepare fish protein concentrate	Protease	Fungal/ Bacterial
Distilled Beverages			
Thick mash	Thins mash and accelerates saccharification	Amylase	Bacterial Malt
Chill haze	Chillproofs beer	Protease	Fungal Bacterial Papain

TABLE 30.2

COMMERCIAL APPLICATIONS OF ENZYMES IN FOOD PRODUCTION (*Continued*)

Processing Difficulty or Requirement	Enzyme Function	Enzyme Used	Enzyme Source
Low runoff of wort	Assists in physical disintegration of resin and improves runoff of wort	Amylase	Fungal Malt
Fruit Products and Wines			
Apple juice haze	Clarifies apple juice	Pectinase	Fungal
High viscosity due to pectin	Reduces viscosity by hydrolyzing the pectin	Pectinase	Fungal
Slow filtration rates of wines and juices	Accelerates rate of filtration	Pectinase	Fungal
Low juice yield	Facilitates separation of juice from the fruit, thus increasing yield	Pectinase	Fungal
Gelled purée or fruit concentrate	Prevents pectin gel formations and breaks gels	Pectinase	Fungal
Poor color of grape juice	Improves color extraction from grape skins	Pectinase	Fungal
Fruit wastes	Produces fermentable sugars from apple and grape pomace	Cellulase	Fungal
Sediment in finished product	Helps prevent precipitation and improves clarity	Pectinase	Fungal
Syrups and Candies			
Controlled level of dextrose, maltose, and higher saccharides	Controls ratios of dextrose, maltose, and higher saccharides	Amylase	Bacterial Fungal Malt
High viscosity syrups	Reduces viscosity	Amylase	Bacterial Fungal
Sugar loss in scrap candy	Facilitates sugar recovery from scrap candy by liquefaction of starch content	Amylase	Bacterial
Filterability of vanilla extracts	Improves filterability of vanilla extracts	Cellulase	Fungal
Miscellaneous			
Poor flavors in cheese and milk	Improves characteristic flavors in milk and cheese	Lipase	Fungal
Tough cooked vegetables and fruits	Tenderizes fruits and vegetables prior to cooking	Cellulase	Fungal
Inefficient degermination of corn	Produces efficient degermination of corn	Cellulase	Fungal
High set times in gelatins	Reduces set times of gelatin without significantly altering gel strength	Protease	Fungal
Starchy taste of sweet potato flakes	Increase conversion of sweet potato starch	Amylase	Fungal Bacterial
High viscosity of precooked cereals	Reduces viscosity and allows processing of precooked cereals at higher solid levels	Amylase	Fungal Bacterial

Source: Pulley (1969).

quires short reaction times. In case of lengthy measurements, the reaction velocity is extrapolated to zero time. Concentrations of enzymes should be expressed in units per cubic centimeter; *specific activities* in units per milligram protein.

The methods of assaying proteinase activity were reviewed by Davis and Smith (1955). They include the determination of changes in the modified substrate, or assay of fractions split off from the substrate. In the first category are included the determination of rheological parameters (viscosity, dough consistency, coagulation, solubility, or light dispersion). In the split-off fraction, the carboxylic or amino groups can be titrated, and free amino acids can be determined by specific or nonspecific (ninhydrin) reactions. Soluble proteins, not precipitated with trichloracetic acid, are measured before and after protease action, and the difference gives an estimate of enzymatically modified substrate. Proteolytic enzymes can cause milk coagulation; this property is particularly useful in testing preparations used in milk processing. Finally, chromogenic components of synthetic substrates are cleaved by certain proteolytic enzymes and measured colorimetrically.

Peptidases vary in their specificity. Their activity is assayed by microtitration or colorimetry of amino acids. The activity of certain peptidases, i.e. carboxypeptidases, can be assayed accurately by colorimetrically determining β-naphthol released from a synthetic substrate (naphthoxycarboxylphenylalanine). Urease activity, *per se*, is generally of little importance in food processing. It is used as an index of heat treatment of soy flour toasted in order to improve palatability and nutritional value. Enzymatic hydrolysis of pectins involves several enzymes varying in their effects on viscosity and cleavage of specific groups which can be determined manometrically or titrimetrically (Deuel and Stutz 1958).

Numerous methods are available to measure enzymatic hydrolysis of starch. Most methods for the saccharogenic β-amylase measure the amount of maltose formed. They are used widely in the brewing and malt industries. Determinations of the starch liquefying α-amylase are made by either viscosimetric or dextrinogenic assays. Viscometric methods are discussed in Chap. 28. In determinations of dextrinogenic activity, the rate of dextrinization of a soluble starch solution by an extract of the material under test is measured. The time required to reach a stage at which the color produced on the addition of an iodine solution is reddish-brown instead of blue in undegraded starch is determined. The determination of α-amylase activity is useful in detecting sprouted grain, and in the control of malt production and supplementation. In several European countries, honey is tested for diastatic activity. Honey in which diastase was destroyed by heating is considered adulterated.

The phosphatases of milk are inactivated within the temperature range used for milk pasteurization. They are, therefore, commonly assayed to determine efficiency of pasteurization of milk and milk products. In the commonly used assay, free phenol hydrolysed from phenyl phosphate is determined with 2,6-dibromoquinone-chlorimide by a spectrophotometric procedure. Fat acidity is often used as an index of soundness of stored grain. The acidity increases in stored grain especially if the cereals are stored at high moisture and elevated temperatures. Lipase activity in such grains is generally determined titrimetrically. Testing for the inactivation of peroxidases in the fruit and vegetable industry is run routinely to determine efficiency of blanching. Milk peroxidase is inactivated at 81°–83°C. The test is used to distinguish between milk heated by various processes. Sound milk contains insignificant amounts of catalase. The enzyme can be determined by manometric and electrometric methods. High catalase levels indicate milk from sick cows, presence of colostrum, or bacterial contamination. Bacterial dehydrogenases are assayed to determine the hygienic status of milk. Lakon (1942) studied enzymatic reduction of several tetrazolium salts and found 2,3,5-triphenyltetrazolium chloride the best indicator of seed viability. A review of enzyme reactions related to seed viability was published by Linko (1960).

Determination of Inhibitors

An inhibitor is a compound that causes a decrease in the rate of an enzymatic reaction, by reacting with the enzyme or substrate to form a complex. In general, the initial rate of an enzymatic reaction will decrease with increasing inhibitor concentration linearly at low inhibitor concentrations, and will gradually approach zero at higher concentration. The effects of inhibitors in foods on enzymes were reviewed by Schormuller (1957B, 1958A, B, C) and by Kiermeier et al. (1962). Several analytical procedures have been proposed based on the inhibition of enzymes.

Organophosphorus insecticides inhibit the cholinesterase of animals and insects. This accounts to a large extent both for their effectiveness in the control of harmful insects in agriculture and for the toxicity of their residues in foods to warm-blooded animals. Many of those insecticides are such powerful inhibitors of cholinesterase that very sensitive enzymatic methods for their determination have been developed utilizing this inhibitory property (Augustinson 1959).

Enzymatic methods for the determination or organophosphorus insecticides can be classified as: (1) electrometric; (2) titrimetric; (3) manometric; and (4) colorimetric.   Use of electrometric methods in studies of enzymatic reactions was reviewed by Blaedel and Hicks (1964) and Guilbault (1966B).   The latter described methods based on measurement of changes in potential that occur in enzymatic hydrolysis of a thiocholine ester by cholinesterase.   Organophosphorus compounds inhibit the enzyme and may be determined in 10^{-9} gm concentrations by the technique.   Some colorimetric methods use chromogenic substrates which form colored products on hydrolysis with cholinesterase or related esterases.   With constant substrate concentration, the color intensity depends on the enzymatic activity.

Giang and Hall (1951) described a method of determining insecticide content based on changes in pH.   The sample is extracted with an organic solvent, the solvent is evaporated, and the residue is incubated for 30 min with a known excess of standardized cholinesterase in a buffered solution.   At the end of the inhibition period, a known excess of acetylcholine is added to the reaction mixture. After 60 min, the acetic acid produced by the hydrolysis of acetylcholine is measured by the change in pH.   The less cholinesterase is inhibited by the insecticide, the more acetic acid will be formed in the reaction.

Very small amounts of DDT [di-(p-chlorophenyl)trichloroethane] can be measured accurately with carbonic anhydrase by the inhibitory effect on the enzyme.   Methods of measuring carbonic anhydrase activity were reviewed by Davis (1963).   The enzyme is inhibited by DDT at concentrations at which other inhibitors, with the exception of sulfonamides, are inactive.   As the sample to be examined usually does not contain sulfonamides, a special extraction and purification can be omitted.

Enzymes as Analytical Aids

Exogenous and endogenous enzymes have been employed to a limited extent to liberate proteins from tissues.   With commercial availability of more purified enzymes, the lipases and carbohydrases should have increasing use in the liberation and purification of proteins especially from plants and bacteria (Keller and Block 1960).

The ability of proteolytic enzymes to hydrolyze the peptide bonds formed by specific amino acids gives those enzymes several advantages over acids as hydrolytic agents. Among these are: high yields of peptides (or amino acids), little nonhydrolytic alteration of the products, and the requirement of only catalytic amounts of the enzyme. The only limitation of enzymatic hydrolysis is the production of artifacts through transpeptidation (Sela and Katchalski 1959). Generally proteolytic enzymes have a broad substrate specificity, but none are known which will hydrolyze all of the types of peptide bonds found in proteins (Hill 1965). The *Streptomyces griseus* proteinase (marketed as Pronase), papain, and the subtilisins extensively hydrolyze most proteins with the liberation of free amino acids, but each also leaves some peptide bonds intact. For total enzymatic hydrolysis of protein, it is necessary to employ mixtures of enzymes with several different specificities.

Complete enzymatic hydrolysis has several advantages over acid hydrolytic procedures. The acid-labile amino acids such as asparagine, glutamine, tryptophan, and the phospho- and sulfoesters of certain amino acids are not destroyed. The amino acids serine and threonine, which are destroyed partially by acid, as well as those which are released incompletely by acid hydrolysis, should be present in theoretical yields in enzymatic hydrolysates.

Acid hydrolysis cleaves amide groups and does not permit the direct determination of glutamine and asparagine. The amount of both can be approximated from the amount of ammonia in the acid hydrolysate. The amount of ammonia is, however, affected by the conditions of acid hydrolysis, and especially by the destruction of serine and threonine. The amide content can be determined directly by using enzyme systems which hydrolyze peptide bonds in protein but do not affect amide nitrogen. Tower *et al.* (1962) employed a whole pancreas preparation for enzymatic hydrolysis of proteins and peptides. The residues released into the hydrolyzates were determined by assay with *Clostridium perfringens* L-glutamic acid decarboxylase, L-glutaminase, and L-aspartic acid decarboxylase, and with guinea pig serum L-asparaginase. Lin *et al.* (1964) hydrolyzed proteins with either hydrochloric acid or with Pronase (Nomoto *et al.* 1960A, B). Amino acids in both hydrolyzates were separated by two-dimensional paper chromatography and compared.

In microbiological assays, the vitamin must be extracted first from the sample in water-soluble form and in a state utilizable by the test organism. When the vitamin is stable to acid or alkali at high temperatures, chemical hydrolysis is used. If the vitamins are destroyed by chemical treatment, enzymatic digestion may be used. Enzymatic digestion is essential in the assay of folic acid and pantothenic acid by microbiological assay, and when the concentrations of the various members of the B-complex (thiamine, riboflavin, niacin, biotin, and inositol) are to be determined on the same sample (Assoc. Vitamin Chem. 1951).

A concentrated pectinase enzyme derived from *Aspergillus niger* is used in establishing the relationship between refractometer readings, specific gravity, and total solids for tomato pulp and paste (AOAC 1965). The pectinases, amylases and proteases are useful in the isolation of filth and extraneous matter from food.

Use of Enzymes in Structural Studies

The most useful property of enzymes is their specificity. Lipases from various sources differ in the site on the triglyceride or phospholipid molecule which they attack. Enzymes are useful in studying lipid structure. Thus, pancreatic lipase shows a preferential hydrolysis of the 1 and 3 positions of a triglyceride molecule, the main monoglyceride formed during the hydrolysis having the 2 configuration (Desnuelle 1961; Johnston 1963). Enzymes are also available which distinguish between fatty acids varying in saturation, in chain length, and in stereospecificity (Alford and Smith 1965).

Since the fundamental work of Sanger on the structure of insulin (Sanger, 1952, 1956), an increasing amount of research has been undertaken to determine the amino acid sequences in proteins and peptides. It is useful to degrade the proteins with proteolytic enzymes, since the specificity of some of the enzymes permits the isolation of uniform fragments. The "spectrum" of the peptides obtained in this way is generally characteristic of a protein. The enzymic digests can be separated by ion-exchange column chromatography or characterized by "fingerprinting" (two-dimensional separation on paper; in one direction by conventional paper chromatography and in the other by high-voltage electrophoresis).

A high selectivity of trypsin for specific peptide linkages makes the enzyme particularly useful. However, use of additional enzymes (chymotrypsin, pepsin, and Pronase) has been found useful in some studies.

Enzymatic hydrolysis may be useful in elucidating bonds which are involved in linkages between proteins and prosthetic groups, certain types of inhibitors, and coenzymes. Because of the specificity of most proteinases for bonds formed by amino acids of the L-configuration, enzymatic hydrolysis provides a means for determining the stereochemical homogeneity of polypeptides and proteins

(Hill 1965). Finally, enzymes can be used to specifically modify biologically active substances in such a way that valuable information relating structure to function is obtained.

Nutritional Availability of Amino Acids in Foods

In vitro enzymatic studies have demonstrated that amino acid availability and the amino acid content of food may differ markedly. Comparative measurements of the enzymatic and chemical liberation of amino acids have been made by numerous investigators and were reviewed by Grau and Carrol (1958), Mauron (1961), and Morrison and Rao (1966).

Enzymic *in vitro* analytical procedures were developed by Scheffner *et al.* (1956) and by Mauron *et al.* (1955). The former determined the pattern of essential amino acids released by pepsin. The work involved is quite considerable as ten amino acids have to be determined both in an acid hydrolysate and in the pepsin-digest. The results showed good agreement with biological values obtained by feeding rats. The large amount of work involved makes the usefulness of the method limited to research investigations and of less value in large scale routine determinations.

By confining the evaluation to the release, by pepsin and pancreatin, of the key amino acids tryptophan, methionine, and lysine, which are most likely to be limiting factors in foods, Mauron *et al.* (1955) reduced substantially the work involved. Lysine was determined enzymatically by lysine decarboxylase, and methionine and tryptophan colorimetrically. The procedure was used in the quality control of heat-processed milk and was in good agreement with protein efficiency ratios measured on growing rats (Mauron and Mottu 1958).

Automated Methods

Automated methods for the determination of enzyme activity were reviewed by Schwartz and Bodansky (1963), Blaedel and Hicks (1964), Mason (1965), Frings (1966), and Guilbault (1966A). Use of manometric techniques in such assays was described by Umbreit *et al.* (1957). The assay of many enzyme activities utilizes the measurement at 340 nm of the oxidation or reduction of nucleotides. For the initial stages of the enzyme reaction that are of zero order, the change in absorbance per minute is a measure of the reaction velocity. Many of the experimental difficulties of using enzymes in analysis could be eliminated or lessened by the use of automation. A number of commercially available instruments have been used for automating the preparation of the reaction mixture in various enzyme activity assays. In many of the new instruments, assay results (i.e., absorbance) are transformed automatically by means of computers into units of enzyme activity, or various other parameters based on such activity.

A variety of automatic procedures are available for generating substrate, activator, inhibitor, or pH gradients for the characterization of enzymes (Anon. 1967B). They have a wide range of applicability to enzyme research. Thus, for example, the activation of alkaline phosphatase by magnesium was described. The enzyme is sampled continuously and is mixed with substrate. A constantly increasing gradient of magnesium is prepared by means of addition of concentrated magnesium at a fixed rate. This gradient is then added to the sample-substrate reaction mixture. The resultant stream is incubated in a constant temperature heating bath for a fixed time, and is followed by color development of the reaction product. The rate of reaction is measured by the increase in color intensity. Optimal activation concentration is reached when a plateau appears on the strip chart record.

In a number of recent applications, the natural fluorescence of NADH, when excited at 340 nm, provides an end point for the measurement of enzyme activity. A fluorimeter was designed to handle up to 80 samples per hour. It has a response time of less than $1/2$ sec while maintaining excellent stability from transient noise. The method is made more specific by dialyzing NADH, thereby, eliminating background interferences.

A mixture of enzymes can be fractionated by column chromatography. A small part of the effluent is continuously monitored and assayed. The bulk of the effluent is diverted to a fraction collector for the isolation and characterization of the enzymes. The system permits accurate, reproducible, and automatic separation, detection, quantitation, and collection of enzyme concentrates within a short time. At least 40 specific enzyme assay methods have been automated. Any of these methods can be used to monitor the effluent from a protein—separating column. Often enzymatic assays are made along with total protein determinations by the Lowry modification of the Lowry-Ciocalteau color reaction. In this way, the enzymatic activity per unit protein can be assayed.

Automatic enzyme assay methods are used widely in clinical laboratories. Commercial instruments are available to determine in 2.0 ml serum 12 selected biochemical parameters at a rate of 60 samples per hour. Many of the procedures are finding increasing applications in the food industry. The assays include the determination of glucose by glucose oxidase, assay of free lysine based on the continuous colorimetric determination of carbon dioxide formed during the enzymatic decarboxylation of lysine by L-lysine decarboxylase; determination of cholinesterase activity for the estimation of organic phosphate pesticide residues; and the determination of ethanol based on its oxidation in the presence of alcohol dehydrogenase and NAD. Many hospitals today use automated methods for the assay of enzymes as a tool in the diagnosis of disease and use the information in the treatment of patients. Automated enzyme assays will no doubt be used in the not too distant future for quality control in the food industry. Feedback devices based on enzyme determinations, should make it possible to maintain a uniform product undergoing heat treatment and resulting in enzyme inactivation (i.e. in milk pasteurization, heat treatment in processing of fruits and vegetables, or in toasting of soyflour).

BIBLIOGRAPHY

ACKER, L. 1962. Enzymic reactions in foods of low moisture content. Advan. Food Res. *11*, 263–330.

ALFORD, J. A., and SMITH, J. L. 1965. Production of microbial lipases for the study of triglyceride structure. J. Am. Oil Chemists Soc. *42*, 1038–1040.

AMADOR, E., and WACKER, W. E. C. 1965. Enzymatic methods for diagnosis. Methods Biochem. Analy. *13*, 265–356.

AM. ASSOC. CEREAL CHEMISTS. 1962. Cereal Laboratory Methods. St. Paul, Minn.

AM. SOC. BREWING CHEMISTS. 1958. Methods of Analysis of the Am. Soc. Brewing Chemists. Madison, Wis.

ANDERSON, J. A. 1946. Enzymes and Their Role in Wheat Technology, Monograph Ser., Vol. 1. Am. Assoc. Cereal Chemists. Interscience Publishers, New York.

ANDERSON, J. A., and ALCOCK, H. W. 1954. Storage of Cereal Grains and Their Products. Monograph Ser., Vol. 2. Am. Assoc. Cereal Chemists. St. Paul, Minn.

ANON. 1967A. Standard Analytical Methods of the Member Companies of Corn Industries Research Foundation, 2nd Edition. Washington, D.C.

ANON. 1967B. Completely Automated Laboratory. Technicon Corp., Ardsley, New York,.

ASSOC. OFFIC. AGR. CHEMISTS. 1965. Official and Tentative Methods of Analysis. Washington, D.C.

ASSOC. VITAMIN CHEMISTS. 1951. Methods of Vitamin Assay. Interscience Publishers, New York.

AUGUSTINSON, KLAS-BERTIL. 1959. Assay methods of cholinesterases. Methods Biochem. Analy. *5*, 1–63.

BECKHORN, E. J., LABBEE, M. D., and UNDERKOFLER, L. A. 1965. Production and use of microbial enzymes for food processing. Agr. Food Chem. *13*, 30–34.

BERGMEYER, H. U. 1963. Methods of Enzymatic Analysis. Academic Press, New York.

BERGSTROM, S., and HOLMAN, R. T. 1948. Lipoxidase and the autoxidation of unsaturated fatty acids. Advan. Enzymol. *8*, 425–495.

BERNFELD, P. 1951. Enzymes of starch degradation and synthesis. Advan. Enzymol. *12*, 379–428.

BLAEDEL, W. J., and HICKS, G. P. 1964. Use of electrochemical methods in study of enzymic reactions. *In* Advances in Analytical Chemistry and Instrumentation, Vol. 3, C. N. Reilley, (Editor). Interscience Publishers, New York.

BOYER, P. D., LARDY, H., and MYRBACK, K. 1959–1963. The Enzymes, Vol. 1–8. Academic Press, New York.

BRUNS, F. H., and WERNERS, P. H. 1962. Dehydrogenases: glucose-6-phosphate dehydrogenase, 6-phosphogluconate dehydrogenase, glutathione reductase, methemoglobin reductase, polyol dehydrogenases. Advan. Clin. Chem. *5*, 238–294.

CASSON, C. B., and GRIFFIN, E. J. 1959. Determination of eggs in certain foods by enzymic hydrolysis of the phospholipids. Analyst *84*, 281–286.

COLOWICK, S. P., and KAPLAN, N. O. 1955. Methods in Enzymology. Academic press, New York.

COOK, A. H. 1962. Barley and Malt, Biology, Biochemistry, and Technology. Academic Press, New York.

CRUESS, W. V. 1943. The role of microorganisms and enzymes in wine making. Advan. Enzymol. *3*, 349–386.

DAVIS, N. C., and SMITH, E. L. 1955. Assay of proteolytic enzymes. Methods Biochem. Analy. *2*, 215–257.

DAVIS, R. P. 1963. The measurement of carbonic anhydrase activity. Methods Biochem. Analy. *11*, 307–327.

DE BECZE, G. I. 1965. Enzymes—industrial. *In* Kirk-Othmer Encyclopedia of Chemical Technology, Vol. 8. John Wiley & Sons, New York.

DESNUELLE, P. 1961. Pancreatic lipase. Adv. Enzymol. *23*, 129–161.

DEUEL, H., and STUTZ, E. 1958. Pectic substances and pectic enzymes. Advan. Enzymol. *20*, 341–382.

DEVILIN, T. M. 1959. Enzymatic methods in analytical chemistry. Anal. Chem. *31*, 977–981.

DICKERMAN, H. W., and CARTER, M. L. 1962. A spectrophotometric method for the determination of lysine utilizing bacterial lysine decarboxylase. Anal. Biochem. *3*, 195–205.

DIXON, M., and WEBB, E. C. 1958. Enzymes. Academic Press, New York.

FORD, F. F., and VITUCCI, J. C. 1948. Certain aspects of microbiological degradation of cellulose. Advan. Enzymol. *8*, 253–298.

FORSYTH, W. G. C., and QUESNEL, V. C. 1963. The mechanism of cacao curing. Advan. Enzymol. *25*, 457–492.

FREE, A. H. 1963. Enzymatic determination of glucose. Advan. Clin. Chem. *6*, 67–96.

FRINGS, C. S. 1966. Analytical Applications of Enzymes. Ph.D. Thesis, Purdue Univ., Lafayette, Indiana.

GALE, E. F. 1945. The use of specific decarboxylase preparations in the estimation of amino acids and in protein analysis. Biochem. J. *39*, 46–52.

GALE, E. F. 1946. The bacterial amino acid decarboxylases. Advan. Enzymol. *6*, 1–32.

GALE, E. F. 1957. Determination of amino acids by use of bacterial amino acid decarboxylases. Methods Biochem. Analy. *4*, 285–306.

GEDDES, W. F. 1946. The amylases of wheat and their significance in milling and baking technology. Advan. Enzymol. *6*, 415–468.

GEDDES, W. F. 1950. The role of enzymes in milling and baking technology. Food Technol. *11*, 441–446.

GIANG, P. A., and HALL, S. A. 1951. Enzymic determination of organic phosphorus insecticides. Anal. Chem. *23*, 1830–1834.

GOLDSTEIN, L., and KATCHALSKI, E. 1968. Use of water-insoluble enzyme derivatives in biochemical analysis and separation. J. Anal. Chem. *243*, 375–396.

GRAU, C. R., and CARROLL, R. W. 1958. Evaluation of protein quality. *In* Processed Plant Protein Foodstuffs, A. M. Altschul (Editor). Academic Press, New York.

GRIFFIN, F. J., and CASSON, C. B. 1961. Enzymic hydrolysis of phospholipids as a means of determining egg in foods. Analyst *86*, 544.

GUILBAULT, G. G. 1966A. Use of enzymes in analytical chemistry. Anal. Chem. *38*, 527R–546R.

GUILBAULT, G. G. 1966B. Symposium on bioelectrochemistry of microorganisms. III. Electrochemical analysis of enzymatic reactions. Bacteriol. Rev. *30*, 94–100.

HALLIWELL, G. 1957A. Cellulolysis by rumen microorganisms. J. Gen. Microbiol. *17*, 153–165.

HALLIWELL, G. 1957B. Cellulolytic preparations from microorganisms of the rumen and from *Myrothecium verrucaria*. J. Gen. Microbiol. *17*, 166–18?.

HALLIWELL, G. 1958. Microdetermination of cellulose in studies with cellulase. Biochem. J. *68*, 605–610.

HALLIWELL, G. 1959. The enzymic decomposition of cellulose. Nutr. Abstr. Rev. *29*, 747–759.

HALLIWELL, G. 1960. Microdetermination of carbohydrates and proteins. Biochem. J. *74*, 457–462.

HILDEBRANDT, F. M. 1947. Recent progress in industrial fermentation. Advan. Enzymology *7*, 557–615.

HILL, R. L. 1965. Hydrolysis of proteins. Advan. Protein Chem. *20*, 37–107.

HLYNKA, I. (Editor). 1964. Wheat Chemistry and Technology, Monograph Ser., Vol. 3, Am. Assoc. Cereal Chemists, St. Paul, Minn.

HOLMAN, R. T. 1955. Measurement of lipoxidase activity. Methods Biochem. Analy. *2*, 113–119.

INTERN. UNION BIOCHEM. 1961. Report of the Commission on Enzymes of the Intern. Union Biochem. Pergamon Press, Oxford, England.

JOHNSON, J. A., and MILLER, B. S. 1953. Enzyme systems important in breadmaking. Wallerstein Lab. Commun. *16*, 105–117.

JOHNSTON, J. M. 1963. Recent developments in the mechanism of fat absorption. Advan. Lipid Res. *1*, 105–131.

JOSLYN, M. A. 1949. Enzyme activity in frozen vegetable tissue. Advan. Enzymology *9*, 613–652.

JOSLYN, M. A. 1964. Enzymes in food processing. *In* Food Processing Operations, Vol. 2, M. A. Joslyn, and J. L. Heid (Editors). Avi Publishing Co., Westport, Conn.

JOSLYN, M. A., and PONTING, J. D. 1951. Enzyme-catalyzed oxidative browning of fruit products. Advan. Food Res. *3*, 1–44.

KAY, G. 1968. Insolubilized enzymes. Process Biochem. *3*, No. 8, 36–39.

KELLER, S., and BLOCK, R. J. 1960. Separation of proteins. *In* A Laboratory Manual of Analytical Methods of Protein Chemistry. I. The Separation and Isolation of Protein, P. Alexander, and R. J. Block (Editors). The Macmillan Publishing Co., New York.

KIERMEIER, F., KERN, R., and WILDBRETT, G. 1962. Influence of organic insecticides on enzymes. I. Problems and methodology. Z. Lebensm. Unters. Forsch. *118*, 201–214.

LAKON, G. 1942. Topographical detection of viability of cereal seeds with tetrazolium salts. Ber. Deut. Bot. Ges. *60*, 299–305.

LEE, E. Y. C., and WHELAN, W. J. 1966. Enzymic methods for the microdetermination of glycogen and amylopectin and their unit-chain lengths. Arch. Biochem. Biophys. *116*, 162–167.

LIN, F. M., POMERANZ, Y., and SHELLENBERGER, J. A. 1964. Determination of protein-bound glutamine and asparagine. Proc. 6th Intern. Congr. Biochem. *2*, 117.

LINKO, P. 1960. The biochemistry of grain storage. Cereal Sci. Today *10*, 302–306.

LUNDQUIST, F. 1959. The determination of ethyl alcohol in blood and tissues. Methods Biochem. Analy. *7*, 217–251.

MACGEE, J. 1959. Enzymic determination of polyunsaturated fatty acid. Anal. Chem. *31*, 298.

MASON, W. B. 1965. Bioanalytical techniques. Rep. 18th Annual Analytical Chemistry Summer Symposium. Anal. Chem. *37*, 1755–1758.

MAURON, J. 1961. The concept of amino acid availability and its bearing on protein evaluation. *In* Progress in Meeting Needs of Infants and Preschool Children. Natl. Acad. Sci. Natl. Res. Council, Publ. *843*. Washington, D. C.

MAURON, J., and MOTTU, F. 1958. Relation between *in vitro* lysine availability and *in vivo* protein evaluation in milk powders. Archives Biochem. Biophys. *77*, 312–327.

MAURON, J., MOTTU, F., BUJARD, E., and EGLI, R. H. 1955. The availability of lysine, methionine, and tryptophan in condensed milk and milk powder. *In vitro* digestion studies. Arch. Biochem. Biophys. *59*, 433–451.

MOHLER, H. 1958. The significance of enzymes in food industries. Mitt. Gebiete Lebensmittunters. Hyg. *49*, 406–422.

MORRISON, A. B., and RAO, N. 1966. Measurement of the nutritional availability of amino acids in foods. Advan. Chem. Ser. *57*, 159–177.

NATL. CANNERS ASSOC. 1954. Bull. *27-L*. Berkeley, Calif.

NEISH, W. J. P. 1957. α-Keto acid determinations. Methods Biochem. Analy. *5*, 107–179.

NOMOTO, M., NARAHASHI, Y., and MURAKAMI, M. 1960A. A proteolytic enzyme of *Streptomyces griseus*. VI. Hydrolysis of protein by *Streptomyces griseus* protease. J. Biochem. (Tokyo) *48*, 593–602.

NOMOTO, M., NARASHASHI, Y., and MURAKAMI, M. 1960B. A proteolytic enzyme of *Streptomyces griseus*. VII. Substrate specificity of *Streptomyces griseus* protease. J. Biochem. (Tokyo) *48*, 906–918.

PEYNAUD, E., BLOUIN, J., and LAFON-LAFOURCADE, Y. 1966. Review of applications of enzymatic methods to the determination of some organic acids in wines. Am. J. Enol. Viticul. *17*, 218–224.

PLESNER, P., and KALCKAR, H. M. 1956. Enzymic microdeterminations of uric acid, hypoxanthine, xanthine, and adenine, and xanthopterine by UV spectroscopy. Methods Biochem. Analy. *3*, 97–109.

POMERANZ, Y. 1964A. Lactase (beta-D-galactosidase). I. Occurrence and properties. Food Technol. *18*, 88–93.

POMERANZ, Y. 1964B. Lactase (beta-D-galactosidase). II. Possibilities in the food industries. Food Technol. *18*, 96–103.

POMERANZ, Y. 1966. The role of enzyme additives in breadmaking. Brot Geback *20*, 40–45.

POMERANZ, Y. 1968. Relation between chemical composition and breadmaking potentialities of wheat flour. Advan. Food Res. *16*, 335–455.

POMERANZ, Y. (Editor). 1971. Wheat: Chemistry and Technology, Second Edition. Am. Assoc. Cereal Chemists, St. Paul, Minn.

PULLEY, J. E. 1969. Enzymes simplify processing. Food Eng. *41*, No. 2, 68–71.

REED, G. 1966. Enzymes in Food Processing. Academic Press, New York.

SANGER, F. 1952. The arrangement of amino acids in proteins. Advan. Protein Chem. *7*, 2–67.

SANGER, F. 1956. The structure of insulin. *In* Currents in Biochemical Research, D. E. Green (Editor). Interscience Publishers, New York.

SCHEFFNER, A. L., ECKFELD, G. A., and SPECTOR, H. 1956. The pepsin-digest-residue (PDR) amino acid index of net protein utilization. J. Nutr. *60*, 105–120.

SCHORMULLER, J. 1957A. Analysis of fermentation in food research. Nahrung *1*, 40–52. (German)

SCHORMULLER, J. 1957B. Inhibition of fermentation as a basis of toxic activities. I. Principles, metabolites as inhibitors, protein and hormone antagonists. Z. Lebensm. Unters. Forsch. *106*, 372–392. (German)

SCHORMULLER, J. 1958A. Inhibition of fermentation as a basis of toxic activities. II. Antibiotics, animal poisons, halogen, and phosphorus-containing antagonists. Z. Lebensm. Unters. Forsch. *107*, 40–56. (German)

SCHORMULLER, J. 1958B. Inhibition of fermentation as a basis of toxic activities. III. Pesticides, surface active compounds and heavy metals. Z. Lebensm. Unters. Forsch. *107*, 257–264. (German)

SCHORMULLER, J. 1958C. Inhibition of fermentation as a basis of toxic activity. IV. Disinfectants, irradiation, and drugs. Z. Lebensm. Unters. Forsch. *107*, 312–368. (German)

SCHORMULLER, J. 1968. Significance and limits of enzymatic methods in food analysis. Z. Anal. Chemie. *243*, 613–624. (German)

SCHULTZ, H. W. 1960. Food Enzymes. Avi Publishing Co., Westport, Conn.

SCHULTZ, H. W. 1962. Lipids and Their Oxidation. Avi Publishing Co., Westport, Conn.

SCHULTZ, H. W., and Anglemier, A. F. 1964. Proteins and Their Reactions. Avi Publishing Co., Westport, Conn.

SCHWARTZ, M. K., and BODANSKY, O. 1963. Automated methods for determination of enzyme activity. Methods Biochem. Analy. *11*, 211–246.

SELA, M., and KATCHALSKI, E. 1959. Biological properties of poly-α amino acids. Advan. Protein Chem. *14*, 391–478.

SOC. CHEM. IND. 1961. Production and Application of Enzyme Preparation in Food Manufacture, Monograph No. *11*. London, England.

TALALAY, P. 1960. Enzymic analysis of steroid hormones. Methods Biochem. Analy. *8*, 119–143.

TAUFEL, K. 1965. The perspectives of enzymes in foods and nutrition. Nahrung *9*, 265–285.

TOWER, D. B., PETERS, E. L., and WHERRETT, J. R. 1962. Determination of protein-bound glutamine and asparagine. J. Biol. Chem. *237*, 1861–1869.

UMBREIT, W. W., BURRIS, R. H., and STAUFFER, J. F. 1957. Manometric Methods. Burgess Publishing Co., Minneapolis, Minn.

CHAPTER 31

Analytical Microbiology

Principles of Microbiological Assays

Based on the similarity in nutritive requirements of microorganisms and experimental animals, it is possible to use microorganisms to determine quantitatively many of the substances which are known to be essential constituents of all living cells (Snell 1945). The requirements of certain microorganisms for specific nutritional factors reflect a loss of their ability to synthesize those factors (Snell 1946). At present, microorganisms are known which require each of the water-soluble vitamins (with the possible exception of ascorbic acid), and each of the amino acids required by higher animals. The fundamental similarity in the metabolic requirements of various organisms has facilitated greatly the identification of nutritional essentials of both microorganisms and animals. Similarly, the application of microorganisms to the quantitative determination of the vitamins (Peterson and Peterson 1945; Koser 1948; Snell 1948, 1949, 1950), and the amino acids, purines, and pyrimidines (Snell 1945, 1948, 1949; Hendlin 1952) which they require is rapidly extending our knowledge of the distribution and importance of those substances in nature. The recognition that the effectiveness of antibiotic agents may sometimes be due to their interaction with essential cellular metabolites, and that antibacterial agents can often be fashioned by varying the structure of known essential metabolites in a suitable manner has served further to intensify interest in microbial nutrition (Knight 1945; Snell 1946).

Analytical microbiology has been defined as the branch of microbiology in which microorganisms are used as reagents for the quantitative determination of certain chemical compounds (Gavin 1956). As an analytical tool, it has proved invaluable for the assay of vitamins, amino acids, growth factors, and antibiotics. Microbiological assays have been reviewed in several publications (Peterson and Peterson 1945; Snell 1945, 1948, 1950, 1952; Hendlin 1952; Kersey and Fink 1954; Gavin 1956, 1957A, B, 1958, 1959; Hutner *et al.* 1958; Sokolski and Carpenter 1959; Kavanagh 1960; Oberzill 1968A, B). Methods used in microbiological assay of vitamins in clinical chemistry were reviewed by Baker and Sobotka (1962). Use of statistics in microbiological assays was discussed by Knudsen (1950) and Bliss (1956). A comprehensive book on theoretical and applied aspects of analytical microbiology was edited by Kavanagh (1963).

A microbiological assay depends upon the observation that in the presence of limiting amounts of certain compounds, the amount of growth is a function of the amount of these compounds. Although test conditions vary with the different assay methods, the fundamental techniques in most of the assays are the same. The test substance is added to a liquid or gel medium, inoculated with the microorganism, and growth stimulation or growth inhibition is measured. The response measured depends on the biochemical effect of the substance on the metabolism of the organism. It may be a growth response (generally positive in the assay of nutrients and negative in antibiotics) that is measured by numerical counts, optical density, weight, or area; the growth response may be a definite end point, or an all-or-none response. In the metabolic response (either positive or negative), metabolic products or change in some function may be measured. Among the measurable metabolic responses are acid production, carbon dioxide production, oxygen uptake, reduction of nitrates, hemolysis of red blood cells, antiluminescent activity, or inhibition of spore germination (Gavin 1956). Cancellieri and Morpurgo (1962) proposed an interesting new method for the determination of amino acids, vitamins, and purine and pyrimidine bases. The growth of wild, protrophic strains of bacteria is inhibited by substances that are analogous and antagonistic to the regular metabolites. By adding graded amounts of the nutrilite to reactivate the growth, the nutrilite content can be assayed. The usefulness of this method has yet to be confirmed.

The Microorganism

The ideal test organism should: (1) be sensitive to the substance being assayed; (2) be easily cultivated; (3) have some metabolic function or response that is readily measurable; and (4) not be susceptible during the assay to variation in either sensitivity or phase (Gavin 1956). It is also desirable that the organism be nonpathogenic and have reasonable specificity. Rapid growth is often advantageous in the speed of obtaining the results and increasing the precision. For convenience, an organism should be suitable for the assay of several compounds, although the most exacting organisms are often either less stable or more difficult to cultivate. For the

examination of the amino acid content of unknowns, particularly un-
purified materials, it is advantageous to introduce into assay tubes
as little extraneous material as possible.   Consequently, sensitivity
of the organism for an assayed compound may be important
(Kavanagh 1963).

The microorganisms used for assay include bacteria, yeasts, fungi,
and protozoa.   The use of bacteria generally poses fewer problems
than the use of other microorganisms.   They have been used for the
assay of proteins, amino acids, carbohydrates, vitamins, and for the
evaluation of antiseptics, disinfectants, and chemotherapeutic
agents.   Lactic acid bacteria (including the genera *Lactobacillus*,
Streptococcus, and *Leuconostoc*) are equal or surpass in complexity of
nutritional requirements all other groups of microorganisms.   The
complex nutritional needs of the lactic acid bacteria for one or more
of the B vitamins, purines, pyrimidines, and an array of amino acids,
as well as their ability to dissimilate carbohydrates to an easily
measurable end point make them exceptionally useful in microbiological assays (Hendlin 1952).   Assay procedures using yeasts
offer the advantages of simplicity and speed, and have been developed for some vitamins.   Yeasts thrive under aerobic conditions
and require the use of shakers.   Fungi do not have as wide an application as assay organisms as the bacteria.   Several methods have
been developed for the assay of antibiotics, trace metals, and
vitamins.   However, fungal growth is generally slow.   Fungi are
used in a number of procedures where no other organism is available,
or where fungicidal or fungistatic evaluation is desired.

Protozoa have a variety of nutritional requirements including
amino acids, vitamins, nucleic acid derivatives, lipid factors, and
hormones.   The proteolytic enzymes elaborated by some protozoans
make them useful in the assay of proteins and of the biological
availability of the proteins.

One of the perennial problems in any work with microorganisms is
the preservation of the culture with a particular physiology (Kavanagh 1960).   The culture must be maintained in such a manner
that it will give a similar response each time it is used in an assay
procedure.   The simplest way to preserve a bacterial culture is in
stabs or agar slants with transfer at monthly intervals.   Another
popular method of preservation is by means of lyophilization.   Over
a long period, a culture preserved this way may differ from the
material lyophilized, because differential death may reduce the
distribution of organisms in the population.

The times of incubation for an assay range from several hours for turbidimetic assays of antibiotics to 7–8 days for assays with protozoans.   In assays that involve short incubation periods, bacterial cultures should be used in the log phase of their growth; for longer incubation periods, the age may vary from 18 to 24 hr.   Similarly, the size of inoculum is more critical in short- than in long-term incubations.   Inoculum size affects antibacterial potency of some antibiotics, and is generally more important in diffusion than in other methods.   Size of bacterial inoculum can be controlled by standardizing with a colorimeter the turbidity of the diluted inoculum.

The selected temperature for incubation should give good, but not necessarily maximum growth.   It is essential, however, that the selected temperature be maintained with little variation.

The Medium

An ideal culture medium for use in analytical procedures should meet the following requirements (Gavin 1956):

(1)   It must contain all of the factors necessary to support growth.   The test organism must be supplied with an energy source, nitrogen source, mineral salts, and in synthetic media with growth factors.

In the assays of growth factors in which a positive growth response is obtained, the culture medium must be prepared so that all of the necessary nutrients are available to the test organism with the exception of the one that will control growth.   The limiting factor is being assayed, and is supplied in graded amounts.   In the negative growth response, lack of growth must be due only to the presence of an inhibitory substance in a medium containing all the nutrients essential for normal growth and development of the test organism.

(2)   Neither the medium nor the food extract should contain stimulants, substitutes, or antagonists to the assayed material.   The water used should be demineralized, and the compounds used in preparing the medium should be pure.

(3)   The medium should have a pH which will be compatible with the activity of the assayed substance and the growth of the microorganism.

Microbiological Methods

There are four main methods of microbiological assay by which the potency of samples and standard solutions can be compared. These are *diffusion, turbidity or dilution, gravimetric,* and *metabolic response* methods.

(1)   In the diffusion method, a zone of growth or inhibition of the test organism is formed around an application point or area.   The assayed substance is allowed to diffuse through solid inoculated culture media.   The test organisms may be bacteria, bacteriophage, fungi, protozoa, or algae; and the zones may be of growth as in the assay for vitamins, amino acids, etc., or of inhibition as in the assay of antibiotics.   The size of the zone is a function of the concentration, and in certain instances of the amount of the assayed substance.   This function can be expressed as a linear relationship between the size of the zone and the logarithm of the concentration of the substance.   By measuring the distance the substance diffuses, and comparing it with that of a known standard preparation, the potency of the sample may be assayed.   Diffusion may be of two types: linear; i.e., by bringing the substance in contact with a column of seeded agar in a capillary or test tube; and radial, around a suitable reservoir on a seeded agar plate.   The linear diffusion method, although theoretically sound, is seldom used because of practical considerations.

There are two methods employing the horizontal type of diffusion —those in which the diameter of the zone of inhibition or growth depends upon the concentration, and those dependent upon the amount of substance being analyzed.   Among those depending on concentration, the *cylinder* and *cup plate methods* are used most commonly.   The small cylinders (from glass, porcelain, or metal) can be embedded to a fixed height in the agar; or in the cup plate method, the depression is formed by removing a small object previously placed in the molten agar, after the agar has solidified, or by removing a slug from solidified agar with a sterile cork borer. The analyzed substance placed in the cylinder or depression, diffuses and the diameter of the zone of inhibition or exhibition is measured and compared to a series of standard concentrations of the assayed substance.

The two methods depending upon the amount of material include the drop plate method and the paper disc method.   The tested material is placed either directly on the agar or indirectly on the agar by means of a filter paper disc.

According to Gavin (1957A) the cylinder and cup method are well-adapted to rapid assays and to testing dilute solutions; the drop and paper pad method are simple, convenient, rapid, and can be used in assaying very small samples.   The accuracy and reproducibility are the same as for the cylinder or cup plate method.

Diffusion assays can be used only for substances that diffuse into agar and are generally less precise than other microbiological methods.   Decreasing the thickness of the agar layer increases sensitivity; too moist agar will cause streaks and indistinct obscure zones; amount, quality, and origin of the agar may affect the results.   The pH of the agar and assay solutions affects the activity and stability of the tested compound and assay results.   Similarly length and temperature of incubation must be rigidly standardized.   Diffusion methods are used widely in the assay of antibiotics.   Their sensitivity is relatively low, and they are used less in the assay of amino acids and vitamins.   In general, cup plate methods require 10 to 2000 times the vitamin levels required by the conventional tube method (Hendlin 1952).   Bolinder *et al.* (1963A, B) investigated the various factors affecting the determination of amino acids (i.e., lysine) by the cup plate assay method, and recommended a procedure for increasing the assay sensitivity and reproducibility.

(2)   The distinction between dilution and turbidimetric methods is that the former gives an all-or-none end point in broth or agar, whereas the latter measures graded growth or metabolic responses. The serial dilution method for assay of antibiotics is quite important for food analysts.   It is simple in principle, easily performed, and of considerable utility (Kavanagh 1960).   Several dilutions of the tested substance in small tubes are inoculated with a test organism, incubated, and the lowest concentration of the substance which causes apparently complete inhibition of growth of the organism is taken to be the minimum inhibitory concentration.

In the turbidimetric tube assay, graded concentrations of the tested substance are added to a series of test tubes or flasks containing a liquid nutrient medium.   The medium is inoculated with the test organism and incubated for a suitable time.   The response of the test organism is measured in a photometer, the scale readings of which may be converted by a calibration curve prepared with graded amounts of pure substance to determine the potency of the assayed sample.   The turbidimetric method of assay is used for determining antibiotics, vitamins, amino acids, and other growth substances. When large numbers of samples of a known compound of approximately known concentration, and uncontaminated with interfering substances are to be assayed, then a turbidimetric method is one of choice.

There probably are fewer important factors to control in a turbidimetric than a diffusion method assay.   Turbidimetric methods are

(1) more sensitive to low dilutions than diffusion methods, (2) can be adapted to rapid assays, and (3) the results obtained at several concentration levels can be analyzed statistically. The assayed solutions must be sterile—especially with long incubation tests; a standard curve must be prepared each time an assay is run. The methods are affected by highly colored substances, though the effects can be allowed for by running uninoculated blanks at each dilution level. Some organisms clump during growth, and it is difficult to suspend them uniformly for turbidimetric measurement. When this method is used, one must be certain that the organisms are uniformly suspended by shaking before measurement.

Sterilization of the culture media can have adverse effects on turbidimetric assays. Heat sterilization may cause color changes, lower pH, and destroy certain nutrients due to amino acid-sugar interactions. Lactobacilli are relatively immune to interference from contaminants due to the rapidity of growth, essentially anaerobic conditions, and immediate acidification of the medium. With slow growing organisms such as *Tetrahymena*, with an incubation time of 6 to 8 days, special precautions must ge taken to avoid contamination. Separate sterilization of glucose solution or reduction of the glucose content will aid in overcoming the problem of color changes.

The limits of error reported in the literature for turbidimetric assays range from ± 2 to $\pm 20\%$. The majority of reports indicate that $\pm 10\%$ is average for this procedure, and that an experienced analyst exercising rigid control of the factors which contribute to variation can reduce the error to $\pm 5\%$.

(3) In gravimetric methods, the response of the test organism to graded concentrations of the analyzed substance is determined after a suitable incubation time by measurement of the amount of growth in terms of dry cell weight. Under the conditions of assay, this weight is proportional to the concentration of the limiting factor. The majority of gravimetric methods used mutants of *Neurospora* as test organisms. The methods are simple, precise, reliable, and inexpensive. Results with some mutants are very specific and can be used for assays of colored solutions. The assays are, however, time-consuming, not well-adapted to large numbers of sample, and require large incubation space. In addition, the requirements for nitrogen metabolism by the test organisms are complex and may be affected by a variety of nonspecific nitrogenous compounds. With *Neurospora* mutants, a separate strain is required for each particular assay, thus, requiring maintenance of a large number of test organisms. Occasional major changes in metabolism may occur as a

result of a single gene mutation with the resultant possibility of adaption (Gavin 1958).

(4) In the metabolic response method, the response of the test organism to various concentrations of the assayed substance is evaluated after a suitable incubation period, as a change in a specific measurable metabolic parameter. Several parameters can be measured; however, acid production is the only one used widely. Rates of growth and acid production do not run exactly parallel courses. In general, growth (as measured turbidimetrically) reaches a maximum considerably before maximum acid production has occurred. The amount of acid eventually produced is, however, closely related to the number of total cells present. Some of the precautions necessary when turbidimetric measurement is used are unnecessary if one titrates acid. Color of the sample, clumping, and turbidity development are of little consequence. With some of the heterofermentative lactic acid bacteria and with *Escherichia coli* which produce smaller amounts of acid, turbidimetric measurement of growth is normally used. With homofermentative lactobacilli, a titration is generally the preferred procedure.

Titrimetric methods require sterile samples and media, are somewhat time-consuming and tedious, and the limiting buffering capacity of the medium poses problems of limits for optimum and maximum acid production. The precision is generally $\pm 10\%$, but $\pm 3\%$ or lower can be attained with careful control of all variables.

Calculation and Reliability of Microbial Assays

Bliss (1956) described in detail the precision of assays and calculations of potency of microbiological assays. Four practices which increase the precision and reliability of many microbial assays are: (1) selecting dosage levels and a function of the growth response that lead to a straight or log-dose response line; (2) randomizing the location of culture tubes within racks, during sterilization, inoculation and incubation; (3) following an objective probability rule in identifying and rejecting supposedly aberrant observations; and (4) basing assayed levels on the means of two or more independent assays.

According to Snell (1945), the best and simplest criterion for establishing reliability of a microbiological assay method is agreement with other procedures. In running an assay, a standard curve is obtained in which growth response is plotted against concentration of the assayed compound. The sample is also assayed at several concentration levels; calculated to yield an amount of growth that falls upon the standard curve. The percent thus found should be constant regardless of the portion of the standard curve from which it was calculated. Repeated assays should give consistent values. When a known amount of a pure compound is added to a sample, it must be quantitatively recovered. High or low recoveries indicate stimulation or supression of response; in either case, the assay value is suspect. If two or more organisms which vary in other nutritive requirements yield the same value on a sample, the probability that the assay value is correct is greatly increased. It is highly unlikely that two or more different organisms would respond to the interfering materials in exactly the same manner and to exactly the same extent. Finally, the specificity of response should always be investigated prior to use of a new medium or organism for assay.

Assays of Vitamins, Amino Acids, and of Nucleic Acids and Their Derivatives

In general, the essential steps involved in the assay are: (1) preparation of media, carrying stock cultures and maintenance of the cultures; (2) preparation of the nutrilite-deficient medium; (3) preparation of the inoculum medium and inoculum culture; (4) extraction of the nutrient from the sample prior to assay (generally, various concentrations of HCl at 15 lb in an autoclave); (5) setting up the assay; (6) sterilization of the assay tubes and media; (7) inoculation with the test organism; (8) incubation; (9) determination of response to the nutrient and nutrient-containing extract; and (10) calculation of results.

The procedure used in each assay is similar. The essential steps common to all methods, but modified in detail according to the test organism being used and the nutrient to be determined were described by Snell (1945, 1948, 1950), Barton-Wright (1963, 1967), and Miller (1958). Details of the methods are reproduced in the Difco Manual (Anon. 1951); manual of methods of vitamin assay (Assoc. Vitamin Chemists, 1951); the comprehensive book on analytical microbiology edited by Kavanagh (1963); the classical reference book on cereal chemistry (Kent-Jones and Amos 1957); and several of the general official handbooks of food analysis (Am. Assoc. Cereal Chemists, 1962; Assoc. Offic. Agr. Chemists, 1965).

For assay with lactobacilli of vitamins, amino acids, and nucleic acids and their derivatives, the selected test organism is carried as stab cultures by monthly transfer in a yeast-dextrose agar. Such cultures are incubated at 37°C for 24–48 hr or until good growth is visible, and are then held in the refrigerator for the remainder of the interval between transfers. About 24 hr before an assay is to be made, a transfer is made from this culture to a tube of sterile inoculum medium. In the assay or riboflavin, this consists of the riboflavin deficient basal medium supplemented with riboflavin (1μg/-10 cc). This inoculum culture is then incubated at 37°C until used. The selected medium deficient in a single nutrilite (the one to be assayed) is prepared at twice its final concentration. The assay is generally carried out in 16 × 180 mm lipless test tubes held in metal racks which are easily autoclaved. To one series of tubes, increasing amounts of a standard solution of the assayed nutrilite are added.

To a second series of tubes, increasing amounts of neutralized extracts of the various tested samples are added. Amounts and concentrations of the nutrient in the standard solutions and in the sample extracts are calculated to produce a gradation in growth of the test organism between no growth and maximum possible growth. Contents of all tubes are diluted with water to the same volume (generally 5 cc), and an equal volume of the double strength basic medium is added. The tubes are then plugged with cotton or capped, sterilized by autoclaving, cooled to room temperature, and each tube is inoculated with a drop of the washed inoculum culture. The racks are incubated at constant temperature near the optimum of the organism used. After a sufficiently long incubation period, growth present in each tube is determined by acidimetry or turbidimetry. The growth response to increasing concentrations of the nutrients is then plotted against the concentration of the nutrient in the standard solution. The amount of nutrient present in the various samples is determined by interpolating the growth response in the sample tubes onto the standard curve.

Snell and co-workers (Snell 1945) devised on the basis of studies of the nutritional requirements of the lactic acid bacteria many of the synthetic laboratory media used today in the assay of water-soluble vitamins, amino acids, and purines and pyrimidines. To prevent the pH from shifting to levels which interfere with growth, the media are buffered. Acetate is the buffer of choice; when citrate buffers are used it is necessary to increase the metal concentration of the medium because of the strong chelating action of citrate. Generally, a fixed mixture of minerals is used though variation in the composition of the salt mixture may improve growth of some microorganisms. Though the microorganisms do not require all the water-soluble vitamins and growth factors, even the nonessential nutrilities often aid early growth. Thus, *L. arabinosus* grows well without vitamin B_1 but its development is greatly enhanced by pyridoxine, pyridoxamine, or pyridoxal. For most lactobacilli, adenine, guanine, and xanthine are interchangeable, but various organisms differ in the ease with which they utilize the individual compounds. The effect of thymine is shown only in a folic-free medium. Thymine plus purine bases replace folic acid more or less for certain organisms. Any synthetic medium for microbiolgical assay of amino acids should contain a complete quota of water-soluble vitamins and growth factors and their physiological equivalents. In assays of vitamins, either a mixture of the essential amino acids or a hydrolysate of a complete protein is used. With the availability of pure amino acids, the nutritional requirements for individual amino acids were determined. This was done by omitting each acid in turn from a mixture which supports growth and noting the effect on growth or acid production. In the Addendum section at the end of this chapter, details of a lysine assay employing a dehydrated medium manufactured commercially (Anon. 1951) are given.

Microbiological Assays and Feeding Tests

Amino Acids.—Microbiological assays of amino acids have certain advantages over other assay methods. They are highly specific and sensitive; they can differentiate between various forms of co-factors, a differentiation which cannot be made by other means. They are admirably adapted to routine work, provided the equipment, techniques and general background of a bacteriological laboratory is available (Tristram 1949). If a complete amino analysis is desired on a limited amount of material, on a more or less continuous basis, ion exchange chromatography is the primary method of choice where the personnel and equipment are available. If, however, the content of one or several amino acids is desired (rather than a complete analysis), if the configuration of the amino acid is of interest, and if facilities for column chromatography are not available, properly selected and executed microbiological assays are quite simple, specific, and accurate. Use of microbiological assays in the determination of amino acids in foods and feeds was described by Williams (1955) and Lyman et al. (1956).

Vitamins.—Microbiological methods of vitamin assay can provide useful information on the total vitamin content, aid in the identification of a compound, and aid in establishing the presence of various active forms of the nutritional cofactors. It should be realized, however, that the potency of the test material, determined by microbiological assay, may not be identical with the amount available for metabolism in man or animal. Melnick and Oser (1947) discussed the limitations of the nonbiological methods for determining vitamins. The emphasis on the part of the worker using microbiological assays has been directed toward the development of extraction and hydrolysis procedures which tend to yield maximal figures for the material being assayed. Certainly, autoclaving for 60 min at 15 lb and hydrolysis with $6N$ H_2SO_4 (recommended for biotin assays) have no counterpart in the animal organism themselves. They measure total and not available vitamin contents. The problem of adequate liberation of vitamin B_6 from its multiple bound forms in biological materials is the crux of any valid method of determination (Storvick and Peters 1964; Storvik et al. 1964). Vitamin B_6 exists in nature both in a "bound" and in a "free" state. The vitamin must first be liberated before accurate microbiological and chemical assays can be made. Biological assays are not dependent on this step, since the rat is able to utilize both bound and free forms of the vitamin. None of the several enzymic or chemical methods for hydrolysis or extraction is completely successful for liberation of vitamin B_6 from the multiple substances of complex nature encountered in foods and tissues. The estimation of the vitamin B_6 content of foods and biological materials is further complicated by the existence of several closely related active compounds (Snell 1958). In biological assays, all the forms of the vitamin (pyridoxine, pyridoxal, and pyridoxamine either free or combined with other substances such as phosphate) have similar growth-promoting activities for the rat. However, in microbiological assays, differences in organism response to the various forms are an important consideration. *Saccharomyces carlsbergensis* has generally been accepted as the organism of choice for the determination of total vitamin B_6. However, in view of the inherent nature of the various forms of the vitamin and of assay microorganisms, the inadequacies of extraction, the presence of growth-stimulating components in food extracts (Toepfer and Polansky 1964), and the interrelationships between vitamin B_6 and other components (i.e. thiamine and alanine in certain lactic acid bacteria), it is not possible to give to date an unqualified recommendation to any single microbiological method for the assay of vitamin B_6 in complex biological substances.

The microbiological assay of vitamin B_{12} has become almost a "science of its own" with a vast literature describing the search for suitable organisms, their specificities, sensitivities, and metabolism; the vitamin B_{12} forms and binding factors liberated during microbial growth; and the vitamin content of body fluids, organs, and foods

(Zucker and Zucker 1950; Jorgensen 1954; Ford and Hutner 1955; Grasbeck 1960; Skeggs 1966). Difficulties encountered in microbiological assays of folic acid resulting from nutritional interrelationships with vitamin B_{12} were reviewed by Girdwood (1960).

Microbiological assays may measure related compounds with similar activity for the microorganism but not for the higher animal. On the other hand, microorganisms may fail to measure vitamin derivatives which nevertheless are active for the animal organism and may, therefore, give erroneously low estimates. Similarly, a food may contain antimetabolites which inhibit enzymatic systems in microbiological assays, but are rendered inactive in the animal.

In summary, although the microbiological procedures can, if the proper organism is selected and the necessary precautions are used, give useful information of the vitamin content of foods it should be realized that microbiological assays generally measure total rather than available vitamin content. Feeding trials with strains of some microorganisms as the experimental animal offer the advantages of speed, precision, and small requirements of labor, space, and materials. Actual feeding tests with animals are, however, the final criterion of specificity and potency. The validity of conclusions drawn from microbiological assays depends on the extent to which they give results in agreement with feeding tests.

Microbiological Methods for Assessing the Nutritional Value of Proteins

There is little doubt today that protein quality is primarily governed by its essential amino acid content. Consequently, much basic and useful information has been gained by development of chemical and microbiological methods of amino acid analysis.

Nevertheless, evaluation of the nutritive value of a protein from knowledge of its amino acid composition still leaves much to be desired. Two reasons are responsible for this (Mauron 1961; Bender 1969); first, although we know the approximate amino acid needs of several species, these needs are variable according to the metabolic state of the subject. Second, amino acid content as revealed by classical amino acid analyses does not necessarily reflect amino acid availability to the organism. The basic assumption implied in all methods of chemical scoring for evaluation of protein value is that the total amount of amino acid as determined by classical methods is available to the organism. There is, however, good experimental evidence that in certain foods a proportion of the analytically determined amino acids may not be available for assimilation. Thus, Gupta *et al.* (1958) found that lysine availability to the weanling rat was only about 50% for corn, 70% for wheat, 85% for rice, 90–95% for spray-dried milk powder, and 68% for a roller-dried milk sample. Several workers found that lysine and, to a lesser extent, methionine and tryptophan, are made unavailable when proteins are submitted to severe heat treatment. Even under less severe conditions, as encountered in milk processing or milk storage, in extraction of oil seeds, and in industrial manufacture of fish flour, some of the amino acids essential and limiting in foods of plant origin may be made less available.

The accepted methods for measuring the nutritional value of proteins involve the use of common laboratory animals—mice, rats, chicks, dogs, and others. A description of these methods is outside the scope of the book. The interested reader may consult for additional information on this subject several excellent reviews and books (Allison 1949; Albanese 1959; Campbell 1961; Mauron 1961; Morrison and Rao 1966). Methods, nomenclature, and definitions used in such tests were described recently by Henry (1965). The following are some of the commonly used definitions:

$$\text{Biological value (BV)} = \frac{\text{Retained N} \times 100}{\text{Absorbed N}}$$

$$\text{True digestibility (TD)} = \frac{\text{Absorbed N} \times 100}{\text{N intake}}$$

$$\text{Protein efficiency ratio (PER)} = \frac{\text{Gain in body-weight}}{\text{Protein intake}}$$

$$\text{Net protein utilization (NPU)} = \frac{\text{Body N content with test protein} - \text{body N content with N-free diet}}{\text{N intake}}$$

$$\text{Net protein ratio (NPR)} = \frac{\text{Gain in body-weight with test protein} + \text{loss in body-weight with N-free diet}}{\text{Protein intake}}$$

There are many difficulties in determining *in vivo* availability of individual amino acids. These were reviewed and summarized by Mauron (1961).

Several authors have, therefore, tried to measure amino acid availability *in vitro*. This may be performed by chemical, enzymatic, and microbiological methods. The only useful chemical method developed to measure amino acid availability is that employing Sanger's reagent for determining available lysine in foods. The method is based on the assumption that the reduced availability of lysine seems to be largely due to its ϵ-NH_2 group combining with other active groups under conditions of moist heat to form a linkage that resists hydrolysis by enzymes but not by acids (Carpenter 1958). Only lysine molecules with reactive ϵ-NH_2 groups are nutritionally available, react with fluorodinitrobenzene (FDNB), and yield a colored ϵ-DNP compound which can be measured colorimetrically after acid hydrolysis. Enzymic *in vitro* methods are discussed elsewhere in this book.

The microbiological methods for assessing the nutritional value of proteins can be divided into those employing protozoans and those measuring growth of bacteria.

Protoza.—Some interesting studies have been made of the possibilities of the use of ciliated protozoa of the genus *Tetrahymena* in studies of the nutritional quality of proteins. Those protozoans are known to possess proteolytic enzymes and are able to utilize intact proteins as sources of amino acids (Grau and Carroll 1958). Studies of the nitrogen metabolism of the protozoan *T geleii* H have shown that the amino acid requirements of the organism resemble in many respects those of higher animals. The resemblance in requirements, the rapid growth, and small size (together with the ability of *Tetrahymena* to hydrolyze proteins) suggested that they would be useful in protein evaluation.

Studies by Rosen and co-workers have shown that the protozoan *Tetrahymena pyriformis* has a general nutritional and metabolic pattern similar to that of higher animals. The similarity comprises protein quality ratings, protein/energy, protein/vitamin, and protein/amino acid relationships, as well as the ability to detect thermal damage to proteins and to grow on multicomponent protein mixtures such as compounded animal feeds (Rosen 1958). Teunison (1961) used *T. pyriformis* W to evaluate the relative nutritional quality of ten protein concentrates and a mixture of amino acids. The reproducibility of the results at several nitrogen levels of the reference standard, "vitamin-free" purified casein, was established by statistical studies. Nutritional quality was estimated as an index based on the ratio of the regression coefficient of growth on the nitrogen content of the test material to the regression coefficient of growth on the nitrogen content of casein. These were estimated for the ranges 0–1.0 and 0–1.7 mg nitrogen/milliliter. The ranges were adopted because there was better growth at these concentrations for most of the nitrogen sources studied. Assays at higher nitrogen concentrations yielded further information on the nutritional quality, particularly, for diets containing mixed proteins. The procedure used by Teunison (1961) was based on a modified procedure described by Rosen et al. (1962). The basal medium contains vitamins, minerals, carbon and energy sources, purines, and pyrimidines, and the tested sample serves as a nitrogen source. Stott et al. (1963) introduced several simplifications and modifications in experimental procedure which included: (1) the use of "count" to assess growth response and compute relative nutritive values; (2) elimination of routine ammonia-nitrogen determination; (3) reduction of phosphate in the basal medium; and (4) use of the Fuchs-Rosenthal hemocytometer (2-mm depth).

The simplified procedure was useful in evaluating the relative nutritive value of proteins, the effect of thermal damage to protein and effects of amino acid composition, and supplementation with essential and limiting amino acids. Similar results were described by Baum (1966A, B). The application of Tetrahymena assays to the measurement of available lysine, methionine, arginine, and histidine was described by Stott and Smith (1966).

Despite apparent advantages in simplicity and convenience, the biological methods involving protozoans have not been widely accepted. Ford (1960) pointed out that the procedures are unfamiliar and complex compared with usual microbiological methods and not entirely suitable as analytical tools. In a comprehensive study of methods, Bunyan and Price (1960) found difficulty in obtaining replicates using protozoans. There is no doubt, however, that good replicability can be obtained with protozoans, and that collaborative studies in which biological value of proteins and availability of amino acids are evaluated under comparable conditions could bring out the merits of the various assay procedures (Boyne et al. 1967).

Bacteria.—Of bacteria, Streptococcus faecalis, S. zymogenes, and Leuconostoc mesenteroides were investigated. For those organisms, the test proteins require an in vitro predigestion with single or successive enzymes or chemical agents; the hydrolysates serve as nitrogen sources for the organism.

Halevy and Grossowicz (1953) tested 2 day pancreatin digests of various proteins as the sole nitrogen source for a strain of S. faecalis requiring 10 common essential amino acids, and determined the quantity needed to achieve half-maximum growth (as determined by photometric measurement of culture turbidities after incubation for 48 hr). The hydrolysates were further tested to determine which of the ten amino acids believed to be essential for the organism was actually responsible for the observed limitation of growth. This information was obtained by supplementation of the basal medium with 9 of the 10 essential acids and retesting the hydrolysate with the supplemented medium. The amino acid which, on omission from the basal medium, failed to allow an increased growth response to the the hydrolysate was chosen as the most deficient. Despite general agreement with rat growth tests, the validity of the microbiological test is questionable as egg albumen was found to be deficient in essential amino acids, not limiting for higher animals fed this protein. In addition, under the conditions of the test, gelatin as sole protein gave an appreciable response although it has practically no tryptophan.

L. mesenteroides P-60 was used by Horn *et al.* (1954) to determine the effects of heat treatment on the nutritive value of cottonseed proteins. The organism is known to require for growth the simultaneous presence of 15 different amino acids, including all those known to be indispensable to the growing rat or chick. *In vitro* enzyme digestions were carried out on successive days with pepsin, trypsin, and hog mucosa. The samples were also hydrolyzed with acid to obtain, for comparative purposes, hydrolysates representing complete liberation of the amino acids actually present in the sample. Abilities of the enzyme and acid hydrolysates to support growth of the organism were compared after three days of incubation by measurement of acid production. Resultant computations of "indices of protein value" agreed fairly well with values determined by rat growth, but the scope of the assay for other protein materials was not investigated.

In comparison with higher animals, possible limitations of most bacterial assays are inability of the organisms to utilize intact protein, their requirement for amino acids nonessential for higher animals, and the probable influence of peptides, stimulatory for bacteria but not for higher animals (Rosen 1958). Rogers *et al.* compared (1959) results of bacterial assays with P.E.R. estimations of several foods. With *S. faecalis* ATCC 9790, autolysis occurred in media containing hydrolysates of proteins deficient in lysine, and erratic results were encountered. Bunyan and Price (1960) compared microbiological methods with N.P.U. determinations and found no correlations between the methods on a large series of meat meals.

Ford (1960, 1962) used *Streptococcus zymogenes*, a vigorously proteolytic microorganism, to measure the availability of methionine, leucine, isoleucine, arginine, histidine, tryptophan, and valine in different food proteins. The foods were hydrolyzed with papain before assay with the microorganism. In 12 whale-meat meals, differences in nutritive quality reflected corresponding differences in the biological availability of several of the constituent amino acids. Values obtained for available methionine, tryptophan, leucine, and arginine were closely correlated with each other and with rat-assay and the "available" lysine values obtained for these materials by Bunyan and Price (1960). The results of microbiological assays were affected by fineness of grinding and conditions of enzymic predigestion of the test samples (Ford 1964; Morrison and Rao 1966). According to Morkowska-Gluzinska (1966), there is good agreement in proteins of animal origin between P.E.R. or N.P.U. values and microbiological assays by the Ford method. The usefulness of the microbiological method in assaying the nutritive value of several plant proteins was, however, very low. According to Campbell (1961), although microbiological assays for protein quality have been found useful in certain limited applications, they have inherent shortcomings which limit their general applicability as screening procedures for predicting the quality of protein in a wide range of foodstuffs.

Potentially, the most useful applications of the microbiological approach lie in the determination of limiting amino acids, effects of supplementation of plant proteins, and changes in availability of specific amino acids as a result of processing. The details of assays of available methionine with *S. zymogenes* and of available lysine and methionine with *T. pyriformis* W were described by Boyne *et al.* (1967).

Assay of Antibiotics

Although chemical methods of estimation have been worked out for most of the common antibiotics, it is often necessary to use biological methods because they are usually more sensitive and may be applied to both known and unknown antibiotics or to chemically heterogenous materials without preliminary fractionation (Kersey and Fink 1954). Microbiological assay methods have been devised for almost every antibiotic in use (Loy and Wright 1959). Many of the procedures have been standardized into precise quantitative methods that have been described in official compendia (such as Official Methods of Analysis of AOAC, Federal Antibiotic Regulations of the Food and Drug Administration, the National Formulary, and U.S. Pharmacopeia). Microbiological assay methods for antibiotics were reviewed by Sokolski and Carpenter (1959). Kirshbaum and Arret (1959) outlined details for assaying the commonly used antibiotics. Detailed information on the procedures for performing the assays can be found in the Code of Federal Regulations (Anon. 1959) and the book by Grove and Randall (1955).

One of the main problems encountered in the analysis of antibiotics is the frequent use of mixtures of 2 or more antibiotics in 1 product. Test organisms are seldom so selective that they are susceptible to one antibiotic only. However, the level at which the organism is inhibited by one antibiotic may be such that it is not affected by commonly encountered amounts of a second antibiotic. To clarify this problem, Arret et al. (1957) determined the "interference thresholds" for 10 antibiotics in 2 widely used tube methods.

Several methods are available for microbiological assaying mixtures of antibiotics. It is possible to develop in certain bacteria artificial resistance to a given antibiotic by growing it in the presence of increasing subinhibitory concentrations. Such artificial resistance is produced most easily with streptomycin. A culture made artificially resistant may lose its resistance unless it is maintained in a medium containing the particular antibiotic. However, when it is lyophilized, it cannot be cultured directly from the lyophilized state in a medium that contains the antibiotic. It must be grown first in an antibiotic-free medium before transfer to the antibiotic-containing medium (Loy and Wright 1959). In the analysis of mixtures of antibiotics, it is also possible to inactivate an interfering component or to separate the antibiotics. Penicillin is inactivated by the specific enzyme penicillinase. Other antibiotics can be inactivated selectively by chemical methods. Streptomycin can be inactivated with carbazide, dihydrostreptomycin with barium hydroxide, the tetracyclines by heating a solution at pH 8 at 100°C for 30 min, and erythromycin by heating a solution at pH 2 at 37°C for 3 hr. Separation by differential solubility is successful in some instances (Weiss et al. 1957).

Automation of the Microbiological Assay

The use of automatic methods of analysis has made significant contributions to microbiology. Several methods have been proposed for removing the tedium involved in many microbiological techniques and increasing the objectivity and precision of assays.

By feeding continuously streams of inoculum and nutrient medium and by periodically introducing antibiotic solutions, Gerke *et al.* (1960) were able to obtain dose response curves for several antibiotics. The assays measured turbidity and respiration, and were comparable with manual agar diffusion assays (Haney *et al.* 1962). Shaw and Duncombe (1963, 1965) described a method depending on measuring the carbon dioxide produced by free respiration of a bacterium during a fixed incubation time, and the depression of respiration by graded concentrations of the antibiotic. Operated at a speed of 20 instrumental responses per hour, with duplicate recording for each sample and the necessary standards, an overall throughput of six samples per hour was attained. Usefulness of turbidimetric automated assays of antibiotics was described by Platt *et al.* (1965).

Automation in assays of antibiotics by a turbidimetric method was broken down to the following steps (McMahan 1965):

A. Preparation of standard
   (1) Weighing
   (2) Diluting

B. Preparation of sample
   (1) Weighing
   (2) Recording weight
   (3) Extracting
   (4) Diluting
   (5) Recording dilution

C. Test procedure
   (1) Pipetting standard and sample solution
   (2) Inoculation
   (3) Incubation
   (4) Reading turbidity
   (5) Recording turbidity
   (6) Preparation of standard curve
   (7) Comparison of sample to standard
   (8) Calculation of results
   (9) Reporting results

An automated assay of 50 samples (400 tubes) per hour was attained. DiCuollo *et al.* (1965) described a semiautomated instrumental system for performing large plate diffusion assays. The system (1) semiautomatically dilutes the antibiotic samples; (2) applies simultaneously and quantitatively large numbers of sample solutions to the assay plate; (3) prepares a permanent photographic record of the assay results; and (4) automatically reads and records the zone sizes on digital printout and computer punch type for automated or manual computation.

A series of basic studies on application of automated analysis to the study of bacterial growth (Ferrari *et al.* 1965; Gerke 1965; Watson 1965) demonstrated the feasibility of using the available instruments for studying bacterial growth. The studies laid foundations for investigations of broader aspects of cell physiology as related to continuous flow analysis. Future developments will no doubt encompass automated microbiological assays of amino acids, vitamins, and other growth factors.

ADDENDUM

Amino Acid Assay Media[1]

Media for the microbiological assay of the amino acids, leucine, methionine, lysine and isoleucine have been prepared. A complete description of each medium is given in the individual discussions in the Difco Manual (Anon. 1951). The formulation of the media per liter is as follows:

	Gm		Gm
Bacto-dextrose	50	Guanine hydrochloride	0.02
Sodium acetate	40	Uracil	0.02
Ammonium chloride	6	Xanthine	0.02
DL-Alanine	0.4	Thiamine hydrochloride	0.001
L-Arginine hydrochloride	0.484	Pyridoxine hydrochloride	0.002
Bacto-asparagine	0.8	Pyridoxamine	
L-Aspartic acid	0.2	hydrochloride	0.0006
L-Cystine, Difco	0.1	Pyridoxal hydrochloride	0.0006
L-Glutamic acid	0.6	Calcium pantothenate	0.001
Glycine	0.2	Riboflavin	0.001
L-Histidine hydrochloride	0.124	Nicotinic acid	0.002
DL-Phenylalanine	0.2	*p*-Aminobenzoic acid,	
L-Proline	0.2	Difco	0.0002
DL-Serine	0.1	Biotin	0.000002
DL-Threonine	0.4	Folic acid	0.00002
DL-Tryptophan	0.08	Monopotassium phosphate	1.2
L-Tyrosine	0.2	Dipotassium phosphate	1.2
DL-Valine	0.5	Magnesium sulfate	0.4
Adenine sulfate	0.02	Ferrous sulfate	0.02
		Manganese sulfate	0.04
		Sodium chloride	0.02

---

[1] Courtesy, Difco Laboratories, Inc., Detroit, Mich.

Bacto Lysine Assay Medium (B422) Dehydrated

The composition of Bacto-Lysine Assay Medium is as shown above, but in addition it contains 0.2 gm of DL-methionine, 0.5 gm DL-isoleucine and 0.5 gm DL-leucine. Bacto-Lysine Assay Medium, prepared according to the formula given by Steel, Sauberlich, Reynolds and Baumann[2] is a complete dehydrated medium for the assay of L-lysine. It is free from lysine but contains all the other growth factors and amino acids necessary for the growth of *Leuconostoc mesenteroides* P-60 ATCC 8042. The addition of L-lysine in specified increasing concentrations gives a linear growth response by *L. mesenteroides* P-60, which may be measured acidimetrically or turbidimetrically.

The following procedure is recommended for the use of Bacto-Lysine Assay Medium:

Stock cultures of the test organism, *L. mesenteroides* P-60 are prepared by stab inoculation of Bacto-Micro Assay Culture Agar. After 24–48 hr incubation at 35°-37°C, the tubes are stored in the refrigerator. Transplants are made at monthly intervals in triplicate.

Inoculum for assay is prepared by subculturing from a stock culture to 10 ml of Bacto-Micro Inoculum Broth. After 16–24 hr incubation at 35°–37°C, the cells are centrifuged under aseptic conditions, and the supernatant liquid decanted. The cells are resuspended in 10 ml sterile isotonic sodium chloride solution. The cell suspension is diluted 5–100 with sterile isotonic sodium chloride. One drop of the latter suspension is used to inoculate each of the assay tubes (10 ml).

It is essential that a standard curve be constructed each time an assay is run, since conditions of autoclaving, temperature of incubation, etc., which influence the standard curve readings, cannot be duplicated exactly from time to time. A standard curve is obtained by using L-lysine at levels of 0, 30, 60, 90, 120, 150, 180, 240, and 300 μg per assay tube (10 ml).

The concentrations of lysine required for the preparation of the standard curve may be prepared by dissolving 6.0 gm L-lysine in 1000 ml distilled water. This is the stock solution (6000 μg/ml). Dilute the stock solution by adding 1-ml to 99-ml distilled water. Use 0.0, 0.5, 1.0, 1.5, 2.0, 2.5, 3.0, 4.0 and 5.0 ml per tube The stock solution is stable for 2 months when stored at 2°–6°C under toluene.

Bacto-Lysine Assay Medium may be used for both turbidimetric and acidimetric analyses. Turbidimetric readings should be made

[2] J. Biol. Chem. *177*, 533 (1949).

after 16–20 hr at 35°–37°C; acidimetric determinations are made after 72 hr incubation. The most effective assay range, using Bacto-Lysine Assay Medium, has been found to be between 30 and 240 μg of L-lysine.

To rehydrate the medium, suspend 105 gm of Bacto-Lysine Assay Medium in 1000 ml distilled water, and heat to boiling for 2–3 min. The slight precipitate which forms should be evenly distributed by shaking. Five milliliters of the medium is added to each tube in the preparation of the tubes for the standard curve and to each tube containing material under assay. For the assay, each tube must contain 5 ml of rehydrated medium, increasing amounts of the standard or the unknown, and sufficient distilled water to give a total volume of 10 ml per tube. The tubes are autoclaved for 10 min at 15 lb pressure (121°C). Oversterilization of the medium will give unsatisfactory results.

One hundred grams of Bacto-Lysine Assay Medium will make 1.9 liters of final medium.

Bacto Micro Assay Culture Agar (B319) Dehydrated

	Gm
Bacto-Yeast extract	20
Proteose peptone No. 3, Difco	5
Bacto-dextrose	10
Monopotassium phosphate	2
Sorbitan monooleate complex	0.1
Bacto-agar	10

To rehydrate the medium, suspend 47 gm of Bacto-Micro Assay Culture Agar in 1000 ml of cold distilled water, and heat to boiling to dissolve the medium completely. The medium is then distributed in 10-ml quantities in tubes of 16–20 mm diam. Sterilize in the autoclave for 15 min at 15 lb pressure (121°C). The tubes are then allowed to cool in an upright position. Final reaction of the medium will be pH 6.7.

One hundred grams of Bacto-Micro Assay Culture Agar will make 2.1 liters of medium.

Bacto Micro Inoculum Broth (B320) Dehydrated

Same as Bacto Micro Assay Culture Agar, except that it contains no agar.

To rehydrate the medium, dissolve 37 gm of Bacto-Micro Inoculum Broth in 1000 ml distilled water. Distribute in 10-ml quantities in tubes of 16–20 mm diam. Sterilize in the autoclave for 15

min at 15 lb pressure (121 °C). Final reaction of the medium will be pH 6.7.
One hundred grams of Bacto-Micro Inoculum Broth will make 2.7 liters of medium.

BIBLIOGRAPHY

ALBANESE, A. A. 1959. Protein and Amino Acid Nutrition. Academic Press, New York.
ALLISON, J. B. 1949. Biological evaluation of proteins. Advan. Protein Chem. 5, 155–200.
AM. ASSOC. CEREAL CHEMISTS. 1962. Cereal Laboratory Methods, 7th Edition. Am. Assoc. Cereal Chemists, St. Paul, Minn.
ANON. 1951. Difco Manual, 3rd Edition. Difco Laboratories, Detroit.
ANON. 1959. Compilation of regulations for tests and methods of assay and certification of antibiotic and antibiotic-containing drugs. Federal Register 141, No. 21CFR. Washington, D.C.
ARRET, B., WOODWARD, M. R., WINTERMERE, D. M., and KIRSHBAUM, A. 1957. Antibiotic interference thresholds of microbial assays. Antibiot. Chemotherapy 7, 545–548.
ASSOC. OFFIC. AGR. CHEMISTS. 1965. Official Methods of Analysis of the Association of Official Agricultural Chemists, 10th Edition. Assoc. Offic. Agr. Chemists, Washington, D.C.
Assoc. Vitamin Chemists. 1951. Methods of Vitamin Assay. Interscience Publishers, New York.
BAKER, H., and SOBOTKA, H. 1962. Microbiological assay methods for vitamins. Advan. Clin. Chem. 5, 147–235.
BARTON-WRIGHT, E. C. 1963. Practical Methods for the Microbiological Assay of the Vitamin B complex and Amino Acids. United Trade Press, London.
BARTON-WRIGHT, E. C. 1967. The microbiological assay of certain vitamins in compound feeding stuffs. J. Assoc. Public Analysts 5, 8–23.
BAUM, F. 1966A. Determining the biological value of proteins with Tetrahymena pyriformis W. II. The protein value of foods and feeds. Die Nahrung. 10, 453–459.
BAUM, F. 1966B. Determining the biological value of proteins with Tetrahymena pyriformis W. III. The effects of amino acid supplementation and storage on protein value of cereal flours. Die Nahrung 10, 571–580.
BENDER, A. E. 1969. Newer methods of assessing protein quality. Chem. Ind. (London), 904–909.
BLISS, C. I. 1956. The calculation of microbial assays. Bacteriol. Rev. 20, 243–258.
BOLINDER, A. E., LIE, S., and ERICSON, L. E. 1963A. Plate assay methods for amino acids. I. A sensitive cup plate assay for lysine with Streptococcus faecalis ATCC 6057. Biotechnol. Bioeng. 5, 131–146.
BOLINDER, A. E., LIE, S., and ERICSON, L. E. 1963B. Plate assay methods for amino acids. II. Factors affecting the cup plate assay for lysine with Streptococcus faecalis ATCC 6057. Biotechnol. Bioeng. 5, 147–165.
BOYNE, A. W., PRICE, S. A., ROSEN, G. D., and STOTT, J. A. 1967. Protein quality of feeding stuffs. IV. Progress report on collaborative studies on the microbiological assay for available amino acid.. Brit. J. Nutr. 21, 181–206.
BUNYAN, J., and PRICE, S. A. 1960. Studies on protein concentrates for animal feeding. J. Sci. Food Agr. 11, 25–37.
CAMPBELL, J. A. 1961. Methodology of protein evaluation—A critical appraisal of methods for evaluation of protein in foods. Nutrition document R. 10/Add 37 WHO/FAO/UNICEF—PAG.

CANCELLIERI, M. F. P., and MORPURGO, G. 1962. A new microbiological method for amino acid assay. Sci. Rept. Ist, Super Sanita 2, 336–344.

CARPENTER, K. J. 1958. Chemical methods of evaluating protein quality. Proc. Nutr. Soc. 17, 91–100.

DI CUOLLO, C. J., GUARINI, J. R., and PAGANO, J. F. 1965. Automation of large plate agar microbiological diffusion assays. N.Y. Acad. Sci. Ann. 130, 672–679.

FERRARI, A., GERKE, J. R., WATSON, R. W., and UMBREIT, W. W. 1965. Application of automated analysis to the study of bacterial growth. I. The instrumental system. N.Y. Acad. Sci. Ann. 130, 704–721.

FORD, J. E. 1960. A microbiological method for assessing the nutritional value of proteins. Brit. J. Nutr. 14, 485–497.

FORD, J. E. 1962. A microbiological method for assessing the nutritional value of proteins. II. The measurement of "available" methionine, leucine, isoleucine, arginine, histidine, tryptophan, and valine. Brit. J. Nutr. 16, 409–425.

FORD, J. E. 1964. A microbiological method for assessing the nutritional value of proteins. III. Further studies on the measurement of available amino acids. Brit. J. Nutr. 18, 449–460.

FORD, J. E., and HUTNER, S. H. 1955. Role of vitamin B₁₂ in the metabolism of microorganisms. Vitamins Hormones 13, 101–136.

GAVIN, J. J. 1956. Analytical microbiology. I. The test organisms. Appl. Microbiol. 4, 323–331.

GAVIN, J. J. 1957A. Analytical microbiology. II. The diffusion methods. Appl. Microbiol. 5, 25–33.

GAVIN, J. J. 1957B. Analytical microbiology. III. Turbidimetric methods. Appl. Microbiol. 5, 235–243.

GAVIN, J. J. 1958. Analytical microbiology. IV. Gravimetric methods. Appl. Microbiol. 6, 80–85.

GAVIN, J. J. 1959. Analytical microbiology. V. Metabolic response methods. Appl. Microbiol. 7, 180–192.

GERKE, J. R. 1965. Application of automated analysis to the study of bacterial growth. II. Continuous analysis of microbial nucleic acids and proteins during normal and antibiotic inhibited growth. N.Y. Acad. Sci. Ann. 130, 722–732.

GERKE, J. R., HANEY, T. A., and PAGANO, J. F. 1960. Automation of the microbiological assay of antibiotics with an Autoanalyzer instrumental system. N.Y. Acad. Sci. Ann. 87, 782–791.

GIRDWOOD, R. H. 1960. Folic acid, its analogs and antagonists. Advan. Clin. Chem. 3, 236–297.

GRASBECK, R. 1960. Physiology and pathology of vitamin B absorption, distribution, and excretion. Advan. Clin. Chem. 3, 299–366.

GRAU, C. R., and CARROLL, R. W. 1958. Evaluation of protein quality, In Processed Plant Protein Foodstuffs, A. M. Altschul (Editor). Academic Press, New York.

GROVE, D. C., and RANDALL, W. A. 1955. Assay Methods of Antibiotics. Medical Encyclopedia, New York.

GUPTA, J. D., DAKROURY, A. M., HARPER, A. E., and ELVEHJEM, C. A. 1958. Biological availability of lysine. J. Nutr. 64, 259–270.

HALEVY, S., and GROSSOWICZ, N. 1953. A microbiological approach to nutritional evaluation of protein. Proc. Soc. Exptl. Biol. Med. 82, 567–571.

HANEY, T. A., GERKE, J. R., MADIGAN, M. E., and PAGANO, J. F. 1962. Automated microbiological analyses for tetracycline and polyenes. N. Y. Acad. Sci. Ann. 93, 627–639.

HENDLIN, D. 1952. Symposium on lactic acid bacteria. III. Use of lactic acid bacteria in microbiological assays. Bacteriol. Rev. 16, 241–246.

HENRY, K. M. 1965. A comparison of biological methods with rats for describing the nutritive value of proteins. Brit. J. Nutr. *19*, 125–135.

HORN, M. J., BLUM, A. E., and WOMACK, M. J. 1954. Availability of amino acids to microorganisms. II. A short microbiological method of determining protein value. J. Nutr. *52*, 375–381.

HUTNER, S. H., CURY, A., and BAKER, H. 1958. Microbiological assays. Anal. Chem. *30*, 849–867.

JORGENSON, E. H. 1954. Microbiological assay of vitamin B_{12}. Methods Biochem. Analy. *1*, 81–113.

KAVANAGH, F. 1960. A commentary on microbiological assaying. Advan. Appl. Microbiol. *2*, 65–93.

KAVANAGH, F. 1963. Analytical Microbiology. Academic Press, New York.

KENT-JONES, D. W., and AMOS, A. J. 1957. Modern Cereal Chemistry, 5th Edition. The Northern Publishing Co., Liverpool, England.

KERSEY, R. C., and FINK, F. C. 1954. Microbiological assay of antibiotics. Methods Biochem. Analy. *1*, 53–79.

KIRSHBAUM, A., and ARRET, B. 1959. Outline of details for assaying the commonly used antibiotics. Antibiot. Chemotherapy *9*, 613–617.

KNIGHT, B. C. J. G. 1945. Growth factors in microbiology; some wider aspects of nutritional studies with microorganisms. Vitamins Hormones *3*, 105–228.

KNUDSEN, L. F. 1950. Statistics in microbiological assay. N. Y. Acad. Sci., Ann. *52*, 889–902.

KOSER, S. A. 1948. Growth factors for microorganisms. Ann. Rev. Microbiol. *2*, 121–142.

LOY, H. W., and WRIGHT, W. W. 1959. Microbiological assay of amino acids, vitamins, and antibiotics. Anal. Chem. *31*, 971–974.

LYMAN, C. M., KUIKEN, K. A., and HALE, F. 1956. Essential amino acid content of farm feeds. J. Agr. Food Chem. *4*, 1008–1013.

MAURON, J. 1961. The concept of amino acid availability and its bearing on protein evaluation. *In* Progress in Meeting Needs of Infants and Pre-school Children, Publ. *843*. Natl. Acad. Sci., Natl. Res. Council, Washington, D.C.

McMAHAN, J. R. 1965. A new automated system for microbiological assays. N.Y. Acad. Sci. Ann. *130*, 680–685.

MELNICK, D., and OSER, B. L. 1947. Physiological availability of the vitamins. Vitamins Hormones *5*, 39–92.

MILLER, H. K. 1958. The microbiological assay of nucleic acids and their derivatives. Methods Biochem. Analy. *6*, 31–62.

MORKOWSKA-GLUZINSKA, W. 1966. Use of the *Streptomyces zymogenes* microbiological method in evaluating the nutritional value of proteins from selected products. Roczniki PZH *17*, 467–475.

MORRISON, A. B., and RAO, M. N. 1966. Measurement of the nutritional availability of amino acids in foods. Advan. Chem. Ser. *57*, 159–177.

OBERZILL, W. 1968A. Microbiological assay methods of vitamins and amino acids. I. General principles. Sci. Pharm. *36*, 30–49.

OBERZILL, W. 1968B. Microbiological assay methods of vitamins and amino acids. II. Special methods for vitamins. Sci. Pharm. *36*, 199–218.

PETERSON, N. H., and PETERSON, M. S. 1945. Relation of bacteria to vitamins and other growth factors. Bacteriol. Rev. *9*, 49–109.

PLATT, T. B., GENTILE, J., and GEORGE, M. J. 1965. An automated turbidimetric system for antibiotic assay. N.Y. Acad. Sci. Ann. *130*, 664–671.

ROGERS, C. G., McLAUGHLAN, J. M., and CHAPMAN, D. G. 1959. Evaluation of protein in foods. III. A study of bacterial methods. Can. J. Biochem. Physiol. *37*, 1351–1360.

ROSEN, G. D., 1958. The microbiological assay of protein quality. *In* Proceedings of International Symposium on Microchemistry, D. Wilson (Editor). Pergamon Press, London.

ROSEN, G. D., STOTT, J. A., and SMITH, A. 1962. Microbiological assays of amino acids and intact proteins. Cereal Sci. Today 7, 36–39.

SHAW, W. H., and DUNCOMBE, R. E. 1963. Continuous automatic microbiological assay of antibiotics. Analyst 88, 694–701.

SHAW, W. H., and DUNCOMBE, R. E. 1965. Continuous automatic microbiological assay of antibiotics. N.Y. Acad. Sci. Ann. 130, 647–656.

SKEGGS, H. R. 1966. Microbiological assay of vitamin B_{12}. Methods of Biochem. Analy. 14, 53–62.

SNELL, E. E. 1945. The microbiological assay of amino acids. Advan. Protein Chem. 2, 85–118.

SNELL, E. E. 1946. Growth factors for microorganisms. Ann. Rev. Biochem. 15, 375–391.

SNELL, E. E. 1948. Nutritional requirements of the lactic acid bacteria. Wallerstein Lab. Commun. 11, 80–104.

SNELL, E. E. 1949. Nutritional requirements of microorganisms. Ann. Rev. Microbiol. 3, 97–117.

SNELL, E. E. 1950. Microbiological methods in vitamin research. In Vitamin Methods, Vol. 1, P. Gyorgy (Editor). Academic Press, New York.

SNELL, E. E. 1952. Symposium on lactic acid bacteria. II. The nutrition of lactic acid bacteria. Bacteriol. Rev. 16, 235–241.

SNELL, E. E. 1958. Chemical structure in relation to biological activities of vitamin B_6. Vitamins Hormones 16, 77–125.

SOKOLSKI, W. T., and CARPENTER, O. S. 1959. Microbiological assay. Progr. Ind. Microbiol. 1, 93–135.

STORVICK, C. A., BENSON, E. M., EDWARDS, M. A., and WOODRING, M. J. 1964. Chemical and microbiological determination of vitamin B_6. Methods Biochem. Analy. 12, 183–276.

STORVICK, C. A., and PETERS, J. M. 1964. Methods for the determination of vitamin B in biological materials. Vitamins Hormones 22, 833–854.

STOTT, J. A., and SMITH, H. 1966. Microbiological assay of protein quality with Tetrahymena pyriformis W. IV. Measurement of available lysine, methionine, arginine, and histidine. Brit. J. Nutr. 20, 663–673.

STOTT, J. A., SMITH, H., and ROSEN, G. D. 1963. Microbiological evaluation of protein quality with Tetrahymena pyriformis W. III. A simplified assay procedure. Brit. J. Nutr. 17, 227–232.

TEUNISON, D. J. 1961. Microbiological assay of intact proteins using Tetrahymena pyriformis W. I. Survey of protein concentrates. Anal. Biochem. 2, 405–420.

TOEPFER, E. W., and POLANSKY, M. M. 1964. Recent developments in the analysis for vitamin B_6 in foods. Vitamins Hormones 22, 825–832.

TRISTRAM, G. R. 1949. Amino acid composition of purified proteins. Advan. Protein Chem. 5, 83–153.

WATSON, R. W. 1965. Application of automated analysis to the study of bacterial growth. III. Regulation of growth processes in Escherichia coli by sulfur compounds. N.Y. Acad. Sci. Ann. 130, 733–743.

WEISS, P. J., ANDREW, M. L., and WRIGHT, W. W. 1957. Solubility of antibiotics in 24 solvents; use in analysis. Antibiot. Chemotherapy 7, 374–377.

WILLIAMS, H. H. 1955. "Essential" amino acid content of animal feeds. Memoir 337. Cornell Univ., Agr. Expt. Sta., Ithaca, N.Y.

ZUCKER, T. F., and ZUCKER, L. M. 1950. "Animal protein factor" and vitamin B_{12} in the nutrition of animals. Vitamins Hormones 8, 1–54.

Applications and Chemical Composition

CHAPTER 32

General Remarks

The determination of food composition is fundamental to theoretical and applied investigations in food science and technology, and is often the basis of establishing the nutritional value and overall acceptance from the consumer standpoint. The analysis may establish the amounts of one or several components or the overall composition.

The constituents to be detected or determined in food analysis may be elements, radicals, functional groups, compounds, groups of compounds, or phases. Few of the reactions used in chemical, physical, or physiochemical methods of food analysis are completely specific or selective. Sometimes, careful adjustment of pH, oxidation or reduction, or complexing of certain groups or elements make determinations possible. Mostly, however (except for specific biological methods—enzymatic, immunological), fractionation procedures must be used. Many such procedures have been developed in recent years. Advances in food analysis in the last two decades have resulted from the development of many instrumental methods, but primarily from improvements in separation methods (mainly chromatography) that, in turn, utilize more and more instrumental techniques. Those methods were described in Section II of this book.

The food analyst of today must have a knowledge of chemical, physical, and physiochemical properties of foods and of the methods used in their analyses. Such knowledge is essential to the selection or modification of analytical methods in order to meet requirements of speed, precision, and accuracy.

Generally, the food analyst knows the nature and qualitative composition of a sample. Often he also knows the approximate ranges of constituents to be determined. In all those cases, there is no need for a qualitative analysis. On few occasions will the analyst be asked to analyze a sample of totally unknown composition; in some cases, the range of a certain component may not be known, or the presence of a certain minor component (i.e., mineral) may have to be established. In all such cases, a qualitative test, that generally gives information of a semiquantitative nature, is required.

In order for an analysis to provide useful information, several criteria must be met. The information obtained must be arrived at by use of reproducible, empirically-tested (preferably standardized and based on collaborative investigations), and scientifically-sound techniques. In the selection of an appropriate technique one should consider first the information that is to be obtained to solve a specific problem. This selection will often depend on the nature of the food examined, availability of equipment, required speed and precision, and background information. The latter are, however, of secondary importance. The main goal is providing specific information for a specific purpose. Using a semiquantitative, rapid test or a simplified procedure instead of an appropriate instrumental assay cannot be justified if the substitute procedures cannot provide the information that is required. At the same time, using expensive equipment that requires costly maintenance cannot be justified if the same information can be obtained by a simple and rapid procedure. In any analysis, the number of assays should be reduced to a minimum of most informative tests. Scores of uniformative data are not only costly, but also complicate evaluation.

In determining the overall composition of a food sample or content of a specified component, four steps are involved: (1) obtaining a representative subsample for analysis, (2) converting the component(s) into a form that permits an assay, (3) the assay, and (4) calculation and interpretation of data.

The analyst often assumes that the sample he is about to analyze is homogenous. It is advisable that before starting a determination, the whole sample be remixed to eliminate heterogeneity—mainly in particle size and moisture distribution. In some foods, such as concentrated sugar solutions, honey, etc., the sample must be heated carefully to dissolve sugar crystals. Some foods pose special problems of sample preparation. The difficulty of obtaining a representative sample from a heterogeneous substance such as a meat carcass is well-known. In addition, incorporation of fatty tissue with the lean meat to obtain a homogenous mixture requires efficient mincing, mixing, and subdivision of the total sample. In some cases, complete homogenization is actually undesirable. Thus, in determining the composition of an ice cream base mix, insoluble particles (fruit, nuts, confectionery) are removed from the melted sample on a sieve. Similarly, in analyzing chocolate lipids it is essential to remove nut particles.

In many foods the composition is determined in the edible portion and the inedible part is carefully removed, weighed, and often reported. It is often impossible or undesirable to incorporate and distribute all impurities uniformly in the whole sample. Thus, in grain analyses, large-sized impurities (straw, stones) are removed by sieving or mechanical cleaning from a large sample, and the partly purified sample is then subdivided for further analysis.

Once a suitable sample is obtained, a representative part of it is to be prepared for the actual assay. This stage permits a rapid and simple critical examination of the food. In evaluating cereal grains, for example, many old and simple tests are used. They include: (1) *feeling* the temperature of the grain in a sack or bin to determine whether the grain is heating; (2) *visually inspecting* a handful of grain to assess its purity and presence of foreign or objectionable material; (3) *smelling* the grain to establish off flavor or off odor; and (4) *chewing and masticating* the grain to get some indication of its vitreousness, hardness, and approximate protein-gluten contents.

Admittedly, the above "tests" cannot establish the precise composition, potential use, and optimum processing. Yet, they continue to have a place in many evaluation schemes. Those simple tests provide useful and important supplementary information. They are good screening tests and often can save costly, lengthy, and laborious analyses.

For a complete characterization of a totally unknown food sample, the following sequence of treatments is recommended: sublimation, steam distillation, and extraction with petroleum ether followed by ether, chloroform, ethanol (absolute or 80%), and water (cold, warm, acidified, and alkaline). Steam distillation is carried out on a separate sample aliquot. For fresh tissues, dehydration with ethanol may be required for efficient fat extraction. The residue from all treatments represents the insoluble that can be subsequently hydrolyzed for detailed determination. Generally, a determination of mineral components is made on the original untreated material. Every assay should include a moisture determination.

The solid and solute of most extracts are separated by filtration.

To minimize retention of solute, suction, pressure, or centrifugation can be used. Extraction by percolation gives the most complete extraction of solute from the solid.

Whatever the extractant, the extract is generally evaporated to dryness or concentrated to reduce the bulk of solvent before subsequent fractionation and characterization. If the solution does not splash or froth, evaporation can be accomplished by simply boiling down the solution. If frothing is excessive, evaporation may be carried out at temperatures below boiling point by heating from below in the usual manner, or by radiant heating from above. The rate of evaporation can be increased by blowing a current of warm air over the surface of the fluid to remove the vapor-saturated air. Preheating the air is often effective provided the flow of air is carefully controlled to avoid splashing, and provided no water-impermeable layer of lipids or crystallized solids is formed. The extract can be dried at relatively low temperatures without danger of dust contamination, by placing it into washed cellophane (dialysis) tubing that is hung up to dry in a well-aerated and dry area. The rate of evaporation is proportional to the exposed surface; consequently, narrow tubes give best results.

Distillation *in vacuo* is the most commonly used method for reducing the bulk of the extract. To get a high rate of distillation at a low temperature, it is essential to have a good vacuum, wide vapor paths, and efficient cooling of the distillate. Apparatus for continuous distillation *in vacuo* is available commercially.

When an aqueous extract freezes, only water solidifies, and the solutes are concentrated in the unfrozen phase. The water can then be removed by lyophilization. The process is particularly useful for heat-labile substances.

By selecting a suitable membrane, it is possible to concentrate a solute by ultrafiltration. If the solute has a high molecular weight and is water soluble, the separation by commercially available membranes can be quite rapid and efficient. With small molecules (molecular weight below 5000) the membranes must be so impervious that filtration is slow, and high pressures must be used to overcome the osmotic pressure caused by the retained solute.

Semipermeable membranes used for dialysis or for ultrafiltration are basically molecular sieves. The separation attained during the dialysis depends on one hand on the opening in the membrane; and on the other hand on the molecular weight, shape, and charge or charge distribution on the solute molecule. Generally, particles with molecular weights above 10,000 are retained. Almost all dialysis is now done with the commercially available cellophane sheets or tubes. For cellulose-containing extracts, membranes of nitrated cellulose or protein (such as gut or bladder) are suitable.

With small amounts of fluids, dialysis is relatively simple. The fluid to be dialysed is put into a bag folded from cellophane sheet or into a tube (tied on both ends) and immersed into water. To increase the rate of dialysis of dilute solutions, the fluid inside and outside the tubing is mixed mechanically. Less efficient mixing can be maintained by bubbling air slowly through the outside vessel or by a dropwise water current. If the concentration of solute is high, there is no need for mechanical mixing. A dialysis tube suspended near the top of a jar of water establishes a convective system in which the dense solution of diffusible solute moves to the bottom of the jar and remains there undisturbed. Consequently, dialysis takes place against a more dilute solution than if there had been mixing.

Characterization of extracts, following dialysis, centrifugation, or concentration is accomplished today most commonly after separation by any of the available, appropriate fractionation procedures.

In the actual assay, two basic approaches can be used. The desired constituent is isolated by physical, chemical, or physico-chemical methods to separate it from interfering substances. Ideally, the substance to be assayed is isolated as a single component or its derivative in a specific fraction. In practice, it suffices if the fraction does not contain any uncontrollable source of interference with the subsequent measurement.

The selection of a fractionation procedure for a specific assay depends on many factors. They include: nature and composition of the food; availability, speed, specificity, efficiency, and accuracy of separation; and recovery of minimally or optimally modified components. Generally, neither can a single fractionation procedure be used for separation or isolation of a specified component in various foods; nor can a single procedure be used to separate all components of a food. Consequently, the food analyst must be aware of the principles and limitations of the separation techniques and of the nature of the food examined so that the best separation technique can be selected or devised.

An assay that requires no fractionation is, of course, preferable. The assay may be highly selective or specific as in enzymatic or immunochemical methods. Sometimes conditions can be made for the assay of a specified component as in certain colorimetric tests for carbohydrates or amino acids. Alternatively, the selectivity of an assay can be attained by immobilization or sequestration of interfering substances. Any physical or chemical property can be used for the identification or quantitative determination provided that property is (or can be made) specific for the intended determination.

Once the analysis is completed, the results must be expressed in such a manner that their meaning and significance can be clearly understood and the information used for solving the specific problem. The food analyst is generally better qualified (than the recipient of the analysis) to evaluate the significance and limitations of his data. The analyst is expected to help interpret the results and draw appropriate conclusions. Such conclusions can be best drawn after the background information on the nature of the sample is known. Interpretation of the analytical data has been greatly facilitated in recent years by the availability of mechanical means of data processing, and by the development of statistical techniques both in devising and designing analytical procedures and in assaying the value of analytical results. The analyst is assisted in his interpretation by consulting available comprehensive books and tables on food composition. He will generally depend to a large extent on consultation with other chemists in the interpretation of results and ultimately on his own judgment and experience.

CHAPTER 33

Determination of Moisture

INTRODUCTION

Moisture determination is one of the most important and most widely used measurements in the processing and testing of foods (Makower 1950). Since the amount of dry matter in a food is inversely related to the amount of moisture it contains, moisture content is of direct economic importance to the processor and the consumer. Of even greater significance, however, is the effect of moisture on the stability and quality of foods. Grain that contains too much water is subject to rapid deterioration from mold growth, heating, insect damage, and sprouting (Martens and Hlynka 1965). Small differences in moisture content may be responsible for unexpected cases of spoilage in commercially stored grain (Christensen and Linko 1963). The rate of browning of dehydrated vegetables and fruits, or oxygen absorption by egg powders increases with an increase in moisture content. Moisture determination is important in many industrial problems, i.e. in the evaluation of materials' balance or of processing losses. We must know the moisture content (and sometimes its distribution) for optimum processing of foods, i.e. milling of cereals, mixing of dough to optimum consistency, and for producing bread with the best grain, texture, and freshness retention. Moisture content must be known in determining the nutritive value of a food, in expressing results of analytical determinations on a uniform basis, and in meeting compositional standards or laws. And finally, it is often desirable to weigh out samples for analytical determinations on a given moisture basis. This is especially important if the measured analytical parameter does not vary in a linear or simple manner with an increase in dry matter content.

The moisture contents of foods vary widely. Fluid dairy products (whole milk, nonfat milk, and buttermilk) contain 87–91% water; various dry milk powders contain about 4% water. Cheeses have intermediate water contents (ranging from about 40% in cheddar to 75% in cottage); the water content of butter is about 15%, of cream 60–70%, and of ice creams and sherbet around 65%. Pure oils and fats contain practically no water; but the processed lipid-rich materials may contain substantial amounts of water (from about 15% in margarine and mayonnaise to 40% in salad dressings). Some fresh fruits contain more than 90% water in the edible portion. Melons contain 92–94%, citrus fruits 86–89%, and various berries 81–90% water. Most raw tree and vine fruits contain 83–87% water; the water content of ripe guavas is 81%, of ripe olives 72–75%, and of avocado 65%. After commercial drying, fruits contain up to 25% water. Fresh fruit juices and nectars contain 85–93% water; the water content is lowered in concentrated or sweetened products. Cereals are characterized by low moisture contents. Whole grains designed for long-term storage have 10–12% water. The moisture content of breakfast cereals is below 4%, of macaroni 9%, and of milled grain products (flour, grits, semolina, germ) 10–13%. Among baked cereals, pies are rich in water (43–59%); bread and rolls are intermediate (35–45%, and 28%, respectively); and crackers and pretzels are relatively dry (5–8%). Ripe raw nuts generally contain 3–5% water, or less, after roasting. Fresh chestnuts contain about 53% water. The moisture contents of meat and fish depend primarily on the fat contents, and vary to a lesser degree with the age, source, and growth season of the animal or fish. The moisture contents range from 50 to 70%, but some organs may contain up to 80% water. The moisture content of sausages varies widely. Water in poultry meats is 50% (in geese) to 75% (in chicken). Fresh eggs have about 74%, and dried eggs about 5% water. White sugar (cane or beet), hard candy, and plain chocolate contain 1% or less of water. In fruit jellies, jams, marmalades, and preserves up to 35%; in honey 20%; and in various syrups 20–40% is water. Sweet potatoes contain less water (69%) than white potatoes (78%). Radishes have most (93%) and parsnips least (79%) water among the common root vegetables. Among other vegetables, a still wider range is found. Green lima beans have about 67%, and raw cucumbers over 96% water. Dry legumes contain 10–12% water, and the water content of commercially dried vegetables is 7–10% (preferably below 8).

The rapid and accurate determination of water in foods varying so widely in texture, overall composition, and moisture content continues to present many problems. Many workers have stressed the complexity of analytical procedures for the determination of water in foods. Makower (1950) pointed out that though the literature is replete with methods of moisture determination, we have no methods that are both accurate and practical. According to Matchett and von Loesecke (1953), accurate, rapid, and simple methods of moisture determination applicable to all types of food materials are continuously sought, but it seems doubtful that such a goal will ever be attained.

In practice, the guiding principle has been to prefer the method that gives the highest moisture values, provided decomposition of organic components and volatilization of compounds other than water are negligible, or that such losses can be compensated by incomplete removal, under fixed experimental conditions, of strongly absorbed water (Hlynka and Robinson 1954). Generally, the reproducibility and practicability of a method (simplicity, convenience of apparatus, and rapidity) have been the important factors in the selection of an analytical method for water determination. Less emphasis is placed on the accuracy of a water determination. Admittedly, reproducibility is of major importance in the control of processing procedures and in the standardization of commercial products. Yet, accuracy is of great significance in establishing conditions that govern food stability. The usefulness and validity of simple and rapid moisture determinations depend on their calibration against standard and accurate reference methods. The difficulties encountered in developing such reference methods can best be understood by considering the manner in which water is held by various food components.

Some Basic Considerations

The nature of the forces between water and the component substances of foods was reviewed by Ward (1963) and Hamm (1963). Water may occur in foods in at least three forms. A certain amount may be present as free water in the intergranular spaces and within the pores of the material. Such water retains its usual physical properties and serves as a dispersing agent for the colloidal substances and as a solvent for the crystalline compounds. Part of the water is absorbed on the surface of the macromolecular colloids (starches, pectins, cellulose, and proteins). That water is closely associated with the absorbing macromolecules by forces of absorption which are attributed to van der Waals forces or to hydrogen bond formation. Finally, some of the water is in a bound form—in combination with various substances, i.e. as water of hydration. The above classification, though convenient is quite arbitrary. Attempts to determine quantitatively the amounts of various forms of water in foods have been unsuccessful. Unlike in some inorganic compounds which show a distinct discontinuous sorption isotherm of various levels of crystallization water, the sorption isotherm of water in foods shows a continuous spectrum of the types of water binding. Consequently, the terms free, absorbed, and bound are relative, and as the true moisture content is not known, the conditions selected for moisture determination are arbitrary.

Methods for the determination of moisture can be divided into drying methods, distillation procedures, chemical assays, and physical procedures. Principles of the various methods are outlined and examples are given in this chapter.

Drying Methods

The procedures for determining the moisture content specified in food standards generally involve thermal drying methods. The material is heated under carefully specified conditions and the loss of weight is taken as a measure of the moisture content of the sample. The determination of moisture from the loss of weight due to heating necessarily involves an empirical choice of the type of oven and temperature and length of drying. Hence the values obtained for moisture depend on the arbitrarily selected conditions. Some of the methods provide, therefore, approximate rather than accurate moisture values. Drying methods, however, are simple, relatively rapid, and permit the simultaneous analyses of large numbers of samples. They continue to be the preferred procedures by many food analysts.

In an ideal procedure for the determination of water, weight losses should result from quantitative and rapid volatilization of water only. In practice, heating of a moist organic substance causes, in addition, volatilization of other absorbed material and of gaseous products formed by irreversible thermal decomposition of organic components. Further, weight changes resulting from oxidation phenomena (i.e. of oils) occur. These changes do not begin at a specified or fixed temperature, and occur at all temperatures at widely varying rates. The rate at which moisture can be removed from the surface of a solid phase is a function of the water vapor pressure and of the drying temperature. Water can be determined at any temperature provided the partial vapor pressure in the air above the solid phase is lower than the vapor pressure of the water in the sample. Thus, we can have thermal drying below the freezing point of water as in lyophilization. For accurate moisture measurement, the tendency is to dry foods at the lowest possible temperatures.

Practical considerations dictate, however, selecting temperatures at which the decomposition of organic compounds is minimized, and

yet the time required for quantitative drying at the selected temperature not unduly prolonged.

Several investigators studied factors affecting the precision of
moisture measurements by drying methods. According to Koster
(1934) and Oxley and Pixton (1961), the accuracy of moisture determinations is affected by the drying temperature, temperature and
relative humidity of the drying chamber, relative humidity and air
movement in the drying chamber, vacuum in the chamber, depth
and particle size of samples, drying oven construction, and the
number and position of samples in the oven. Guilbot (1955) emphasized that the surface of the material being dried and the rate of
diffusion of water vapor in the drying substance also affect results.
Mitchell (1950 A, B) evaluated the effects of the following variables
on the rate of evaporation of water in steam and vacuum ovens:
diameter, depth, and material of the container; number of containers
and their position in the oven; rate of conduction of heat to container (i.e. from different types of shelf material); and ventilation
and temperature of the oven. The rate of evaporation was higher
in aluminum than in glass or porcelain dishes, higher in vacuum than
in steam ovens, and higher in shallow than in deep dishes. The rate
of heat supply to the bottom of the dish was the most important
factor.

Hlynka and Robinson (1954) listed general sources of errors in the
determination of moisture in cereals. In addition to sampling
errors (of the bulk lot or at the laboratory subsampling stage), moisture may change during subsequent storage of the samples. If the
material for a moisture determination must be ground, loss of moisture to, or gain of moisture from, the air may take place. For
damp grain a two-stage drying method has been recommended.
A sample of 100 gm or more is first allowed to equilibrate with the air,
and the moisture loss of the damp wheat is determined. A subsample is then ground for the determination of the remaining moisture. In two-stage drying,

$$\text{total moisture} = A + (100 - A)B/100 \qquad (331)$$

where $A = \%$ moisture lost in air drying

$B = \%$ moisture in the air-dry sample as determined by
oven drying

The 2-stage method gives generally higher results (0.2–0.6%) than
the 1-stage procedure, but prolongs the determination. To prevent

moisture losses after oven drying, moisture dishes with tight covers should be promptly closed after the oven is opened, rapidly transferred from the oven to a good desiccator, and weighed rapidly after they have attained room temperature. Common desiccators and desiccants are of limited usefulness, and unless properly maintained can cause erratic results from pickup of moisture during cooling.

Solid materials must be pulverized under conditions that minimize compositional changes. Maximum particle size and distribution of particle size are of great significance.

In drying liquids, it is essential to spread the material over a large surface. The liquid is preferably evaporated first on a water bath, and then drying is completed in an oven. The dried residue should weigh 1–2 gm. Similarly, the loss of moisture in cheese is most rapid when the final drying in a water or vacuum oven is preceded by heating on a water bath for about 1 hr. To increase the area of dried cheese, it can be smeared on a glass rod, mixed with inert material, or placed between two previously weighed sheets of aluminum.

Hunt and Neustadt (1966) recommended grinding grain samples for moisture determination in a Wiley laboratory mill (intermediate model). The Wiley mill was recommended because the sample is subjected to a minimum of heating during grinding, and because it is protected from contact with the atmosphere. With most cereal grains, the 2-stage drying method was required for samples containing above 16% moisture. For soybean samples with more than 10% moisture and rough rice samples with more than 13% moisture, the 2-stage method must be used. Cereals should be ground to pass an 18-mesh screen; grinding through a 40-mesh sieve was recommended for other foods; the rotor and stator blades of the mill must be sharpened periodically depending on use. Minimum pressure should be applied in feeding the sample through the mill and the sample weight should be limited to 5 gm. Standard aluminum dishes are recommended for cereals (55 mm diam and 15 mm high, with slightly tapered sides and tightly fitted slip-in covers designed to fit under the dish when they are placed in the oven).

The drying temperature used in water determination ranges—depending on the tested material—from 70° to 155°C. Yet, some moisture is retained by biological systems at least to temperatures as high as 365°C, the critical temperature of water (Nelson and Hulett 1920). The average time of drying is from below 1 hr to 6 hr or more. Foods can be dried for moisture determination either for a selected period of time or until two successive weighings show a negligible loss in weight (generally less than 2 mg for a 5-gm sample, at 1-hr intervals). The drying time is inversely related to drying temperature. However, increasing the temperature increases loss of weight to a level that cannot be attained even by prolonged heating at lower temperatures. Measurements of carbon dioxide evolved during drying indicate that some weight losses at temperatures over 80°C result from decomposition of the product. Consequently, the use of high temperatures to obtain some rapid results can be justified on pragmatic grounds only (Hlynka and Robinson 1954). In foods susceptible to decomposition, drying temperatures can be reduced by using vacuum ovens, and drying times can be shortened by using drying agents or by passing dry air over the samples.

A common source of error in a moisture determination is the formation of a crust that is impervious to evaporation of moisture from the center of a dried sample. In drying samples rich in sugars, the effects of crust formation can be reduced by moistening with water and thorough mixing with sand or asbestos to increase the exposed surface or by top drying of thin layers under infrared heat lamps. In some plant materials, drying at relatively low temperatures (to prevent crust formation), and completing the drying at higher temperatures is the solution. Losses of water through chemical reaction in dextrinization or hydrolysis are accelerated at elevated temperatures and high moisture levels. To reduce such losses, predrying at low temperatures followed by final drying at a recommended temperature is helpful (Willits 1951).

The sensitivity of certain sugars (especially fructose) to decomposition at elevated temperatures rules out determining moisture by drying in air ovens in such foods as honey and fruit syrups. Mitchell (1950 A,B) found that the drying of fructose, glucose, or sucrose solutions was faster and the tendency to decompose was less if the pH was below 7. Fructose solutions decompose at temperatures above 70°C; glucose is relatively stable at 98°C. Sucrose solutions dry very slowly in the steam oven. In the vacuum oven, slightly acidic fructose solutions on pumice adsorbant dried in 4 hr at 70°C; and at 60°C (preferable temperature) in 7 to 9 hr.

Conventional or standard procedures were developed for most foods. Details of such procedures are described in handbooks or standard reference manuals. Most methods employ conditions that were agreed upon after collaborative tests. The conditions were selected to simplify the analytical determination and to increase the reproducibility.

Air-Oven Methods.—Because of their convenience, air-oven methods of various types are widely used in control laboratories for the determination of moisture. Ovens ranging from simple water or steam jacketed types to elaborate equipment with forced air circulation and built-in balances are in wide use.

Either convection-type ovens or forced-draft ovens can be used for moisture determinations. Forced-draft ovens are preferred, since they generally accommodate more samples and attain the desired temperatures more rapidly after the samples are inserted in the oven. The principal criterion of the suitability of an oven is precision of control and uniformity of temperature at different positions in the oven. A good oven should have a thermoregulator of ±0.5°C or less, that will maintain its setting without requiring constant adjustment. Heating units should give an on-and-off cycle of 15 min or less for an oven with a normal load (Hlynka and Robinson 1954). Variations of temperature (±3.0°C or more) with position in the oven is a major problem in convection ovens, can be somewhat reduced by the proper distribution of heating elements, adequate insulation, and drying limited numbers of samples placed on one shelf in a central, uniformly-heated area.

Variations of temperature with position in the oven can be minimized by mechanical circulation of the air in the oven. The Brabender, semiautomatic moisture tester (Fig. 33.1) uses a small fan to circulate air in the oven and has a built-in automatic balance. The oven holds ten 10-gm samples in flat, tared dishes. After a specified drying period, each dish is weighed (without removing from the oven or cooling) and the moisture is indicated on an illuminated scale. The precision, relative speed, and convenience have made the oven popular for the determination of moisture in various foods. The instrument is useful also for determining the drying rate of foods at various temperatures. Cereals are generally dried in the Brabender oven for 1 hr at 130°C. The Carter-Simon oven is used to determine moisture in cereal products by drying at 155°C for 15 min. Three samples can be tested at 5-min intervals. The instrument gives results that compare with the standard air-oven procedure (130°C), but requires constant attention. The Chopin instrument uses drying temperatures up to 200°C. The vaporized water is passed through a calcium carbide container fitted with a flame jet on top. As long as water vapor is evolved, the generated acetylene is burning. Drying is completed when the flame is reduced to a specified size. The sample is cooled and weighed. Actual drying times are about 5 min for flour and 7 min for ground grain.

Courtesy C. W. Brabender Instruments, Inc.

FIG. 33.1. BRABENDER RAPID MOISTURE OVEN

Vacuum-Oven Methods.—Vacuum-oven determinations are generally the standard and most accurate procedures of moisture determination for most foods. Although it is impossible to obtain an absolute moisture level by drying methods, moisture determination by drying in a vacuum oven is a close and reproducible estimate of true moisture content. In many biological materials, most of the water can be removed with relative ease, but removing the residual 1% (or so) is difficult. This residual water can be removed easier, more rapidly, and with less overall change of the organic components by drying under vacuum than by other methods.

The rate of drying can be increased through lowering the vapor pressure in the air by using vacuum. If during drying no air were let into the vacuum oven, the pressure of water vapor in the oven would eliminate usefulness of the vacuum oven, especially in samples with high moisture content. In addition, the efficiency of the oven is limited by the rate of water diffusion into the pump. A small amount of air, preferably dried by passing through concentrated sulfuric acid is, therefore, let in to continuously sweep the oven.

The usual vacuum-oven method involves drying to constant weight at a pressure below 50 (preferably 25) mm of mercury. Drying to constant weight requires rather long periods for most foods. In cereals, practically constant weight can be attained within 16 hr. Drying temperatures specified for most vacuum oven methods are 98°–102°C, indicating an accuracy of ±2°C. Mercury-in-glass thermoregulators with suitable relays provide much better accuracy ($\pm 0.1\,^\circ$C) maintain their setting, and give trouble-free operation (Hlynka and Robinson 1954). Foods rich in levulose (i.e. fruits) must be dried at 70°C or below. Using a 0.5 in. aluminum shelf with good contact with the walls improves substantially the important uniformity of temperature and heat transfer in vacuum ovens.

Other Drying Methods.—Drying by conventional methods involves heat conductivity and convection, and drying times are long. Infrared drying is more effective as it involves penetration of heat into the sample being dried. Infrared drying can shorten the drying time to $^1/_3$–$^1/_8$ of that required in conventional drying. In infrared drying, a 250–500-watt lamp is used, the filament of which developes a temperature of 2,000° to 2,500°K; 1,000°K is best for drying animal and plant material (Schierbaum 1957). The distance of the infrared source from the dried material is the critical parameter as close proximity may cause substantial decomposition. Distance of about 10 cm is recommended. The thickness of the dried material should not exceed 10–15 mm. Drying times under optimum conditions are up to 20 min in meat products, and up to 25 min in baked products. In ground grain, 10 min is generally satisfactory. Sample size should be 2.5–10 gm, depending on food and moisture content. Infrared drying instruments are available equipped with forced ventilation, or connected to torsion balances with indicator scales to read directly moisture content.

Methods of drying foods by desiccation in an evacuated desiccator over dehydrating or water absorbing chemicals, as sulfuric acid, freshly ignited powdered lime, phosphorus pentoxide or lumps of calcium carbide at room temperature, are very slow. Some foods cannot be dried completely even after several months. The use of vacuum at somewhat higher temperatures (50°C) is generally more satisfactory.

Drying at low temperatures has been particularly popular with researchers interested in developing reference, standard methods for moisture determination. The method for moisture determination in dried vegetables by Makower and Nielsen (1948) involves the addition of a large amount of water to a weighed sample, freezing and drying to the frozen stage (lyophilization); and completion of the drying in a vacuum oven or vacuum desiccator in the presence of an efficient water absorbent. The last step can be completed in a relatively short time at, or slightly above, room temperature because of a marked increase in drying rate brought about by lyophilization.

True moisture contents have also been determined by several other methods. The reversibility method of Sair and Fetzer (1942) measures the extent of decomposition during drying. Makower *et al.* (1948) developed the redrying technique to establish empirical conditions (time and temperature) for a true moisture determination. Oxley and Pixton (1960), however, have shown that with oven-drying methods, the ability to measure the recovery of added water varies with the food to which water was added. The isotopic dilution method of Brand and Kassell (1942) is another primary method for the determination of water. A known amount of heavy water is added to the system, and the total amount of water in the food can be computed from the heavy water content in a portion of the total water.

In addition to thermal drying methods, distillation procedures and the Karl Fischer titration procedure can be used to determine the water content of some foods. These methods are described in detail in subsequent sections of this chapter.

Distillation Methods

Distillation methods have been used for moisture determinations for almost 70 yr. There are two main types of distillation procedures. In one, water is distilled from an immiscible liquid of high boiling point. The sample suspended in a mineral oil having a flash point much above the boiling point of water is heated to a predetermined temperature in a suitable apparatus. The water that distills off condenses and is collected in a suitable measuring cylinder. In the second type, the mixture of water and an immiscible solvent (i.e. xylene, toluene) distills off, and is collected in a suitable measuring apparatus in which water separates and its volume can be measured. Distillation with an immiscible solvent under a refluxing type of condenser is the method most commonly used.

Distillation with a boiling liquid provides an effective means of heat transfer, the water is removed rapidly, and the test is made in an inert atmosphere that minimizes danger of oxidation. Distillation methods cause less decomposition in some foods than drying at elevated temperatures. However, chemical reactions produced by heat are reduced but not eliminated. Adverse effects of heat can be reduced still further by selecting organic solvents with a boiling point below that of water, such as benzene. Such a choice, however, lengthens the distillation time. The liquids most commonly used for distillations of the second type are xylene, toluene, and tetrachloroethylene. If the boiling liquid is lighter than water, the collecting trap usually contains a tube, sealed at the bottom, and calibrated upward. This form requires reading only one meniscus in measuring the amount of collected water. The calibrated portion of the tube may be cooled by a water bath to a standard temperature; or as the water layer is covered by a layer of the immiscible liquid, the whole trap may be removed for temperature adjustment (Fetzer 1951). If a liquid heavier than water is used, two menisci must be read, the condensing liquid must pass through the water, formation of an emulsion is enhanced, and the temperature adjustment somewhat complicated. On the other hand, recommended liquids with a high specific gravity (tetrachloroethylene, carbon tetrachloride) eliminate fire hazard, and reduce the danger of overheating or charring, as the sample floats on top of the liquid.

The literature on distillation procedures is extensive and has been reviewed by Fetzer (1951), and in annual reviews on food analysis in *Analytical Chemistry.* Consequently, only two of the most commonly used distillation procedures will be discussed.

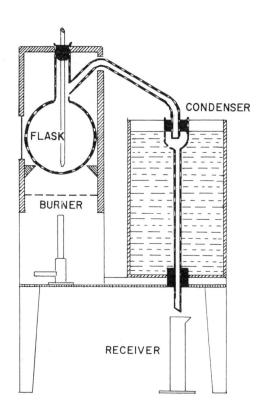

FIG. 33.2. BROWN AND DUVEL APPARATUS

The Brown-Duvel method was developed in 1907 and was used with certain modifications from 1912 till 1959 for moisture determinations in cereals by the Board of Grain Commissioners for Canada, a Federal Government branch responsible for quality control of Canadian Grain. A drawing of the apparatus is shown in Fig. 33.2. For routine tests, units for heating 2 to 6 flasks simultaneously are available. A 100-gm sample of whole grain is heated in the flask with 150 ml of nonvolatile oil to a specified cutoff temperature (180°C) for wheat. The amount of water that is distilled into the graduated cylinder is read in milliliters and reported as percent moisture. The determination takes about 1 hr and, with a suitable bank of equipment, 1 man can make about 12 to 18 determinations per hour (Hlynka and Robinson 1954). The method can be taught to inexperienced persons, and the required equipment is reasonably rugged and easily replaced.

Development of the Dean and Stark (1920) distilling receiver tube which permits continuous refluxing and separation of water provided a means of efficiently removing the water with limited amounts of solvent. In 1925, Bidwell and Sterling improved the Dean and Stark tube by placing a small reservoir above the calibrated tube. The tubes are shown in Fig. 33.3. Figure 33.4 shows an apparatus designed by the Corn Industries Research Foundation for the de-

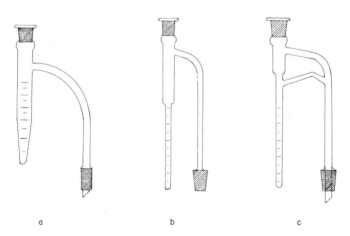

FIG. 33.3. TRAPS FOR MOISTURE DETERMINATIONS BY DISTILLATION PROCEDURES

(A) Dean and Stark (B) Bidwell and Sterling, and (C) Modified Bidwell and Sterling

termination of moisture in corn products. The apparatus basically uses the Bidwell and Sterling design. For accurate results, standard apparatus and careful attention to specified procedures are essential. Different solvents have been recommended for various corn products, and required distillation times range from 6 to 48 hr.

Many difficulties may be encountered in the determination of moisture by the distillation method. These include: relatively low precision of the receiving measuring device, difficulties in reading the meniscus, adherence of moisture droplets to the glass, overboiling (especially with xylol), solubility of water in the distillation liquid, incomplete evaporation of water and underestimation of moisture contents, and distillation of water-soluble components. Adherence of water to the walls of the condenser tube or sides of the receiving tubes can be generally remedied by using thoroughly cleaned glassware. However, in distilling some materials a small amount of fat or wax may be carried to the reservoir trap or the drip tip of the con-

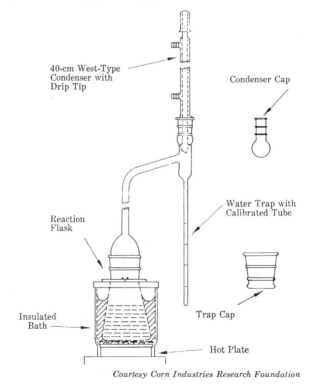

40-cm West-Type
Condenser with
Drip Tip

Condenser Cap

Water Trap with
Calibrated Tube

Reaction
Flask

Insulated
Bath

Trap Cap

Hot Plate

Courtesy Corn Industries Research Foundation

FIG. 33.4. CIRF APPARATUS FOR DETERMINATION OF
MOISTURE BY AZEOTROPIC DISTILLATION

denser.   Removing the lipid material mechanically releases water adhering to walls and gives better reading of the meniscus.   Use of a small amount of wetting agent will also improve meniscus reading. Incomplete recovery of water due to the formation of an emulsion can sometimes be remedied by adding small amounts of amyl or isobutyl alcohol.   To improve the moisture distillation, using wide-mouthed boiling flasks is recommended; dispersing the tested material on diatomaceous earth or on Filter-Cel is useful with many viscous foods rich in sugar or protein.   Foods in powder form (cereals, flours, starches) tend to bump during the distillation through overheating on the bottom of the flask.   This can be largely overcome through the introduction of a small amount of dry short-fiber asbestos (Fetzer 1951).   If the material is heat-sensitive and decomposition of the food with the formation of water is suspected, effective techniques of food dispersion and low boiling point liquids should be used.

The main objection to distillation procedures is that they are not adaptable to routine testing.   The *Tenth Edition of Official Methods of Analysis* of the Assoc. Offic. Agr. Chemists (1965) lists use of distillation with toluene only for testing of grain and stock feeds, and even this on a provisional basis only.   It is doubtful whether 48-hr distillation procedures are actually used in the corn milling industry for routine testing.   It is, therefore, not surprising that Fetzer (1951) concluded that the distillation procedure for moisture determination should not be used if the conventional oven methods are applicable.   A distillation procedure for rapid, approximate, determination of moisture and simultaneous removal with n-butyl ether of fat from meat and meat products was described by Davis *et al.* (1966).

Chemical Methods

Karl Fischer Reagent Titration—The wartime need for dehydrated foods stimulated the search for more rapid and accurate methods for determining moisture.   The Karl Fischer reagent has proved to be quite adaptable for this purpose (Oser 1949).   The method is particularly applicable to foods which give erratic results when heated or submitted to a vacuum.   The Karl Fischer titration has been found the choice method for determination of water in many low-moisture foods such as dried fruits and vegetables, candies, chocolate, roasted coffee, oils, and fats.   Superiority of the method was demonstrated in determining moisture in sugar-rich foods (sugars, honey), or foods rich both in reducing sugars and proteins. The procedure has been applied also to foods with intermediate moisture levels (bakery doughs, baked products, fat-rich cake mixes), and to foods with high levels of volatile oils. The method is seldom used in the determination of water in structurally heterogenous, high-moisture foods such as fresh fruits and vegetables. The Karl Fischer method for moisture determination is based on the reaction described by Bunsen in 1853 involving the reduction of iodine by sulfur dioxide in the presence of water:

$$2H_2O + SO_2 + I_2 \rightarrow H_2SO_4 + 2HI \qquad (33.2)$$

Karl Fischer (1935) modified the procedure and established the conditions for quantitating the modification. Methanol and pyridine were used in a four-component system to dissolve iodine and sulfur dioxide.

The basic reaction takes place in two steps (Mitchell (1951):

$$C_5H_5N \cdot I_2 + C_5H_5N \cdot SO_2 + C_5H_5N + H_2O \rightarrow$$
$$2C_5H_5N \cdot HI + C_5H_5N \cdot SO_3 \qquad (33.3)$$

and

$$C_5H_5N \cdot SO_3 + CH_3OH \rightarrow C_5H_5N(H)SO_4CH_3 \qquad (33.4)$$

As shown by the above reactions, for each mole of water 1 mole of iodine, 1 mole of sulfur dioxide, 3 moles of pyridine, and 1 mole of methanol are required. In practice, an excess of sulfur dioxide, pyridine, and methanol is used, and the strength of the reagent depends on the concentration of iodine. For general work, a methanolic solution containing other components in the ratio of 1 iodine: 3 sulfur dioxide:10 pyridine, and at a concentration equivalent to about 3.5 mg of water per milliliter of reagent is used. Other compositions may be preferable for various foods. Numerous variations have been proposed for the preparation of the Fischer reagent. In laboratories that prepare and consume large amounts of the reagent, a stable stock solution of iodine dissolved in pyridine and then diluted with methanol is prepared. Liquid sulfur dioxide is added to portions of the stock solution a day or two before use. To minimize losses of active reagent from side reactions that consume iodine, many of the laboratory supply houses market the reagent as two solutions: a solution of iodine in methanol, and sulfur dioxide in pyridine. The solutions are mixed shortly before use. The sample in which water is to be determined, is dispersed in an appropriate solvent (i.e. dry methanol) and the complete, four-component, reagent is added to the sample.

Alternatively, a two-mixture Karl Fischer reagent can be used. The sample is dissolved in a mixture of sulfur dioxide-pyridine-methanol, and titrated with a solution of iodine in methanol. The iodine in methanol solution is stable and needs standardization only rarely, provided the solution is protected against moisture.

In the titration with the Karl Fischer reagent, iodine and sulfur dioxide are added in the appropriate form to the water-containing food. The excess of iodine that cannot react with water is in a free form. The amount required for the titration can be determined visually, till a yellow-mahogony brown color is seen. Adding to the system a few drops of methylene blue gives a green end point. Reagents, titrating flask, and buret must be protected from atmospheric moisture.

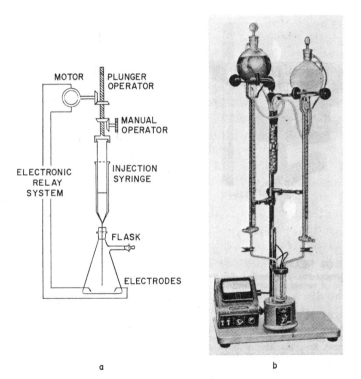

FIG. 33.5. APPARATUS FOR MOISTURE DETERMINATION BY THE KARL FISCHER METHOD

(A) Schema of apparatus, (B) Commercial equipment.

Instead of visual observations, one can use a photometric end-point determination. In highly colored solutions, electrometric titration by the dead-stop technique is most often used. It is based on the fact that in applying a small potential difference to two equal platinum electrodes immersed in a redox system, the potential formed by polarization is compensated and the current flow is interrupted. During the last phase of the redox titration—at the end-point—polarization or depolarization of an electrode to a complete depolarization or polarization of both electrodes takes place. That change is accompanied by a large deflection of a galvanometer needle or change in a magic eye. Instruments based on this principle are sold by many laboratory supply houses. An outline of an apparatus for automatic titrations and a commercial assembly are shown in Fig. 33.5. Units that automatically perform the Karl Fischer water determination by the conductometric method are also available. The units are equipped with electromagnetic, glass-sealed titrating valves for either direct or back titration. Visual determinations of end points are sensitive to differences of less than 0.5 mg water. The sensitivity is increased (to 0.2 mg) if endpoints are determined electrometrically.

Theoretically, the Karl Fischer reagent may be used for the determination of water in liquids, gases, and solids. Basically, the method may be considered as a primary standard method for determining water content. In most biological materials, it must be calibrated and standardized against another reference method. Most organic compounds do not interfere with the determination. Some exceptions have been observed. Thus, ascorbic acid is oxidized essentially quantitatively to dehydroascorbic acid, and the Fischer reagent determines the sum of the moles of water plus ascorbic acid. Most carbonyl compounds interfere. Aldehydes and ketones tend to react with the methanol of the Fischer reagent, form acetals, and release water. The interference may result in a fading endpoint. Similarly, mercaptans, diacylperoxides, thio acids, and hydrazines may interfere though usually methods are available to eliminate such interferences. Inorganic compounds which affect the water titration include metal-oxides, hydroxides, carbonates, bicarbonates, chromates, dichromates, borates, and sulfides.

The determination of moisture is carried out in a nonaqueous system. This requires preparing samples for analysis under conditions that minimize changes in water content. Fluids are delivered best with an automatic pipet or syringe. Viscous fluids or pastes are generally homogenized with a solvent. Solids are either homogenized with the solvents or titrated as suspensions. For granular products that must be pulverized, titration is often carried out in a grinding container. Hart and Neustadt (1957) described as accurate method for determining moisture in seeds involving an efficient grinding-extraction procedure that takes place in a completely enclosed cup. Methanol present in the cup extracts water from the grain as it is being ground. After suspended matter has settled from the extract, an aliquot is titrated with the Karl Fischer reagent. Fats and waxes are spread out or made into a paste with an appropriate fluid. Oil-rich foods (oil seeds, germ of cereals) must be ground to a fine powder. If the material is pulverized in a high-speed grinder, the temperature increases and moisture may be lost.

Applications of the Fischer reagent to the determination of water in foods have been covered adequately in the literature (Schroeder and Nair 1948; Oser 1949; Jones 1951; Frediani 1952; Matchett and von Loesecke 1953; von Loesecke 1957; Morgareidge 1959; Cavagnol 1961; Borker and Sloman 1963; Sloman and Borker 1965). Satisfactory techniques have been developed for most foodstuffs. The simplest technique involves direct titration of the sample with the Karl Fischer reagent. In order to titrate all of the water in a sample, it may be necessary to extract the water by refluxing the sample with an appropriate solvent (i.e. methanol). Sometimes excess reagent is added, and back titrated with a standard water solution. Early investigations have clearly shown that the preparation of the sample is important. Thus Fosnot and Haman (1945) found that in cereals and cereal products, completeness of the reaction between the water in the sample and the reagent varied with the kind of material and fineness of the grind. Frediani et al. (1952) found that for best results, cereals, dried yeast, etc., should be ground to pass a 40-mesh sieve. On the other hand, extraction of water from dehydrated fruits or vegetables in boiling methanol is rapid. The determination can be completed in less than $1/_2$ hr, and the results are less dependent on the fineness of grinding than with most other common methods (Makower 1950). Yet, in some dried vegetables, water is bound very tenaciously, diffuses slowly, and the apparent moisture changes considerably with an increase in extraction time. To obtain agreement between the Karl Fischer titration and a vacuum oven method, it is necessary to extract dry potatoes for 6 hr and carrots for less than $1/_2$ hr.

Various modifications have been proposed for the determination of water in oils and fats. Francois and Sergent (1950) suggest boiling a larger sample in methanol, and titrating an aliquot of the cooled solution.   Meelheim and Roark (1953) distilled water from oils and fats with toluene into a receiving titration flask containing methanol and excess Karl Fischer reagent.

The main difficulty in using the Karl Fischer method arises from the lack of complete water extraction.   McComb and Wright (1954) found formamide to be a more rapid and versatile extractant of water from foods, than methanol.   Modification of the extraction procedure is exemplified by a method for the water determination in dairy products, wherein xylene or carbon tetrachloride is employed in mixed solvent systems with alcohol (Morgareidge 1959).

Other Chemical Methods.—When foods are mixed with powdered calcium carbide, the water reacts to produce acetylene.   The quantity of acetylene can be measured by the loss in weight of the mixture, pressure or volume of the gas produced in a closed system, and the formation of copper acetylide (gravimetric, titrimetric, or colorimetric).   Accuracy of the methods depends on the fineness of grinding of the interacting components.   Gas production can be doubled by using calcium hydride in place of calcium carbide.   None of the methods has found wide application.

Water in sugar products has been determined by an elegant method whereby the sugar is ground with cobaltous bromide in dry chloroform (Gardiner and Keyte 1958).   Combination of the salt with water causes it to precipitate from solution; it is separated subsequently by filtration and the cobaltous bromide is weighed after evaporation of the solvent.   Semiquantitative estimates of free water in food can be obtained by measuring the rate of change in the color of filter paper impregnated with cobaltous chloride.   The dry paper is blue and becomes pink when moist.

Mixing sulfuric acid with food results in heating, and the increase in temperature under fixed conditions is roughly proportional to the water content of a food.   Launer and Tomimatsu (1952) described a procedure in which nonaqueous organic matter is determined by an oxidation with a dichromate solution in sulfuric acid, and the moisture is measured by difference.   The method is rapid (5 to 10 min exclusive of sample preparation) and relatively simple. Its usefulness is somewhat limited, and the use of corrosive reagents makes it unattractive for use in routine quality control.

Determinations of water by the acetyl chloride method (Smith and Bryant 1935) is generally useful in determining moisture in products that give satisfactory results with the Karl Fischer reagent. The determination involves titration of the acid formed in the following reactions.

$$H_2O + CH_3COCl \rightarrow CH_3COOH + HCl \qquad (33.5)$$

$$ROH + CH_3COCl \rightarrow CH_3COOR + HCl \qquad (33.6)$$

The method is useful in determining moisture in oils, butter, margarine, dry spices, and many low-moisture foods. The acetyl chloride reagent is generally dissolved in toluol and the food is diluted or dispersed in pyridine. The method has found little acceptance.

Physical Methods

Infrared Determination.—The infrared determination of water is based on measuring the absorption at wavelengths characteristic of the molecular vibration in water (Kaye 1954). The most useful wavelengths are 3.0 and 6.1 μm, corresponding to fundamental vibrational modes of the water molecule, the combination absorption band at 1.93 μm, and the 1.45 μm first overtone of the OH stretching. Measurement of the water content using infrared absorption techniques can provide analyses with a sensitivity of a few parts per million in a wide range of organic and inorganic materials. High photometric sensitivity to 1 part in 5,000 is possible by use of a tungsten light source and a lead sulfide detector with carefully designed electronic circuitry. Wavelength isolation is obtained by using narrow-band interference filters or a simple grating monochromator. Use of glass or quartz windows and lens materials simplifies maintenance and permits measurements at elevated temperatures. The 1.45, 1.91, and 2.83 μm water absorption regions have been used for water analyses at sensitivities ranging from 1 to 5 ppm to several percent (Trippeer 1965). Hart *et al.* (1962) and Norris and Hart (1965) investigated the spectral region from 0.7 to 2.4 μm for measuring the moisture content of grain and seeds. The spectral absorbance curve for a thin layer of ground wheat showed minimum interference for the 1.94 μm band. Using a 2-gm sample of ground material mixed with 1.5–2.0 ml of carbon tetrachloride in a 4.4-cm diam cell, the transmittance values of a large number of wheat, soybean, wheat flour, and wheat bran samples are measured at 1.94 and 2.08 μm. From these data, the optical density difference ΔOD (1.94 μm and 2.08 μm) was computed for each sample and related to the moisture content as determined by standard procedures. Calibration curves for each of the 4 materials showed standard deviations from 0.28 to 0.37% moisture for the moisture range from 0 to 20%. The water absorption band at 0.97 μm was measured on individual intact peanuts and related to the moisture content. A measurement within ±0.7% moisture content was obtained using ΔOD (0.97–0.90 μm) as the measured value.

A near infrared method has been reported for the analysis of fruit and vegetables (Gold 1964). In collaborative work the near infrared method was found to be as accurate as, and more rapid and specific than, the vacuum-oven method for the determination of moisture in dried vegetables and spices. The usefulness and limitations of the near infrared spectral absorption method for determining moisture (and fat) in meat products were described by Ben-Gera and Norris (1968).

Gas Chromatographic Methods.—The methods are based on extracting the moisture with an organic solvent and determining water in the extract by gas chromatography. The procedure has general application provided (1) water is extracted effectively, and (2) the extracts contain no substances that coincide with either the solvent or water peaks on the chromatogram (Weise *et al.* 1965). In a typical analysis, 15 gm of sample is blended with 100 ml of absolute methanol and several milliliters of secondary butanol (depending on the expected moisture content of food). The mixture is allowed to settle for 15 sec and a sample of two microliters is removed from the clear supernatant and tested. The gas liquid chromatography analysis takes about 5 min. The method is useful for a variety of foods varying in moisture content from 8 to 65% (Schwecke and Nelson 1964). Gas chromatography has been reported to be a reliable method for determining the moisture content of a number of foods such as cereals, cereal products, fruits, and fruit products (Brekke and Conrad 1965).

Nuclear Magnetic Resonance.—When a sample containing nuclei is placed simultaneously in a fixed magnetic field and a radio frequency magnetic field of the proper frequency, the phenomenon of nuclear magnetic resonance (NMR) occurs (Rollwitz 1965). The theory, effects of food components, physical properties, and uniformity of test specimens on moisture determination by NMR were discussed by Shaw *et al.* (1953). In NMR instruments, the frequency of the magnetic field is kept constant and the strength of the field is varied over a narrow range. The peak-to-peak amplitude of the signal of a sample in the field can be obtained from a calibration curve of amplitude plotted against moisture content. The method depends on the absorption of radio-frequency energy by the hydrogen nucleus, and is therefore not specific for water. With most materials, however, it is easy to distinguish between the signals contributed by the hydrogen of absorbed water and that which is present in more hindered forms (Conway *et al.* 1957). Elsken and Kunsman (1956) reported that NMR methods are satisfactory for moisture determinations in foods with limited amounts of soluble solids, as the hydrogen nuclei of the solutes absorbed energy in the NMR spectrometer in a manner undistinguishable from those of water. In the case where the soluble solids were constant and preferably small, a calibration curve may be made. A limiting factor in the high-moisture range was the nonuniform character of many biological materials in their natural state.

Although the technique of NMR requires somewhat elaborate equipment, the method has sufficient advantages to have prompted extensive investigations. The usefulness of the technique in food analysis was described by Shaw *et al.* (1953), Conway *et al.* (1957), and Rubin (1958). NMR measurements are rapid (actual measurement requires 1 min), nondestructive, accurate, and for many foods more precise than drying techniques. Calibrated standards can be kept sealed in a glass tube and used repeatedly over a period of time. The relatively large sample size reduces difficulties caused by nonuniformity of specimen. The wide band NMR commercial instruments can be used for both moisture and lipid determination.

Electrical Methods.—The relatively long time required to carry out most moisture determinations has prompted the development of numerous rapid electrical methods. The most commonly used ones are based on the determination of conductivity or capacitance. The usefulness and limitations of electrical methods were outlined by Schupp and Mecke (1948). The main difficulty stems from the fact that moisture determinations by electrical methods in foods are affected by many factors including the texture of the food, packing, mineral contents, temperature, and moisture distribution. The relation between the moisture content and the measured parameter is empirical and assumes that the effects of other factors have been minimized or accounted for by proper calibration.

In conductivity measurements, the conductivity or resistance of a food introduced into an electrical circuit is measured. The measurement is most commonly made on instruments using a modification of the Wheatstone bridge.

Several instruments, widely used by the industry for rapid, routine determination of moisture by electrical methods, are based on measuring resistance. These include the Universal meter, Marconi, and Tag-Heppenstall meter. In the Universal meter, a hand-driven megger establishes a voltage across a sample of 20 gm pressed in a steel cup to a specified thickness. The electrical resistance is indicated on an ohmmeter of the dynamometer type. A correction scale for temperature is provided. The instrument requires neither batteries nor power supply. The Marconi instrument is battery-operated. A sample is compressed in a small cell, the bottom of which is equipped with electrodes consisting of two circular rings. The sample need not be weighed accurately, as the current penetrates the sample only to a depth of the order of the electrode separation. For many years, the Tag-Heppenstall (also known under the name of the Weston Moisture Meter) was widely used in the determination of moisture in cereal grains. In that instrument, the electrical resistance of the grain is measured as it passes between 2 corrugated steel rolls (1 motor driven and 1 stationary) that serve as electrodes. The grain is crushed to provide good electrical contact for measuring the conductance or resistance of the grain as it passes between the rolls. Galvanometer readings and temperature corrections are used in reading moisture contents from calibration charts. Moisture determinations by the Tag-Heppenstall are rapid (10–20 sec) and precise (standard error of estimation about ± 0.2%), but it is difficult to maintain the calibration, as wearing of the rolls and bearings increases the spacing between the rolls, giving lower results.

In dielectric type meters, the determination is based on the fact that the dielectric constant of water at 20°C (80.37) is higher than that of most fluids and solvents. The heart of the instrument is a capacitance cell in which two metal plates of given size are at a fixed distance. Measurements can be made over the range −20°C to 180°C. The two plates have equal, but opposite charges that are reversed at fixed frequencies to give an alternating current field. If a material is introduced between the plates, polarization takes place. With nonpolar substances, induced dipoles are formed in the electrical alternating field. Polar substances, with permanent dipoles, are oriented in the alternating field. This increases the charge of the plates and their capacitance. Consequently, the determination is based on measuring the capacitance, and the dielectric constant that it measures is an index of the increase in capacitance. It is determined by comparing the capacitance C of a capacitance cell filled with the tested dielectric substance to the capacitance C_o of an empty (vacuum) condenser

$$C = \epsilon \cdot C_o \qquad (33.7)$$

ϵ is a dimensionless constant that is affected strongly by temperature and frequency. The dielectric-type meters most widely used in testing cereals are the Steinlite and the Motomco moisture meters. In the Steinlite meter, dry grain is introduced into a narrow cell of the meter, the long walls of which are plates of a parallel plate condenser. Galvanometer readings at an appropriate radio frequency range, temperature corrections, and apparent specific gravity are included in computing moisture contents from calibration charts. In determining moisture with the Steinlite in samples with high water content (i.e. sweet corn for canning), the samples are wet ground with a hygroscopic solvent. The filtered solvent containing the moisture from the corn sample is used as the dielectric constituent in a coaxial cylindrical condenser. Changes in the dielectric characteristics of the solvent-corn extract mixture can be calibrated in terms of the moisture content of the sample (Geiser *et al.* 1951).

The Motomco electric moisture meter was made official in 1959 for most grain in Canada (Martens and Hlynka 1965), and has been used widely in the United States since 1963 (Hunt and Neustadt 1966). The principle of the instrument operation in the balancing of two oscillating circuits as indicated by a milliammeter. One oscillator is fixed to oscillate at 18 Mc, while the other that contains the test cell for holding the grain can be adjusted to balance with the fixed oscillator by means of a variable condenser coupled to a graduated dial.

In addition to conductivity and capacitance measuring instruments, other types of electric meters are available. The signal strength from a radio transmitter falls off as the distance increases. In air, this decrease follows the inverse square law. If water or a material containing water is inserted between the transmitter and the receiver, an additional factor affects the results. The absorption of electromagnetic energy by water forms the basis of a method of measuring moisture content by microwave absorption (Watson 1965). Moisture measurements by high frequency offer the advantage that they can be made by means of electrodes that do not come in contact with the material. The measurements can be made in three ways: (1) measurements of the capacitance of a flat capacitor in which the material to be measured acts as the dielectric, (2) measurement of the resistance of the material at high frequency, and (3) measurement of the impedance of an induction in which the material acts as a core (Leroy 1965).

For industrial processing, the methods of continuous moisture measurement that have been most commonly applied include resistance, capacitive reactance, power loss at microwave and radio frequencies, temperature difference due to evaporation, thermal neutron capture by hydrogen, neutron reflection by hydrogen, and nuclear magnetic resonance (Green 1965). Of these methods, capacitive reactance and resistance have been the most widely applied. The equipment employed for the two methods is generally more simple and less expensive than for the others.

Electrical instruments for moisture determination have the important advantage of speed. Some of the instruments give average accurate results, if the meters are properly calibrated and checked for uniform performance. All instruments are affected by distribution of moisture in a food. For precise measurements, essentially complete equilibration of moisture is essential. Readings on all meters are also affected by temperature; in some, correction for temperature is so high that a serious error in temperature measurement introduces a serious error in moisture determination. Moisture determinations by capacitance measurements are theoretically more satisfactory than by conductivity measurements. The latter measure free water only, and assume a constant ratio between free and bound water. Dielectric moisture meters measure both free and bound water, though the great difference in the dielectric value of the two types of water may introduce errors in the measurements when the ratio of the two shifts. Conductivity measurements are much more affected than dielectric constants by the composition of soluble solids. To reduce that effect, Hancock and Burdick (1956) proposed a modified indirect conductivity method for determining water in foods and feeds. Adding a large amount of sodium chloride to an alcohol-acetone extraction mixture tends to mask the effect of other electrolytes that may be present in a sample.

Particle size and tightness of packing in the measuring cell affect all electrical measurements of moisture content. Hughes et al. (1965) showed that a measurement of capacitance with a correction based on the simultaneous determination of density improved correlation with moisture content.

Miscellaneous Physical Methods.—Many densimetric, refractometric, and polarimetric methods have been devised for the rapid determination of total solids in foods. The methods are discussed in detail in Chap. 25, 26, and 27. In all those methods, calibration curves are available or must be prepared to correlate soluble solids with the selected physical parameter. To determine the total solids contents (or moisture by difference), the insoluble solids must be determined or assumed.

Densimetric methods (by pycnometer, specific gravity balance, or various types of hydrometers) are the most commonly used routine tests to determine dry solids in milk (in combination with the fat determination), sugar solutions (including fruit juices, syrups, concentrates), fruit products (particularly tomatoes), beverages (alcoholic and malts), and salt solutions (used by the pickling industry). Refractive index measurements give the most reliable, reproducible, and rapid values for estimating the soluble solids contents of sucrose solutions, fruits, fruit products (jellies, jams, juices, tomato products) and dried fruit (Bolin and Nury 1965), corn syrups, honey, candy, and various carbohydrate-rich products. The solids content can be determined rapidly and precisely by refractometry in milk products, protein solutions, and oil-rich foods. Polarimetric methods are widely used to determine the identity and concentrations of sugars.

The freezing point of unadulterated milk is close to $-0.550°C$; it is lowered approximately $0.0055°C$ for each 1% of added water. The percentage of added water can be determined by a cryoscopic method.

The methods described thus far were designed to measure (directly or indirectly) the total moisture content of foods. In many instances the total moisture content may be of limited significance. Moisture within a food is not distributed uniformly. Deterioration of a food might be a function of the moisture content of the component most susceptible to deterioration. Under those conditions the vapor pressure of water above (or relative humidity of the atmosphere in equilibrium with) the food may provide more meaningful information (Makower 1950). In foods, with a relatively uniform distribution of water, relative humidity elements can be used to determine moisture content rapidly and precisely (Tessem et al. 1965).

BIBLIOGRAPHY

Assoc. Offic. Agr. Chemists. 1965. Official Methods of Analysis of the Assoc. Agr. Chemists. 10th Edition. Assoc. Offic. Agr. Chemists, Washington, D.C.

Ben-Gera, I., and Norris, K. H. 1968. Direct spectrophotometric determination of fat and moisture in meat products. J. Food Sci. 33, 64–67.

Bidwell, G. L., and Sterling, W. F. 1925. Preliminary notes on the direct determination of moisture. J. Ind. Eng. Chem. 17, 147–149.

Bolin, H. R., and Nury, F. S. 1965. Rapid estimation of dried fruit moisture content by refractive index. J. Agr. Food Chem. 13, 590–591.

Borker, E., and Sloman, K. J. 1963. Food. Anal. Chem. 35, 62R–77R.

Brand, E., and Kassell, B. 1942. Analysis and minimum molecular weight of β-lactoglobulin. J. Biol. Chem. 145, 365–378.

Brekke, J., and Conrad, R. 1965. Gas liquid chromatography and vacuum oven determination of moisture in fruits and fruit products. J. Agr. Food Chem. 13, 591–593.

Bunsen, R. 1953. A volumetric method of wide application. Ann. Chem. 86, 265.

CAVAGNOL, J. C. 1961. Food. Anal. Chem. *33*, 50R–61R.

CHRISTENSEN, C. M., and LINKO, P. 1963. Moisture contents of hard red winter wheat as determined by meters and by oven drying, and influence of small differences in moisture content upon subsequent deterioration of the grain in storage. Cereal Chem. *40*, 129–137.

CONWAY, T. F., COHEE, R. F., and SMITH, R. J. 1957. NMR moisture analyzer shows big potential. Food Eng. *29*, No. 6, 80–82.

DAVIS, C. E., OCKERMAN, H. W., and CAHILL, V. R. 1966. A rapid approximate analytical method for simultaneous determination of moisture and fat in meat and meat products. Food Technol. *20*, No. 11, 94–98.

DEAN, E. W., and STARK, D. D. 1920. A convenient method for the determination of water in petroleum and other organic emulsions. J. Ind. Eng. Chem. *12*, 486–490.

ELSKEN, R. H., and KUNSMAN, C. H. 1956. Further results of moisture determination of foods by hydrogen nuclei magnetic resonance. J. Assoc. Offic. Agr. Chemists *39*, 434–444.

FETZER, W. R. 1951. Determination of moisture by distillation. Anal. Chem. *23*, 1062–1069.

FISCHER, K. 1935. New methods for the quantitative determination of water in fluids and solids. Angew. Chem. *48*, 394.

FOSNOT, R. H., and HAMAN, R. W. 1945. A preliminary investigation of the application of Karl Fischer reagent to determination of moisture in cereals and cereal products. Cereal Chem. *22*, 41–49.

FRANCOIS, M. Th., and SERGENT, A. 1950. Determination of moisture in oil seeds by the method of Fischer. Bull. Mens. ITERG *4*, 401–404.

FREDIANI, H. A. 1952. Automatic Karl Fischer titration. Apparatus using dead-stop principle. Anal. Chem. *24*, 1126–1128.

FREDIANI, H. A., OWEN, J. T., and BAIRD, J. H. 1952. Application of an automatic Karl Fischer titrator to moisture determinations in food products. Trans. Am. Assoc. Cereal Chemists *10*, 176–180.

GARDINER, S. D., and KEYTE, H. J. 1958. A study of some methods for determining water in refined sugars, including the newly devised cobaltous bromide method. Analyst *83*, 150–155.

GEISER, C. E., HOMEYER, P. G., and FISCHER, R. G. 1951. A comparison of three methods for determination of moisture in sweet corn. Food Technol. *5*, 250–253.

GOLD, H. J. 1964. General application of near-infrared moisture analysis to fruit and vegetable materials. Food Technol. *18*, No. 4, 184–185.

GREEN, R. M. 1965. Continuous moisture measurement in solids. *In* Humidity and Moisture, Measurement and Control in Science and Industry, A. Wexler (Editor). Van Nostrand-Reinhold Publishing Co., New York.

GUILBOT, I. 1955. Determination of moisture by the oven drying method. Proc. Intern. Bread Congress, Hamburg, Germany.

HAMM, R. 1963. The water imbibing powder of foods. *In* Recent Advances In Food Science, Vol. 3, J. M. Leitch, and D. N. Rhodes (Editors). Butterworths, London.

HANCOCK, C. K., and BURDICK, R. L. 1956. Modified indirect conductivity method for determining water in cottonseed meal. Agr. Food Chem. *4*, 800–802.

HART, J. R., and NEUSTADT, M. H. 1957. Application of the Karl Fischer method to grain moisture determination. Cereal Chem. *34*, 26–37.

HART, J. R., NORRIS, K. H., and GOLUMBIC, C. 1962. Determination of the moisture content of seeds by near infrared spectroscopy of their methanol extracts. Cereal Chem. *39*, 94–99.

HLYNKA, I., and ROBINSON, A. D. 1954. Moisture and its measurement. *In* Storage of Cereal Grains and Their Products, J. A. Anderson, and A. W. Alcock (Editors). Am. Assoc. Cereal Chem., St. Paul, Minn.

HUGHES, E. J., VAALA, J. L., and KOCH, R. B. 1965. Improvement of moisture determination by capacitance measurement through density correction. *In* Humidity and Moisture, Measurement and Control in Science and Industry, A. Wexler (Editor). Van Nostrand-Reinhold Publishing Co., New York.

HUNT, W. H., and NEUSTADT, M. H. 1966. Factors affecting the precision of moisture measurement in grain and related crops. J. Assoc. Offic. Agr. Chemists *49*, 757–763.

JONES, A. G. 1951. A review of some developments in the use of the Karl Fischer reagent. Analyst *76*, 5–12.

KAYE, W. 1954. Near infrared spectroscopy. I. Spectral identification and analytical applications. Spectrochim. Acta *6*, 257–287.

KOSTER, A. 1934. Moisture determination by the oven-drying method. (German) Muhle *71*, 1303.

LAUNER, H. F., and TOMIMATSU, Y. 1952. Rapid method for moisture in fruits and vegetables by oxidation with dichromate. I. Potatoes and peas. Food Technol. *6*, 59–64.

LEROY, R. P. 1965. Moisture measurements by high frequency currents. *In* Humidity and Moisture, Measurement and Control in Science and Industry, A. Wexler (Editor). Van Nostrand-Reinhold Publishing Co., New York.

MAKOWER, B. 1950. Determination of water in some dehydrated foods. *In* Advances in Chemistry Series 3. Am. Chem. Soc., Washington, D.C.

MAKOWER, B., CHASTAIN, S. M., and NIELSEN, E. 1948. Moisture determination in dehydrated vegetables, vacuum method. Ind. Eng. Chem. *38*, 725–731.

MAKOWER, B., and NIELSEN, E. 1948. Use of lyophilization in determination of moisture content of dehydrated vegetables. Anal. Chem. *20*, 856–858.

MARTENS, V., and HLYNKA, I. 1965. Determination of moisture in Canadian grain by electric moisture meter. *In* Humidity and Moisture, Measurement and Control in Science and Industry, A. Wexler (Editor). Van Nostrand-Reinhold Publishing Co., New York.

MATCHETT, J. R., and VON LOESECKE, H. W. 1953. Food. Anal. Chem. *25*, 24–30.

McCOMB, E. A., and WRIGHT, H. W. 1954. Application of formamide as an extraction solvent with Karl Fischer reagent for the determination of moisture in some food products. Food Technol. *8*, 73–75.

MEELHEIM, R., and ROARK, J. N. 1953. Determination of moisture in oils and greases. Anal. Chem. *25*, 348–349.

MITCHELL, J. 1951. Karl Fischer reagent titration. Anal. Chem. *23*, 1069–1075.

MITCHELL, T. J. 1950A. The rate of evaporation in the determination of water. Chem. Ind. 751.

MITCHELL, T. J. 1950B. The determination of water in sugar solutions by desiccation at room temperature. Chem. Ind. 815.

MORGAREIDGE, K. 1959. Food. Anal. Chem. *31*, 691–696.

NELSON, O. A., and HULETT, G. A. 1920. The moisture contents of cereals. Ind. Eng. Chem. *12*, 40–45.

NORRIS, K. H., and HART, J. R. 1965. Direct spectrophotometric determination of moisture content of grain and seeds. *In* Humidity and Moisture, Measurement and Control in Science and Industry, A. Wexler (Editor). Van Nostrand-Reinhold Publishing Co., New York.

OSER, B. L. 1949. Food. Anal. Chem. *21*, 216–227.

OXLEY, T. A., and PIXTON, S. W. 1961. Determination of moisture content in cereals. II. Errors in determination by oven drying of known changes in moisture content. J. Sci. Food Agr. *11*, 315–319.

OXLEY, T. A., PIXTON, S. W., and HOWE, R. W. 1960. Determination of moisture content in cereals. I. Interaction of type of cereal and oven method. J. Sci. Food Agr. *11*, 18–25.

ROLLWITZ, W. L. 1965. Nuclear magnetic resonance as a technique for measuring moisture in liquids and solids. *In* Humidity and Moisture, Measurement and Control in Science and Industry, A. Wexler (Editor). Van Nostrand-Reinhold Publishing Co., New York.

RUBIN, H. 1958. New tool for moisture analysis. Nuclear magnetic resonance. Cereal Sci. Today *3*, 240–243.

SAIR, L., and FETZER, W. R. 1942. The determination of moisture in the wet milling industry. Cereal Chem. *19*, 633–692, 714–720.

SCHIERBAUM, F. 1957. Determination of moisture by infrared in food analysis. Dtsch. Lebensmittel-Rdsch. *53*, 173–178.

SCHROEDER, C. W., and NAIR, J. H. 1948. Determination of water in dry food materials. Karl Fischer method. Anal. Chem. *20*, 452–455.

SCHUPP, R., and MECKE, R. 1948. Precise dielectric measurements of associating materials. Z. Elektrochem. *51*, 40–44, 54–60.

SCHWECKE, W. M., and NELSON, J. H. 1964. Determination of moisture in foods by gas chromatography. Anal. Chem. *36*, 689–690.

SHAW, T. M., ELSKEN, R. H., AND KUNSMAN, C. H. 1953. Moisture determination of foods by hydrogen nuclei magnetic resonance. J. Assoc. Offic. Agr. Chemists *36*, 1070–1076.

SLOMAN, K. G., and BORKER, E. 1965. Food. Anal. Chem. *37*, 70R–86R.

SMITH, D. M., and BRYANT, W. M. D. 1935. Titrimetric determination of water in organic liquids, using acetyl chloride and pyridine. J. Am. Chem. Soc. *57*, 841–845.

TESSEM, B. M., HUGHES, F. J., PEARCY, G., and TSANTIR, K. 1965. Description and test evaluation of the Honeywell relative humidity flour moisture meter. Cereal Sci. Today *10*, 50–52, 62.

TRIPPEER, A. 1965. Infrared analysis of water. *In* Humidity and Moisture, Measurement and Control in Science and Industry, A. Wexler (Editor). Van Nostrand-Reinhold Publishing Co., New York.

VON LOESECKE, H. W. 1957. Food. Anal. Chem. *29*, 647–656.

WARD, A. G. 1963. The nature of the forces between water and the macromolecular constituents of food. *In* Recent Advances in Food Science, Vol. 3. J. M. Leitch, and D. N. Rhodes (Editors). Butterworths, London.

WATSON, A. 1965. Measurement and control of moisture content by microwave absorption. *In* Humidity and Moisture, Measurement and Control in Science and Industry, A. Wexler (Editor). Van Nostrand-Reinhold Publishing Co., New York.

WEISE, E. L., BURKE, R. W., and TAYLOR, J. K. 1965. Gas chromatographic determination of moisture content of grain. *In* Humidity and Moisture, Measurement and Control in Science and Industry, A. Wexler (Editor). Van Nostrand-Reinhold Publishing Co., New York.

WILLITS, C. O. 1951. Methods for determination of moisture. Oven drying. Anal. Chem. *23*, 1058–1062.

CHAPTER 34

Ash and Mineral Components

INTRODUCTION

Ash is the inorganic residue from the incineration of organic matter. Its content and composition depend on the nature of the food ignited and on the method of ashing. Most fluid dairy products contain 0.5 to 1.0% ash; the ash content increases to 1.5% in evaporated milk, and to almost 8% in nonfat dry milk. Ash in cheese depends on their water content and on the presence of mineral additives. Pure fats, oils, and shortenings contain practically no mineral components. The main mineral component in butter, margarine, mayonnaise, and salad dressing is sodium chloride. Ash in fresh fruits ranges from 0.2 to 0.8% and is generally inversely related to moisture content. Some dried fruits (i.e. apricots) may contain as much as 3.5% ash. The white endosperm fraction obtained during milling of wheat contains less than 0.5% ash; the bran, aleurone layer, and germ are rich in minerals. The ash of baked products depends mainly on their salt content. Most nuts contain 1.5 to 2.5% ash. Fresh meat and poultry contain around 1% mineral components, but the ash of processed meats may be as high as 12% (in dried and salted beef). The ash content of the edible portion of fresh fish ranges from 1 to 2%. The yolk contains almost 3 times as many mineral components as the white of eggs (respectively, 1.7 and 0.6%). Pure sugar, candy, honey, and syrups contain trace amounts to 0.5% ash; but sweets made with brown sugar or chocolate contain more inorganic components. The ash content of most vegetables is around 1%, and is generally higher than of fruits. Pickles, sauerkraut, and other processed vegetables are rich in salt and their ash content is high. Beans contain up to 4% ash.

Among the mineral components, calcium is present in relatively high concentrations in most dairy or dairy-containing products, in cereals, nuts, some fish, eggs, and certain vegetables. Small concentrations of calcium are present in practically all foods, except pure sugar, starch, and oil. Phosphorous-rich foods include most dairy products, grains and grain products, nuts, meat, fish, poultry, eggs, and legumes. Smaller concentrations of phosphorus are present in most other foods. Iron is present in relatively high concentrations in most grains and grain products, in flours, meals and other farinaceous materials, and in baked and cooked cereals (especially enriched ones). Most nut and nut products, meat, poultry, seafoods, fish and shell fish, eggs and legumes are good sources of iron. Smaller amounts of iron are present in most dairy products, fruits and vegetables, and in some sweets. Salt is the main source of sodium in all salted foods. Most dairy products, fruits, cereals and processed cereals, nuts, meat, fish, poultry, eggs, and vegetables contain substantial amounts of potassium. Magnesium can be found in relatively high concentrations in nuts, cereals, and legumes; manganous-rich foods include cereals, vegetables, and some fruits and meats. In addition to some seafoods and liver, cereals and vegetables are good sources of copper. Sulfur is well-distributed among most protein-rich foods and some vegetables; vegetables and fruits are good sources of cobalt. Some seafoods are particularly rich in zinc, that is present in lesser amounts in most classes of foods.

The determination of mineral constituents in foods can be divided into two classes: assay of ash (total, soluble, insoluble), and the determination of individual components. Total ash is a widely accepted index of refinement of foods, such as wheat flour or sugar. The objectives of flour milling are to separate starchy endosperm from bran and germ and, subsequently, to reduce endosperm particles to flour. Since the mineral content of the bran is about 20 times that of the endosperm, the ash test fundamentally indicates the purity of the flour or thoroughness of the separation of bran and germ from the rest of the wheat kernel. The ash test has assumed greater importance in the milling trade than any other test for the control of the milling operation. In refining cane sugar, excessive amounts of minerals interfere with processing (decolorization and crystallization). On the other hand, adequate levels of total ash are indicative of functional properties of some foods products (i.e. gelatin). In fruit jellies and marmalades, ash is determined to estimate the fruit content of the product; total electrolyte content can be used to determine adulteration of some juices and beverages. Levels of ash and ash alkalinity are useful parameters in distinguishing fruit vinegar from synthetic vinegar. Ashing of tissue slices is useful in histological identification. Total ash content is a useful parameter of the nutritional value of some foods and feeds. High levels of acid-insoluble ash indicate the presence of sand or dirt.

Mineral components present in biological systems can be divided into those that are indispensable for normal metabolism and generally constitute essential dietary elements, and those that have no known function or are even deleterious. The latter may gain access from the soil, from spraying of plant materials, or from industrial processes. In addition to the nutritional significance of mineral elements, their physiological and technological aspects should be considered. Some metallic residues (i.e. lead or mercury) may have toxic effects. The oxidation of ascorbic acid and the stability of fruit juices are affected by copper. Certain mineral components enhance and others impede panary fermentation and the quality of fermented beverages. Mineral components may affect cold hardiness of cereal grains, and storability of fruits and vegetables.

Determination of Ash

Ash in foods is determined by weighing the dry mineral residue of organic materials heated at elevated temperatures (around 550°C). The details of ashing procedures are described later in this chapter (dry ashing in the assay of mineral components). The form of the mineral constituents in ash differs considerably from the form of those components in the original food. Thus, calcium oxalates are converted into carbonates, and upon further ashing to oxides. Some trace minerals, linked to biologically active systems, are converted to inorganic components. Yet, the total mineral content is a meaningful analytical parameter, and dry ashing is often used prior to elementary analysis.

In addition to the determination of total mineral content by ashing, indirect methods are available to determine the total electrolyte content of foods. Conductometric methods provide a simple, rapid, and accurate means of determining the ash content of sugars (Gillet 1949). Such foods are generally low in minerals and direct ashing requires incineration of large samples rich in strongly foaming carbohydrates. Conductometric methods are based on the principle that in a solution of the sugar, the mineral matter that constitutes the ash dissociates; whereas the sucrose, a nonelectrolyte, does not dissociate. The conductance of the solution is therefore an index of the concentration of the ions present and of the mineral or ash content. Numerous papers have been published on the subject and the methods are described in detail in a number of handbooks (Browne and Zerban 1941; Bates 1942).

Results of conductometric ash determinations in sugar products are affected somewhat by nonelectrolytes, and an experimental correction factor—depending on the assayed food—is determined. In certain foods conductivity is measured in a medium acidified to displace all the weak acids of the salts. The assay requires two conductivity measurements (before and after acid addition). In syrups or molasses, a second conductivity determination in an alkaline medium gives more precise values.

Electrolyte concentration of sugar products can be determined also by an ion-exchange method (Pomeranz and Lindner 1954). The determination is based on the fact that if a solution containing several salts is passed through a cation exchange column (in hydrogen form), the effluent contains a quantity of acid equivalent to the original salt content. By titration of the liberated acid, the total electrolyte content of the tested solution can be determined.

Water-soluble ash is sometimes used as an index of the fruit content of jelly and fruit preserves. Acid insoluble ash is a useful index of mineral matter (dirt or sand in spices), efficiency of wheat washing prior to milling, talc in confectionery, coating of rice, or surface contamination of fruits and vegetables. Acid insoluble ash is determined after digesting the total ash in 10% hydrochloric acid.

The ash of fruits and vegetables is alkaline in reaction, that of meat products and of certain cereals is acid. Alkaline ash is ascribed to the presence of the salts of organic acids that are converted on ashing into the corresponding carbonates. Alkalinity of ash was used for many years to analyze fruit juices. In foods rich in fruit acids or salts, ash alkalinity is an index of fruit content. The results are, however, affected by the presence of phosphates, and the effect of the latter varies with the ashing procedure. The significance of ash alkalinity determination has decreased with the availability of excellent methods for the determination of individual organic acids. The assay is occasionally made to determine the relative amounts of cream of tartar in grape products, in detecting adulteration of foods with minerals, and in determining the acid-base balance of foods.

Salt-free ash is especially important in seasoned foods. It is determined as the difference between total ash and sodium chloride in the ash (as assayed titrimetrically in ash dissolved in dilute nitric acid). Salt in foodstuffs can be determined without ashing by a potentiometric titration of chloride with silver nitrate, with a silver-silver chloride electrode as an indicator (Cole 1967).

Assay of Mineral Components

Sampling and Contamination Problems.—The problem of contamination in elemental analysis has been summarized in an excellent review by Thiers (1957). Trace element analysis and contamination are unavoidably linked. The contamination can be minimized by stringent precautions at all stages of the work—sampling, storage of samples, preparation for analysis, and actual assay.

Foods, and especially plant materials, are subject to surface contamination because of their format and habitat (Mitchell 1960). The impurities must be removed. The foods must be mixed thoroughly. Lipids are practically free of minerals, and as animal foods may vary widely in their lipid content, such variations may cause large differences in mineral content. Smith *et al.* (1968) reported that the grinding of plant materials may result in a segregation into fractions of varying particle size and elemental composition.

Wet or liquid samples are usually dried in ovens. Conditions for corrosion are ideal in such ovens and the perforated metal shelves assist in making them a likely source of contamination. To prevent contamination, the samples are often covered with fluted watch glasses so as not to interfere with drying. Such precautions have been found insufficient to prevent contamination particularly in fume hoods or in ovens with forced air circulation (Thiers 1957). Individual small glass chambers are preferable. Contamination may come from the container in which samples are dried at elevated temperatures, i.e. arsenic, zinc, lead, and other metals in borosilicate glass. Hardened steel mortars, widely used for powdering hard and brittle materials, result in severe contamination. It is practically impossible to make a correction for such a contamination, because of its extremely erratic nature. Grinding by hand with a mortar and pestle is relatively safe. However, for very precise work, a new mortar may be required to eliminate contamination from previous samples. Metallic sieves for classifying ground material may cause problems; plastic (i.e. nylon) sieves are safer. In wet digestion, contamination from reagents and containers can cause erratic results. Contamination from the interior walls and ceiling of the muffle furnace present problems in dry ashing.

Dry Ashing.—In dry ashing, the sample is weighed into a dish and the organic matter is burned off without flaming, and heated either for a fixed period of time or to constant weight. The residue must be free from carbon. The dish containing the residue is cooled in a desiccator and the amount of total ash is determined by weighing. The selection of ashing dishes (crucibles) depends on the nature of the food analyzed, and on the analyses that are to be performed on the ash. The materials used include quartz, Vycor, porcelain, steel, nickel, platinum, and a gold-platinum alloy. Quartz dishes, smooth on the inside, are resistant to halogens, neutral solutions, and acids (except hydrogen fluoride and phosphoric acid) at most concentrations and temperatures. Resistance to alkali is relatively poor. The crucibles are stable at high temperatures (up to 1100°C for routine work) and can be cleaned with hot dilute hydrochloric acid. Porcelain crucibles resemble quartz in chemical and physical properties. Temperature resistance is even higher— unglazed crucibles can withstand up to 1200°C for routine work. The crucibles retain their smooth surface and are easy to clean with dilute hydrochloric acid. They are widely used because of their good weight constancy and relatively low price, but are, however, susceptible to alkali and crack from sudden large temperature changes. This latter property makes their use somewhat troublesome. Steel is sometimes used for ashing large samples. The low price and relatively good resistance of some alloys (i.e. steel containing 18% chromium and at least 8% nickel) to acid and alkali make their use attractive. The crucibles are cleaned mechanically with fine sand or steel wool. Nickel crucibles are used little as they deteriorate due to nickel carbonyl formation in reducing atmospheres.

Platinum is the best widely used crucible material, but is too expensive for routine ashing of large numbers of samples. Platinum has a high melting point (1773°C), good heat conductivity, and high chemical inertness. Platinum dishes can corrode and their high price makes it mandatory to eliminate conditions conducive to such corrosion. Platinum dishes should be cleaned thoroughly as dirt-containing organic matter has a reducing-corroding action. Corrosion may be caused by reduction to metals of oxides of iron, lead, and tin. The elements lead, arsenic, antimony, silicium, and bismuth and their compounds are considered platinum poisons. Corrosion from heavy metals leads to pitting and formation of holes. Some elements give phosphides and silicides that form low melting point eutectic mixtures, weaken the platinum crucible or dish, and cause cracks. Platinum crucibles should be touched with platinum-tipped tongs and placed after ashing on clean porcelain, asbestos, or marble surfaces. They can be cleaned by boiling with water or acids, but not with aqua regia or hydrochloric acid in the presence of strong oxidants. Silicic acid residues can be removed with dilute hydrofluoric acid. Cleaning by melting (if necessary repeatedly) with potassium pyrosulfate is often effective. Mechanical cleaning should be avoided and performed, if necessary, carefully with acid-washed clean sea sand. Dishes from gold-platinum (90:10) melt at 1100°C, but are superior to pure platinum dishes in resistance to phosphoric acid and alkali melting.

For total ash determinations, where recovery and determination of individual metals is not necessary, ashing in porcelain crucibles at temperatures ranging from 400° to 700°C (most commonly around 550°C) is satisfactory. If the ash components are to be determined, the biological material that is to be ashed and the elements that are to be determined must be considered individually. According to Grant (1951), iron, aluminum, copper, tin, silicon, and magnesium can be determined as oxides in platinum crucibles. Porcelain crucibles have been suggested for the determination of chromium as the metal. Sodium, potassium, lithium, magnesium, calcium, strontium, barium, cadmium, manganese, and lead can be determined in platinum crucibles as sulfates. In the case of lead compounds, nitric acid must be added to avoid the reduction to metallic lead and possible crucible damage. Ashing over an open flame requires constant attention, and except in cases of ashing samples that froth and bubble excessively, only the initial stage of ashing is done with an open flame. A furnace with a rheostat for temperature control is used for most routine work.

If prolonged ashing fails to give a carbon-free ash, the residue should be moistened, dried, and reheated until a white-gray ash remains. In some cases, it may be necessary to dissolve the ash in a small amount of water, filter the carbon-containing residue through a small low-ash filter paper, dry the two parts, and ash separately. If water fails to break up the material the residue may be treated with a few drops of hydrogen peroxide, nitric acid and/or sulfuric acid; but in the latter cases the composition of the ash is changed and special precautions must be taken to report the correct type of ash (Dunlop 1961).

Dry ashing for the destruction of organic matter prior to the determination of trace elements is not used extensively because it is generally believed that losses occur from volatilization. According to Lynch (1954), dry ashing is the most satisfactory method if no loss occurs at temperatures up to about 500°C. The method cannot be used for arsenic and mercury; its usefulness for lead is uncertain; and iron can sometimes be troublesome owing to the difficulty of getting the metal into solution after ashing is completed. Thiers (1957) summarized reported losses during dry ashing (Table 34.1). Most of the losses can be minimized if proper ashing conditions are used. Thiers (1957) recommended a dry ashing procedure for biological material that involves drying and preashing the sample in a special apparatus with the aid of a hot plate and an infrared lamp (Fig. 34.1). The temperature of the hot plate surface is gradually raised to about 300°C. This gives the sample a charred appearance. Ashing is completed in a muffle furnace with a controlled temperature that starts at 250°C, and is raised within about 1 hr gradually to 450°C where it is held until the decomposition of organic matter is complete.

TABLE 34.1

SUMMARY OF REPORTS ON LOSSES DURING DRY ASHING

Metal	Conditions
Arsenic	Loosely bound as in blood; may volatilize as unknown compound at 56°C
Boron	Volatilizes with steam from acid solutions
Cadmium	Volatilizes, possibly as the chloride, or metal, between 400° and 500°C
Chromium	Volatilizes as chromyl chloride at low temperatures under oxidizing conditions
Copper	Volatilizes as porphyrin compounds when petroleum samples are burned
	Volatilizes from vinegar, possibly as copper acetate at low temperatures
	Reduces to metal which is not dissolved by hydrochloric acid
Iron	Volatilizes as ferric chloride at 450°C
	Volatilizes as porphyrin compounds when petroleum samples are burned
	When materials with a high phosphorus-to-iron ratio are ashed, an unidentified compound is formed which resists solution or hydrolysis, causing low results
Lead	Volatilizes from blood or petroleum unless sulfate is present
Mercury	Volatilizes as metal below 450°C
Nickel	Volatilizes as porphyrin compounds when petroleum samples are burned
Phosphorus	Volatilizes, presumably as one of the oxy acids, especially when sulfate is present, except in the presence of excess magnesium
Vanadium	Volatilizes as porphyrin compounds when petroleum samples are burned
	Volatilizes as the chloride below 450°C
Zinc	Volatilizes, presumably as the chloride, above 450°C

Gorsuch (1959) found that dry ashing of cocoa at 550°C gives satisfactory recovery of antimony, chromium, cobalt, iron, molybdenum, strontium and zinc; but recoveries of arsenic, cadmium, copper, mercury, and silver were not satisfactory. Lead was recovered completely after ashing at 450°C, but recovery was questionable at 550°C. To overcome losses at high temperature, a low temperature dry asher has been developed and is available commercially (Anon. 1967). The instrument uses a high purity stream of excited oxygen, produced by a high frequency electro-

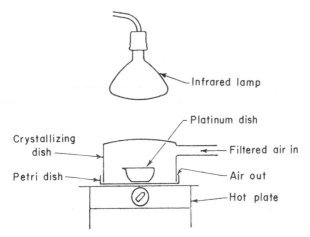

FIG. 34.1.   CHAMBER FOR DRYING OR PREASHING

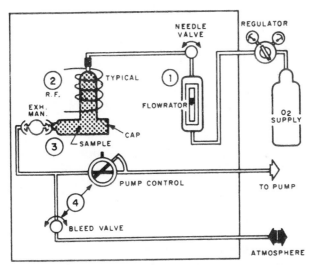

Courtesy Trapelo Lab., Waltham, Mass.

FIG. 34.2.   PRINCIPLE OF OPERATION OF LOW-TEMPERA-
TURE DRY ASHER

Oxygen is admitted through a flowrator (1) and passes
through the radio-frequency electromagnetic field (2) which
produces excited oxygen species that attack the sample in
the oxidation chamber (3).   Exhaust vapors are removed
by a mechanical vacuum pump which is controlled by the
valves (4).

magnetic field as the only oxidizing agent (Gleit and Holland 1962). Figure 34.2 illustrates the principle of operation of the instrument.

Modified Dry Ashing Procedures.—In addition to the simple ashing procedure, certain modifications and additives have been proposed to accelerate the process, to prevent overall losses of minerals, or to improve the retention of critical components.

Liquids and moist materials should be dried prior to ashing. The drying is generally done for moisture determination. It has been suggested that ashing dry materials can be better controlled when filter paper is used to line the dish. Protein-rich materials ash slowly, particularly in salt-rich foods (i.e. salted meat). Ashing at elevated temperatures causes large losses of salt. The use of fixatives and separate ashing of the water-insoluble food and salt-rich water extract may be useful. To carbohydrate-rich materials, that swell and foam excessively, one can add a few drops of pure olive oil after drying and before ashing. The sample is heated over a small flame till foaming ends, and the ashing is completed in a furnace. Fermenting the carbohydrates prior to ashing will generally eliminate foaming difficulties, but is too lengthy and complicated for routine tests. Whenever possible, mineral constituents are estimated in carbohydrate-rich materials by indirect methods (i.e. conductivity). Lipid-rich materials ash rapidly. Schneider (1967) recommends heating such foods till they catch fire and allowing the fat to burn-off. As pure fat is practically ash-free, it may be advisable to determine the ash after an extraction of the fat with an organic solvent. This is the generally recommended procedure if both ash and fat are determined. According to Davidsohn (1948), difficulties due to spattering in determining the ash content of various moist materials, such as jams, may be overcome by adding about 0.2 gm of pure cottonwool of known ash content. The correction to be made is very small.

To accelerate ashing, the addition of small amounts of pure glycerin or alcohol has been suggested. In regular dry ashing, atmospheric oxygen serves as oxidant; chemical oxidants (i.e. hydrogen peroxide) may be added sometimes to accelerate the process. Ammonium carbonate may be useful as an aid, even if ashing is followed by ash analysis. Ammonium nitrate is not recommended as an ashing aid, as it causes puffing and ash losses (Schneider 1967). According to Zonneveld and Gersons (1966), ashing is accelerated by adding an aluminum chloride solution. An accelerated method of ash determination in cereals recommends moistening 3–5 gm of the material with 5 ml of an alcoholic solution of magnesium acetate and incineration at 700°C in a muffle oven. In both the aluminum chloride and magnesium acetate methods, a blank must be subtracted from the ash.

In dry ashing samples rich in silicon and aluminum, an insoluble residue may form. If this occurs, the sample should be fused with a small amount of sodium carbonate. In the cases of silicates, the silica may be volatilized by careful treatment with sulfuric and hydrofluoric acids, taking special precautions to remove fluoride ions before subsequent determinations are made. Special cases of dry ashing may require the addition of a fixative. Six such procedures were described by a committee on analytical methods for the destruction of organic matter (Anon. 1960). A comprehensive survey of ashing procedures was published by Middleton and Stuckey (1953). For the determination of chlorine and boron, it is necessary to ash under alkaline conditions. In the determination of fluorine in foods, the sample is ashed in the presence of calcium oxide. A mixture of aluminum nitrate and calcium nitrate is recommended as an ash aid in the determination of metals. It has been suggested to use magnesium chloride to moisten the charred mass before final ashing of biological material rich in phosphate. As some lead losses occur in regular dry ashing above 550°C, dry ashing with sulfuric acid as an ashing aid permits increasing temperature to 650°C with little lead loss (Gorsuch 1959). The addition of acid slows down the oxidation rate and lengthens the ashing time required. Dry ashing with magnesium nitrate or acetate, or with nitric acid as an ashing aid had little advantage.

In addition to regular furnaces, special tube and combustion furnaces are available for the decomposition of organic compounds on a microscale or prior to an elemental determination. The oxygen bomb method, in which oxygen under pressure replaces air, is widely used for the determination of sulfur and halogens. Elvidge and Garratt (1954) reported that complete combustion can be achieved in a commercial bomb calorimeter in an atmosphere of oxygen without loss of the more volatile components. With light and bulky foods, preliminary compression improves combustion. Wet materials must be dried before combustion. About 3 to 4 gm of material can be burned in one step under an initial pressure of 30 atm of oxygen. An excellent flask combustion (Fig. 34.3) method has been proposed by Schoniger (1955). The method uses a combustion procedure in which the weighed sample is rolled in a small piece of filter paper, and placed in a platinum cup which is suspended from the stopper of a 250-ml conical flask, previously filled with oxygen. The filter paper is ignited, plunged back into the flask, where it burns brightly for a few seconds. The products of combustion are absorbed in a few milliliters of water in the bottom of the flask, the final determination being made on that solution. Corner (1959) described the use of the procedure for the rapid microdetermination of organically-bound halogens, arsenic, phosphorus, and boron.

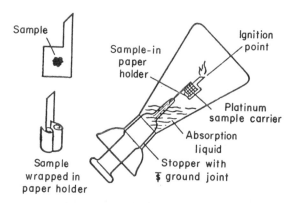

FIG. 34.3. EQUIPMENT FOR OXYGEN FLASK COMBUSTION

A method intermediate between dry and wet ashing was proposed by Wirthle (1900). The sample is digested with sulfuric acid until a porous mass is formed. Digestion of the partly decomposed material is completed by ashing after adding soda and sodium nitrate.

Wet Ashing.—Wet ashing is used primarily for the digestion of samples for determining trace elements and metallic poisons.

The use of a single acid is desirable, but usually not practical for the complete decomposition of organic material. Sulfuric acid is not a strong oxidizing agent and the time required for decomposition is long. Adding a salt, i.e. potassium sulfate, raises the boiling point of the acid and accelerates decomposition. The technique is particularly useful for samples in which adding nitric acid causes the formation of insoluble oxides. Nitric acid alone is a good oxidant, but it usually boils away before the sample is completely oxidized. Middleton and Stuckey (1954) described a method of destroying organic matter at temperatures below 350°C by digestion with nitric acid as the only major reagent. Small amounts of sulfuric acid are added at the initial charring stage to prevent the ignition of fat-rich materials when only nitric acid is added. Gorsuch (1959) found the method generally satisfactory, though recoveries were slightly lower than with conventional wet oxidation methods. Selenium and mercury were lost almost completely. The procedure is somewhat tedious and time-consuming.

Mixed acids are the usual reagents for the decomposition of organic material. The use of a mixture of sulfuric and nitric acids is recommended by many workers, and is the most acceptable procedure. Suggested quantities of each acid, order, and rate of addition vary with different biological materials and investigators.

To avoid excessive foaming in the digestion of fat or sugar-rich materials, it may be advisable to add sulfuric acid, allow it to soak in, add nitric acid in small portions, with heating in between. In later digestion stages, hydrogen peroxide may be added to complete digestion.

The use of perchloric acid with nitric acid or with nitric-sulfuric acid mixtures has been suggested by Smith (1953) for the rapid decomposition of many organic compounds that are difficult to oxidize. Perchloric acid is an excellent oxidant at elevated temperatures and above 60%, but can be explosive and very dangerous if improperly used. It is used routinely by many laboratories. Five grams of wheat can be completely digested in 10 min with HNO_3 + 70% $HClO_4$ (1:2) compared with 8 hr using the usual HNO_3 + H_2SO_4 method. Digestion with perchloric acid should be performed under a special hood containing no plastic ingredients and no glycerol-containing caulking substances. Figure 34.4 shows the apparatus for doing $HClO_4$ digestions in an ordinary hood.

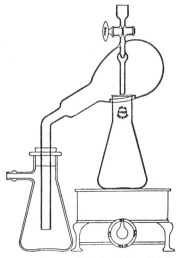

FIG. 34.4. WET OXIDATION APPARATUS

Courtesy G. F. Smith Chemical Co.

Dry Ashing vs Wet Digestion.—Dry ashing is the most commonly used procedure to determine the total mineral content of foods. It is used to determine water soluble, water insoluble, and acid insoluble ash. Dry ashing is applicable also to the determination of most common metals. Dry ashing requires no attention, is simple, and is well-suited to handle routinely large numbers of samples. Generally no reagents are added and no blank substraction is required. Dry ashing takes a long time, but can be shortened by accelerated methods or taken care of by ashing overnight. High temperatures and the relatively expensive equipment limit somewhat the usefulness of the method. The main objection is, however, the interaction between components themselves or the receptacle material. In the estimation of certain trace elements in foods the use of either silica or porcelain crucibles leads to the absorption of the element by the vessels. If the temperature of ashing is not excessively high, the absorption loss may be greater than the volatilization loss. Excessive heating makes certain metallic compounds insoluble (i.e. those of tin). Foods with a high phosphorus to base ratio fuse to a dark melt in which carbon particles are trapped and will not burn. Foods with a high alkaline balance show progressive decomposition of the carbonates and the volatilization of chlorides. High hygroscopicity, lightness, and fluffiness of the ash may sometimes present problems in determining precisely the total ash content or in handling the mineral residue for subsequent analysis. In wet oxidation, relatively low temperatures and liquid conditions are maintained; the apparatus is simple and oxidation is rapid. On the other hand, the procedures require large amounts of corrosive reagents and the correction from the reagents. Handling routinely large numbers of samples by wet digestion is difficult and the required operators' time is large.

Elemental Analysis.—This chapter includes no detailed description of the methods to determine individual mineral components. Such procedures are described in general textbooks of inorganic analysis, standard reference books on food analysis, and specialized textbooks on the determination of minerals in biological materials. A partial list of the latter includes a comprehensive treatise in 13 volumes on the chemistry and the determination of individual elements (Kolthoff and Elving 1961–1966), several general books on trace analysis (Monier-Williams 1953; Yoe and Koch 1955; Sandell 1959; Koch and Koch-Dedic 1964; Morrison 1965; and Bowen 1966). Complete monographs were devoted to the analysis of such biologically important elements as cobalt (Nemodruk and Karalova 1965) and boron (Young 1966). Determination of mineral elements in biochemical systems was recently covered by Stewart and Frazer (1963), Bertrand (1964), and Alcock and MacIntyre (1966).

Element Enrichment.—The determination of trace elements often requires enrichment of the elements, and/or the separation of many elements at the trace level from large amounts of major elements. The most useful technique for element enrichment is ion exchange. Ion exchange has proved to be a valuable tool in the concentration, isolation, and recovery of ionic materials present in a solution in trace amounts. Ion exchange chromatography on an ion exchange resin can be also used for fractionation, separation, and the elimination of interfering ions (i.e. phosphates). The use of ion exchange chromatography for enrichment was described by Samuelson (1963), and is reviewed in annual reviews of *Analytical Chemistry*. A special preconcentration procedure of trace elements by precipitation ion exchange was described by Tera *et al.* (1965). Enrichment by extraction with organic solvents was described by Gorbach and Pohl (1951), and enrichment problems in microanalysis were reviewed by Schulek and Laszlovsky (1960).

Instrumental Methods.—The use of instrumental methods of analysis for metallic ions in foods has increased tremendously during recent years. The principles of instrumental methods used in the determination of mineral components and trace elements were described earlier. This chapter is primarily concerned with the applications of those principles to food analysis.

Emission Spectroscopy.—Emission spectroscopy is the oldest among analytical methods for trace analysis. It depends on observing and measuring the radiation emitted by atoms of the various elements when planetary electrons displaced from their orbits by various means fall back to the original (or a lower) level. For each element there is a pattern of wavelengths characteristic of the element when excited in a particular way. When several elements are present, each emits its own wavelengths. By identifying the wavelengths in the spectrum, the sample can be analyzed. Emission spectroscopy is sensitive but the precision is rather low. Use of emission spectroscopy in biochemical analyses were summarized by David (1962).

Flame photometry.—The original studies of Herschel (1848), Alter (1854), and of Kirchhoff and Bunsen (1860) on qualitative differentiation of salts depending on their emission in a flame were extended by Janssen (1870), Champion *et al.* (1873) and Lundegardth (1929, 1934), who respectively suggested using the phenomenon in quantitative analysis, developed the prototype of today's flame photometer, and devised a satisfactory method of introducing solutions into a flame at a constant rate. In subsequent years, many workers introduced important modifications and improvements (Margoshes and Vallee 1956; MacIntyre 1961; Pungor 1966).

The flame photometer (see Chap. 10) consists essentially of an atomizer, a burner, some means of isolating the desired part of the spectrum, a photosensitive detector, sometimes an amplifier, and finally a method of measuring the desired emission by a galvanometer, null meter, or chart recorder. The instruments are used primarily to determine calcium, sodium, and potassium.

Atomic Absorption Spectroscopy (See Chap. 10).—Less than 15 yr ago the term atomic absorption spectroscopy was familiar to only a small group of physicists and to few chemists. Within the last few years the analytical technique has found enthusiastic acceptance by science and industry (Zettner 1964). It is estimated (Walsh 1966) that during the year 1967 about 1000 laboratories adopted atomic absorption spectroscopy to chemical analysis. Review issues of *Analytical Chemistry* indicated that hundreds of papers are published annually on basic research, instrumentation, specific analytical methods, and practical applications of atomic absorption spectroscopy. Several comprehensive reviews and books deal with the principles and application of atomic absorption spectroscopy in testing biological materials (David 1960; Allan 1962; Willis 1963; Zettner 1964; Ellwell and Gidley 1966; West 1967).

Atomic absorption spectroscopy is not quite as free from inter-element effects as was originally expected, but is by far better in this respect than any form of emission spectrography. It is quite sensitive; the limit of detection ranges from 0.01 ppm for magnesium to 5.00 ppm for barium. The method is rapid; about 1000 determinations can be made per week. The equipment is relatively inexpensive (about $5,000), only $1/_{10}$ the cost of X-ray fluorescence equipment. The limiting factor is the need for cathode lamps for each element or several combinations. Recent major advances in the technique promise to broaden greatly the scope of atomic absorption spectroscopy.

In atomic fluorescence spectroscopy, atoms are generated in the same way as in atomic absorption spectroscopy, except that a cylindrical flame is used. The flame is irradiated by resonance radiation from a powerful spectral source, and the fluorescence which is generated in the flame is measured at right angles to the incident beam of radiation. This is done to minimize the contamination of the fluorescence signal by light from the source. Atomic absorption spectroscopy can be used in the ppm range (about 10^{-6} M solutions); atomic fluorescence spectroscopy in the ppb range (about 10^{-9} M solutions). The analytical potentialities of both methods were summarized by West (1967).

Activation Analysis.—In activation analysis, a weighed sample together with a standard which contains a known weight of the element sought are exposed to nuclear bombardment. The radioactivity of the element in the sample is then compared with the radioactivity in the standard. Generally, a chemical separation is required to purify the radioisotopes of the element sought, and remove all other induced radioactivity. The quantity of the element in the sample is then calculated from the ratio of the separated activities. It some instances, the final measurement of activity can be made on the intact sample. If the background remains inactive during nuclear bombardment or if the energies of the emitted radiations differ widely, a direct measurement of trace elements is possible. Also, if the trace element has a substantially longer half life than the other induced activities, the interfering materials may be allowed to decay and the radio-assay completed when the interference is insignificant (Loveridge and Smales 1957). Results obtained by neutron activation may be expected to lie within 5% of the true value, and replicate analyses may lie under favorable conditions within 2 or 3% of the mean.

The attractive features of neutron activation analyses are wide applicability, high sensitivity (0.001–1 ppm), and satisfactory accuracy and precision. There have been numerous applications of activation analysis in botany and agriculture (Bowen 1967). The determination in foods of toxic residues containing such elements as arsenic, bromine, chlorine, mercury, or nickel is of great interest. At least 36 elements were detected or measured in foods from animal sources between 1962 and 1967 (Spronk 1967).

X-ray Spectroscopy.—There are currently three uses of X-rays in chemical analysis. Absorption methods are of limited practical application because the adjustment of wavelength is most critical. X-ray diffraction is useful in crystallography, and in establishing the complicated structure of biological molecules. The use of X-rays for the identification of chemical components is based on emission methods, involving secondary or fluorescent emission. Measurement of the intensity and wavelength of fluorescence radiation is now a well-established method of analysis, and has been applied to the determination of the elements from ^{11}Na to ^{92}U in powder, liquid, or metal samples. Coefficients of variation of about 1% in the concentration range 5–100%, and of 5% in the 0.1–1.0% range can be obtained. In some instances determinations in the ppm range can be made (Brown 1959). The method is rapid (1–4 min), independent of the chemical combination of the element, and nondestructive in the sense that the specimen examined is not destroyed, though some specimen preparation may be required. Instrumentation for X-ray spectroscopy is quite expensive. Liebhafsky *et al.* (1960) are authors of a book on uses of X-ray absorption and emission in analytical chemistry; a comprehensive review of the literature on X-ray spectrometry was prepared by Buwalda (1965). Determination of elements by X-ray emission spectrometry in biological systems was described by Natelson (1964), and in foods by Tuchscheerer (1965).

Glass Electrodes.—When a thin membrane of glass is interposed between two solutions, an electrical potential difference is observed across the glass. The potential depends on the ions present in the solutions. Depending on the composition of the glass, the response may be to the hydrogen ion chiefly, to other cations, or even to organic cations (Eisenman 1965). Cation sensitive electrodes are now available commercially, and can be used to determine concentrations of sodium, potassium, silver, ammonium, thallium, lithium, and other ions in biological fluids or ash solutions. The electrodes are unaffected by oxidants and reducing agents, and are affected little by anions (except fluoride), and by high concentrations of proteins and amino acids. The biochemical application of cation-sensitive glass electrodes has been the subject of many papers (D'Eustachio 1968).

Miscellaneous.—Trace elements are determined in many laboratories by specific colorimetric and turbidimetric methods, by fluorescence analysis, and by polarography. The methods were described in detail earlier. The use of infrared spectroscopy in determining polyatomic ions was described by Miller and Wilkins (1952). Relatively simple chromatographic methods for rapid routine evaluation of trace elements in crops and foods were described by Duffield (1958) and Coulson *et al.* (1960). Connoly and Maguire (1963) described a quantitative paper chromatographic procedure for the determination of copper, cobalt, nickel, molybdenum, and manganese in foods. The elements are separated from the ashed materials as their chlorides on slotted chromatography papers, and identified colorimetrically. The concentrations are determined by reflectance measurements on the papers and the results are calculated with the aid of prepared standard graphs. As little as 0.05γ of the element can be determined.

Impressive advances have been made in recent years in developing instruments that permit an essentially complete elemental analysis to be performed *in situ* on the structures observed in the tissues of thin sections prepared by standard histological methods. The electron probe microanalyzer or electron probe X-ray scanning microscope (Birks 1960) can perform nondestructive elemental chemical analyses on localized regions with diameters as small as 1 μ, and volumes of a few cubic microns. The limit of detectability is about 0.1%, and many inorganic elements can be measured. The method recently has been extended to analyses and scanning biological specimens (Tousimis 1963; Mellons 1964; Andersen 1967).

Another promising new technique that has been adapted to microanalysis of inorganic elements is the laser microprobe. In that procedure, a laser beam is flashed through the optics of a regular microscope set to analyze a very small arc. The instrument is attached to a sensitive spectrograph (Glick and Rosan 1966).

Finally, mention should be made of biological (noninstrumental) methods of trace analysis. The principle of some of the procedures was described in the chapter on microbiological assays. The techniques were recently reviewed by Nicholas (1966).

Comparison of Methods of Elemental Analyses.—Bowen (1967) described the results of elemental analyses of a standard plant material (dried kale powder) analyzed for 40 elements by 29 laboratories. The techniques used were activation analysis, atomic absorption spectroscopy, a catalytical technique, colorimetry, flame photometry, turbidimetry, and titrimetric analysis. Consistent results were obtained by more than one laboratory for Au, B, Ba, Br, Ca, Cl, Co, Cr, Fe, Ga, I, Mn, Mo, N, P, Rb, S, Sc, and W. Small differences in results obtained by different techniques were found for Cu, K, Mg, Na, P, Se, Sr, and Zn. Of these, the most significant were: flame photometry gave high results for Na, activation analysis without chemical separation was unreliable for determining K and Mg, and atomic absorption spectrometry gave high results for Cu and Sr. Gross discrepancies were found in the results reported for Al, As, Hg, Ni, and Ti. Similar anomalies have been reported in the analyses of mammalian blood. Precisions of the various techniques varied widely for the studied elements.

A comparison of instrumental methods for trace analysis in foods is summarized in Table 34.2 (from Ames 1966).

TABLE 34.2

COMPARISON OF ANALYTICAL METHODS FOR TRACE ANALYSIS

	Several Elements Simultaneously	Sensitivity (Ppm)	Specificity	Accuracy	Freedom from Contamination, Reagent Blanks, etc.	Possibility of Overcoming Surface Contamination
Activation analysis	In some cases	Very high (0.001–1)	Good	Good	Good	Good
Atomic absorption and flame	No	0.01–5	Good	Good	Bad	Good
Emission spectroscopy	Yes	0.1	Good	Reasonable	Bad	Bad
X-ray spectroscopy	Yes	10–100	Good	Needs standards	Good	Possible by pre-cleaning
Mass spectrometry vacuum spark	Yes	0.01	Good	Needs standards	Good	Good
Gas analysis	Yes	Recently improved	Fair (N₂ CO)	Good	Good	Good

Bibliography

ALCOCK, N. W., and MACINTYRE, I. 1966. Methods of estimating magnesium in biological materials. Methods Biochem. Analy. *14*, 1–52.

ALLAN, J. E. 1962. A review of recent work in atomic absorption spectroscopy. Spectrochim. Acta *18*, 605–614.

ALTER, D. 1854. Cited by Margoshes and Vallee. Am. J. Sci. Art *18*, 55.

AMES, R. 1966. New instrumental procedures for determination of trace elements. Wallerstein Lab. Commun. *29*, No. 100, 107–113.

ANDERSEN, C. A. 1967. An introduction to the electron probe microanalyzer and its application to biochemistry. Methods Biochem. Analy. *15*, 147–270.

ANON. 1960. Analytical Methods Committee. Methods for the destruction of organic matter. Analyst *85*, 643–656.

ANON. 1967. LTA-600 low temperature dry asher. Trace-lab, Richmond, Calif.

BATES, F. J. 1942. Bureau of Standards Circular *C-440*.

BERTRAND, D. 1964. Chemical methods of mineral assays; biological applications Ann. Nutr. (Paris) *18*, H1; Al-A69.

BIRKS, L. S. 1960. The electron probe: an added dimension in chemical analysis. Anal. Chem. *32*, 19A-24A.

BOWEN, H. J. M. 1966. Trace Elements in Biochemistry. Academic Press, London.

BOWEN, H. J. M. 1967. Comparative elemental analyses on standard plant material. Analyst *92*, 124–131.

BROWN, F. 1959. X-ray fluorescence analysis—a review. Analyst *84*, 344–355.

BROWNE, C. A., and ZERBAN, F. W. 1941. Sugar Analysis. John Wiley & Sons, New York.

BUWALDA, J. 1965. X-ray spectrometry: Review of Literature. Phillips Co., Eindhoven, Holland.

CHAMPION, P., PELLET, H., and GRENIER, M. 1873. Spectrophotometry. Compt. Rend. Acad. Sci. *76*, 701–711.

COLE, S. J. 1967. Potentiometric determination of salt in foodstuffs. Food Technol. *21*, 302–304.

CONNOLY, J. F., and Maguire, M. F. 1963. An improved chromatographic method for determining trace elements in foodstuffs. Analyst *88*, 125–130.

CORNER, M. 1959. Rapid microdetermination of organically bound halogens, arsenic, phosphorus and boron. Analyst 84, 41–46.
COULSON, C. B., DAVIES, R. I., and LUNA, C. 1960. Quantitative paper chromatography of inorganic ions in soils and plants. Analyst 85, 203–207.
DAVID, D. J. 1960. The application of atomic absorption to chemical analysis. Analyst 85, 779–791.
DAVID, D. J. 1962. Emission and atomic absorption spectrochemical methods. In Modern Methods of Plant Analysis, M. V. Tracey, and H. F. Linskens (Editors). Springer, Berlin.
DAVIDSOHN, A. 1948. Aid for the determination of ash in matter containing high percentages of water. Analyst 73, 678.
D'EUSTACHIO, A. J. 1968. Biochemical analysis. Anal. Chem. 40, 19R–33R.
DUFFIELD, W. D. 1958. A system for the determination of certain trace elements in foods. Analyst 83, 503–508.
DUNLOP, E. C. 1961. Decomposition and dissolution of samples: Organic. In Treatise on Analytical Chemistry, Vol. 2, I. M. Kolthoff, and P. J. Elving (Editors). Interscience Publishing Co., New York.
EISENMAN, G. 1965. The electrochemistry of cation—sensitive glass electrodes. Advan. Anal. Chem. Instr. 4, 213–369.
ELLWELL, W. T., and GIDLEY, J. A. F. 1966. Atomic Absorption Spectrophotometry. Pergamon Press, Oxford, England.
ELVIDGE, D. A., and GARRATT, D. C. 1954. A note on a bomb technique for preparing samples for determination of lead in foodstuffs. Analyst 79, 146–147.
GILLET, T. R. 1949. Conductometric measurement of ash in white sugars. Anal. Chem. 21, 1081–1084.
GLEIT, C. E., and HOLLAND, W. D. 1962. Use of electrically excited oxygen for the low temperature decomposition of organic substances. Anal. Chem. 34, 1454–1457.
GLICK, D., and ROSAN, R. C. 1966. Laser microprobe for elemental microanalysis, application in histochemistry. Microchem. J. 10, 393–401.
GORBACH, G., and POHL, F. 1951. Enrichment and spectral analytical assay of trace elements; extraction with organic solvents. Mikrochem. 38, 258–267.
GORSUCH, T. T. 1959. Radiochemical investigations on the recovery for analysis of trace elements in organic and biological materials. Analyst 84, 135–173.
GRANT, J. 1951. Qualitative Organic Microanalysis, 5th Edition. Blakiston Publishing Co., Philadelphia.
HERSCHEL, J. F. 1848. Treatises on sound and light. In Encyclopedia Metropolitana; or System of Universal Knowledge on a Methodical Plan Projected by Samuel Taylor Coleridge. J. J. Griffin, London, England.
JANSSEN, J. 1870. Spectroscopie—Sur l'analyses spectrale quantitative. Compt. Rend. 71, 626–629.
KIRCHHOFF, G., and BUNSEN, R. 1860. Chemische Analyse durch Spektralbeobachtungen. Pogg. Ann. 110, 161–189.
KOCH, O. G., and KOCH-DEDIC, G. A. 1964. Handbook of Trace Analysis. Springer-Verlag, Berlin. (German)
KOLTHOFF, I. M., and ELVING, P. J. 1961–1966. Treatise on Analytical Chemistry, Part II. In Analytical Chemistry of the Elements. Interscience Publishing Co., New York.
LIEBHAFSKY, H. A., PFEIFFER, H. G., Winslow, E. H., and Zemany, P. D. 1960. X-ray Absorption and Emission in Analytical Chemistry. John Wiley & Sons, New York.
LOVERIDGE, B. A., and SMALES, A. A. 1957. Activation analysis and its application in biochemistry. Methods Biochem. Analy. 5, 225–272.

LUNDEGARDTH, H. 1929. The Qualitative Spectral Analysis of the Elements. G. Fischer Publishing Co., Jena, Germany.

LUNDEGARDTH, H. 1934. The Quantitative Spectral Analysis of the Elements. G. Fischer Publishing Co., Jena, Germany.

LYNCH, G. R. 1954. The destruction of organic matter. Analyst 79, 137.

MACINTYRE, I. 1961. Flame photometry. Advan. Clin. Chem. 4, 1–28.

MARGOSHES, M., and VALLEE, B. L. 1956. Flame photometry and spectrometry. Principles and applications. Methods Biochem. Analy. 3, 353–407.

MELLONS, R. C. 1964. Electron probe microanalysis. I. Calcium and phosphorus in normal human cortical bone. Lab. Invest. 13, 183–195.

MIDDLETON, G., and STUCKEY, R. E. 1953. The preparation of biological material for the determination of trace metals. I. Critical review of existing procedures. Analyst 78, 532–542.

MIDDLETON, G., and STUCKEY, R. E. 1954. The preparation of biological material for the determination of trace metals. II. A method for the destruction of organic matter in biological material. Analyst 79, 138–142.

MILLER, F. A., and WILKINS, C. H. 1952. Infrared spectra and characteristic frequencies of inorganic ions. Anal. Chem. 24, 1253–1294.

MITCHELL, R. L. 1960. Contamination problems in soil and plant analysis. J. Sci. Food Agr. 11, 553–560.

MONIER-WILLIAMS, G. W. 1953. Trace Elements in Foods. Chapman and Hall, London, England.

MORRISON, G. H. 1965. Trace Analysis—Physical Methods. Interscience Publishing Co., New York.

NATELSON, S. 1964. Determination of elements by X-ray emission spectrometry. Methods Biochem. Analy. 12, 1–68.

NEMODRUK, A. A., and KARALOVA, Z. K. 1965. Analytical Chemistry of Boron. Israel Program for Scientific Translation, Jerusalem.

NICHOLAS, D. J. D. 1966. Microbiological techniques as analytical and purification tools with special reference to trace metals. Ann. N. Y. Acad. Sci. 137, 217–231.

POMERANZ, Y., and LINDNER, C. 1954. The determination of the total electrolyte concentration of sugar products. Anal. Chim. Acta 11, 239–243.

PUNGOR, E. 1966. Flame Photometry Theory. D. Van Nostrand Co., London, England.

SAMUELSON, O. 1963. Ion Exchange Separations in Analytical Chemistry. John Wiley & Sons, New York.

SANDELL, E. B. 1959. Colorimetric Determination of Traces of Metals, 3rd Edition. Interscience Publishing, New York.

SCHNEIDER, E. 1967. Minerals; preparation of samples, determination of total ash, detection and determination of individual components. In Analysis of Foods, Vol. II, W. Diemair (Editor). Springer, Berlin.

SCHONIGER, W. 1955. A rapid micro-analytical determination of halogen in organic substances. Mikrochim. Acta 1, 123–129. (From Anal. Abstr. 2, 1816, 1955).

SCHULEK, E., and LASZLOVSKY, J. 1960. Problems of destruction and enrichment in microanalysis. Microchim. Acta, 485–501.

SMITH, G. F. 1953. The wet ashing of organic material employing hot concentrated perchloric acid. The liquid fire reaction. Anal. Chim. Acta 5, 397–421.

SMITH, J. H., CARTER, D. L., BROWN, M. J., and DOUGLAS, C. L. 1968. Differences in chemical composition of plant sample fractions resulting from grinding and screening. Agron. J. 60, 149–151.

SPRONK, N. 1967. Nuclear activation in the animal sciences. In Nuclear Activation Techniques in the Life Sciences. International Atomic Energy Agency, Publ. SM-91/16. Vienna, Austria.

STEWART, C. P., and FRAZER, S. C. 1963. Magnesium. Advan. Clin. Chem. *6*, 29–65.

TERA, F., RUCH, R. R., and MORRISON, G. H. 1965. Preconcentration of trace elements by precipitation exchange. Anal. Chem. *37*, 358–360.

THIERS, R. E. 1957. Contamination in trace element analysis and its control. Methods Biochem. Analy. *5*, 273–335.

TOUSIMIS, A. J. 1963. Scanning electron probe microanalysis of biological specimens. Biomed. Sci. Instr. *1*, 249–261.

TUCHSCHEERER, Th. 1965. Determination of inorganic food components by roentgen fluorescence. Z. Lebensmittel Unters. Forsch. *127*, 185–194.

WALSH, A. 1955. Application of atomic absorption spectra to chemical analysis. Spectrochim. Acta *1*, 108–117.

WALSH, A. 1966. Some recent advances in atomic absorption spectroscopy. J. New Zealand Instr. Chem. *30*, 7–21.

WEST, T. S. 1967. Atomic analysis in flames. Endeavour *26*, 44–48.

WILLIS, J. B. 1963. Analysis of biological materials by atomic absorption spectroscopy. Methods Biochem. Analy. *11*, 1–67.

WIRTHLE, F. 1900. Chem. Ztg. *24*, 263. *Cited by* Middleton and Stuckey (1954).

YOE, J. H., and KOCH, H. J. 1955. Trace Analysis. John Wiley & Sons, New York.

YOUNG, R. S. 1966. The Analytical Chemistry of Cobalt. Pergamon Press, Oxford, England.

ZETTNER, A. 1964. Principles and applications of atomic absorption spectroscopy. Adv. Clin. Chem. *7*, 1–62.

ZONNEVELD, H., and GERSONS, L. 1966. A rapid dry ashing method. Z. Lebensmittel Unters. Forsch. *131*, 205–207.

CHAPTER 35

Carbohydrates

Composition and Occurrence

Carbohydrates are the most abundant and widely distributed food component. *Carbohydrates* include (a) monosaccharides (polyhydroxy aldehydes or ketones) among which are 5-carbon compounds, such as xylose or arabinose, and 6-carbon compounds, such as glucose and fructose, (b) oligosaccharides in which a hydroxyl group of 1 monosaccharide has condensed with the reducing group of another monosaccharide; if 2 sugar units are joined in this manner, a disaccharide results; a linear array of 3 to 8 monosaccharides joined by glycosidic linkages gives oligosaccharides, and (c) polysaccharides that may be separated roughly into 2 broad groups, the so-called structural polysaccharides (i.e. cellulose, hemicellulose, lignin) that constitute or are part of rigid, mechanical structures in plants, and nutrient polysaccharides (i.e. starch, glycogen) that are metabolic reserves in plants and animals.

Perhaps the most important of the known monosaccharides is D-glucose. It is found as such in the blood of animals, in the sap of plants, and in many fruit juices. It also forms the structural unit of the most important polysaccharides. It is probably produced in all green plants, though its conversion into starch, cellulose, and other polysaccharides may prevent its detection. Fructose is found in fruit juices and in honey. An abundant source of both glucose and fructose is the disaccharide sucrose that can be hydrolyzed to yield one mole of each of them. Other disaccharides include the milk sugar, lactose, that constitutes about 5% of cow's milk and about 6% of human milk, and that yields on hydrolysis, glucose and galactose; maltose, a disaccharide in which 1 molecule of glucose is joined through a 1,4-α-glycosidic linkage to a second molecule of glucose, the disaccharide being formed abundantly by amylolytic breakdown of polysaccharides during malting or digestion in the animal body; and cellobiose, a degradation product of cellulose resembling maltose except that the 2 glucose units are joined through a β-glycosidic linkage. The most important freely occurring trisaccharide is the sugar beet raffinose, in which galactose is linked to a sucrose unit.

In addition to their nutritional and metabolic function, carbohydrates are important as natural sweeteners, raw materials for various fermentation products including alcoholic beverages, and the main ingredient of cereals. Carbohydrates govern the rheological properties of most foods of plant origin. The involvement of carbohydrates in the browning reaction is known to improve or impair consumer acceptance and the nutritional value of many foods (Stadtman 1948; Coulter *et al.* 1951; Danehy and Pigman 1951; Hodge 1953).

In food composition tables, the carbohydrates content has been usually given as total carbohydrates by difference, i.e. the percentage of water, protein, fat, and ash substracted from 100. Another widely used term is nitrogen-free extract, calculated as components other than water, nitrogenous compounds, crude fiber, crude fat, and minerals. The increasing awareness that specific carbohydrates play significant metabolic and functional roles, and the availability of analytical tools to determine individual components aroused interest in investigations on their distribution in many foods.

Fruits are a rich source of mono- and disaccharides; dates contain up to 48.5% sucrose, and dried figs contain a mixture of 30.9% fructose and 42.0% glucose (Hardinge *et al.* 1965). The sucrose content of most fruits and fruit juices is low, though some varieties of melon, peaches, pineapple, and tangerine contain 6–9% sucrose, and mango 11.6% sucrose. Reducing sugars (primarily a mixture of fructose and glucose) are the main soluble carbohydrate of most fruits, and account for 70% of the seedless raisins. Partly-ripe bananas are relatively rich in starch (8.8%), uncooked prunes in cellulose and hemicellulose, and citrus fruits in pectins. Vegetables contain substantially less glucose and fructose than fruits, and the only significant source of sucrose is sugar beets. Fresh corn and white and sweet potatoes contain about 15% or more of starch. Practically the only carbohydrate present in unsweetened milk and milk products is lactose. Nuts are generally a poor source of mono- and disaccharides; chestnuts contain up to 33% starch. The main component of cereals and cereal products is starch. In milled products, the starch content increases with the degree of refinement, and is about 70% in white flour compared to about 60% in whole grain. The increase in starch is accompanied by a parallel decrease in cellulose, hemicellulose, and pentosans. Spices contain 9.0 to 38.6% reducing sugars, in cloves and black pepper, respectively, and the latter has as much as 34% starch. White commercial sugar contains 99.5%

(or more) sucrose; corn sugar has about 87.5% glucose; honey has about 75% reducing sugars, a mixture of fructose and glucose; and most syrups and sweets have various amounts of sucrose (up to about 65%), reducing sugars (up to 40%), and dextrins (up to 35%).

Water-Soluble and Water-Insoluble Solids

The analyses of syrups, fruit preserves, malted products, and many other foods include the determination of the water-soluble and water-insoluble fractions. The insoluble fraction is of value in the determination of the fruit content of jams and preserves. The determination involves heating with boiling water, separating— by filtration—the soluble components, and drying the insoluble fraction to constant weight.

Total water-soluble solids can be determined directly by evaporating an aliquot of the extract and drying the residue in a vacuum oven at 70°C. In the indirect determination, the water-soluble solids are calculated as the difference between total solids (as determined from loss on drying) and the water-insoluble solids. If only an estimate of the water-soluble solids is required, determining the specific gravity or refractive index is rapid, simple, and reliable.

Removing Interfering Substances.—In determining the composition of the carbohydrates in the water extract, it is essential to remove interfering materials. Solid foods must be ground under conditions that cause little change in moisture content, and do not significantly affect the composition and properties of the foods. Lipids and chlorophyll are generally removed by extraction with petroleum-ether in which the carbohydrates are practically insoluble. Extraction is generally carried out at 40°–50°C; higher temperatures may solubilize starch components. To avoid hydrolysis and inversion of sucrose by organic acids during extraction at elevated temperatures, the addition of calcium carbonate for neutralization has been recommended. This addition is inadvisable if the extract contains large amounts of reducing sugars (Streuli and Stesel 1951). In enzymatically active extracts, it is important to prevent the hydrolysis of sugars during the extraction and storage of samples. This may be accomplished by the addition of mercuric chloride (Hadorn and Jungkunz 1952). Enzymatic modifications can be eliminated by the extraction of sugars with an aqueous solution of ethanol, or dropping the finely divided material into boiling 80% ethanol. The standard AOAC procedure for fresh plant materials, recommends dropping into twice distilled ethanol neutralized with calcium carbonate. The amount of ethanol is selected to give a concentration of 80% with the water extracted from the sample. The sample in ethanol is heated on a water bath for 30 min. The alcohol is generally evaporated at a low temperature, and the excess of calcium ions is removed during clarification.

Clarifying Agents.—Water extracts of most foods are clarified prior to a sugar determination. Turbidity caused by proteins and soluble starch affects polarimetric assays, and end-point determinations are masked in highly colored solutions. The color of the solution may not interfere in some reductometric methods provided the coloring substances do not react with the sugar reagents. Proteins precipitate copper in copper-reduction methods.

Clarification of water extracts is based on the principle that heavy metals precipitate colloidal substances, i.e. proteins, or that precipitates formed through the action of heavy metals, i.e. zinc ferrocyanide, combine with and coprecipitate the proteins (Acker 1967).

Clarification agents should meet several requirements (Streuli and Stesel 1952). They should remove the interfering substances completely without adsorbing or modifying the sugars. A reasonable excess of clarifying agent should not affect the assay. The precipitate should be small, and the precipitation procedure should be relatively simple.

Different clarifying agents meet those requirements to a varying extent. The use of a specific agent will, therefore, depend on the analyzed food, on the kind and amount of interfering substances, and on the proposed assay method. Extracts for polarimetric assay should be clear, practically colorless, and free of optically active substances other than sugars (i.e. amino acids, tannins, glycosides) or substances that influence the optical rotation of sugars (i.e. acid salts). Ferricyanide and copper reducing methods are more sensitive to soluble nonsugar, reducing compounds than gravimetric copper reducing procedures. In the latter, it is essential to remove completely colloidal matter that might be coprecipitated with the Cu_2O.

Most clarifying agents also have a decolorizing action through the adsorption of coloring substances by the precipitate, or through precipitation of natural chromogens of the polyphenol type by lead salts. Lead salts, especially the basic lead acetate, precipitate also optically active organic acids and are therefore useful in polarimetric assays. Aluminum hydroxide and kieselguhr have a limited clarifying power. Generally, an increase in decolorization is accompanied by increased adsorption of reducing sugars. This correlation is particularly noticeable in activated carbons.

In the assay of sugars in colorless or slightly colored, protein-rich solutions, the Carrez precipitation method gives excellent results (Acker 1967). The precipitant is less satisfactory for solutions of plant origin that are rich in gums, pectins, and acidic colloids. The Carrez solution involves the consecutive addition of equal solutions containing per liter, respectively 150 gm $K_4Fe(CN)_6 \cdot 3H_2O$, and 300 gm $ZnSO_4 \cdot 7H_2O$. The zinc is in excess but generally does not interfere, except in the complexometric determination of sugars according to Potterat and Eschman (1954), where excess zinc is removed after precipitation with $1.0N$ NaOH. Also in the determination of sugars by fermentation methods, the excess zinc must be removed.

Neutral lead acetate is the most commonly used clarifying agent both for chemical and polarimetric determinations. Its main limitation is the low decolorizing power; consequently, it is not suited for polarimetry of dark colored solutions. Excess lead acetate must be removed with a sodium sulfate or phosphate solution, solid sodium oxalate, or a mixture of disodium phosphate and oxalate. Basic lead acetate was used widely for deproteinization, but has been largely replaced by the Carrez reagent. Its use is limited today to the clarification of highly colored solutions rich in organic acids. Whereas a slight excess of neutral lead acetate has little effect, excess basic lead acetate precipitates and occludes reducing sugars, and affects the specific rotation of sugars. Excess basic lead acetate is removed as excess of the neutral salt. Aluminum hydroxide (alumina cream) is sometimes used for clarifying slightly colored sugar solutions. It is efficient in removing flocculate colloids, but not noncolloidal materials.

Additional clarifying agents include a mixture of barium hydroxide and zinc sulfate (Somogyi 1945B); dialyzed ferrioxychloride for the precipitation of slightly alkaline solutions; a mixture of copper sulfate and alkali for the precipitation of protein in milk; and mercuric nitrate in conjunction with an alkali for animal tissue extracts. Proteins can be removed by the general precipitants, trichloroacetic acid or phosphotungstic acid.

The clarifying agent is added either as a saturated aqueous solution, and the sugar solution is made to volume after clarification, or as a powder after the solution is made to volume.

Even with safe clarifying agents such as neutral lead acetate, a large excess should be avoided as it affects the polarization of sugars, and in heating, the interaction between the lead and sugar may result in some destruction of the latter.

Compounds added to remove excess clarifying agents should not be added in large excess. For certain methods of sugar determination, treatment with special reagents is mandatory. Oxalate is oxidized by ceric sulfate and must be replaced by disodium phosphate if ceric sulfate is to be used for the titration of ferrocyanide. In iodometric assays, the use of neutral lead acetate and sodium oxalate is satisfactory.

In certain assays, passing the solution through a mixed-bed ion-exchange column, for the purpose of desalting, is recommended (Wiseman et al. 1960). Amino acids present a special problem as they interfere with the determination of sugars, and are not separated by common protein precipitants (Hadorn and Biefer 1956). Some separation of amino acids from sugars may be effected by ion-exchange chromatography.

Detection of Carbohydrates

Qualitative tests for sugars are based on (1) color reactions effected by the condensation of degradation products of sugars in strong mineral acids with various organic compounds; (2) the reducing properties of the carbonyl group; (3) and on oxidative cleavage of neighboring hydroxyl groups. Many qualitative tests are determined on fractions separated by paper, thin-layer, or column chromatography. Many of the qualitative tests have been adapted for the quantitative determinations. The action of strong mineral acids (sulfuric, hydrochloric, and phosphoric) on carbohydrates leads to the formation of colored decomposition products. Aldohexoses and ketohexoses give as one of the main decomposition products hydroxymethyl furfural that in acid solution further decomposes to levulinic acid. Pentoses and hexuronic acids (after decarboxylation) give furfural; methyl pentoses produce methyl furfural.

The observations of Bandow (1937) on the reactions of carbohydrates in concentrated sulfuric acid have been developed into quantitative methods by Ikawa and Niemann (1949) and Bath (1958). Scott et al. (1967) described a sensitive and rapid ultraviolet spectrophotometric method for the determination of hexoses, pentoses, and uronic acids after their reactions with concentrated sulfuric acid. A more distinctive coloration is obtained when to the decomposition products of sugars in acid, some organic compounds are added (Dische 1962; Stanek *et al.* 1963; Acker 1967). The compounds include phenols, aromatic amines, thio compounds, urea, anthrones, and others, and gave rise to the Molisch reaction (α-naphthol), Selivanoff test (resorcinol), Bial procedure (orcinol), the naphthoresorcinol and phloroglucinol tests of Tollens, the Dische reaction (diphenylamine), and the Tauber test (benzidine). Some of the reagents are selective for certain sugars only, some show wide application, and some give various colored reactions depending on the sugar present.

Two of the reagents (anthrone and phenol) have found very wide use both in qualitative and quantitative analysis, especially in determining the concentration of the fractions separated by column and paper or thin-layer chromatography (Hodge and Hofreiter 1962).

Anthrone (9,10-dihydro-9-oxoanthracene), a reduction product of anthraquinone, was first recognized by Dreywood (1946) to react specifically with many carbohydrates in concentrated sulfuric acid solutions to produce a characteristic blue-green color. The color has been attributed to the reaction product of hydroxymethyl furfural or furfural, and anthrone. Carbohydrates and their derivatives that do not yield these substances display a wide range of different colors. The differences preclude the use of anthrone in the determination of total carbohydrate in sugar mixtures. However, in other instances this property has been used in the differential analysis of mixtures. Anthrone gives the best results when applied to pure solutions of hexose sugars or their polymers which produce the characteristic blue-green color. The phenol-sulfuric acid (Dubois *et al.* 1956) is a simple, rapid, sensitive, accurate, specific, and widely applicable method for carbohydrates. Virtually all classes of sugars, including sugar derivatives, oligosaccharides, and polysaccharides can be determined. The reagents are inexpensive, readily available, and stable. A stable color is produced, and the results are reproducible. The method is excellent for determining sugars separated by chromatography. In the direct determination of lactose in milk and cheese, normal amounts of casein, amino acids, and organic acids do not interfere (Barnett and Tawab 1957).

Color reactions based on the reducing properties of mono- and short-chain oligosaccharides are nonspecific, and can be used only after the removal of other reducing organic compounds. The reactions are, however, sensitive and well-suited to routine assays (Dische 1962). Several types of such reactions are useful.

In the reaction of reducing sugars with arsenomolybdate, copper[2+]

salts are reduced to copper$^+$ oxide which in turn reduces arseno-molybdate to molybdene blue. The absorbance of the latter is a measure of the concentration of the sugar. Ferricyanide at pH above 10.5 can be reduced by sugars to ferrocyanide that produces Prussian blue. If an organic compound containing an easily oxidizable group (i.e. reducing sugar) is heated with a solution of a tetrazolium salt at pH 12.5, a red, violet, or blue color—turning into a precipitate—is formed. Intensification of the color may be achieved by the addition of acetone. The test is highly sensitive. The most commonly used salt is 2,3,5-triphenyltetrazolium chloride or bromide.

In the periodate oxidation test, neighboring hydroxyl groups are oxidized and the aldehyde formed is detected by a fuchsin-sulfurous acid reaction.

Reducing sugars contain an aldehyde or keto group and, therefore, reduce in alkaline solution copper, silver, bismuth, and mercury salts to compounds of lower valence or to a metallic state. The best known reagent, based on the reduction of copper, is the Fehling solution. It is prepared by mixing before use two solutions, one containing cupric sulfate and one containing sodium potassium tartrate and sodium hydroxide. Depending on the concentration of sugars in a solution, heating with the Fehling solution gives a yellowish orange to red colored solution or precipitate. Some monosaccharides, i.e. glucuronic acid, react in the cold.

The Tollens reagent is based on the oxidative effect of the complex ion $[Ag(NH_3)_2]^+$. It is the most sensitive of all reagents utilizing the reduction of metal ions by sugars. The reaction is, however, not specific and is exhibited also by other easily oxidizable organic compounds such as polyhydric phenols, amino phenols, and alde-hydes. Some sugars without a free hemiacetal group, i.e. sucrose, also give a positive reaction.

The oxidation of sugars by an alkaline solution of trivalent bismuth in the presence of potassium-sodium-tartrate is the basis of the Nylander reaction. Crystalline phenylhydrazones with specific melting points are obtained when a cold aqueous sugar solution is treated with 1 vol of phenylhydrazine, 1 vol of 50% acetic acid, and 3 vol of water (Acker 1967). The phenylhydra-zones are somewhat soluble in water and for identification, sub-stituted phenylhydrazines such as bromophenyl-, nitrophenyl-, or 2,4-dinitrophenyl-hydrazine are preferred. The original sugar can be recovered from the phenylhydrazones by treatment with benzaldehyde.

Osazones are obtained by the interaction of 3 moles of phenylhydrazine and 1 mole of sugar. They are less useful for the identification, as epimer sugars (glucose, mannose, and fructose) give the same osazone. Identification of sugars by their crystalline derivatives has been largely replaced by various techniques of partition chromatography.

Determination of Mono- and Oligosaccharides

The available assay methods of mono- and oligosaccharides include chemical, colorimetric, chromatographic, electrophoretic, optical, and biochemical procedures. Today more and more assay techniques involve preliminary separation by chromatographic and electrophoretic techniques prior to actual assay by classical chemical procedures or colorimetric tests. Optical tests are useful as identifying aids in the determination of total solubles, and for the determination of specific sugars (generally in combination with chemical or enzymatic pretreatment). Microbiological assays of carbohydrates have found relatively little application. The use of enzymes as aids in sugar analysis or in actual assays is gaining significance with the commercial availability of pure, selective, and stable preparations.

Chemical Procedures.—An excellent and detailed review on selected methods for determining reducing sugars was published by Hodge and Davis (1952). Comprehensive and detailed reviews on the analysis of carbohydrates were prepared by Bates (1942) and Browne and Zerban (1941), in the series of monographs edited by Whistler and Wolfrom (1962–1964), and by Acker (1967). Reviews of the specialized aspects of carbohydrate analysis were presented in several issues of *Advances in Carbohydrate Chemistry and in Methods of Biochemical Analysis.*

(1). *Copper Methods.*—Probably no other analytical method has been utilized in so many modifications as the assay of reducing sugars for their oxidation by copper ions. In all copper methods, the reduction of copper and the oxidation of sugars are not stoichiometric. Yet, the reaction conditions can be adapted to give quantitative reproducible results, the amount of reducing sugars being determined from calibration tables.

The oxidation of reducing sugars by alkaline copper solutions was first proposed by Trommer in 1841. The assay procedure was improved by Barresvil who proposed, 3 yr later, to add potassium tartrate to prevent the precipitation of cupric hydroxide. Details of the method were worked out by Fehling in 1848 and reevaluated critically by Soxhlet in 1878. Two solutions are prepared for the determination of reducing sugars by the Fehling-Soxhlet methods. They contain per 500 ml respectively, 34.64 gm of $CuSO_4 \cdot 5H_2O$, and 173 gm of Rochelle salt $(NaKC_4H_4O_6 \cdot 4H_2O)$ and 50 gm of sodium hydroxide.

When a reducing sugar is treated with alkali at elevated temperatures, the sugar is degraded, and some of the degradation products reduce the cupric ions in the solution to a cuprous oxide precipitate. For quantitative determinations the method is varied as to the composition of the alkaline copper solution and the details of the assay procedure.

Several procedures have been suggested in which the volume of sugar solution required to reduce a definite amount of alkaline copper reagent is measured. The only direct titrimetric procedure that has found wide acceptance in Europe is that proposed by Lane and Eynon (1923). It is used to a limited extent in the United States for determining the dextrose equivalent of starch syrups as the official analytical method of the Corn Industries Research Association. The *dextrose equivalent* is defined as reducing sugars, calculated as dextrose, in the dry substance of starch syrups. Use of the Lane-Eynon method in the analysis of starch hydrolysates by the International Commission for Unified Methods of Sugar Analysis was described by Heyns (1959). In principle, the sugar solution is added slowly from a buret to a vigorously boiling mixture (1:1) of the two Fehling-Soxhlet solutions. Close to the end point, the sugar solution is added dropwise in the presence of 1 ml of a 2% aqueous solution of methylene blue, which changes from blue to white by an excess of the reducing sugar. The determination is then repeated and the amount of sugar solution (less 0.5 ml) determined in the preliminary assay is added at once, and the titration is continued to the end point.

In most methods, an excess of alkaline-tartrate cupric sulfate is added to a sugar solution, the solution boiled under specified conditions, and the amount of precipitate formed is determined. The precipitate can be weighed as Cu_2O, transformed to CuO or Cu, determined by titration after dissolving the Cu_2O precipitate, or determined by measuring the unreacted cupric ion complex.

In the gravimetric methods, the precipitate is filtered through a glass or porcelain filter stick with a fritted insert or asbestos layer, washed, dried, and weighed. The method most commonly used in the United States is the modified Munson-Walker (1906) procedure of the AOAC. It uses a unified procedure for all sugars, but the results are calculated for each sugar from empirical tables. The tables were computed to allow for the presence of sucrose along with reducing sugars (glucose, fructose, maltose, lactose, and their mixtures). The Munson-Walker procedure was evaluated critically by Hammond (1940) who studied the accuracy of the procedure, developed a refined method, and prepared more concise tables for dextrose, levulose, invert sugars, and invert sugar-sucrose mixtures. The precision and errors of the Munson and Walker's method were studied also by Jackson and McDonald (1941). Wise and McCammon (1945) extended the use of the Munson-Walker method by preparing tables for some of the less common sugars.

Several titrimetric methods for the indirect determination of the reduced copper are available. The titrimetric permanganate methods are based on the procedure proposed by Bertrand in 1906 (Acker 1967). The washed precipitate is dissolved in an acidified $Fe_2(SO_4)_3$ solution, and the amount of ferrous ions from the reduction by the cuprous ions is determined with standard permanganate. In the AOAC modification, the cuprous oxide is dissolved in neutral ferric ammonium sulfate, and the acid is added shortly before titration. The end-point determination is sharpened by adding phenanthroline as an indicator. Acid ferric sulfate dissolves cuprous oxide faster than does the neutral solution, but leads to low results as some of the ferrous sulfate formed is oxidized by air under acid conditions.

In the titrimetric iodide method, the precipitate is filtered, washed, and oxidized to cupric nitrate with nitric acid. The excess of nitric acid is removed by boiling, and after acidification with strong acetic acid, 10 ml of a 30% potassium iodide solution is added. The liberated iodine is titrated with standard sodium thiosulfate.

The complexometric titration of cupric ions (from the Cu_2O precipitate dissolved by boiling nitric acid) with Na_2EDTA in the presence of indicators was proposed by Potterat and Eschmann (1954). The indicators change color when bound to cupric ions, but are less stable than the cupric chelates. At the end of the titration, cupric ions are removed by the chelating agent from the indicator Cu-complex and the color of the indicator changes. An indicator commonly used is murexide (ammonium salt of purpuric acid) that is blue-violet at pH 10, and yellow as a copper-complex. The titration is carried out in an ammoniacal medium. The reagent (1000 ml) is prepared by adding a solution containing

25 gm $CuSO_4 \cdot 5H_2O$ to 500 ml of a solution containing 286 gm $Na_2CO_3 \cdot 10H_2O$ and 38 gm Na_2EDTA. The reagent is stable for many months and even boiling for 24 hr affects it little.

The direct iodometric determination of Shaffer and Hartmann (1921) is based on the fact that in the presence of oxalate a complex is formed with the cupric ion in solution, and the precipitated cuprous oxide can be titrated after acidification with an iodate-iodide solution without filtering. A considerable saving of time for each determination is effected. Shaffer and Hartmann employed the alkaline copper reagent and heating conditions specified by Munson-Walker, so that the tables worked out by the latter can be used. The usefulness of the method was confirmed by Hadorn and Fellenberg (1945). Apparently, the speed is achieved at some sacrifice of accuracy.

Somogyi (1945A) modified the Shaffer-Hartmann procedure for the determination of micro quantities of reducing sugars. The alkaline reagent developed by Somogyi is buffered with phosphates and includes potassium iodate as a source of iodine for the oxidation of the cuprous ion. The inclusion of 18% sodium sulfate is claimed to eliminate back oxidation of the cuprous ions by air. The presence of inorganic halides and nitrates depresses the reducing value, apparently by partly solubilizing the cuprous oxide and enhancing its oxidation in air. The titrimetric method has a precision of $\pm 2\%$ for the range of 0.3–3.0 mg of glucose. In a colorimetric modification, reduced copper is determined by reacting it with phospho- or arsenomolybdate color-forming reagents (Nelson 1944). The colorimetric method is as precise as the titrimetric; the useful range is 5–600 γ.

Iodometric titration of excess copper sulfate has several advantages over the titration of reduced Cu_2O. The procedure is simple, requires no filtration, and one method can be used for all reducing sugars. This method was developed by Schoorl and Regenbogen (1917). It employs the Fehling-Soxhlet solutions; the reduced copper is determined indirectly by iodometric titration of the unreduced copper salt remaining after the oxidation of sugars. The accuracy of the method was confirmed by collaborative studies (Flohil 1933).

The highly alkaline Fehling-Soxhlet solution causes strong and not always reproducible degradation of various sugars, and either lowers the assay precision or requires strict adherence to experimental conditions. The use of alkaline carbonate solutions has been the basis of a procedure known in Europe as the Luff-Schoorl method, and in the United States as the Benedict reagent. The latter employs sodium carbonate instead of potassium hydroxide, and sodium citrate in place of tartrate. Determinations of reducing sugars in carbonate-buffered water solutions were reviewed by Heidt and Colman (1952).

Barfoed proposed as early as 1873 (Hodge and Davis 1952) to use copper acetate to differentiate between reducing mono-saccharides and disaccharides, as the latter are not oxidized appreciably by the reagent. The reagent was adapted for quantitative use by Steinhoff (1933) and by Sichert and Bleyer (1936), and is used for the determination of glucose in the presence of maltose and dextrin. In the modified procedures, sodium acetate was substituted for acetic acid and the reduced copper determined iodometrically, or by dissolving it in a ferric sulfate solution and titrating the resulting ferrous ion with ceric sulfate.

(2). *Oxidation by Alkaline Ferricyanide Solutions.*—The alkaline ferricyanide reagent was developed as an analytical procedure for the determination of sugar in blood by Hagedorn and Jensen (1923), and modified for food analysis by Hanes (1929). The method is based on the reduction of alkaline ferricyanide to ferrocyanide in the presence of a reducing sugar, the amount of reduction being a measure of the amount of sugar the sample contains. The amount of reduced ferrocyanide is determined as the difference between that added, and that after reduction. The changes occurring in the presence of potassium iodide are $2K_3Fe(CN)_6 + 2KI \rightarrow 2K_4Fe(CN)_6 + I_2$. In the presence of zinc ions, the ferrocyanide formed is precipitated as a zinc complex and the equilibrium is shifted to the right: $2K_4Fe(CN)_6 + 3ZnSO_4 \rightarrow K_2Zn_3[Fe(CN)_6]_2 + 3K_2SO_4$. The liberated iodine is titrated with standard thiosulfate. Direct titration of the ferrocyanide can be carried out with ceric sulfate and phenanthroline as the indicator (Whitmoyer 1934; Hassid 1936). The ferricyanide can also be used in the direct titrations of the sugars in the presence of an indicator (picric acid or methylene blue), or the ferrocyanide produced estimated colorimetrically as Prussian blue.

The most commonly used are the procedures involving the titration of excess ferricyanide. Numerous modifications of the end-point determination have been proposed, including titrimetric, colorimetric, and potentiometric procedures. Ferricyanide reduction methods are popular because the reduction and subsequent determination of reduced ferricyanide can be carried out in one reaction vessel. The ferricyanide reagent is stable in the alkaline solution used, and the ferrocyanide formed is more stable than the Cu_2O formed in copper reduction methods. The reaction is reproducible and well-suited to routine analyses. However, the oxidizing action of ferricyanide is not as specific as that of copper-reducing methods, since ferricyanide is reduced easier by substances other than sugars. As in the copper-reduction methods, the oxidation yields a variety of partly unstable oxidation products. Consequently, no stoichiometric relationship exists for the sugar oxidation.

Following the investigations of Blish and Sandstedt (1933) and Kneen and Sandstedt (1941), the ferricyanide method is used extensively to determine reducing sugars, diastatic activity, and the β-amylase activity of wheat flour, and the saccharifying activity of enzyme preparations. It is particularly useful in the determination of small amounts of maltose in starch hydrolyzates.

(3). *Iodometric Methods.*—Iodine in an alkaline medium is converted rapidly into hypoiodite that can oxidize aldoses; ketoses are oxidized little. The method, originally proposed by Willstatter and Schudel (1918), is applicable to aldoses alone or in a solution with other carbohydrates, provided no interfering iodine-consuming compounds are present. The dissolved sample is treated with an excess of dilute iodine, and sodium hydroxide is added and mixed rapidly (otherwise the iodine may be oxidized to iodate that does not react on sugars in alkaline solutions). After the solution is acidified with hydrochloric or sulfuric acid, and a few minutes of standing, the excess of the standard iodine solution is titrated with a standard thiosulfate solution. Unlike in the copper and ferricyanide methods, the oxidation of aldoses by iodine approaches the stoichiometric reaction, $RCHO + I_2 + 3NaOH \rightarrow RCOONa + 2NaI + 2H_2O$. Various modifications of the iodine method concern primarily the optimum alkalinity of the medium to realize the stoichiometric relation and eliminate the interference of ketoses. Ethanol, acetone, and other substances that react with iodine must be absent. Under optimum conditions only about 1% of the ketoses are oxidized. Other substances that consume small amounts of iodine include mannitol, glycerin, sodium lactate, sodium formate, urea, and sucrose (Hodge and Davis 1952). The modified procedures of Hinton and Macara (1924) and Lothrop and Holmes (1931) are accurate to within $\pm 0.5\%$ in samples high in aldoses. Fructose can be determined (after oxidation of glucose with iodine in an alkaline medium) by copper reduction methods, provided the excess iodine is titrated in the glucose determination with sulfurous acid rather than with thiosulfate (Acker 1967).

By replacing the iodine solution with solutions of potassium iodide and chloramine T (the sodium salt of N-chloro-p-toluene sulfonamide), many of the limitations of the iodine method are overcome. Chloramine T hydrolyzes slowly, producing in a slightly alkaline medium sodium hypochlorite which reacts with potassium iodide to slowly release the hypoiodite oxidizing agent. The oxidation of aldoses proceeds slowly and the danger of a side reaction in minimized.

Dextrose, maltose, lactose, and invert sugar can be analyzed satisfactorily. The method has been used in the determination of lactose in milk products (Hinton and Macara 1927). The determination of lactose is unaffected by the presence of sucrose; small amounts of fructose do not affect the determination of glucose. Substances that form iodoform with hypoiodite affect the results.

(4). *Cerimetric Methods.*—The cerimetric determinations of sugars can be carried out in two general ways (Stanek *et al.* 1963). Either the consumption of cerium perchlorate is estimated titrimetrically with nitroferroin as an indicator, or a sugar solution is boiled with a solution of cerium sulfate in dilute sulfuric acid and the excess of cerium salts is back titrated with ferrous salts. Under these conditions, glucose yields formic acid as the highest oxidation product, while ketoses are partially oxidized to carbon dioxide. By adding chromic ions to the cerium sulfate solution, glucose is also oxidized to carbon dioxide.

Chromatographic and Electrophoretic Methods.—Chromatographic methods have been used to fractionate, isolate, identify, and determine carbohydrates in complex mixtures. The methods range from paper chromatographic procedures for the identification of components on a microscale to large-scale column separations for the isolation of relatively large amounts of pure compounds for further identification and study.

(1). *Paper and Thin-Layer Chromatography.*—The separation of sugars by paper chromatography has been developed since the pioneering investigations of Partridge (1948) into one of the most versatile forms of qualitative and quantitative methods of carbohydrate microanalysis (Cramer 1953; Hough 1954; Kowkabany 1954; Lederer and Lederer 1957; Whistler and DeMiller 1962).

Qualitative paper chromatography is the best and simplest method to distinguish between various forms of sugars present in foods along with a mixture of various other compounds. In the case of a complex mixture, separation by paper chromatography may be supplemented by the use of specific sprays, additional identification of separated spots, use of additional separation methods, and—in all cases—comparison with R_f values of mixtures of pure sugars chromatographed along with the investigated food extracts.

In the chromatography of pure sugar solutions, no special preparation is required. For food extracts, precipitation of proteins or other interfering substances by the conventional sugar precipitants is advisable. To avoid tailing, the removal of inorganic compounds by passing the solution through a mixed bed of ion exchangers prior to paper chromatography is advisable. It should be noted, however, that strongly basic anion exchangers in the hydroxyl form may adsorb or modify sugars, and that strong acidic cation exchangers may hydrolyze sensitive oligosaccharides (Jayme and Knolle 1960).

Separations of 1% sugar solutions containing a maximum of 60γ of an individual sugar give best results. The solvent system butanol-acetic acid-water (4:1:5) gives good resolution, but requires long development times. Resolution is more rapid with phenol-water or collidin-water systems (Acker 1967). Other useful systems include acetic acid-pyridine-water and n-propanol-acetic acid-water. The selection depends also on the expected mixture of sugars. Whereas a mixture of glucose, sucrose, and lactose cannot be separated well by phenol-containing systems, the separation is satisfactory in n-propanol-acetic acid-water systems.

Generally, the higher the sugar in a homologous series, the lower its R_f value, i.e. trisaccharides have lower R_f values than disaccharides, and monosaccharides have still higher mobilities. Pentoses have higher R_f values than hexoses; and among hexoses, ketoses are faster moving than aldoses. There are exceptions to these rules; in phenol-containing systems fructose moves faster than xylose. Disaccharides with 1,4 linkages migrate faster than those with 1,6 linkages; and α-D-glucosides faster than β-D-glucosides (Isherwood and Jermyn 1951). For better resolution, 2-dimensional paper chromatography can be used; generally however, separation in 1 dimension with several solvent systems is preferred.

For the identification of the separated components, many spraying reagents are available. Aniline phthalate gives a color reaction with reducing sugars only; fructose is, however, less reactive. Spraying with m-phenylenediamine gives fluorescing acridine derivatives

(Chargaff *et al.* 1948).   Silver nitrate in ammonia is a useful general reagent for reducing sugars.   For nonreducing sugars, various phenols in acid, i.e. naphthoresorcinol in syrupy phosphoric acid is recommended.   The phloroglucinol-hydrochloric acid reagent of Borenfreund and Dische (1957) gives a purple color with aldopentoses, dark green with ketopentoses, yellow brown with ketohexoses, and green with methyl pentoses.

Qualitative determinations of sugars by paper chromatographic methods are useful primarily in the separation of sugars on a microscale, and in the identification of components of a complex mixture that cannot be resolved easily by other available procedures. They are used to determine oligosaccharides in malt extracts, the composition of starch hydrolyzates, or small amounts of raffinose in sucrose solutions.   The two main limitations of the quantitative determinations of carbohydrates by paper chromatography are high blanks, and the lack of a linear relation between sugar concentration and a measured analytical parameter.   One must prepare each time a calibration curve covering the expected range. The precision is generally only $\pm 5\%$ though some methods give, with proper precautions, an accuracy of $\pm 2\%$ (Whistler and Hickson 1955).

Quantitation of the separated carbohydrates can be accomplished in several ways.   The stained spots can be determined by direct photometry (Jayme and Knolle 1960).   Some colored sugar derivatives, i.e. triphenyltetrazolium, can be eluted quantitatively and the concentration measured in solution (Schoenemann *et al.* 1961).   Most commonly, the sugars are separated by paper chromatography, their position determined, and they are eluted for subsequent determination by colorimetric methods (Dimler *et al.* 1952).   Many methods have been developed for the quantitative determination of carbohydrates in a mixture after separation by paper chromatography.   Of the microtitrimetric methods, the alkaline ferricyanide reagent of Hagedorn and Jensen (1923) is one of the more useful.

Colorimetric methods are usually more sensitive than titrimetric methods.   Two of the more useful are the phenol-sulfuric acid and anthrone methods.   They were described earlier in this chapter. In both, it is essential that the extract be free of cellulose fibers.

Sugars can be separated by thin-layer chromatography, the advantage of which is speed and good resolution (Stahl and Kaltenbach 1961, 1965;   Weill and Hanke 1962;   Scherz *et al.* 1968). Huber *et al.* (1966) described a rapid method for the determination of saccharide distribution of corn syrups by direct densitometry of thin-layer chromatograms. The results were of comparable accuracy with those obtained by gravimetric paper chromatography.

(2). *Electrophoresis.*—The application of zone electrophoresis (on a strip of filter paper or other supporting medium) for the separation of carbohydrates is relatively new, but has found wide application (Foster 1957; Weigel 1963). Separations of carbohydrates are generally made with borate derivatives. The method cannot separate borate-complex epimers such as fructose, glucose, and mannose.

(3). *Gas-Liquid chromatography.*—Gas chromatographic analysis of carbohydrates has received much attention as a result of the pioneering work of Sweeley *et al.* (1963) with trimethylsilyl derivatives. Other workers (Alexander and Garbutt 1965; Kagan and Mabry 1965; and Sawardeker and Sloneker 1965) have followed the silyl ether derivative approach to the quantitative analysis of sugar samples. The determination of the carbohydrates by gas chromatography of the alditol acetates was reviewed by Crowell and Burnett (1967). Gas-liquid chromatography of carbohydrates was reviewed by Bishop (1964) and Geyer (1965). For the separation of carbohydrates by gas-liquid partition chromatography, free reducing groups of sugars must be blocked; excellent resolution and quantitation can be achieved.

(4). *Ion-Exchange Chromatography.*—Sugars, being weak electrolytes, have little tendency to react with ion-exchange resins. However, it has been known for a long time that certain polyhydroxy compounds react with the borate ion to form complexes of negatively charged ions. Khym and Zill (1952) first introduced the technique of separating a mixture of sugar-borate complexes on a column of strong-base anion exchange in the borate form. Later, Zill *et al.* (1953) extended their work to include the separation of related compounds. The sugars were eluted by a stepwise gradient of borate and pH, and the elution required up to 60 hr. Since the initial reports of Khym and Zill, a variety of ion exchange matrices and modifications have been proposed. The procedures that were developed permitted separations of 15 sugars in an automated procedure (Green 1966). Accelerated (7.5 hr) procedures for automated separations were reported recently (Anon. 1966; Ohms *et al.* 1967). The sugars are eluted from a strongly anionic styrene-divinylbenzene column with a borate buffer of gradually increased chloride concentration and pH, or with a linear gradient of borate buffer. Sugar in the eluate was determined by the orcinol-sulfuric acid

method.   Further development of this technique in the separation
and determination of mixtures of mono-, di-, and trisaccharides was
described by Kesler (1967).

(5). *Other Column-Chromatographic Methods.*—Extrusion chromatography involves development of the column to approximately a
predetermined extent, extrusion of the column from the tube,
location and sectioning of the zones, and extraction of the adsorbed
material with a highly polar solvent in which the material is soluble.
The method is fast and eliminates the need to evaporate and handle
large amounts of solvents.   The method has not been used, thus
far, extensively.

Carbon column chromatography was devised by Whistler and
Durso (1950) for the separation of oligosaccharides into classes according to the degree of polymerization (mono-, di-, and tri-, and
oligosaccharides).   Either carbon alone, or carbon-Celite mixtures
are used.   The capacity of the columns is high and quantitative
separations of large amounts of material are possible (Hoover *et al.*
1965).   In automated elution and recording procedures (French *et al.*
1966), gradient elution of charcoal columns using aqueous tertiary
butanol was effective in separating starch oligosaccharides in the
range up to 15 D-glucose units.   Branched, as well as linear, oligosaccharides could be separated.   Resolution is improved if mixtures
of charcoal-Celite are treated with stearic acid (Alm, 1952; Miller,
1960).   Mixtures that can be resolved on paper chromatography
using a developing system that is not completely miscible with water,
can usually be separated on Celite columns with the same solvent
system.   Cellulose column chromatography has also found wide
application in preparative carbohydrate chemistry (Hough 1954;
Brinkley 1955).

C. Optical Methods.—Optical methods were discussed in
detail in several chapters in this book.   The following brief outline
deals specifically with special aspects of carbohydrate analysis.

(1). *Refractometry.*—Refractometers are widely used in the
sugar industry to measure the content of dissolved solids in sugar
solutions.   Generally, a sugar industry laboratory may be expected
to have several refractometers, sometimes including a variety of
models (Charles and Meads 1962).   Refractometers are located at
plant operating stations and in control laboratories to provide information for prompt process or product control.   They also find
application in research laboratories for process study and other investigations.

(2). *Polarimeters.*—Carbohydrates are optically active and can be assayed polarimetrically. Optical rotations can be measured by means of visual polarimeters, visual saccharimeters, and photoelectric spectrophotometers (Whistler and Wolfrom 1962–1964). Visual polarimeters measure the angle of rotation directly on a circular scale. A monochromatic light source must be used. Saccharimeters are specially designed for the determination of optical rotation of sucrose-containing solutions. They are generally of the quartz-wedge compensating type; permitting illumination with white light in conjunction with a dichromate filter. Photoelectric spectropolarimeters are high-precision polarimeters in which the intensity of monochromatic light is measured by photoelectric cells instead of the human eye.

Polarimetric assays of sugar are nondestructive and rapid. They are accurate provided (1) the solution is clear and colorless or only slightly colored, (2) the concentration of tested sugars is within an optimum range of the instrument, and (3) the solution contains no interfering optically active compounds. The use of clearing agents removes turbid materials and part of the coloring substances. To account for the presence of optically active substances, other than sugars, determination before and after inversion (as in sucrose analysis) is useful. According to the Biot law, the rotating capacity of individual sugars is proportional to the concentration of the solution and to the length of the liquid column. A measure of this capacity is the specific rotation

$$[\alpha] = \frac{100 \cdot \alpha}{lc} = \frac{100 \cdot \alpha}{lpd} \tag{35.1}$$

where $[\alpha]$ specific rotation, and α is the angle of rotation of a solution of specific gravity d, containing p gm of active substance per 100 gm solution (or a concentration of c gm/100 ml of solution) in a tube of l dm length. The specific rotation depends on the temperature and wavelength of the rotated light. Generally, measurements are made at 20°C with a sodium light. The green mercury light is used in light-electric polarimeters because of its high intensity.

For a sucrose solution of $c = 26.016$ gm (corresponding to 26.000 gm weighed in air) in 100 ml, employing the sodium light, at 20°C and 760 Torr, $[\alpha] = 66.523°$, and

$$c = \frac{100}{[\alpha]} \cdot \frac{\alpha}{l} = 1.5032 \frac{\alpha}{l} \tag{35.2}$$

[α] varies with concentration, but the variation is small and can be neglected. The temperature correction is

$$[\alpha]_t^D = [\alpha]_{20}^D \, [1-0.000184 \, (t - 20)] \qquad (35.3)$$

A rotation of polarization of 1° corresponds to 0.7519 gm/ml sucrose. Corresponding values for other sugars are 0.9470, 0.5405, 0.9524, and 0.3623, respectively for glucose, fructose (levorotatory), lactose, and maltose.

According to decisions of the International Commission for Unified Methods of Sugar Analysis, saccharimeters should be calibrated according to the International sugar-scale of ICUMSA, in sugar degrees. The 100° point on the international scale is defined as the polarization of a normal solution of pure sucrose (26.000 gm per 100 ml) at 20°C in a 200-mm long tube with regular light and a dichromate filter, a solution of $K_2Cr_2O_7$ of such concentration that the percentage of dichromate times the width of solution is 9 (i.e. 6% dichromate and filter width of 1.5 cm). The details of the calibration of the saccharimeter are given in AOAC.

Products containing reducing sugars show mutarotation involving the establishment of an equilibrium state of the α and β forms, and the intermediate open-chain form. Such solutions must be allowed to stand overnight for analysis. If the determination is to be made immediately, the neutral solutions can be heated to boiling, a few drops of ammonia added before bringing to volume, or dry sodium bicarbonate can be added to a solution adjusted to final volume till clearly alkaline.

Sucrose can be determined directly by polarimetry-saccharimetry, in the presence of reducing sugars after the latter are eliminated by heating with alkali (i.e. barium hydroxide for 1 hr at 70°–80°C). The method is used in testing chocolate (Thaler 1940). Precise and reproducible chemical and polarimetric methods of sucrose determination are based on the fact that sucrose as a fructofuranoside is hydrolyzed by acids faster than other disaccharides and most oligosaccharides (except those containing fructofuranosides, i.e. raffinose). On hydrolysis, 1 part of sucrose is hydrolyzed to 1.053 parts of an equimolar mixture of glucose and fructose known as invert sugar, because the hydrolysis is accompanied by a change in optical rotation (from +66.5 to −20.00 at 20°C).

In the absence of other optically active substances, the sucrose content of food solutions can be determined by direct polarimetry. In the presence of other sugars, the sucrose content of a food solution can be determined from the change in optical rotation (or reducing power, in chemical determinations) after inversion. It is thus necessary that sucrose alone be hydrolyzed and that the rotation (or reducing power) of no other compounds be changed.

Hydrolysis with invertase is more specific than acid hydrolysis. In the optical rotation procedures, it is important to hydrolyze and polarize the samples under specified conditions. Both acids and salts affect the specific rotation, and their effects may be accounted for by blank determinations.

(3). *Infrared Spectroscopy.*—Infrared spectroscopy of simple sugars is complicated by the fact that they are practically insoluble in the organic solvents commonly used in such determinations. Consequently, the potassium bromide disc and paraffin suspension methods are used. Infrared spectroscopy has been widely used in the studies of carbohydrate structure (Whistler and House 1953; Solms *et al.* 1954; Baker *et al.* 1956; Brock-Neely 1957; White *et al.* 1958; Spedding 1964). Investigations of Parker (1960) concerned an infrared method of studying mutarotation; Underwood *et al.* (1961) used infrared spectroscopy in his investigations of browning in foods; Lin and Pomeranz (1965, 1968) found the technique useful in the detection of carbohydrates in wheat protein preparations and in investigations of the water-soluble wheat flour pentosans. The determination of lactose in milk products by infrared spectroscopy was described by Dyachenko and Samsonov (1964) and by Biggs (1967).

D. Biochemical Methods.—Biochemical methods of carbohydrate analysis can be classified into two groups. In microbiological assays, yeast strains varying in their fermentative capacity are selected. Such yeasts are particularly useful in differentiating between pentoses and hexoses (or their polymers). Baker's yeast can also be used to differentiate between nonfermentable lactose and fermentable sucrose or maltose. In enzymatic methods, selective cleavage to monosaccharides and/or enzymatic assay of the monosaccharide is specific and widely used. The use of amylase or amyloglucosidase is an example of the first type; determination of glucose with glucose oxidase of the second type. In addition, microorganisms and enzymes can be used in the pretreatment of substrate prior to chemical assay. The various types of biochemical methods are outlined in detail in the respective chapters of this book.

Hexosamines

The amino sugars constitute the building stones of many biologically important substances (Balazs and Jeanloz 1965). Although many types of amino sugars are known, only a few types are known to occur naturally. Of these, the most common are 2-desoxy-2-amino-D-glucose (known as D-glucosamine or chitosamine) discovered by Ledderhose in 1878, and 2-desoxy-2-amino-D-galactose (D-galactosamine or chondrosamine) discovered by Levene in 1914 (Gardell 1958). In their reactions the hexosamines have the properties of both hexoses and primary amines.

In the hydrolysis of amino sugar containing polysaccharides, the monomers must be liberated from the polysaccharides. In general, the amino sugar glycosides are very resistant to acid hydrolysis and special precautions must be taken to secure complete hydrolysis. Gardell (1958) recommends heating a 1–2% solution for 8 hr on a boiling water bath with 6N HCl. According to Pusztai (1965) most glucosamine-containing constituents of the seeds of kidney beans can be hydrolyzed with 0.5N HCl for 16 hr, or with 2N HCl for 2 hr at 100°. Some acid-stable polysaccharide fractions require, however, hydrolysis with 6N HCl.

The isolation, crystallization, and subsequent identification of the 2-amino sugar components of mucopolysaccharides, in which they are present in relatively large proportions is readily accomplished (Foster and Stacey 1952). In biological materials that contain 2-amino sugars in minor proportion or only in trace amounts, other methods are required. Such methods include specific colorimetric tests, isolation of crystalline derivatives, and chromatographic procedures.

The most widely used of the colorimetric tests is based on the observation that glucosamine pentacetate, after being warmed with dilute alkali and then treated with p-dimethylaminobenzaldehyde (Ehrlich reagent) gives an intense reddish-purple coloration. The original procedure of Zuckerhandl and Messiner-Klebermass (1931) has undergone many modifications.

In the original procedure, free 2-amino sugars were detected and estimated colorimetrically as N-acetates after careful acetylation. The procedure was inaccurate and the color produced was not stable. Elson and Morgan (1933) and Morgan and Elson (1934) made use of the reaction noted by Pauly and Ludwig (1922) in which acetylacetone was condensed with hexosamines under alkaline conditions to yield a product that gave a stable reddish color with an acid solution of Ehrlich's reagent. In the first part of the test, the reaction of acetylacetone with glucosamine under alkaline conditions, 4 compounds are formed, 2 or 3 of which are chromogenic. The determination is, thus, highly empirical and the indiscriminate use of the so-called Elson-Morgan reaction may lead to confusing results. Neutral sugars (ketoses more than aldoses) and certain amino acids (mainly lysine, and to a lesser extent glycine and arginine) may react to yield products which give a color with Ehrlich's reagent that is indistinguishable from that produced by glucosamine. The sugar-amino reaction may be differentiated from the hexosamine reaction in several ways. The optimum pH for the hexosamine color is 9.5 and that for the amino acid-sugar complex color 10.8–11.2. Glucosamine does not give a color with Ehrlich's reagent after being heated with a carbonate buffer alone in the first part of the test, whereas the sugar-amino acid complex gives a red color. The color intensity with glucosamine is proportional to the acetylacetone concentration, whereas a maximum intensity is obtained with the sugar amino acid complex.

The test for N-acetyl-2-amino sugars involves warming the hexosamine-N-acetate or pentacetate with dilute alkali followed by treatment with an acid solution of Ehrlich's reagent. A reddish-purple color develops. Amino acids and neutral sugars apparently do not interfere if the test is conducted under special, rigidly controlled conditions.

The sensitivity of the original Elson-Morgan test is 20 γ. Dische and Borenfreund (1950) developed a method for determining 5 γ of the amino sugar. The hexosamines are deaminated with nitrous acid, and the derivative is reacted with skatole in dilute hydrochloric acid to yield stable characteristic colors well-suited to a quantitative colorimetric determination.

The identity of the amino sugar cannot be established by the above colorimetric tests, since they do not differentiate among the 2-amino sugars. The ideal means for the identification of a hexosamine is the isolation of a well-characterized crystalline derivative. Numerous compounds have been recorded in the literature for this purpose. The most satisfactory derivatives are the condensation products with aldehydes, the Schiffs' bases (Jolles and Morgan 1940). The p-nitrobenzylidene, 2-hydroxy-1-naphthylidene, and 3-methoxy-4-hydroxybenzylidene derivatives have solubility properties well-suited for the isolation of small quantities of amino sugars. Large quantities of neutral sugars do not interfere, and amino acids give condensation products that can be extracted with chloroform.

Glucosamines can be separated on ion-exchange resin columns, starch columns, and by paper or thin-layer chromatography. The most significant advance came with the introduction of 1-fluoro-2,4-

dinitrobenzene as a reagent for condensation with the basic groups of amino acids to give well-defined, colored 2,4-dinitrophenyl derivatives well-suited to chromatography (Annison *et al.* 1951; Kent *et al.* 1951).

Hexosamines and hexosamine uronic acids can be separated by paper electrophoresis, in a borate buffer, of their deamination products (Williamson and Zamenhof 1963), or as molybdate complexes (Mayer and Westphal 1968). The separation and determination of hexosamines by gas-liquid chromatography were described by Sweeley *et al.* (1963) and by Radhakrishnamurty *et al.* (1966).

Adding borate to the acetylation mixture in the Elson-Morgan method causes a depression of the color formation. With borate, glucosamine gives only 20–30% and galactosamine about 50% of the color formed without borate. Tracey (1952) used this principle in elaborating a method for the qualitative and semiqualitative determination of the glucosamine and galactosamine in mixtures. A similar method was described more recently by Good and Bessman (1964).

Glucosamine is transferred to glucosamine-6-phosphate by ATP and hexokinase. This phosphorylating system, however, has little or no action on galactosamine. The glucosamine-6-phosphate can be precipitated from the reaction mixture with zinc sulfate and barium hydroxide. If the Elson-Morgan reaction is carried out before and after the enzymic reaction, the amino sugar composition of a mixture can be determined (Slein 1952). A similar procedure based on the specificity of yeast hexokinase for D-glucosamine, and of the yeast acetylating enzyme for D-glucosamine 6-phosphate was described by Luderitz *et al.* (1964).

Polysaccharides

A. **Starch.**—Next to cellulose, starch is the most abundant and widely distributed substance of vegetable life. Starch occurs in the form of granules as reserve food in various parts of most plants: in the seeds of cereal grains, the roots of tapioca, the tubers of potatoes, and—in small amounts—in the stem-pith of sago, and fruit of the banana. The shape and size of the starch granules are characteristic for each plant. The appearance of the granules under microscopic examination is used to identify starches from various sources (Fig. 35.1).

Extremes in the size of starch granules are shown by taro or dasheen root (1 μ in diameter) and by root of the edible type of canna (150–200 μ in diameter).

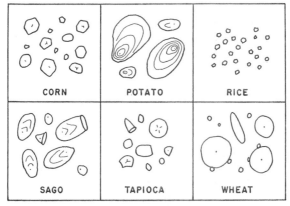

From T. J. Schoch

Fig. 35.1. Microscopic Appearance of Various Granular Starches

Under the polarizing microscope, the granular starches have a characteristic birefringence, with spherical granules giving a *Maltese cross* interference pattern and others a differently formed cross, indicating an organized spherocrystalline structure. Starch is a natural high polymer of D-glucose units; the polymer is formed through successive condensation of glucose units involving the enzyme systems of a developing plant. Complete, chemical or enzymatic hydrolysis of starch yields only glucose units.

Ordinary starches contain two types of glucose polymers. The linear fraction, called amylose, is a linear chain of some 500 to 2000 glucose units. Corn starch amylose contains an average of about 500 glucose units and has a molecular weight of about 80,000. The individual units in amylose are connected by alpha-1,4 linkages, i.e. from the aldehydic carbon 1 of each glucose to carbon 4 of the preceding glucose molecule.

The branched fraction, amylopectin, contains several hundreds of short linear branches, with an average branch length of 25 glucose units. The average molecular weight of amylopectin is at least 1,000,000. The interglucose linkages in each linear portion are alpha-1,4 (as in amylose); the branch points are through alpha-1,6 linkages.

Common starches (corn, wheat, tapioca, and potato) contain amylose and amylopectin in (generally) fixed proportions. Starches in certain varieties of corn, sorghum, and rice (but not of wheat or common root starches) are composed entirely of amylopectin and are known as waxy starches. On the other hand, wrinkled-seeded, garden-type varieties of peas and some recently developed corn varieties (amylomaize) have starches that contain from 70 to over 80% amylose. Waxy starches have certain unique properties which make them preferred for various uses. The linear fraction in cooked starch tends to deposit in the form of discrete insoluble particles as a result of starch retrogradation; amylopectin shows little tendency to retrograde.

Starches can be fractionated into amylose and amylopectin by gelatinization in water at elevated temperatures and pressures. The dissolved starch granules are cooled in the presence of polar agents such as butyl alcohol, amyl alcohol, thymol, and nitroparaffins. A complex precipitate of the polar components with amylose can be isolated by ultracentrifugation. Amylose similarly treated with fatty acids forms a complex that separates at 90° ±5°C as a pulverulent precipitate. Commercially, corn and potato starch can be fractionated into amylose and amylopectin by several processes.

The intense blue color that starches give with iodine is due solely to the amylose fraction; amylopectin gives a red or violet-red color.

The presence of starch can be established by the sensitive iodine reaction; 0.002 mg/ml can be positively detected.

To determine the starch content of foods, several methods are available. The starch can be extracted and dispersed into a colloidal solution which can be separated from extraneous matter. The starch content of the dispersion can be determined by precipitation and gravimetric or titrimetric assay of the precipitate; by a polarimetric method; colorimetrically; or as glucose after chemical or enzymatic hydrolysis.

In fat- and protein-rich foods (sausages, cheese), the starch can be reacted with alkali to form alcohol-insoluble complexes that can be readily separated from other components. The method is not suitable for the determination of starch in plant materials that contain a mixture of various polysaccharides.

Although starch can be solubilized in boiling water, its complete extraction from a plant tissue is difficult to achieve because of its high molecular weight and colloidal properties. Perchloric acid is an efficient extractant. The method of Pucher et al. (1948) consists of extracting the soluble sugars from dried plant tissue with 80% ethanol. The sugar-free residue is then treated with a perchloric acid solution; the extracted starch is precipitated with iodine and the starch-iodine complex is decomposed with alkali. The liberated starch is then determined colorimetrically with the anthrone reagent.

Starch can be extracted with a hot, concentrated solution of calcium chloride, and the extract assayed by various methods. To eliminate the interference from other extractable components, the method has undergone several modifications (Hadorn and Doevelaar 1960; Zelenka and Sasek 1966). From the carbohydrates dissolved in the calcium chloride solution, a starch-iodine complex is precipitated. Under proper conditions, precipitation of other carbohydrates is eliminated. The starch content can be determined gravimetrically; or the amount of iodine absorbed can be measured titrimetrically or colorimetrically; or the starch-iodine complex can be decomposed with sodium thiosulfite, the starch redissolved in calcium chloride and measured polarimetrically. Starch has a high specific optical rotation. Depending on the extractant composition and extraction method, $[\alpha]_D = +190°$ to 203°. Hemicelluloses, that may be solubilized, are substantially less optically active. Dextrins have a high optical rotation and must be separated.

Methods based on titrimetric or colorimetric determination of the absorbed iodine are highly empirical, as the relation between color and starch concentration is not stoichiometric. The results are affected by the method of starch extraction, and depend on the source of the starch and the ratio of amylose and amylopectin.

In several methods, starch is determined polarimetrically in dilute hydrochloric extracts (Winkler and Lukov 1967). The acid extract follows the extraction of soluble carbohydrates for the determination of a blank. The method is inapplicable to foods containing pre-gelatinized or otherwise modified starches. Such starches are partly soluble (Ulman and Richter 1961). Difficulties encountered in determining starch in such foods were recently evaluated in collaborative studies, based on the investigations of Friedman et al. (1967) and Friedman and Witt (1967). A proposed modification of the classical Maercker method consists of the extraction of nonstarchy components with isopropanol, enzymatic hydrolysis, clarification of the digest with zinc hydroxide, and determination of the starch by titration or spectrophotometry after ferricyanide reduction. Eheart and Mason (1966) have shown that in cereal products, reproducible starch determinations can be run by extracting the starch with a calcium chloride solution, precipitation of proteins with uranyl acetate, and a polarimetric assay of starch in the solution.

A complete acid hydrolysis of starch yields glucose that can be determined by the usual chemical or physicochemical methods. Methods that involve acid hydrolysis are subject to error caused on one hand by hydrolysis of nonstarch polysaccharides, and on the other hand by destruction of dextrose by acid during hydrolysis. To overcome the difficulties encountered in polarimetric methods or in acid hydrolysis several procedures employing a combination of acid and enzymatic hydrolysis or hydrolysis by several enzymes were proposed (Thivend *et al.* 1965; Lee and Whelan 1966; Donelson and Yamazaki 1968). Heat gelatinized starch in wheat products can be partially hydrolyzed with heat-resistant bacterial alpha-amylase, and the hydrolysis to dextrose completed with amyloglucosidase. The glucose can be assayed by glucose oxidase. In all stages of the assay, only starch and its hydrolysis products are analyzed. No interference from nonstarch components is expected and the elimination of such components is therefore not required (Pomeranz 1966). An excellent discussion of the scope and limitations of starch determinations was presented by Richter and Schierbaum (1968).

B. **Dextrins.**—Dextrins are a mixture of carbohydrates that are products of partial acid or enzymatic hydrolysis of starch. The dextrins include glucose polymers above hexoses and below those colored blue with iodine, though the higher homologues that give a brown-yellow to brown-red color are included.

Dextrins dissolved in water give a milky suspension upon the addition of a tenfold excess of ethanol. If other materials that might be precipitated are present, they can be removed by precipitation with a zinc acetate or potassium ferrocyanide solution. Yet some interference may be caused by pectins and water-soluble gums.

A relatively simple, rapid, and effective separation of dextrins can be made by thin-layer chromatography (Diemair and Kolbel 1963). In quantitative determinations, it must be realized that the assay involves an undefined mixture. Consequently, the methods are highly empirical. In precipitation with alcohols, starch and proteins (and under certain conditions pectins) are coprecipitated. In removing the interfering substances with lead acetate, some losses of dextrins are encountered. A mixture of acetone-ethanol (1:1) containing 2 gm calcium chloride per liter is a more useful precipitant (Thaler 1967).

Dextrins can be determined chemically. First glucose is determined by the Barfoed method, then total reducing sugars are assayed by the Fehling procedure, and finally dextrins and oligosaccharides are acid-hydrolyzed and determined as glucose by the Fehling method. Such a procedure has been developed for dextrins in syrups and bread (Rotsch 1947). A fermentation procedure with baker's yeast was developed by Taufel and Muller (1956). If the dextrins have been separated from the starch, their contents can be determined polarimetrically. Van der Bij (1967) described the usefulness of thin-layer chromatography, infrared spectroscopy, and gas-liquid chromatography in testing dextrins.

C. Glycogen.—Glycogen is a polysaccharide, generally, of animal origin that is stored in muscles and mainly in the liver. It resembles plant amylopectin, except that the number of glucose units per chain is smaller (about 12). It is water soluble and colors red-brown after iodine addition.

Most of the current methods for the determination of glycogen are based on a Pfluger (1909) method as modified by Good et al. (1933). The glycogen is isolated by precipitating with alcohol after heating the tissue in strong potassium hydroxide. It must be emphasized (Stetten et al. 1958) that as a result of alkaline extraction the glycogen with a molecular weight of 10^8 is reduced to 10^5. In addition, degradation products that are not precipitated with 70% ethanol, are formed. After isolation, the glycogen can be measured directly with anthrone or indirectly by hydrolysis followed by a glucose determination (Seifter et al. 1950; Van Handel 1965).

Analytical procedures based on the color that glycogen gives with iodine are little used because they lack sensitivity and are affected by temperature. Krisman (1962) developed a sensitive and relatively specific procedure for the colorimetric estimation of glycogen based on the effects of salts on color intensity and stability.

An enzymic method for measuring glycogen has been described by Passonneau et al. (1967). Glycogen plus inorganic phosphate and TPN$^+$ are converted in one analytical step to 6-P-gluconolactone and TPNH. The TPNH is measured by its fluorescence or ultraviolet absorption. The method uses commercially available enzymes: phosphorylase, P-glucomutase, and glucose-6-P-dehydrogenase. Most commercial preparations contain enough transglucosylase and glucosidase for the complete degradation of glycogen. The method is specific, sensitive (0.05 γ can be detected), and applicable to the analyses of whole tissues.

D. Cellulose and Other Cell Wall Components.—The determination of this group of materials is wrought, even today, with many difficulties. The materials contain many undefined polymers varying in size and in composition. During isolation of the various cell wall components, modifications—including degradation—take place. Part of the cellulose is soluble in alkali. The insoluble fraction is called α-cellulose. The fraction in the alkaline extract that is precipitated upon acidification is β-cellulose, and the fraction remaining in solution is γ-cellulose containing mainly hemicelluloses.

Hemicelluloses comprise a mixture of alkali-soluble polysaccharides, the amount and composition of which vary with the extraction procedure. These include polymers of mannose, galactose, xylose, and arabinose as well as polymers of uronic acids and of methyl- or acetyl-substituted monoses. Cellulose and hemicelluloses are among the most abundant and widely distributed carbohydrates in plant materials. They are the main cell wall constituents of fruits, vegetables, and cereals. Their content increases with maturity; they are highly concentrated in pericarp tissues of cereal grains and present in low concentration in the starchy endosperm.

Lignin is present in trace amounts in high quality fruits and vegetables, but is present in the bran of cereal grains and in some spices.

Cellulose and hemicellulose are generally determined in defatted foods following the extraction of soluble carbohydrates. To eliminate the effects of enzymes on degradation and solubilization, prior storage at low temperature and enzyme inactivation at the beginning of the analysis are important. In addition, proteins and lignin must be removed. Jermyn (1955) described in detail a procedure for the determination of celluloses and hemicelluloses in plant material.

After defatting the powdered material with ether and ethanol, it is boiled under reflux with 85% ethanol to extract some soluble carbohydrates. Water-soluble carbohydrates—starches and pectins— are extracted after vigorous boiling in water under reflux. Lignins are extracted by heating at 70–75°C for 4 hr with an acidified sodium chlorite solution. The residual material, holocellulose, that is practically lignin-free is digested with alkali and the cellulose and hemicellulose are determined. The cellulose content in the alkali-insoluble fraction can be determined after a complete hydrolysis with sulfuric acid, and determination of the reducing sugars.

Enzymatic methods for the assay of cellulose and other cell wall components are described in Chap. 30.

Mannans (i.e. in coffee or yeast) are generally determined by precipitation with an alkaline copper solution (Thaler 1967). The mannans are generally in the hemicellulose fraction from which they are extracted with alkali.

Uncertainty regarding the detailed chemistry of lignin has precluded establishment of a generally satisfactory method for its determination. Most methods are based on the insolubility of lignin in concentrated mineral acids (i.e. 72% sulfuric acid) in which most cell constituents are soluble. The residue may, however, contain some humins as a result of the action of acids at elevated temperatures. It is, therefore, recommended to remove potentially interfering substances by an extraction with water and dilute acid. In addition, defatted, air-dried materials are generally determined (Freudenberg and Ploetz 1940; Ellis *et al.* 1946).

Pentosans are polymers of pentoses and methyl pentoses. The pentoses can be calculated from the pentose contents of pentosan hydrolyzates. Either the reducing power after removal of fermentable hexoses, or the pentose content of chromatographically fractionated hydrolyzates can be determined. The pentose or pentosan content is often determined by measuring the amount of furfural formed after a reaction with hydrochloric acid. The latter method is highly empirical and significantly affected by the assay conditions and the presence of nonpentosan components that yield furfural (or hydroxymethyl furfural) under test conditions. In most methods, including the official AOAC procedure, the furfural is distilled and precipitated with phloroglucinol; in other procedures barbituric or thiobarbituric acid is used.

The furfural can also be titrated with bromine, 1 molecule of furfural giving 2 molecules of hydrogen bromide. Finally, the furfural can be determined spectrophotometrically, based on its absorbance in the ultraviolet (275 nm for furfural and 284 nm for hydroxymethyl furfural) (Thaler 1967).

E. Pectins.—Pectins are important as jelling and thickening agents. The pectins are present, bound to calcium mainly in the middle lamella, in growing tissues of many higher plants, and bound to cellulose in the primary cell membrane. The pectins are generally obtained commercially by processing the parenchymatic tissues of plant materials. During fruit ripening, the cell pectins are solubilized enzymatically and are present in fruit juices. They serve as the cementing agents of the cell and regulate the water content.

Chemically, pectins are polymers of galacturonic acid connected by alpha-1,4-linkages to long chains. They can be divided and defined (Gudjons 1967) as:

Pectic substances—materials comprising all polygalacturonic acid containing materials

Protopectins—water insoluble materials, in bound form, yielding pectins upon hydrolysis

Pectin—partly esterified polygalacturonic acids; generally methyl esters, in some (rapeseeds and beets) also esterified with acetic acid. They can be divided into low methoxy and high methoxy pectins, depending whether they contain less or more than 7% methoxy esterified galacturonic acids. A completely (100%) esterified polygalacturonide has theoretically 16.3% methoxyl groups (Baker 1948)

Pectinic acids—have all carboxyl groups in the free form and are water-insoluble. Salts of pectinic acids are water-soluble.

Pectic substances isolated from plant materials often contain various arabans and galactans. The isolated substances are a heterogenous mixture that varies (depending on the source, method of extraction, and subsequent treatment) in molecular weight, degree of esterification, and araban and galactan content. In view of the heterogenous nature of the pectic substances, their analysis is based either on empirical procedures designed to evaluate their usefulness for specific purposes, or on the assay of specific constituents.

The solubility of pectins increases with an increase in the degree of esterification and with a decrease in molecular weight. The less soluble a pectin is in water, the easier it can be precipitated by adding an electrolyte. Pectins with an esterification grade of up to 20% are precipitated by sodium chloride solutions, with 50% by calcium chloride solutions, and with 70% by aluminum chloride or copper chloride solutions. Completely esterified pectins are not precipitated by electrolytes. The pectins can also be precipitated with organic solvents, such as acetone, methyl-, ethyl-, or propyl alcohol. The concentration of alcohol that is required increases with the degree of esterification.

The usefulness of pectins stems mainly from their capacity to form stable gels or films, and increase the viscosity of acidified, sugar-containing solutions. Completely esterified pectins can be jelled without acid or electrolyte addition. Pectins with a high esterification grade jell rapidly at a relatively low acidity. With a decrease in esterification, the rate of gel formation decreases and the required pH optimum decreases. The jelling capacity is greatly enhanced by the presence of calcium salts, which also increase the stability and decrease the dependence on pH and sugar concentration. Viscosity increases, at a given esterification grade, with an increase in molecular weight; decreased esterification lowers viscosity. The effects on viscosity depend also on the total concentration and the electrolytes present. The stability of gels is affected by numerous factors—including temperature, the presence of acids or alkali, enzymes, and mechanical treatment. The pectin content and its quality (determined as jelly grade, i.e. pounds of sugar that 1 lb of pectin would set to a gel under standardized conditions) are important in the manufacture of jellies, jams, and preserves.

For the detection of pectins, the McCready and Reeve (1955) method is used. To pectin and hydroxamic acid in an alkaline solution, a ferric compound is added, and a water-insoluble red complex is formed. Numerous colorimetric and paper chromatographic methods are available to detect and estimate the components after hydrolysis with 2% hydrochloric acid or 4% sulfuric acid under reflux (Gudjons 1967).

To determine the identity and purity of a pectin preparation, the galacturonide content, neutralization equivalent, degree of esterification, and physical properties of the pectic substances are determined. The precision and significance of the determination are largely dependent on the amount and nature of nonpectic compounds, and on the difficulty of their separation.

Pectin can be determined gravimetrically in aqueous, ammonium citrate, or dilute hydrochloric acid extracts, by a procedure involving precipitation with alcohol or acetone, saponification of the precipitate with cold alkali, acidification, boiling with acid and conversion into a precipitate of pectic acid (Wichmann 1922, 1923). More commonly, the saponified pectate is precipitated as calcium pectate (Carre and Haynes 1922; Emmett and Carre 1926; Griebel and Weiss 1929; Hinton 1939).

In the titrimetric method of Deuel (1943), the pectin is saponified with sodium hydroxide, slightly acidified, precipitated with 96% ethanol, washed, dissolved with excess standardized alkali, and the excess back-titrated.

As uronic acids are decarboxylated by boiling with mineral acids, the amount of carbon dioxide evolved can be used to determine the pectin contents (Conrad 1931; McCready et al. 1940). A colorimetric procedure, based on the carbazole reaction of uronic acids has been proposed by McComb and McCready (1952). The number of free carboxyl groups can be determined by titration. Numerous methods of organic analysis can be used to determine the methyl or acetyl groups linked to uronic acids in pectic substances. For general information on the role, composition, and assay of pectic substances, the interested reader may consult several books (Kertesz 1951; Hottenroth 1951) or excellent and comprehensive reviews (Hinton 1939; Anderson and Sands 1945; Hirst and Jones 1946; Joslyn and Phaff 1947; Baker 1948; Deuel and Stutz 1958).

Crude Fiber

In the analysis of cellulose-containing foods, the determination of crude fiber is widely used.   Crude fiber includes, theoretically, materials that are indigestible in the human and animal organism. It is determined as material insoluble in dilute acid and dilute alkali under specified conditions.

The present method is based on a procedure developed by Hennenberg, Stohmann, and Rautenberg in 1864 in the Agricultural Experiment Station in Weende (near Gottingen) in Germany.   In the determination, 2 gm of material are defatted with petroleum ether and boiled under reflux for exactly 30 min with 200 ml of a solution containing 1.25 gm sulfuric acid per 100 ml solution.   The solution is filtered through linen or several layers of cheese cloth on a fluted funnel and washed with boiling water until the washings are no longer acid.   The residue is tranferred quantitatively to a beaker, boiled for 30 min with 200 ml of a solution containing 1.25 gm of carbonate-free sodium hydroxide per 100 ml.   The final residue is filtered through a thin but close pad of washed and ignited absestos in a Gooch crucible, dried in an electric oven, weighed, incinerated, cooled, and weighed again.   The loss in weight in incineration is taken as crude fiber.

Crude fiber determinations are greatly affected by manipulations and procedures.   Particle size is very important;   the finer the material is ground, the lower the determined crude fiber content. Apparent crude fiber is lowered by defatting, though the low lipid content of some foods (i.e. white flour) affects the results little.   The rate of heating to the boiling point and rate of boiling must be controlled.   Filtering after each digestion must be completed within a given time;   delays in filtering after acid or alkali digestion generally lower the results.

To reduce evaporation losses during boiling, and concomitant changes in concentration of digestion solutions, specially constructed apparatus—with condensers—is available commercially (Fig. 35.2).   Most difficulties are encountered in determining the crude fiber content of protein-rich foods;   predigestion by proteolytic enzymes has been recommended to accelerate filtration after digestion.

The residue from a crude fiber determination contains about 97% cellulose and lignin.   It does not represent, however, all the cellulose and lignin present initially.   Thus, the crude fiber content of whole wheat is about 2%, whereas the kernel has about 7% of pericarp, most of which cannot be digested by humans and nonruminant animals.   In addition, the crude fiber is a mixture of cellulosic materials and does not represent any specific compound or group of compounds. Yet, the crude fiber is a useful parameter in food and feed analyses. Crude fiber is commonly used as an index of the feeding value of poultry and stock feeds; seeds high in crude fiber content are low in nutritional value. A determination of crude fiber is used in evaluating the efficiency of milling and separating bran from the starchy endosperm. It is a more lengthy, laborious, and complicated but more direct index of flour purity than color or ash. Crude fiber is also useful in the chemical determination of succulence of fresh vegetables and fruits; overmature products have increased levels of crude fiber. Generally, however, substantial changes in texture and consumer acceptance are accompanied by relatively small changes in crude fiber content.

There are at least 100 modifications of the original crude fiber determination, including addition of oxidizing agents, use of special solvents (phenol or trichloroacetic acid), and various concentrations of acid and alkali. Some of the methods simplify, some shorten, and some improve reproducibility of the determination. Most improve little the accuracy of the actual determination of indigestible materials (Tunger 1964; Hallab and Epps 1963).

Kurschner and Hanak (1930) proposed the use of dilute nitric acid for digestion. The hemicelluloses are hydrolyzed, and lignin

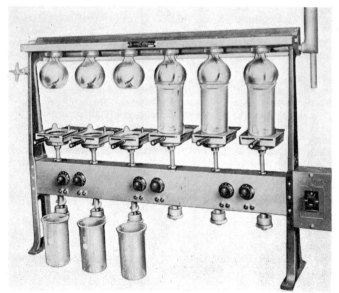

Courtesy Labconco Co

FIG. 35.2. CRUDE FIBER APPARATUS

is oxidized and nitrated.   The dual action of nitric acid is improved
by dissolving it in 80% acetic acid rather than in water.   Digestion
is improved by the addition of trichloroacetic acid (Scharrer and
Kurschner 1932).   The method has been used in cocoa products
(Fincke and Thaler 1942).

Van Soest (1967) reviewed the limitations of the crude fiber assay
in terms of the validity and usefulness of the information it provides.
A rapid procedure was developed (Van Soest and Wine 1967) for
determining the fiber insoluble in a neutral detergent.   The method
was compared and standardized against determinations of vegetable
matter that was indigestible by proteolytic and diastatic enzymes,
and that could not be utilized except by microbial fermentation in
the digestive tracts of animals.

BIBLIOGRAPHY

ACKER, L.   1967.   Carbohydrates—water-soluble compounds (total extract); detection and
    determination of mono- and oligosaccharides.   In Handbook of Chemistry, Food Analysis,
    Vol. 2.   W. Diemair (Editor).   Springer-Verlag, Berlin.   (German)
ALEXANDER, R. J., and GARBUTT, J. T.   1965.   Use of sorbitol as internal stan-
    dard in determination of D-glucose by gas liquid chromatography.   Anal.
    Chem. 37, 303–305.
ALM, R. S.   1952.   Gradient elution analysis.   II.   Oligosaccharides.   Acta
    Chem. Scand. 6, 1186–1193.
ANDERSON, E., and SANDS, L. A.   1945.   Discussion of methods of value in
    research in plant uronides.   Advan. Carbohydrate Chem. 1, 329–334.
ANNISON, E. F., JAMES, A. T., and MORGAN, W. T. J.   1951.   Separation and
    identification of small amounts of mixed amino sugars.   Biochem. J. 48,
    477–482.
ANON.   1966.   The Technicon Sugar Chromatography System.   1564-10-6.
    Technicon Chromatography Corp., Ardsley, N.Y.
ASSOC. OFFIC. AGR. CHEMISTS.   1965.   Official Methods of Analysis, 10th
    Edition.   Assoc. Offic. Agr. Chemists, Washington, D.C.
BAKER, E. L.   1948.   High polymer pectins and their de-esterification.   Advan.
    Food Res. 1, 395–424.
BAKER, S. A., BOURNE, E. J., and WHIFFEN, D. H.   1956.   Use of infrared
    analysis in the determination of carbohydrate structure.   Methods Biochem.
    Analy. 3, 213–245.
BALAZS, E. A., and JEANLOZ, R. W. (Editors).   1965.   The Amino Sugars—The
    Chemistry and Biology of Compounds Containing Amino Sugars, Vols. 1, 2, 3,
    4.   Academic Press, New York.
BANDOW, F.   1937.   Absorption spectra of organic substances in concentrated
    sulfuric acid.   I.   Experiments on carbohydrates.   Biochem. Z. 294, 124–137
    (Chem Abstr. 32, 11819).
BARNETT, A. J. G., and TAWAB, G. A.   1957.   Determination of lactose in milk
    and cheese.   J. Sci. Food Agr. 8, 437–441.
BATES, F. J.   1942.   Polarimetry, Saccharimetry, and the Sugars.   Natl. Bur.
    Std. Circ. 440.
BATH, I. H.   1958.   The ultraviolet spectrophotometric determination of sugars
    and uronic acids.   Analyst 83, 451–455.

BIGGS, D. A. 1967. Milk analysis with the infrared milk analyzer. J. Dairy Sci. *50*, 799–803.

BISHOP, C. T. 1964. Gas liquid chromatography of carbohydrate derivatives. Advan. Carbohydrate Chem. *19*, 95–147.

BLISH, M. J., and SANDSTEDT, R. M. 1933. An improved method for the estimation of flour diastatic value. Cereal Chem. *10*, 189–202.

BORENFREUND, E., and DISCHE, Z. 1957. A new spray for spotting sugars on paper chromatograms. Arch. Biochem. Biophys. *67*, 239–240.

BRINKLEY, W. W. 1955. Column chromatography of sugars and their derivatives. Advan. Carbohydrate Chem. *10*, 55–94.

BROCK-NEELY, W. 1957. Infrared spectra of carbohydrates. Advan. Carbohydrate Chem. *12*, 13–33.

BROWNE, C. A., and ZERBAN, F. W. 1941. Physical and Chemical Methods of Sugar Analysis, 3rd Edition. J. Wiley & Sons, New York.

CARRE, M. H., and HAYNES, D. 1922. The estimation of pectin as calcium pectate and the application of this method to the determination of the soluble pectin in apples. Biochem. J. *16*, 60–69.

CHARGAFF, E., LEVINE, C., and GREEN, C. 1948. Techniques for the demonstration by chromatography of nitrogenous lipide constituents, sulfur-containing amino acids and reducing sugars. J. Biol. Chem. *175*, 67–71.

CHARLES, D. F., and MEADS, P. F. 1962. Refractive Index. *In* Methods in Carbohydrate Chemistry, R. L. Whistler, and M. L. Wolfrom, (Editors). Academic Press, New York.

CONRAD, C. M. 1931. Decarboxylation studies on pectins and calcium pectates. J. Am. Chem. Soc. *53*, 1999–2003.

COULTER, S. T., JENNESS, R., and GEDDES, W. F. 1951. Physical and chemical aspects of the production, storage and utility of dry milk products. Advan. Food Res. *3*, 45–106.

CRAMER, F. 1953. Paper Chromatography, 2nd Edition. Chemie Publishing Co., Weiheim, Germany. (German)

CROWELL, E. P., and BURNETT, B. B. 1967. Determination of carbohydrate composition of wood pulps by gas chromatography of alditol acetates. Anal. Chem. *39*, 121–124.

DANEHY, J. P., and PIGMAN, W. W. 1951. Reactions between sugars and nitrogenous compounds and their relationship to certain food problems. Advan. Food Res. *3*, 241–281.

DEUEL, H. 1943. Pectin as high molecular electrolyte. Z. Lebensmittelunters. Hyg. (Bern) *34*, 41. (German).

DEUEL, H., and STUTZ, E. 1958. Pectic substances and pectic enzymes. Advan. Enzymol. *20*, 341–382.

DIEMAIR, W. and KOLBEL, R. 1963. Detection and determination of dextrins. Z. Lebensmittelunters. Forsch. 124, 1–14, 157–179. (German).

DIMLER, R. J., SCHAEFER, W. C., WISE, C. S., and RIST, C. E. 1952. Quantitative paper chromatography of D-glucose and its oligosaccharides. Anal. Chem. *24*, 1411–1414.

DISCHE, Z. 1962. Color reactions of carbohydrates. *In* Methods in Carbohydrate Chemistry, R. L. Whistler, and M. L. Wolfrom (Editors). Academic Press, New York.

DISCHE, Z., and BORENFREUND, E. 1950. A spectrophotometric method for the determination of hexosamines. J. Biol. Chem. *184*, 517–522.

DONELSON, J. R., and YAMAZAKI, W. T. 1968. Enzymic determination of starch in wheat fractions. Cereal Chem. *45*, 177–182.

DREYWOOD, R. 1946. Qualitative test for carbohydrate material. Ind. Eng. Chem. Anal. Edition *18*, 499.

DUBOIS, M. *et al.* 1956. Colorimetric method for determination of sugars and related substances. Anal. Chem. *28*, 350–356.

DYACHENKO, P., and SAMSONOV, Y. 1964. Infrared spectra of dried milk and its components. Molochn. Prom. *25*, No. 7, 11–13. (From Chem. Abstr. *61*, 15263 f).

EHEART, J. F. and MASON, B. S. 1966. Assay methodology studies of carbohydrate fractions of wheat products. J. Assoc. Offic. Agr. Chemists *49*, 907–912.

ELLIS, G. H., MATRONE, G., and MAYNARD, L. A. 1946. A seventy-two percent sulfuric acid method for the determination of lignin and its use in animal nutrition studies. J. Animal Sci. *5*, 285–297.

ELSON, L. A., and Morgan, W. T. J. 1933. Colorimetric method for the determination of glucosamine and chondrosamine. Biochem. J. *27*, 1824–1828.

EMMETT, A. M., and CARRE, M. H. 1926. A modification of the calcium pectate method for the estimation of pectin. Biochem. J. *20*, 6–12.

FINCKE, H., and THALER, H. 1942. Determination of crude fiber in cocoa and cocoa products. Z. Untersuch. Lebensmittel *84*, 97–134. (German).

FLOHIL, J. T. 1933. Report of the subcommittee on the development of a volumetric copper reduction method for sugar determinations. Cereal Chem. *10*, 470–476.

FOSTER, A. B. 1957. Zone electrophoresis of carbohydrates. Advan. Carbohydrate Chem. *12*, 81–115.

FOSTER, A. B., and STACEY, M. 1952. The chemistry of the 2-amino sugars (2-amino-2-deoxy sugars). Advan. Carbohydrate Chem. *7*, 247–288.

FRENCH, D., ROBYT, J. F., WEINTRAUB, M., and KNOCK, P. 1966. Separation of maltodextrins by charcoal chromatography. J. Chromatog. *24*, 68–75.

FREUDENBERG, K., and PLOETZ, T. 1940. Quantitative assay of lignin. Ber. Deut. Chem. Ges. *73*, 754–757. (German)

FRIEDMANN, T. E., and WITT, N. F. 1967. Determination of starch and soluble carbohydrates. II. Collaborative study of starch determination in cereal grains and in cereal products. J. Assoc. Offic. Agr. Chemists *50*, 958–963.

FRIEDMANN, T. E., WITT, N. F., and NEIGHBORS, B. W. 1967. Determination of starch and soluble carbohydrates. I Development of method for grain stock feeds, cereal foods, fruits, and vegetables. J. Assoc. Offic. Agr. Chemists *50*, 944–958.

GARDELL, S. 1958. Determination of hexosamines. Methods Biochem. Analy. *6*, 289–317.

GEYER, H. V. 1965. Application of gas chromatography in analysis of sugars and starch hydrolysates. Die Starke *17*, 307–313. (German)

GOOD, C. A., KRAMER, H., and SOMOGYI, M. 1933. The determination of glycogen. J. Biol. Chem. *100*, 485–499.

GOOD, T. A., and BESSMAN, S. P. 1964. Determination of glucosamine and galactosamine using borate buffers for modification of the Elson-Morgan and Morgan-Elson reactions. Anal. Biochem. *9*, 253–262.

GREEN, J. G. 1966. Automated carbohydrate analyzer. Natl. Cancer Inst. Monograph *21*, 447–467.

GRIEBEL, C., and WEISS, F. 1929. The problem of pectin. Z. Unters. Lebensm. *58*, 189–201.

GUDJONS, H. 1967. Pectins. *In* Handbook of Food Chemistry, Food Analysis. W. Diemair, (Editor). Springer-Verlag, Berlin. (German)

HADORN, H., and BIEFER, K. W. 1956. Effects of amino acids on determination of sugars, especially by the method of Potterat and Eschmann and Hadorn and von Fellenberg. Mitt. Lebensmitteluntersuch. Hyg. (Bern) *47*, 4–15.

HADORN, H., and DOEVELAAR, F. 1960. Systematic evaluation of titrimetric, colorimetric and polarimetric methods of starch determinations. Mitt. Lebensmitteluntersuch. Hyg. (Bern) *51*, 1–68. (German)

HADORN, H., and FELLENBERG, T. V. 1945. Reevaluation of iodometric sugar determinations. Lebensmitteluntersuch. Hyg. (Bern)*36*, 359–367. (German)

HADORN, H., and JUNGKUNZ, R. 1952. Analysis and composition of chestnuts (Castanea vesca). Z. LebensmittelUnters. Forsch. 95, 418–429. (German).

HAGEDORN, H. C., and JENSEN, B. N. 1923. Micro assay of blood sugar with ferricyanide. Biochem. Z. 135, 46–58. (German).

HALLAB, A. H., and EPPS, E. A. 1963. Variables affecting the determination of crude fiber. J. Assoc. Offic. Agr. Chemists 46, 1006–1010.

HAMMOND, L. D. 1940. Redetermination of the Munson-Walker reducing-sugar values. J. Res. Natl. Bur. Std. 24, 579–596 (RP 1301).

HANES, C. S. 1929. An application of the method of Hagedorn and Jensen to the determination of larger quantities of reducing sugars. Biochem. J. 23, 99–106.

HARDINGE, M. G., SWARNER, J. B., and CROOKS, H. 1965. Carbohydrates in foods. J. Am. Dietet. Assoc. 46, 197–204.

HASSID, W. Z. 1936. Determination of reducing sugars and sucrose in plant materials. Ind. Eng. Chem. Anal. Edition 8, 138–140.

HEIDT, L. J., and COLMAN, C. M. 1952. Degradation of D-glucose, D-fructose, and invert sugar in carbonate buffered water solution. J. Am. Chem. Soc. 74, 4711–4713.

HEYNS, K. 1959. Standard methods of analysis of starch and starch hydrolysates. ICUMSA Methods. Die Starke 11, 67–74, 215–233. (German).

HINTON, C. L. 1939. Fruit Pectins—Their Chemical Behaviour and Jellying Properties. Dept. Sci. Ind. Res., Food Investigations, Spec. Rep. 48. Her Majesty's Stationery Office, London.

HINTON, C. L., and MACARA, T. 1924. Application of the iodometric method to the analysis of sugar products. Analyst 49, 2–24.

HINTON, C. L., and MACARA, T. 1927. The determination of aldose sugars by means of the chloramine T with special reference to the analysis of milk products. Analyst 52, 668–688.

HIRST, E. L., and JONES, J. K. N. 1946. The chemistry of pectic materials. Advan. Carbohydrate Chem. 2, 235–251.

HODGE, J. E. 1953. Chemistry of browning reactions in model systems. J. Agr. Food Chem. 1, 928–943.

HODGE, J. E., and DAVIS, H. A. 1952. Selected Methods for Determining Reducing Sugars. U. S. Dept. Agr., Agr. Res. Serv. Publ. AIC 333.

HODGE, J. E., and HOFREITER, B. T. 1962. Determination of reducing sugars and carbohydrates. In Methods in Carbohydrate Chemistry, R. L. Whistler, and M. L. Wolfrom (Editors). Academic Press, New York.

HOOVER, W. J., NELSON, A. I., MILNER, R. T., and WEI, L. L. 1965. Isolation and evaluation of the saccharide components of starch hydrolysis. I. Isolation. J. Food Sci. 30, 248–252.

HOTTENROTH, B. 1951. Pektine und Ihre Verwendung. R. Oldenbourg Publ. Co., Munchen, Germany.

HOUGH, L. 1954. Analysis of mixtures of sugars by paper and cellulose column chromatography. Methods Biochem. Analy. 1, 205–242.

HUBER, C. M., SCOBELL, H., and TAI, H. 1966. Determination of saccharide distribution of corn syrup by direct densitometry of thin-layer chromatograms. Cereal Chem. 43, 342–346.

IKAWA, M., and NIEMANN, C. 1949. A spectrophotometric study of the behavior of carbohydrates in seventy-nine percent sulfuric acid. J. Biol. Chem. 180, 923–931.

ISHERWOOD, F. A., and JERMYN, M. A. 1951. Relationship between the structure of the simple sugars and their behavior on the paper chromatogram. Biochem. J. 48, 515–524.

JACKSON, R. F., and MCDONALD, E. J. 1941. Errors of Munson and Walker's reducing sugar tables and the precision of their method. J. Res. Natl. Bur. Std. 27, 237.

614 FOOD ANALYSIS

JAYME, G., and KNOLLE, H. 1960. Quantitative assay of paper chromatograms by direct photometry in the UV. Z. Analyt. Chem. 178, 84–100.
JERMYN, M. A. 1955. Cellulose and hemicellulose. In Modern Methods of Plant Analysis, Vol. 2. K. Paech, and M. V. Tracey, (Editors). Springer-Verlag, Berlin.
JOLLES, Z. E., and MORGAN, W. T. J. 1940. The isolation of small amounts of glucosamine and chondrosamine. Biochem. J. 34, 1183.
JOSLYN, M. A., and PHAFF, H. J. 1947. Recent advances in the chemistry of pectic substances. Wallerstein Lab. Commun. 10 No. 29, 39–56.
KAGAN, J., and MABRY, T. J. 1965. Sugar analysis of flavonoid glycosides. Anal. Chem. 37, 288–289.
KENT, P. W., LAWSON, G., and SENIOR, A. 1951. Chromatographic separation of amino sugars and amino acids by means of the TN-(2,4-dinitrophenyl) derivatives. Science 113, 354–355.
KERTESZ, Z. 1951. The Pectic Substances. Interscience Publishing Co., New York.
KESLER, R. B. 1967. Rapid quantitative anion-exchange chromatography. Anal. Chem. 39, 1416–1422.
KHYM, J. X., and ZILL, L. P. 1952. Separation of sugars by ion exchange. J. Am. Chem. Soc. 74, 2090–2094.
KNEEN, E., and SANDSTEDT, R. M. 1941. Beta amylase activity and its determination in germinated and ungerminated cereals. Cereal Chem. 18, 237–252.
KOWKABANY, G. N. 1954. Paper chromatography of carbohydrates and related compounds. Advan. Carbohydrate Chem. 9, 303–353.
KRISMAN, C. R. 1962. A method for the colorimetric estimation of glycogen with iodine. Anal. Biochem. 4, 17–23.
KURSCHNER, K., and HANAK, A. 1930. Assay of the "so-called" crude fiber. Z. Unters. Lebensm. 59, 484–494.
LANE, J. H., and EYNON, L. 1923. Determination of reducing sugars by means of Fehling's solution with methylene blue as internal indicator. J. Soc. Chem. Ind. 42, 32T–37T.
LEDERER, E., and LEDERER, M. 1957. Chromatography. Elsevier Publishing Co., New York.
LEE, E. Y. C., and WHELAN, W. J. 1966. Enzymic methods for the microdetermination of glycogen and amylopectin and their unit chain lengths. Arch. Biochem. Biophys. 116, 162–167.
LIN, F. M., and POMERANZ, Y. 1965. Characterization of wheat components by infrared spectroscopy. I. Infrared spectra of major wheat flour components. J. Assoc. Offic. Agr. Chemists 48, 885–891.
LIN, F. M., and POMERANZ, Y. 1968. Characterization of water-soluble wheat flour pentosans. J. Food Sci. 33, 599–606.
LOTHROP, R. E., and HOLMES, R. L. 1931. Determination of dextrose and levulose in honey by use of iodine-oxidation method. Ind. Eng. Chem. Anal. Edition 3, 334–339.
LUDERITZ, O., SIMMONS, D. A. R., WESTPHAL, O., and STROMENGER, J. L. 1964. A specific microdetermination of glucosamine and the analysis of other hexosamines in the presence of glucosamine. Anal. Biochem. 9, 263–271.
MAYER, H., and WESTPHAL, O. 1968. Electrophoretic separation of hexosamine and hexatonic acid derivatives as molybdate complexes. J. Chromatog. 33, 514–525. (German)
McCOMB, E. A., and McCREADY, R. M. 1952. Colorimetric determination of pectic substances. Anal. Chem. 24, 1630–1632.
McCREADY, R. M., and REEVE, R. M. 1955. Test for pectin based on the reaction of hydroxamic acids with ferric ion. J. Agr. Food Chem. 3, 260–262.
McCREADY, R. M., SWENSEN, H. A., and MACLAY, W. D. 1940. Determination of uronic acids. Ind. Eng. Chem. Anal. Edition 18, 290–291.

MILLER, G. L. 1960. Micro column chromatographic method for analysis of oligosaccharides. Anal. Biochem. 2, 133–140.

MORGAN, W. T. J., and ELSON, L. A. 1934. A colorimetric method for the determination of N-acetylglucosamine and N-acetyl chondrosamine. Biochem. J. 28, 988–995.

MUNSON, L. S., and WALKER, P. H. 1906. The unification of reducing sugar methods. J. Am. Chem. Soc. 28, 663–686.

NELSON, N. 1944. A photometric adaptation of the Somogyi method for the determination of glucose. J. Biol. Chem. 153, 375–380.

OHMS, J. I., ZEC. J., BENSON, J. V., and PATTERSON, J. A. 1967. Column chromatography of neutral sugars. Operating characteristics and performance of a newly available anion-exchange resin. Anal. Biochem. 20, 51–57.

PARKER, F. S. 1960. Infrared spectra of carbohydrates in water and a new measure of mutarotation. Biochim. Biophys. Acta 42, 513–519.

PARTRIDGE, S. M. 1948 Filter-paper partition chromatography of sugars. I. General description and application to the qualitative analysis of sugars in apple juices, egg white, and foetal blood of sheep. Biochem. J. 42, 238–250.

PASSONNEAU, J. V., GATFIELD. P. D., SCHULTZ, D. W., and LOWRY, O. H. 1967. An enzymic method for measurement of glycogen. Anal. Biochem. 19, 315–326.

PAULY, H., and LUDWIG, E. 1922. Glucosamine as a former of heterocyclic compounds. Z. Physiol. Chem. 121, 170–176. (Chem. Abstr. 16, 4211).

PFLUGER, E. 1909. Arch. Ges. Physiol. 129, 362. Cited by J. V. Passoneau et al. 1967. An enzymic method for measurement of glycogen. Anal. Biochem. 19, 315–326.

POMERANZ, Y. 1966. Unpublished data.

POTTERAT, M., and ESCHMANN, H. 1954. Use of complexing agents in assay of sugars. Mitt. Lebensmitteluntersuch. Hyg. (Bern) 45, 312–329. (French).

PUCHER, G. W., LEAVENWORTH, C. S., and VICKERY, H. B. 1948. Determination of starch in plant tissues. Anal. Chem. 20, 850–853.

PUSZTAI, A. 1965. Studies on extraction of nitrogenous and phosphorus-containing materials from the seeds of kidney beans (Phaseolus vulgaris). Biochem. J. 94, 611–616.

RADHAKRISHNAMURTY, B., DALFERES, E. R., and BERENSON, G. S. 1966. Determination of hexosamines by gas liquid chromatography. Anal. Biochem. 17, 545–550.

RICHTER, M., and SCHIERBAUM, F. 1968. Possibilities and limits in starch assay in cereals. Die Nahrung. 12, 189–197. (German).

ROTSCH, A. 1947. Determination of low-molecular weight carbohydrates in cereals. Getreide, Mehl, Brot 1, 10–13. (German).

SAWARDEKER, J. S., and SLONEKER, J. H. 1965. Quantitative determination of monosaccharides by gas liquid chromatography. Anal. Chem. 37, 945–947.

SCHARRER, K., and KURSCHNER, K. 1932. A new method for crude fiber determination in feeds. Zentralblatt Agrikulturchemie B, 3, 302–310. (German).

SCHERZ, H., STEHLIK, G., BAUCHER, E., and KAINDL, K. 1968. Thin-layer chromatography of carbohydrates. Chromatog. Rev. 10, 1–17. (German).

SCHOENEMANN, K., JESCHEK, G., and FROMMHOLD, K. 1961. A new method for quantitative paper chromatographic determination of sugars and sugar alcohols. Z. Anal. Chem. 181, 338–350. (German).

SCHOORL, N., and REGENBOGEN, A. 1917. Gravimetric sugar assays. Z. Anal. Chem. 56, 191–202. (German).

SCOTT, R. W., MOORE, W. E., EFFAND, M. J., and MILLET, M. A. 1967. Ultraviolet spectrophotometric determination of hexoses, pentoses, and uronic acids after their reactions with concentrated sulfuric acid. Anal. Chem. 21, 68–80.

SEIFTER, S., DAYTON, S., NAVIC, B., and MUNTWEYLER, E. 1950. Estimation of glycogen with the anthrone reagent. Arch. Biochem. Biophys. 25, 191–200.

SHAFFER, P. A., and HARTMANN, A. F. 1921. The iodometric determination of copper and its use in sugar analysis. J. Biol. Chem. 45, 349–390.
SICHERT, K., and BLEYER, B. 1936. Determination of dextrose, maltose, and dextrin in sugar mixtures. Z. Anal. Chem. 107, 328 (Chem. Abstr. 31, 13237).
SLEIN, M. W. 1952. A rapid method for distinguishing D-glucosamine from galactosamine in biological preparations. Proc. Exptl. Biol. Med. 80, 646–647.
SOLMS, J. DENZLER, A., and DEUEL, H. 1954. Amides of polygalacturonic acid. Helv. Chim. Acta 37, 2153–2160.
SOMOGYI, M. 1945A. A new reagent for the determination of sugars. J. Biol. Chem. 160, 61–68.
SOMOGYI, M. 1945B. Determination of blood sugar. J. Biol. Chem. 160, 69–73.
SPEDDING, H. 1964. Infrared spectroscopy and carbohydrate chemistry. Advan. Carbohydrate Chem. 19, 23–49.
STADTMAN, E. R. 1948. Nonenzymatic browning in fruit products. Advan. Food Res. 1, 325–372.
STAHL, E., and KALTENBACH, W. 1961. Thin layer chromatography. VI. Analyses of sugar traces on silica gel plates. J. Chromatog. 5, 351–356. (German)
STAHL, E., and KALTENBACH, W. 1965. Sugars and derivatives. In Thin-layer Chromatography, A Laboratory Handbook. E. Stahl (Editor). Academic Press, New York.
STANEK, J., CERNY, M., KOCUREK, J., and PACAK, J. 1963. The Monosaccharides. Academic Press, New York.
STEINHOFF, G. 1933. Analysis of starch sirup. Determination of glucose in the presence of maltose and dextrin. Z. Spiritus Ind. 56, 64 (Chem. Abstr. 27, 6004).
STETTEN, M. R., KATZEN, H. M., and STETTEN, D. 1958. Comparison of the glycogens isolated by acid and alkaline procedures. J. Biol. Chem. 232, 475–488.
STREULI, H., and STESEL, M. 1951. Digestion of chocolate in sugar analysis. Intern. Fachschr. Schokolade Ind. 6, 200–201. (German)
STREULI, H., and STESEL, M. 1952. Clarifying in assay of sugars in chocolate. Mitt. Lebensmittelunters. Forsch. 43, 417–444. (German)
SWEELEY, C. C., BENTLEY, R., MAKITA, M., and WELLS, W. W. 1963. Gas liquid chromatography of trimethylsilyl derivatives of sugars and related substances. J. Am. Chem. Soc. 85, 2497–2507.
TAUFEL, K., and MULLER, K. 1956. Determination of glucose, maltose, and other fermentable oligosaccharides and dextrins. Z. Lebensmittelunters. Forsch. 103, 272–284. (German)
THALER, H. 1940. Determination of lactose and sucrose in cocoa products Z. Untersuch. Lebensm. 80, 439–450. (German)
THALER, H. 1967. Detection and determination of polysaccharides. Determination of cellulose and other cell wall components. In Handbook of Food Chemistry, Food Analysis, Vol. 2. W. Diemair (Editor). Springer-Verlag, Berlin. (German)
THIVEND, P., MERCIER, Ch., and GUILBOT, A. 1965. Use of glucamylase in starch determination. Die Starke 17, 278–283. (German)
TRACEY, M. V. 1952. Determination of glucosamine by alkaline decomposition. Biochem. J. 52, 265–267.
TUNGER, L. 1964. The definitions of crude fiber and cellulose content. Ernahrungsforschung 9, 57–59. (German)
ULMAN, M., and RICHTER, M. 1961. Review of principles of polarimetric starch assays by the Ewers method. Die Starke 13, 67–75. (German)
UNDERWOOD, J. C., WILLITS, C. O., and LENTO, H. G. 1961. Browning of sugar solutions. VI. Isolation and characterization of the brown pigment in maple sirup. J. Food Sci. 26, 397–400.

VAN DER BIJ, J. R.   1967.   Modern methods of analysis of starch derivatives and starch hydrolysates.   Die Starke 19, 256–263.   (German)
VAN HANDEL, E.   1965.   Estimation of glycogen in small amounts of tissue. Anal. Biochem. 11, 256–265.
VAN SOEST, P. J.   1967.   Development of a comprehensive system of feed analyses and its application to forages.   J. Animal Sci. 26, 119–128.
VAN SOEST, P. J., and WINE, R. H.   1967.   Use of detergents in the analysis of fibrous foods.   The determination of plant cell-wall constituents.   J. Assoc. Offic. Agr. Chemists 50, 50–55.
WEIGEL, H.   1963.   Paper electrophoresis of carbohydrates.   Advan. Carbohydrate Chem. 18, 61–97.
WEILL, C. E., and HANKE, P.   1962.   The thin-layer chromatography of malto-oligosaccharides.   Anal. Chem. 34, 1736–1737.
WHISTLER, R. L., and DE MILLER, J. N.   1962.   Quantitative paper chromatographic determination of carbohydrates.   In Methods in Carbohydrate Chemistry, R. L. Whistler, and M. L. Wolfrom (Editors).   Academic Press, New York.
WHISTLER, R. L., and DURSO, D. F.   1950.   Chromatographic separation of sugars on charcoal.   J. Am. Chem. Soc. 72, 677–679.
WHISTLER, R. L., and HICKSON, J. L.   1955.   Determination of components in corn sirups by quantitative paper chromatography.   Anal. Chem. 27, 1514–1517.
WHISTLER, R. L., and HOUSE, L. R.   1953.   Infrared spectra of sugar anomers. Anal. Chem. 25, 1463–1465.
WHISTLER, R. L., and WOLFROM, M. L. (Editors).   1962–1964.   Methods in Carbohydrate Chemistry, Vols. 1–5.   Academic Press, New York.
WHITE, J. W., EDDY, C. R., PATTY, J., and HOBAN, N.   1958.   Infrared identification of disaccharides.   Anal. Chem. 30, 506.
WHITMOYER, R. B.   1934.   Determination of small amounts of glucose, fructose, and invert sugar in absence and presence of sucrose.   Ind. Eng. Chem. Anal. Edition 6, 268–271.
WICHMANN, H. J.   1922.   Report on determination of pectin in fruit and fruit products.   J. Assoc. Offic. Agr. Chemists 6, 34–40.
WICHMANN, H. J.   1923.   Report on determination of pectin in fruit and fruit products.   J. Assoc. Offic. Agr. Chemists 7, 107–112.
WILLIAMSON, A. R., and ZAMENHOF, S.   1963.   Detection and rapid differentiation of glucosamine, galactosamine, glucosamine uronic acid and galactosamine uronic acid.   Anal. Biochem. 5, 47–50.
WILLSTATTER, R., and SCHUDEL, G.   1918.   Determination of glucose with hypoiodite.   Ber. Deut. Chem. Ges. 51, 780–781.
WINKLER, S., and LUKOV, G.   1967.   Standardization of polarimetric starch assays in dilute hydrochloric acid.   I.   Errors in the Ewers method and reproducibility of acidic digestion of starch.   Die Starke 19, 110–115.
WISE, L. E., and MCCAMMON, D. C.   1945.   Munson-Walker reducing values of some of the less common sugars and of sodium glucuronate.   J. Assoc. Offic. Agr. Chemists 28, 167–174.
WISEMAN, H. G., MALLACK, J. C., and JACOBSON, W. C.   1960.   Determination of sugar in silages and forages.   J. Agr. Food Chem. 8, 78–80.
ZELENKA, S., and SASEK, A.   1966.   Evaluation of analytical methods of starch assay.   Die Starke 18, 77–81.
ZILL, L. P., KHYM, J. X., and CHEMINE, G. M.   1953.   The separation of the borate complexes of sugars and related compounds by ion-exchange chromatography.   J. Am. Chem. Soc. 75, 1339–1342.
ZUCKERHANDL, F., and Messiner-Klebermass, L.   1931.   A method for the demonstration and determination of glucosamine.   Biochem. Z. 236, 19–28 (Chem. Abstr. 25, 4902).

CHAPTER 36

Lipids

INTRODUCTION

Lipids have at least three important functions in foods: culinary, physiological, and nutritional (Kummerow 1960). The ability of lipids to carry odors and flavors, and their contribution to palatability of meats, tenderness of baked products, and richness and texture of ice cream are examples of the first kind. As lipids serve as a convenient means of rapid heat transfer, they have found increasing use in commercial frying operations. Dietary lipids represent the most compact chemical energy available to man. They contain twice the caloric value of an equivalent weight of sugar. They are vital to the structure and biological function of cells. Dietary lipids provide the essential linoleic acid which has both a structural and functional role in animal tissue, and are carriers of the nutritionally essential fat-soluble vitamins.

The term lipid is used to denote fats and fatlike substances, and is synonymous with the terms lipoids or lipins, used in the earlier literature (Feldman 1967). Lipids are usually defined as food components that are insoluble in water and that are soluble in organic fat solvents. In this book, the terms lipids, fats, and oils are used interchangeably to denote such components. Solvents (and mixtures of solvents) used to extract lipids include ether, petroleum-ether, acetone, chloroform, benzene, alcohols (i.e. methanol, ethanol, butanol), and water-saturated butanol. Lipids are chemical constituents of living organisms, or are derived from such constituents, and most commonly possess fatty acids as part of their moiety.

This definition has, however, certain limitations. Thus, sterols, squalene, and carotenoids meet the solubility criteria of lipids but contain no fatty acids. On the other side, gangliosides are soluble in water and alcohol-water mixtures, but insoluble in many of the organic solvents used to extract lipids from their source. Despite these limitations, the definition is useful in describing the general characteristic of a class of compounds.

The nomenclature of lipids has been described (Anon. 1967) by a committee of the Biological Nomenclature Commission of IUPAC and the Commission of Editors of Biochemical Journals of IUB. The proposed rules concern lipids containing glycerol; sphingolipids; neuraminic acid; fatty acids, long-chain alcohols, and amino acid components of lipids; and specific generic terms such as phospholipids and others.

The classification of lipids is difficult because of their heterogenous nature. The system most commonly used despite its limitation is

TABLE 36.1

CLASSIFICATION SCHEME FOR THE LIPIDS

Simple lipids—compounds containing two kinds of structural moieties
  Glyceryl esters—these include partial glycerides as well as triglycerides, and are esters of glycerol and fatty acids
  Cholesteryl esters—esters formed from cholesterol and a fatty acid
  Waxes—a poorly defined group which consists of the true waxes (esters of long chain alcohols and fatty acids), vitamin A esters, and vitamin D esters
  Ceramides—amides formed from sphingosine (and its analogs) and a fatty acid linked through the amino group of the base compound. The compounds formed with sphingosine are the most common

Composite lipids—compounds with more than two kinds of structural moieties
  Glyceryl phosphatides—these compounds are classified as derivatives of phosphatidic acid
  Phosphatidic acid—a diglyceride esterified to phosphoric acid
  Phosphatidyl choline—more descriptive term for lecithin which consists of phosphatidic acid linked to choline
  Phosphatidyl ethanolamine—often erroneously called cephalin, a term referring to phospholipids insoluble in alcohol
  Phosphatidyl serine—also erroneously called cephalin
  Phosphatidyl inositol—major member of a complex group of inositol-containing phosphatides including members with 2 or more phosphates
  Diphosphatidyl glycerol—cardiolipin

Sphingolipids—best described as derivatives of ceramide, a unit structure common to all. However, as in the case of ceramide, the base can be any analog of sphingosine
  Sphingomyelin—a phospholipid form best described as a ceramide phosphoryl choline
  Cerebroside—a ceramide linked to a single sugar at the terminal hydroxyl group of the base and more accurately described as a ceramide monohexoside
  Ceramide dihexosides—same structure as a cerebroside, but with a disaccharide linked to the base
  Ceramide polyhexosides—same structure as a cerebroside, but with a trisaccharide or longer oligosaccharide moiety. May contain one or more amino sugars
  Cerebroside sulfate—a ceramide monohexoside esterified to a sulfate group
  Gangliosides—a complex group of glycolipids that are structurally similar to ceramide polyhexosides, but also contain 1 to 3 sialic acid residues. Most members contain an amino sugar in addition to the other sugars. However, not all gangliosides contain amino sugars

Derived lipids—compounds containing a single structural moiety that occur as such or are released from other lipids by hydrolysis
  Fatty acids
  Sterols
  Fatty alcohols
  Hydrocarbons—includes squalene and the carotenoids
  Fat-soluble vitamins, A, D, E, and K

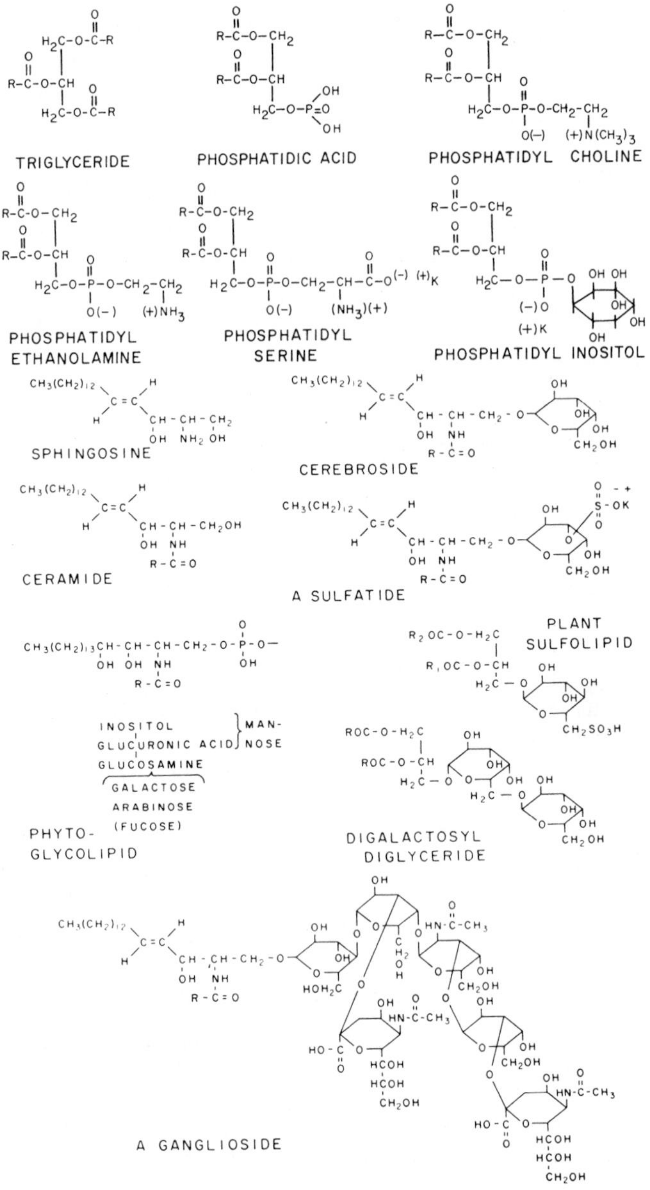

FIG. 36.1. MOLECULAR STRUCTURES OF THE MAJOR LIPID
CLASSES

R represents a fatty acid residue.

that proposed by Bloor (1925) as shown in Table 36.1. Molecular structures of the major lipid classes are given in Fig. 36.1.

Foods vary widely in their lipid content and composition. Lard, shortening, and vegetable or animal cooking fats and oil contain almost 100% lipids. The fat content of butter and margarine is about 81%, and of commercial salad dressings 40–70%. Most nuts are very rich in lipids (almonds, 55; beechnuts, 50; Brazils, 67; cashews, 46; peanuts, 48; pecans, 71, and walnuts up to 64%). The main seeds used for extracting lipids on a commercial scale include (in addition to peanuts) sesame seeds (50% fat), sunflower (47), safflower seeds, hulled (60), and soybeans (18). Among dairy products a wide range is found. Cottage cheese contains 4%, and cream cheese 38% (on an as is basis). Fresh, fluid cow's milk has 3.7% fat, but after drying 27.5% fat. The fat content of cream ranges from 20% in light-coffee to 38% in heavy whipping. Ice cream contains about 12% fat. A very wide range in fat content is encountered in cereal products: grains contain only 3 to 5% but the germ around 10%, bread 3 to 6%; most cookies 15 to 30%, and crackers from 12% in saltines to 24% in chocolate-coated Graham crackers. Raw beef carcass trimmed to retail level contains 16 to 25% fat; sausages 15 to 50%; total edible hens and cocks 25%; and herring 11%. The fat ranges from 4% in pink to 16% in chinook salmon. Raw tuna fish contains only 4% fat, but canned in oil, 21%. The whole edible portion of eggs contains 12% lipids, the yolks alone 29%; after drying the fat content increases to 41% in commercial dried whole eggs, and to 57% in dried yolks. Most fruits and vegetables contain small amounts of lipids (especially when expressed on an as is basis), but avocado contains 16% lipids, and the lipid content ranges from 10% in giant size pickled olives to 36% in salt-dried, oil-coated Greek style olives. Sweet chocolate contains 35% fat, and bitter or baking chocolate 53%; in dry cocoa powders the fat ranges from 8% in low-fat to 24% in high-fat or breakfast types.

The methods of analysis applied to lipids have been collected in several books. Kaufmann (1958) has gathered the chemical and physical methods of analysis of fats and fat products. Mehlenbacher (1960) has collected the methods of analysis of fats and oils emphasizing those for use in industrial control laboratories. Cocks and Van Rede (1966) are the editors of a laboratory handbook for oil and fat analysis, and Boekenogen (1964) has recently initiated a series of volumes on the analysis and characterization of oils, fats, and fat products. Shorter reviews on methods of analysis and separation were published by Fontell *et al.* (1960) and by Mehlenbacher (1958).   General information on lipids is available in several textbooks including those by Williams (1950), Deuel (1951), and Hilditch and Williams (1964).   A series of monographs on progress in the chemistry of lipids is published at irregular intervals.   At least five scientific journals are devoted entirely or almost entirely to lipids (*Lipids; Journal of the American Oil Chemists' Society; Fette, Seifen, Anstrichmittel; Journal of Lipid Research; Chemistry and Physics of Lipids*) and practically all journals concerned with biochemistry, analysis, and food science deal occasionally with various aspects of lipid chemistry and lipid analysis.   Symposia (organized by the American Oil Chemists' Society) on various aspects of lipid chemistry and analysis, and published periodically in their Journal are particularly useful.   The bibliography sections of several journals (including *J. Lipid Research, J. American Oil Chemists' Society*, and *J. Chromatography*) are excellent sources of information on current developments in the analysis of lipids.

The American Oil Chemists' Society has published official and tentative methods of analysis of fats, giving detailed instructions for each method (Anon., 1964).   Similar collections of official methods applicable to fats have been issued by the German Society for Fat Science (Anon. 1950) and the Fat Commission of the International Union of Pure and Applied Chemistry (Anon. 1954).   Standard methods of the Association of Official Analytical Chemists in the United States (Anon. 1965) include sections on fats and oils.   Methods of the Society for Analytical Chemistry (England) and some methods included in British Standards are published in *Analyst.*   Specific methods are included also in *Cereal Laboratory Methods* (Anon. 1962), and in publications of the American Society for Testing Materials (ASTM).   Methods for the analysis of lipids fall into two general groups: those used to determine the lipid content and composition, and those relating to physical and chemical properties of extracted lipids.

LIPID CONTENT AND COMPOSITION

Solvent Extraction of Lipids

As indicated, lipids are characterized by their sparing solubility in water and their considerable solubility in organic solvents—physical properties which reflect their hydrophobic, hydrocarbon nature.   In practice, the wide range of relative polarities of lipids, as a result of their various structures, makes the selection of a single universal solvent impossible. Successful extraction requires that bonds between lipids and other compounds be broken so that the lipids are freed and solubilized. Generally, such solubility is attained when polarities of the lipid and the solvent are similar. Thus, the nonpolar triglycerides are dissolved in nonpolar solvents such as hexane and petroleum ether (a low boiling point distillate from petroleum). Polar compounds, such as glycolipids, are soluble in alcohols. In some cases, solubility is modified as a result of molecular interaction. Thus, phosphatidyl choline (lecithin) behaves as a base because of the quarternary ammonium group of its choline moiety. It dissolves in weakly acidic solvents such as alcohol. Phosphatidyl serine, is structurally similar to phosphatidyl choline, but is a polar, relatively strong acid. It is insoluble in weakly acidic alcohol but dissolves in chloroform that readily associates with acidic polar compounds, even though it is rather a nonpolar solvent. The presence of other lipids also affects solubility. Phosphatidyl serine will be solubilized partly in the presence of phosphatidyl choline, though it is insoluble in alcohol alone. Chloroform is generally a useful solvent, but fails in quantitative extraction of compound lipids (glycolipids and proteolipids).

Preparation of Sample for Lipid Extraction.—There is no single standard method for lipid extraction. The method used depends on the type of analyzed material and nature of the subsequent analytical problem (Marinetti 1962). Thus, extraction of lipids from milk is relatively simple compared to the extraction of lipids from plant or animal tissue. These latter materials require some type of fragmentation, such as mechanical grinding, sonic distintergration, homogenization or compression-decompression. During these steps, it is important to keep the chemical, physical, and enzymatic degradation of lipids to a minimum. This is usually accomplished by the control of temperature, chemical environment, and time of exposure of the material to each solvent.

Entenmann (1961) stressed the fact that if the lipids are incompletely extracted or altered during extraction, the results will be inaccurate regardless of the effort spent later or the precision of the apparatus and techniques used in the analysis. For best results (especially if the extracted lipids are to be characterized) the following requirements must be met: (1) all procedures must be carried out under an atmosphere of nitrogen; (2) the solvent (s) used should be purified and peroxide-free, and be used in the proper solute: solvent ratio; (3) the tissue should be removed from the source and subdivided as soon as possible; (4) heating should be minimized, (5) the extract should be purified to remove nonlipid components, and (6) the purified lipids should be stored under conditions that minimize alteration.

The moisture content is an important factor. Only part of the lipids can be extracted with ether from moist material, as the solvent cannot penetrate the tissues and the extractant becomes saturated with water and inefficient for lipid extraction. Drying at elevated temperatures is undesirable as some lipids become bound to proteins and carbohydrates and are rendered inextractable. Lyophilization affects extractability little and increases the surface area of the sample. On the other hand, more lipid will be extracted from a ground wheat sample containing above 11% moisture than from one that has a very low moisture content.

The extraction of dry materials depends on particle size; consequently, efficient grinding is very important. The classical method of determining fat in oilseeds involves extracting the ground seeds with a selected solvent after repeated grinding in a mortar with sand. Soft materials such as peanuts and copra must be grated, preferably in a grating mill without expressing oil during grinding. Cooling copra to below 0°C minimizes losses. To promote rapid extraction, the sample and solvent are mixed together in a high-speed comminuting device such as a Waring blendor. When the cutting device is operated at high speed, the particle size is rapidly reduced and the fat extraction accelerated.

Pinto and Enas (1949) described a special chopper (similar to a hammer mill but with fixed steel blades equipped with special edges), and a Waring blendor with a special blade assembly for rapid grinding-extraction of copra. Lipids are extracted with difficulty from whole yeast cells because of the limited porosity of the cell wall and its sensitivity to dehydrating agents. The extraction may be accelerated and rendered quantitative if the yeast cells are broken mechanically by shaking with glass beads (Trevelyan 1966). To reduce metabolic modification of lipids during mechanical disintegration, the yeast suspension must be preheated briefly to 90°C. To overcome the tedium involved in extracting lipids from oil seeds, Troeng (1955) proposed milling whole seeds with steel balls and sand with simultaneous extraction. This is the basis of the procedure described by Cocks and Van Rede (1966). The procedure involves preliminary extraction in a vibrating ball mill followed by final extraction for 1 hr in an efficient drip extractor.

In many processed foods; in by-products of the dairy, bread, fermentation, sugar and flour industries; and in animal products, a major part of the lipids is bound to proteins and carbohydrates and direct extraction with nonpolar solvents is inefficient. Such foods must be prepared for lipid extraction by acid hydrolysis or other methods. Two general procedures have been developed. Preliminary extraction of ground material is followed by acid hydrolysis and reextraction after acid treatment and drying. According to the procedure of Campen and Geerling (1954) that has been adopted by the Netherlands Standard Institute, the pre-extraction can be omitted. The sample is predigested by refluxing for 1 hr with 3 N hydrochloric acid; ethanol and solid hexametaphosphate—to facilitate separation—are added, and the fat is extracted with tetrachloromethane.

Solvents.—Ethyl and petroleum ether are the common extraction solvents. There is a growing tendency to use petroleum ether because it is more selective toward true lipids. Ethyl is a better solvent for fat than petroleum ether and will dissolve oxidized lipids. It has, however, several disadvantages; it is more expensive, danger of explosion and fire hazard are somewhat greater; and it picks up water during extraction of a sample and dissolves nonlipid materials (i.e., sugars). Dried ether tends to form peroxides.

Combinations or alternate extraction with ethyl and petroleum ether are used often in extraction of lipid from dairy products. Mixtures of alcohol and ether are employed to remove fat from certain biological materials, although the extract is rich in nonlipid components that must be subsequently separated. Treatment with alcohol facilitates the removal of fat from some materials.

Of a number of solvents tried, water-saturated n-butanol was most effective in extracting lipid material from ground wheat, flour, bran, and gluten (Mecham and Mohammad 1955). Water-saturated butanol has since been used extensively to extract lipids from cereals. Bloksma (1966) extracted flours with various butanol mixtures containing 0 to 17.5% water. With increasing water content, the extracted lipids increased from 1.16 to 1.37%. At the same time, the extracted nonlipid components increased from 0.06 to 0.27%. Water-saturated butanol is an effective lipid extractant. Yet, some of the lipids are released only after acid hydrolysis. In addition, the solvent has a strong odor and requires relatively high temperatures for evaporation, even in vacuo.

A rapid method for lipid extraction was proposed originally by Folch et al. (1957) for isolation and purification of total lipids from animal tissue by means of phase partition of a ternary mixture of

chloroform-methanol-water. Bligh and Dyer (1959) simplified the
method. In the simplified procedure, the sample is homogenized
with a mixture of chloroform and methanol in such proportions that
a miscible system is formed with the water in the sample. Dilution
with chloroform and water separates the homogenate into two layers,
the chloroform layer containing all the lipids and the methanol
layer containing all the nonlipids. A purified extract is obtained by
isolating the chloroform layer. The method has been applied by
the authors to fish muscle and has been adapted by Tsen *et al.* (1962)
for the extraction of lipids from wheat products.

Lee *et al.* (1966) compared the extraction of lipids from fish meal
by 6 methods including: (1) the official AOAC method that involves
a 16-hr extraction with acetone in a Soxhlet extractor, digestion of
the extracted residue with dilute hydrochloric acid, filtration, drying
the residue and reextraction with acetone for 16 hr; (2) extraction
with ether; (3) and (4) two modifications of the Bligh-Dyer method,
(5) an alkaline saponification method; and (6) the Mojonnier
method. The ether and the alkaline saponification methods gave
as expected greatly reduced yields. Also in the Mojonnier method
the yield was slightly reduced. One of the Bligh-Dyer methods
gave practically a theoretical yield.

Two general types of lipid-protein complexes are known: *lipoproteins* and *proteolipids* differing in the relative amounts of lipid and
protein. Lipoprotein is mostly protein, whereas proteolipids contain almost equal amounts of the two and exhibit lipid solubility
characteristics (Feldman 1967). Mixtures of chloroform and
methanol are among the most effective and relatively mild extract-
ants; the proteolipids are solubilized with little damage to most
proteins. The extracted proteolipids can be readily cleaved to
liberate the lipids by repeated evaporation of the extract or by
evaporation to dryness after forming a two-phase system with water.
This cleavage is based on the ratio of chloroform to methanol during
evaporation. The original ratio of 2:1 (v/v), or 1:1 on a molar
basis, is useful in the extraction of intact proteolipids as the high
concentration of chloroform reduces greatly the hydrogen-bonding
capacity of the methanol and prevents disruption of the electrostatic
bonds of the proteolipid. During evaporation, the concentration
of chloroform decreases more rapidly than that of methanol. As
more methanol becomes available to attack and rupture linkages
between the protein and the lipid, the protein is gradually denatured
and precipitated.

Purification of Extracts.—Water-soluble nonlipid substances (carbohydrates, salts, amino acids) are invariably extracted from tissues along with lipids. The nonlipids must be removed for the gravimetric determination of total lipid and for the prevention of contamination during the subsequent fractionation of the total extract. Various partition procedures have been used. The more common ones employ water alone or aqueous salt solutions. The methods fail to remove inorganic phosphate, and with phosphatide-rich extracts may produce emulsions that can be broken only by centrifugation or standing for very long times. Some lipids are appreciably soluble in water, and excessive washing may cause lipid losses.

The removal of contaminants may be accomplished (partly or completely) by evaporation of the extracts to dryness in vacuo under nitrogen, and reextraction with a nonpolar solvent. Separation by dialysis, electrodialysis, electrophoresis, and chromatography are some of the other methods. Wells and Dittmer (1963) suggested that the cross-linked dextran gel, Sephadex, would be useful for the column chromatographic separation of lipids from nonlipid water-soluble substances. Siakotos and Rouser (1965) modified the above procedure for better separation of lipid and nonlipid components.

Apparatus.—Direct extraction is often carried out in a Soxhlet type extractor (see Chap. 23). The sample in an extraction thimble (from filter paper or Alundum) covered with defatted (by soaking in ether) cotton wool to prevent small particles from finding their way into the flask, is placed in the middle part of the apparatus. The flask is filled with solvent; the three parts of the apparatus are assembled; the condenser attached to a tap; and the heating of solvent in the flask started. The condensing vapors fill the middle part containing the sample and carry the dissolved lipid into the flask by a siphoning action each time the height of the siphon is attained. At the end of exhaustive extraction, the apparatus is disconnected, the solvent from the tared flask is evaporated and the weight increases calculated. The most efficient extractors for most materials are the continuous percolator types specified by the methods of the American Oil Chemists' Society. Commercial units of the Goldfish extractors of the percolator type are available (Fig. 36.2). An extraction apparatus recommended by ASTM is shown in Fig. 36.3. It consists of a glass extraction flask, a coiled block tin condenser, polished copper cover (tinned inside), and a glass siphon cup suspended by aluminum wire.

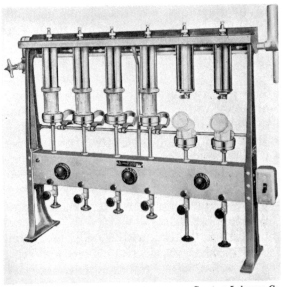

Courtesy Labconco Co.

FIG. 36.2.   LABCONCO-GOLDFISH SOLVENT TYPE EXTRAC-
TOR FOR RAPID DETERMINATION OF FAT AND OIL CONTENT

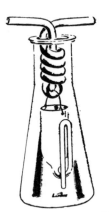

FIG. 36.3.   ASTM EXTRACTION APPARATUS

Courtesy LaPine Scientific Co.

The Mojonnier extraction apparatus (Mojonnier and Troy 1925)
originally devised for fat determination in dairy products, has been
found to be convenient for assay of many other foods when it is
desirable to make a liquid-liquid separation.   The procedure utilizes
a specially designed flask that permits an intimate mixture of fat
and solvent, that is essential in the extraction of fat from a liquid
phase.

Extraction of Lipids in Selected Foods

For extraction of fat from soy flours by the AOCS method, the Butt percolator-type extraction apparatus (Fig. 36.4) is used. The sample is folded in filter paper as shown in Fig. 36.5 and extracted exhaustively with petroleum-ether.

Predried (preferably in a vacuum oven at 95°–100°C under pressure not to exceed 100 Torr for about 5 hr, or in a vacuum desiccator over H_2SO_4 for 24 hr at 10 Torr) and ground grain is extracted exhaustively with dry ether in a Soxhlet, Butt-type, Goldfish, or similar extractor.

For the extraction of crude fat from bread and baked cereal products not containing fruit, the sample (2 gm) is moistened with 2 ml ethanol to prevent lumping on the addition of 10 ml hydrochloric acid (mixed in a ratio of 25:11 (v/v) with water). The material is digested at 70°–80°C in a water bath for 30 to 40 min. After adding 10 ml ethanol and cooling, the contents are extracted 3 times with

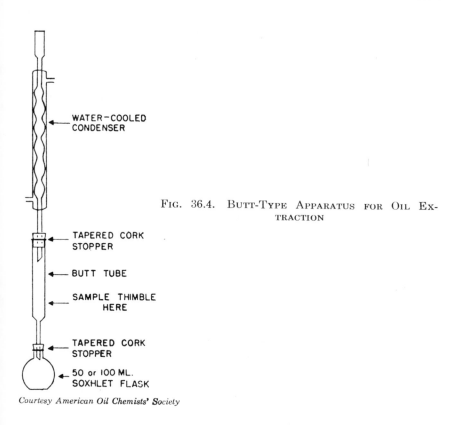

WATER–COOLED
CONDENSER

Fig. 36.4. Butt-Type Apparatus for Oil Extraction

TAPERED CORK
STOPPER

BUTT TUBE

SAMPLE THIMBLE
HERE

TAPERED CORK
STOPPER

50 or 100 ML.
SOXHLET FLASK

Courtesy American Oil Chemists' Society

ether in a Mojonnier type apparatus; the extracts are combined and filtered; the solvent is evaporated and the extracted lipids are weighed. Crude fat in egg yolk and dried whole egg is extracted with a mixture of ether and petroleum ether in a sample pretreated

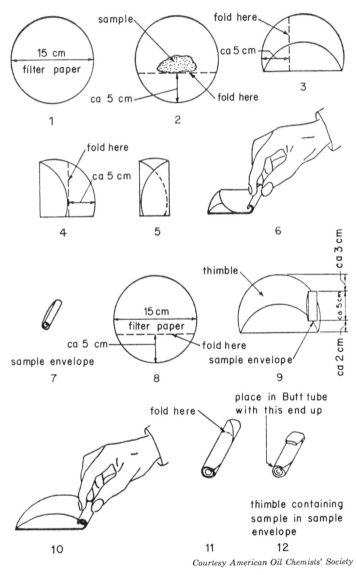

Courtesy American Oil Chemists' Society

FIG. 36.5. METHOD OF FOLDING FILTER PAPER FOR USE IN BUTT-TYPE EXTRACTION APPARATUS

with hydrochloric acid (mixed at a ratio of 4:1 with water) at a rather elevated temperature. On the other hand, crude fat in cocoa can be extracted by simply passing 10 times, 10-ml portions of petroleum ether through a sample (2 gm) in a sintered glass filter, coarse porosity, attached to a suction filtering apparatus.

In milk, fat globules are present as an emulsion of oil in water and are surrounded by a thin protein film. The emulsion must be broken and the protein film removed before the fat can be separated and determined volumetrically. This is accomplished in the Babcock and Gerber methods by using sulfuric acid. In the Babcock method (Anon. 1960), 17.6 ml milk are mixed with 17.5 ml of sulfuric acid of specific gravity 1.80–1.83 in a special bottle, shaken till apparently homogenous, centrifuged and submerged into water at 145°F. The rising fat is determined from the height formed in a graduated neck.

In the Gerber method, 11 ml of milk followed by 1 ml isoamyl alcohol are added to a Gerber glass butyrometer containing 10 ml sulfuric acid (specific gravity 1.82). A lock stopper is inserted and the bottle is shaken until the curd disappears and the contents are homogenous. The butyrometer is centrifuged 4 min, submerged to top of graduate stem in 145°F water for 5 min, and the fat contents are read off.

The Gerber test is simpler and of wider general application than the Babcock test. Charring is generally avoided by the ratio of milk to sulfuric acid. Isoamyl alcohol improves the fat separation and reduces the effects of sulfuric acid.

Phospholipids are not included in the fat determined by either the Gerber or Babcock test (Levowitz 1967). This is not important in milk and cream in which the polar lipids constitute about 1% of total lipids, but is of significance in testing skim milk and buttermilk. The fat of the latter contains up to 24% phospholipids and must be determined by any one of available gravimetric methods.

In ice cream, the presence of stabilizers requires prolonged acid digestion, yet the concentration of the latter must be reduced to avoid excessive charring of the sugar-rich food. Consequently, the accuracy of the Gerber test is reduced. The Gerber test is 2 to 3 times more rapid than the Babcock test. It is much better suited for determining fat in homogenized milk and is official in most European countries. In the United States, the Gerber test is optional in several States, but the Babcock test is used much more. The Gerber test can be used for the assay of fat in cream and cheese provided special butyrometers are employed. The method, with various digesting reagents has been applied satisfactorily to other fat-containing products such as processed meat and fish products. In most of those methods, speed and simplicity are achieved at some sacrifice of precision and accuracy of routine tests.

To overcome the unpleasant corrosive features of sulfuric acid in the Babcock test, Schain (1949) suggested the use of detergents. An anionic detergent, dioctyl sodium phosphate, disperses the protein layer around the fat globule and liberates the fat; a strongly hydrophilic nonionic detergent, polyoxyethylene sorbitan monolaurate, completes the separation.

Over the years, various modifications of the detergent method have been proposed. They include the DPS method of Sager et al. (1955) and the official AOAC Te Sa test. The Te Sa test requires a special two-neck bottle; the Schain and DPS tests are carried out in a regular Babcock milk bottle. In the Te Sa and DPS tests, fat contents is read directly; in the Schain procedure, volumetric readings are converted to fat content by nomograms supplied with each batch of reagent. The rapid detergent test (Te Sa) gives significantly lower fat values in homogenized and chocolate milk than conventional gravimetric methods (Mitchell 1967).

The reference gravimetric methods for fat analysis in dairy products involve successive extractions with ether or mixed ethers after preliminary digestion with acid or alkali. Digestion with hydrochloric acid (as in the Werner-Schmidt or Schmidt-Bondzynski procedures used to a limited extent in Europe) are less suitable than digestion with alkali—especially in sugar-rich foods.

In the Rose-Gottlieb method, the sample is rendered alkaline with ammonium hydroxide; if necessary, diluted with water and heated to facilitate dispersion. The digested mixture is extracted repeatedly with ethanol, ether, and petroleum ether. The combined extracts are dried, purified by extraction with petroleum ether, and weighed. In the Majonnier test, the combined extracts are weighed without purification. Consequently, the results are slightly higher than in the Rose-Gottlieb method.

Indirect Rapid Tests

Various devices and techniques are available for approximate rapid determination of lipid contents. Zimmerman (1962) reported a high ($r = 0.96$) correlation between flaxseed density (as determined by an air-comparison pycnometer) and oil content, and recommended seed density as a criterion for effective screening of flax lines with high oil content.

The recent application of wide-line nuclear magnetic resonance spectroscopy (NMR) to analysis of oil in seeds opens new opportunity for geneticists and plant breeders. The method is non-destructive, rapid, and can be used to determine the oil content of a single seed or of a bulk sample (Bauman *et al.* 1963). The measured NMR value is related to the total hydrogen in the oil fraction of the seed, independent of the hydrogen in the nonoil fraction. The oil content is calculated from calibration tables or curves. Application of the method has been initiated by Conway (1960). The results are affected by moisture content and distribution. Yet, the method has been applied successfully in the assay of oil in seeds such as corn (Alexander *et al.* 1967), soybeans (Collins *et al.* 1967; Fehr *et al.* 1968), and various other oil-containing seeds.

King (1966) reported that NMR may be used to measure instantaneously and continuously moisture and fat in meat products. The two components can be determined separately with high-resolution NMR, since the resonant frequency for protons in water is slightly different than that of protons in fats. The frequency difference is adequate to permit separate analyses of the fat and water contents.

Biggs (1967) and Dyachenko and Samsonov (1964) described the use of infrared spectroscopy in determination of fat in milk products. Fitzgerald *et al.* (1961) described an ultrasonic method for the determination of fat and nonfat milk solids in fluid milk. The method is based on measuring the speed of sound through milk at various temperatures. A turbidimetric procedure for the fat assay in milk was proposed by Hangaard and Pettinati (1959). The milk fat is homogenized to bring it to substantially uniform globules, the protein is chelated, light transmission is measured spectrophotometrically and converted to fat by a chart or graph. Murphy and McGann (1967) found that the turbidimetric method was quite precise and reproducible in the range 0 to 4% fat, and was unaffected by the addition of preservatives. The method uses noncorrosive reagents and permits the testing of 75 samples per hour by an operator.

Stern and Shapiro (1953) described a colorimetric method for oil determination. The extracted lipid is treated with an alkaline solution of hydroxamic acid and allowed to react for a specified time. Upon acidification with hydrochloric acid and addition of ferric chloride, a relatively stable color with an absorbance maximum at 540 nm is formed. The use of the hydroxamic method for the determination of fat in milk was described by Katc *et al.* (1959).

The addition of fat to an organic solvent changes several properties of the mixture. Some of the changes (density and refractive index) are proportional to the concentration of the fat, and large enough for analytical purposes. Standard density-measuring methods such as hydrometers, pycnometers, and the Westphal balance have been used to a limited extent. The refractive index of an oil solution can be used as a measure of the oil content by comparison with calibration graphs or tables. The method has been adopted as an official AOAC procedure for the determination of oil in flaxseed after extraction with a mixture of α-chloronaphthalene and α-bromonaphthalene. The method is used for rapid routine factory control of oilseed extraction. It can be used for cake or expeller meal of copra, palm kernels, peanuts, or soybeans (Cocks and Van Rede 1966). The precision depends primarily on the identity of the tested sample with samples used in establishing the calibration graphs or tables. In favorable cases, the precision is quite high. However, the results are affected by the free fatty acid contents of the sample, that have to be accounted for. Determinations involve extracting 10 gm of the sample with 7.5 ml of tetraline in a vibrating ball mill or mortar if no mill is available. Tetraline (tetrahydronaphthalene) is used as it is a stable solvent with a high refractive index (n_D^{35} not higher than 1.5338 is recommended) and has a large refractive index difference from oils. The extract is filtered and the refractive index is determined at 35°C in a precise Abbe refractometer illuminated by a sodium lamp. The method cannot be used for the lipid determination of extracted soy meals as they absorb excessively high amounts of the solvent, and the final extract that is low in oil gives erratic results.

The refractometric method has been found useful in determining the fat content of vegetable foods, canned meat (Babicheva and Gorelik 1968), and fish and fish products (Schober 1967). The rapid refractometric method is used also for the determination of fat in chocolate-type products (Nadj and Weeden 1966).

Another method of fat determination is based on measuring the change in the dielectric properties of a selected solvent as a result of the presence of fat. The method was originally developed for estimating the oil content of soybeans (Hunt *et al.* 1952), but has since been applied to other oil seeds. The procedure involves grinding a sample with a solvent in a high speed mill. The dielectric measurement is made on the solvent-solute mixture and the reading is referred to a chart that relates dielectric readings with fat content.

Fractionation of Extracted Lipids

Extracted lipids comprise a heterogenous mixture ranging from nonpolar hydrocarbons to the highly polar gangliosides and phyto-glycolipids. This mixture is difficult to analyze, and is most commonly divided into major subgroups.

The classical method of separation by precipitation of phospholipids by acetone from a solution of mixed lipids in ether is inefficient because of the ability of the triglycerides to cosolubilize polar lipids into acetone, and because some phospholipids (i.e. some unsaturated lecithins) and glycolipids are soluble in acetone (Nichols 1964). Dialysis through a rubber membrane of lipid solutions in petroleum ether, whereby the polar lipids are retained and neutral lipids pass through the membrane, is efficient but tedious if the concentration of triglycerides is high. Similarly, fractional crystalization that achieves separation based on differences in solubility is both laborious and unsatisfactory.

Separations have been tremendously improved by the introduction of newer methods. The two methods used most commonly are countercurrent distribution and silicic acid chromatography.

Countercurrent Distribution.—The separation of solutes on the basis of their differential solubilities in two immiscible solvents has been known for some time. Countercurrent distribution is the name given to a particular type of liquid-liquid multiple stage extraction. Although this operation can be carried out as a separatory funnel procedure, the ease and labor-saving advantages of the ingenious laboratory equipment designed by Craig (1944) were most instrumental in the utilization of the method. With a modern automatic countercurrent distribution apparatus as many as 10,000 separations are carried out in 1 hr. The technique holds particular advantage for the lipid chemist for several reasons. The conditions of fractionation are mild and well-suited to the study of labile lipids. Recoveries are practically quantitative. Finally, the partition coefficient of a compound in two solutes is both a characteristic constant and a basis for isolating and characterizing structural features of a molecule.

Separations by countercurrent distribution depend upon the differences which exist in the differential solubility of individual chemical compounds when these compounds are distributed between two immiscible solvents. The differential solubility is described by the partition coefficient and is given by the equation

$$K = C_1/C_2 \qquad (36\text{-}1)$$

where C_1 and C_2 are the concentrations of a given solute, respectively in two solvent layers. The partition coefficients are influenced both by chain length and degree of unsaturation. The effect of increasing the chain length by two methylene groups is opposite and nearly equal to that of one additional double bond. Counter-current distribution has been the main tool of lipid chemists prior to the availability of chromatographic procedures. Today its great-est usefulness is in preparative work. Its main limitation is high price and the maintenance of complicated apparatus. Details of countercurrent distribution apparatus and methodology were de-scribed by Ahrens and Craig (1952), Dutton (1954, 1955), and Casinovi (1963) (see also Chap. 23). Separations include fatty acids, glycerides, phospholipids, bile acids, pigments, fat oxidation products, and many others.

　　Column Chromatography.—The systematic studies of Hirsch and Ahrens (1958) based on the introduction of silicic acid column chromatography by Trappe (1940) emphasized the complexity of factors involved in the separation of various components of lipid extracts. Fractionation on silicic acid columns is achieved by pro-gressively eluting the adsorbed lipids with solvent mixtures of in-creasing polarity (Wren 1961). The neutral lipids (hydrocarbons, glycerides, sterols and sterol esters) are eluted with chloroform; the remaining polar lipids are eluted with methanol. Separations of compounds with widely varying properties within each group can be achieved by stepwise or gradient elutions. To make such separa-tions reproducible, the silicic acid (moisture content, particle size, silicate content, and column preparation) and lipid sample size must be carefully controlled. If those conditions are met, separations of neutral lipids on silicic acid or Florisil (primarily magnesium silicate) columns are highly satisfactory. Column separations of polar lipids were described by Barron and Hanahan (1958) and Lea (1956). Polar lipids are, however, partly adsorbed irreversibly, modified, and separated poorly on silicic acid (or Florisil) columns. Recognizing that no single system was adequate for the complete fractionation of lipid classes, Rouser et al. (1963, 1964, 1965) developed multicolumn schemes for the analysis of lipids. The techniques use DEAE-cellulose, silicic acid, and silicic acid-silicate-water columns. Gen-erally, the combination of columns was to reduce the complexity of the total extract. The DEAE-cellulose columns were introduced specifically to separate acidic lipids. The more homogenous frac-tions were subsequently separated by rapid and efficient 1- and 2-dimensional thin-layer chromatography.

　　Thin-Layer Chromatography.—Thin layer chromatography (TLC) has become in recent years one of the main analytical tools in lipid research and analysis. TLC has largely replaced in lipid analyses the earlier useful but somewhat complicated methods of paper chromatography (Marinetti 1964; Booth 1965). This position of thin-layer chromatography has been attained as a result of its simplicity, speed, sensitivity, and versatility (Mangold 1961 and 1964; Mangold et al. 1964; Nichols 1964; Padley 1964; Blank et al. 1964; Privett and Blank 1964; Privett et al. 1965; Pelick et al. 1965; Pomeranz 1965). The versatility of TLC is exemplified by its application to fractionation of complex lipid mixtures, assay of purity, identification, information on structure (as related to chromatographic mobility in various solvent systems or precoated plates), use in monitoring extractions and separations on columns for preparative work, and in general preparative work.

Practically every known type of lipid can be separated from other lipids and identified by TLC. Each fraction, so separated, generally consists of a whole family of related compounds differing only in the chain length and degree of unsaturation of the component fatty acids, aldehydes, or alcohols. The composition of each lipid with respect to these functional entities can be readily determined by gas-liquid-chromatography after hydrolysis or transesterification. If GLC equipment is not available, separations can be made by thin-layer or column chromatography on impregnated adsorbents. The adsorbents complex or interact selectively with specific functional groups. They include silver nitrate impregnated adsorbents for separation of compounds differing in degree or type of unsaturation, and glycol-complexing agents for the separation of isomeric poly-hydroxy compounds (Morris 1964 and 1966; Schmid et al. 1966; Nutter and Privett 1968).

Methods for the quantitative analysis by TLC may be divided into 2 main groups; (1) methods based on direct analysis of the spots on the chromatoplate, generally by photometric-reflectometric or spectrofluorometric-methods; and (2) methods involving recovery of the separated compounds from the chromatoplate, followed by analyses using conventional procedures: gravimetric, radiometric, spectroscopic for phosphorus determination, ester analysis by the hydroxamic method, oxidation by solutions of chromic acid; and determination of glycerol or carbonyl compound (Privett et al. 1965).

Separations of Lipid Classes.—Separations of lipids by column or a combination of column and thin-layer chromatography by Rouser and co-workers were described earlier in this chapter. Separations of plant lipids were reviewed by Allen et al. (1966). Complete separations by thin-layer chromatography alone were described by Freeman and West (1966) and Skipski *et al.* (1968). A method of separating muscle lipids into phospholipids, free fatty acids, triglycerides, and cholesterol was developed by Hornstein *et al.* (1967). Phospholipids are separated from the total lipid extract by adsorption on activated silicic acid. The free fatty acids in the supernatant are adsorbed on an anion exchange resin. The solution, free of phospholipids and of free fatty acids is saponified with alcoholic potassium hydroxide. After acidification, a hexane extract is obtained containing glycerides, fatty acids, and cholesterol. The fatty acids are adsorbed on an anion exchange resin and the cholesterol remains in solution. Fatty acids are converted to methyl esters directly on the resin; and phospholipids adsorbed in silicic acid are transmethylated to produce the methyl esters of the phospholipid fatty acids. The methyl esters are analyzed by gas chromatography.

Lester (1963) developed a reproducible and sensitive anion-exchange procedure for quantitative analysis of phospholipids. The method was used by Wells and Dittmer (1966) as a basis for the quantitative determination of 24 classes of brain lipids. Selective mild and acid hydrolyses are used to obtain water-soluble phosphate esters characteristic of the diacyl phosphoglycerides and plasmalogens of brain. Those phosphate esters are separated by ion-exchange chromatography and assayed quantitatively. Phospholipids stable to hydrolysis are assayed after fractionation on silicic acid. Gangliosides, neutral lipids, and glycosphingolipids are measured by specific spectrophotometric determination of characteristic components after an initial solvent fractionation and chromatography on Florisil.

PHYSICAL AND CHEMICAL PROPERTIES OF LIPIDS

Physical Characteristics

Physical measurements of fats are useful for identification, for checking purity, and for the control of certain aspects of processing.

Color.—The color of fats and oils is estimated by several methods that vary with the type of examined lipid and with the country (Mehlenbacher 1958). The Wesson method is used in the United States for most edible oils and fats. The color of a 5.25-in. column of oil is compared in a Wesson comparator under specified viewing conditions with Lovibond glasses, using specified yellow ratios. The Lovibond glasses and the Tintometer are used extensively in England and Canada. The FAC method employs standard color solutions (prepared from solutions of inorganic salts) in 10-mm tubes, against which the tested oil in a similar tube is compared. The standards consist of 24 tubes in an odd-numbered series with 3 overlapping series; 1 normal, 1 green, and 1 red (Stillman 1955). Supplemental tubes for closer grading are available in some series. This simple and somewhat imprecise method is used widely for color-grading inedible oils, particularly tallows and greases. The German official method specifies a solution of iodine as the color standard, the comparison being made in a Pulfrich colorimeter. The Gardner color standards are specified by the Methods of the American Oil Chemists' Society and the American Society for Testing Materials for drying-oils. The standards are patterned along the same lines as FAC standards, but are unrelated to the latter numerically. The standards have been revised from time to time and are widely used and accepted.

The photometric method is specified for measuring the color of cottonseed, soybean, and peanut oils. The instrument used is a wide band spectrophotometer. The absorbance of the oil in a 25 mm cuvette is measured at 460, 550, 620, and 670 nm

$$\text{Photometric color} = 1.29\ A_{460} + 69.7\ A_{550} + 41.2\ A_{620} - 56.4\ A_{670}$$

The photometric method was designed by the color committee of AOCS to give values identical with Lovibond color values determined by the Wesson method. There are many shortcomings to this method as it is correlated with purely arbitrary color values. A more logical approach would be the determination of the pigments responsible for oil color (Mehlenbacher 1958).

Melting Point, Solidification, and Consistency.—As fats and oils are a complex mixture of compounds, they have no definite melting point and pass through a gradual softening before becoming liquid. Consequently, the melting point must be defined by the specific conditions of the method. The two methods most commonly used by the shortening and margarine industry include the capillary and Wiley melting point determinations (Smith 1955). The capillary method is essentially the method used by the organic chemist for determining the melting point of pure organic compounds. In this method, a 1-mm (internal diameter) thin-walled capillary tube is filled to the height of 10 mm with melted fat, one end is sealed, and the fat is allowed to stand at 4° to 10°C for 16 hr. The tube is then attached to the thermometer and placed in a bath maintained at 8 to 10°C below the expected melting point. The bath is heated at a rate of 0.5°C per min. The melting point is taken as the temperature at which the fat becomes completely clear.

The Wiley melting point is much more reproducible and reliable. In that method, a disk of the fat ($^3/_8$ in. in diameter and $^1/_8$ in. thick) is solidified and chilled in a metal form for 2 hrs. or more. The disk is then suspended in an alcohol-water bath of its own density and heated slowly while being stirred with a rotating thermometer. The melting point is taken as the temperature at which the fat disk becomes completely spherical. An agreement within 0.2°C between analyses can be generally obtained. The Wiley melting point, in conjunction with the refractive index, has been used for many years in control of hydrogenation. For control purposes the sample is chilled rapidly before assay.

The titer test determines the solidification point of fatty acids. A titer tube is filled with fatty acids (obtained by saponification of oil or fat with potassium hydroxide in glycerol), suspended in an air bath surrounded by a water bath maintained at 15 to 20°C below the titer. The sample is stirred until the temperature begins to rise or remains constant for 30 sec, after which the stirring is stopped, and the maximum temperature which the fatty acids attain as a result of the heat of crystallization is determined. The utility of this method is limited. It is used sometimes in evaluating fats for soap manufacture.

The setting or congeal point determinations that give the solidification point of the fat rather than that of the separated fatty acids are used quite extensively in the margarine or shortening industries. The tests provide useful information on the consistency of a plastic fat and on the performance of a catalytic hydrogenation. In the manufacture of margarine, the solidification point (congeal) and melting point (Wiley method) are kept as close as possible (Smith 1955).

Dilatometry.—Dilatometry is essentially a measurement of changes in specific volume that occurs with change in temperature. It is useful in the field of fats and oils to detect or analyze phase transformations because fats expand when they melt and generally contract when they undergo polymorphic change to a more stable form (Braun 1955). Fat dilatometers vary considerably in size and construction. Basically, they consist of a bulb that is attached to a calibrated capillary tube. The fat in the bulb is confined by a liquid such as a colored water solution or mercury. As the fat expands, the confining liquid is displaced into the capillary tube. When the volume is plotted against temperature a curve is obtained, the initial and final linear portions of which represent the completely solid and liquid states.

Nuclear Magnetic Resonance.—The precise calculation of liquid and solid fats by dilatometric methods is difficult or even impossible in a complex mixture of glycerides. Low-resolution NMR appears to provide a solution to the problem (Chapman *et al.* 1960; Johnson and Shoolery 1962; Ferren and Morse 1963; Taylor *et al.* 1964).

The magnetic field strength at the center of a hydrogen resonance line always has the same value for a given frequency, but the shape of the line is influenced by the chemical and physical state of the sample. The width of the adsorption line is related to the mobility of the hydrogen in the sample or the mobility of the compound containing the hydrogen, and to the field homogeneity. The effective magnetic field strength at a nucleus is the sum of the applied magnetic field plus the field contributed by neighboring nuclei. In a solid where the nuclei are fixed rather rigidly, a nucleus is in a magnetic field significantly higher or lower than the applied field. Consequently, the line width of a solid is relatively wide. In a liquid, the molecules are in a state of thermal agitation, the field contributed by hydrogen nuclei averages out rapidly, and the absorption line is narrow. This difference in widths between solids and liquids can be used to determine the liquid content of fats. Conway and Johnson (1969) have recently shown that high-resolution NMR can be used to determine unsaturation in single corn kernels. The procedures are useful in breeding programs to alter the fatty acid composition of corn oil.

Refractive Index.—The refractive index is constant, within certain limits, for each type of oil or fat. It is, therefore, used in identification. The use of a refractive index determination in the assay of lipid content was described earlier in this chapter. Since the refractive index is related to unsaturation, it can be used in the determination of an iodine value. However, this correlation differs for various types of oils. A simple iodine-number refractometer for testing flaxseed and soybeans was described by Hunt *et al.* (1951).

Infrared Spectroscopy.—Infrared has been used more extensively in the analysis of lipids than of any other food components. In addition to structural analyses, it is useful in the identification of a lipid source; in the detection of adulteration (butterfat by plant lipids, cocoa by hydrogenated fats, durum wheat alimentary pastes by hard wheat semolina); in studies of autoxidation, rancidity, and drying properties of lipids; in following the effects of food processing on lipids; and in nutritional investigations on the effects of food composition and interaction with lipids. Extensive bibliographies on the use of infrared spectroscopy in lipid investigations were prepared by Wheeler (1954); O'Connor (1955, 1961A, 1961B); Chapman (1960, 1965); Schwarz et al. (1957); Freeman (1957); Kaufman (1964); Kohn (1965); Kohn and Laufer-Heydenreich (1966).

Ultraviolet Spectrophotometry.—The common polyunsaturated fatty acids in untreated oils show no absorption peaks in the ultraviolet. If the double bonds in those oils can be rearranged to form a conjugated system, selective absorption will appear in the ultraviolet region, and can be used for analytical purposes. Polyunsaturated fatty acids are converted into conjugated isomers by heating in alkali, and the intensity of the resulting selective absorption is determined. The isomerization is enhanced at elevated temperatures. By using ethylene glycol as a solvent, the temperature can be increased and the assay performed more rapidly (Pitt and Morton 1957).

Mitchell et al. (1943) first published a detailed method by which both linoleic and linolenic acids can be determined. The procedure involved heating 10 ml of ethylene glycol containing 6.5% potassium hydroxide in a test tube in an oil bath. As the temperature of 180°C is reached, 100 mg of fat or fatty acids is added, mixed, heated for 25 min and cooled rapidly. After dilution with ethanol to a suitable volume, the absorbance is measured at 234 and 268 nm. A blank solution in the control cell of the spectrophotometer consists of an alkaline glycol solution heated and diluted as the assayed sample. The extinction at 268 nm is derived from linolenic acid only; the contribution of the linolenic acid at 234 nm can be computed, and the remaining extinction at 234 nm used to calculate the linoleic acid content. The method is empirical and suffers from the limitations of nonstoichiometrical analytical procedures. The ultraviolet method has undergone modifications (included in the latest AOCS methods) to permit also a determination of arachidonic acid in a mixture with other unsaturated fatty acids. Use of ultraviolet absorption spectroscopy in the determination of α-eleostearic acid in tung oil was described by O'Connor (1955).

For many years, differentiation between polyunsaturated fatty acids was based on, generally, nonspecific chemical reactions. The small differences in composition of the fatty acids are, however, expressed in relatively large changes in physical properties. Development of optical equipment and methods based on differences in vapor pressure, partition coefficients, adsorption, pattern of fragmentation by degradative methods, and electron resonance are some of the more powerful, specific, and new tools that provide much more information both in scope and validity (Holman and Rahm, 1966).

Oxidation of polyunsaturated fatty acids is accompanied by increased ultraviolet absorption. The magnitude of the change is, however, not easily related to degree of oxidation, because the effects vary with fatty acids. The test, is, therefore, of a semiquantitative nature (Holman et al. 1945).

Gas-Liquid Chromatography.—The use of gas-liquid chromatography (GLC) was introduced by James and Martin (1952) for the separation of normal saturated carboxylic acids up to 12 carbon atoms in chain length. Cropper and Heywood (1953) extended the method to the separation of the methyl esters of even-numbered fatty acids up to behenic acid. Since then a rapid expansion in the use of this method in lipid investigations and analyses has taken place. Today, methyl esters of fatty acids containing up to 34 carbon atoms, and minor components containing less than 0.05% of the original sample can be detected and estimated reliably. The importance of GLC has increased particularly with the development of polyester stationary liquids that enable separation of methyl esters of fatty acids varying in degree of unsaturation.

Advantages of GLC over other methods of fatty acid analysis have been so great that it has almost entirely replaced them in both research and routine determinations. The advantages include use of small quantities; relative specificity and simplicity; and adaptability to both qualitative and quantitative determinations.

In addition to the determination of total fatty acid composition, GLC can be used to determine the distribution and position of fatty acids in the lipid molecule, in the studies of fat stability and oxidation by chemical and biological agents, in assaying heat or irradiation damage to lipids, and in the detection of adulterants (i.e. hydrogenated fats in cocoa butter). Use of GLC in the analysis of lipids was reviewed by Kaufmann et al. (1961, 1962), Kohn (1964), Horning and Vandenheuvel (1964), and many others.

In most of the published work on the determination of fatty acids, the acids are first converted to the corresponding methyl esters before separation by GLC. Methods for obtaining the methyl esters of the fatty acids can be divided into those involving transesterification of the glycerides in the presence of excess methanol, and saponification of the glycerides with alkali, isolation of the free fatty acids, and esterification of the acids. Transesterification

methods are less time-consuming.   In addition, the conditions normally employed in these procedures cause less isomerization of polyunsaturated fatty acids, than in the saponification-esterification
schemes (Jamieson and Reid 1965).

It is impossible in the limited space available to list the various
uses of gas chromatography in the analysis of lipids in foods.   The
usefulness will be, therefore, illustrated by the description of a few
selected methods for determining fat adulteration.

When the triglyceride type composition of native butter was
known, adulteration with vegetable fat could be detected by GLC
at the 1% level, and with lard at the 3% level (Kuksis and Mc-
Carthy 1964).   Adulteration of unknown butter samples could be
detected at the 5 to 10% level due to considerable variations in
chromatographic patterns of the fatty acids of butters of different
origin.   The ease of detection and identification depended on the
fat added.   Mixtures of coconut and lard, matching closely gas
chromatographic patterns of butter fatty acids, could be prepared.

With the advent of modern processing techniques, animal fats can
be incorporated with vegetable oils and processed into solid shortenings comparable in organoleptic and functional properties with the
all-vegetable shortenings.   To detect the adulteration, Cannon
(1964) developed a procedure for determining vegetable fat in butterfat, involving fat saponification followed by precipitation of the
sterols with digitonin.   The digitonides are acetylated and the sterol
acetates separated by gas chromatography.   The method gives
satisfactory results for lard, soybean oil, and cottonseed oil.   However, digitonides of beef tallow, palm kernel oil, and coconut oil
are difficult to precipitate and acetylate.   The method of Eisner et al.
(1962) and Eisner and Firestone (1963) involved saponification, extraction of the unsaponifiable matter, separation of the latter on
Florisil columns into sterols and other compounds, and gas chromatography of the sterol fraction after acetylation.   The method gives
satisfactory results, but is time-consuming.   Ettinger et al. (1965)
developed a simplified and more rapid procedure.   The total unsaponifiable matter (without separation on Florisil columns) was extracted with ether following saponification, acetylated, and subjected to gas chromatography analysis.   The presence of 2.5%
animal fat in vegetable oil could be detected.

To improve the usefulness of GLC, combined thin-layer and gas-
liquid chromatographic systems (Kuksis 1966) are often used.   Gas
chromatography and mass spectrometry (Ryhage and Stenhagen
1960;   Dutton 1961) have been used respectively for the separation

of complex mixtures and for structural characterization of lipids. In both techniques, analyses of microgram quantities of samples in the vapor phase are performed. Direct combination of both techniques provides an exceptionally rapid, powerful, and versatile tool for the separation, characterization, and structural elucidation of components of complex mixtures. The ability to record rapidly several mass spectra of one emerging chromatographic peak may also be used to determine the efficiency of separation, the presence of impurities, and hydrolytic, thermal, or oxidative decomposition (Leemans and McCloskey 1967). On the other hand, interpretation of the enormous data provided by the combined instruments requires considerable experience, time, and effort.

Various additional attachments to gas chromatographs, (i.e. infrared spectrophotometers) have extended its use as a precise and informative tool. Alternatively, small-scale preparative chromatography can be combined with automated chemical identification.

Chemical Methods

Chemical characteristics of lipids are determined on the entire extract or on water-insoluble fatty acids isolated from a purified extract.

For the preparation of water-insoluble fatty acids, the fat is saponified with a potassium hydroxide solution (in ethanol) or by a glycerol-potassium hydroxide mixture, and the fatty acids in the formed soap are split by acidification with a mineral acid. The fatty acids are extracted with ether in a separatory funnel, washed and dried cautiously. Many methods have been suggested for saponification. Some are rapid and effective, but likely to cause more or less extensive modification of the fatty acids (isomerization, oxidation, volatilization, and degradation), or formation of some mono- and diglycerides. Usefulness of the shortened methods will depend on their agreement with standard procedures.

Determination of Impurities.—Impurities in fats and oils, and fatty acid products are mainly moisture, volatile compounds, insoluble matter, unsaponifiable matter, trace metals, and their soaps. The term MIU (moisture, insoluble, unsaponifiable) is a frequently used group designation for the determination of the nonfatty constituents of crude oils and other fatty acid products where settlement is on the basis of oil or acid content. It also figures predominantly in the trading rules of the various oil trading organizations (Rodeghier 1955). The total MIU is considered valueless material except to those interested in the recovery of sterols and tocopherols from the unsaponifiable fraction of fatty acids split from the soapstock of soybean oils.

The insoluble matter found in fats and oils is dirt, meal, and any other substances insoluble in kerosene and petroleum ether.

Unsaponifiable matter is found dissolved in fats or fatty acids, and is that material that cannot be saponified by potassium hydroxide. The unsaponifiables include sterols, higher alcohols, and some hydrocarbons. Oil-refining removes most of those substances. The analytical determination is made by weighing a 5-gm sample into a 250-ml Erlenmeyer flask. Then, 30 ml of alcohol and 5 ml of a 50% potassium hydroxide solution are added, and the mixture refluxed until saponified. The saponified material is then transferred to an extraction cylinder, the flask is rinsed with alcohol and washed into a cylinder with alcohol, water, and petroleum ether. After cooling and shaking, the petroleum ether extract is siphoned off into a 500-ml separatory funnel, and the extraction repeated. The combined petroleum ether extracts are combined and washed with 10% ethanol, and the petroleum ether extract dried and weighed in a tared beaker as unsaponifiable.

Oxidized fatty acids are determined after acidifying with hydrochloric acid the saponified fat from the assay of unsaponifiable matter. The free fatty acids, that are insoluble in petroleum ether, are considered oxidized fatty acids.

The presence of as little as 0.3 ppm iron is detrimental to the quality and stability of oils. Similarly, other metallic traces are undesirable. The determinations are made by accepted procedures.

Fatty acids react readily with alkali to produce soaps; the soaps are mainly produced during oil refinement. Levels of 5 ppm soap are detrimental to refined oil and impair significantly the quality of hydrogenated oils. In oil processing, soaps are removed by water-washing the neutral oil, followed by thorough drying before bleaching with an adsorbent.

The analytical determination involves alcohol or dilute hydrochloric acid extraction of the soap from the oil followed by quantitative determination of the sodium ion. A rapid conductivity assay of soap in oil is also available.

Determination of Unsaturation.—One of the most important analytical determinations that an oil chemist makes frequently is the measurement of unsaturation of an oil. This determination is important both in classification of fats and oils for trade and use, and for control of manufacturing processes. The generally accepted method of expressing the degree of carbon to carbon unsaturation of a fat, oil, or derivative is as iodine value (or iodine number). This value (number) is defined as the grams of iodine that add to 100 gm of sample (Allen 1955). The results are expressed in terms of iodine whether iodine or some other halogen is actually used. Since halogens all add to carbon-carbon double bonds, most of the methods make use of this property.

Basically, the determination consists of adding a halogen to a weighed quantity of sample and determining the amount of reacting halogen. There are several methods for determining the iodine value, but the most important are the Wijs (developed in 1898) and the Hanus (from 1901). The Wijs method is probably the most widely used and is believed to yield results closer to theoretical values than any other method. Hanus results range from 2 to 5% below Wijs, but the Hanus reagent has the advantage of being more stable (Mehlenbacher 1958).

In the Wijs method, 0.1 to 3.0 gm (depending on the expected iodine value) is dissolved in 15 ml carbon tetrachloride, and 25 ml of the Wijs solution (commercially available, or prepared by dissolving 9 gm iodine trichloride in a mixture of 700 ml glacial acetic acid and 300 ml carbon tetrachloride) is added. The conical flask is closed with a ground glass stopper, mixed, and allowed to stand at about 20°C for 1 hr in the dark. Then, 20 ml of a 10% aqueous potassium iodide solution and about 150 ml water are added, and the unreacted halogen is titrated with an accurately standardized thiosulfate solution in the presence of starch towards the end of the titration.

To shorten the reaction time, a 2.5% mercuric acetate solution in glacial acetic acid is added to the Wijs reaction mixture. The addition of a catalyst shortens the reaction time from 1 hr to 3 min. For high iodine-value fats (above 50), the difference between the 2 methods is negligible. The shortened method gives actually more meaningful (then the Wijs procedure) results in fats with hydroxy-fatty acids, but is somewhat unreliable for Chinese tung oil and other highly conjugated oils (Cocks and Van Rede 1966).

The iodine value (IV) is calculated from the difference in titration of a blank and sample according to the equation

$$IV = \frac{(B - S)N \times 12.692}{\text{weight of sample}} \tag{36.2}$$

where B = titration of blank
S = titration of sample
N = normality of $Na_2S_2O_3$ solution

For a pure oil or fat, the iodine value

$$IV = \frac{2 \times 126.92 \times \text{no. double bonds} \times 100}{\text{molecular weight}} \qquad (36.3)$$

and

$$\text{no. double bonds} = \frac{IV \times \text{molecular weight}}{2 \times 126.92 \times 100} \qquad (36.4)$$

Nondrying oils have an iodine value below 100; drying oils 130 to 200; and semidrying oils have intermediate values. Theoretically, the iodine values of oleic, linoleic, and linolenic acids are respectively 89.9, 181.0, and 273.5. The iodine values of free fatty acids are higher than of glycerides. In a mixture of free fatty acids and glycerides, for each percent of free fatty acids the iodine value increases 0.00045 × iodine value (Cocks and Van Rede 1966).

Thiocyanogen, $(SCN)_2$, adds to double bonds in a manner similar to halogens. However, with polyunsaturated acids the addition is not complete. Thiocyanogen adds 1 mole per mole to dienoic acids, and 2 moles per mole to trienoic acids. This permits the calculation of oleic, linoleic, and linolenic acids from the thiocyanogen value, iodine value, and a separate assay of saturated fatty acids.

Several tests are available for specific polyunsaturated acids, that are based on the formation of crystalline polybromides varying in melting point, when the fatty acids are brominated. The methods vary as to temperature, solvent used to dissolve the fatty acids, and condition of bromination. By careful standardization, the results are reproducible and provide information on composition of some fats.

Determinations of iodine value give a reasonably quantitative measure of unsaturation if the double bonds are not conjugated with each other or with a carbonyl oxygen, and if the determination is carried out under specified conditions as to the excess of halogen reagent, time of reaction, and exclusion of light (Allen 1955).

To overcome some of the limitations of the halogen methods, quantitative hydrogenation can be used. Hydrogenation is used to measure the unsaturation of acetylenic or conjugated double bonds. Although such fats do not absorb halogen readily, the addition of hydrogen can be practically quantitative. Essentially the methods consist of catalytic (nickel, palladium, or platinum) hydrogenation of a heated sample. The volume of hydrogen absorbed is determined from measurements before and after the reaction and reduced to standard conditions. The results can be expressed as mole of hydrogen per mole of sample, or can be calculated to an iodine value basis. The latter is known as the hydrogen-iodine value.

There is, as yet, no standard, rapid, and precise method for the determination of the hydrogen value. Recently, Miwa *et al.* (1966) developed a procedure for the quantitative determination of un-saturation in oils by using an automatic-titrating hydrogenator. The method utilizes a catalyst prepared by *in situ* treatment of platinum salts with sodium borohydride, *in situ* generation of hydro-gen from sodium borohydride, and a valve at the tip of a buret that automatically introduces standardized sodium borohydride solution into the reaction mixture only as long as hydrogenation is proceeding (Fig. 36.6).

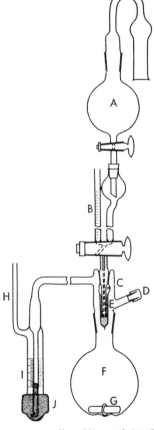

FIG. 36.6. AUTOMATIC-TITRATING HYDRO-GENATOR

(A) 250-ml NaBH$_4$ reagent reservoir, (B) 5 ml precision-bore buret, (C) seal of the automatic valve assembly, (D) 6-mm rubber septum, (E) mercury filled automatic valve, (F) reaction flask, (G) stirring bar, (H) sidearm of bubbler, (I) scale for pressure reading, (J) mercury pool.

From Miwa et al. (1966)

Saponification Value.—The saponification value is a measure of the amount of alkali required to saponify a definite weight of fat. It is expressed as milligrams potassium hydroxide required to saponify 1 gm of fat, i.e. to neutralize the free fatty acids and the fatty acids present in the form of glycerides. The saponification equivalent is the amount of oil or fat saponified by 1 gm equivalent of potassium hydroxide and is equal to

$$\frac{56108}{\text{saponification value}} \qquad (36.5)$$

The saponification value was originated by Koettsdorfer and is sometimes known by this name. The procedure involves saponifying under reflux 4 gm filtered oil with 50 ml of a 0.5 N potassium hydroxide solution in 96% ethanol, for 30 min. Excess potassium hydroxide is determined by back-titration with an aqueous, standardized 0.5 N hydrochloric acid in the presence of phenolphthalein.

The saponification value is an indication of the average molecular weight of fat. For pure fatty acids, the saponification value equals the *acid value*. The *ester value* is the difference between the saponification value and the acid value. In oils and fats, the ester value is a measure of the amount of glycerides present.

Identification.—There is no organized, foolproof, scheme for the qualitative analysis of fats and the problems of identifying individual fats and oils is quite complicated. This is particularly true in the case of mixtures and processed fats. Even with the most sophisticated instruments it is quite possible to encounter mixtures that defy identification of the source of the oil. Admittedly, the availability of more specific, meaningful, and reliable instrumental techniques has simplified identification and detection of adulteration. Some of the methods were described in the discussion of uses of gas-liquid chromatography in lipid analyses. Additional chromatographic procedures and spectrophotometry (especially infrared) are used most commonly in combination with the determination of physical and chemical constants (saponification value, iodine value, melting point, and others) that are constant and typical for individual fats and oils.

In addition, there are several specific tests for individual oils or for certain functional groups. This procedure is much more useful in detecting the presence of a specific adulterant than in ascertaining the purity of an oil. The usefulness of such tests is sometimes limited by the destruction of a specific reaction through processing.

The specific tests are described in detail in books listed at the beginning of this chapter.   They are discussed briefly here.

Squalene, an unsaturated aliphatic hydrocarbon, occurs in higher concentrations in olive (and fish) oil than in most other oils.   The amount of squalene, isolated from the unsaponifiable matter by column chromatography can be used to estimate roughly the olive oil content of vegetable oils.   In the Bellier test for peanut oil, insolubility of arachidic and lignoceric acids in 70% ethanol is measured. The Halphen test is used to detect cottonseed oil.   A reddish color develops on heating the oil with amyl alcohol and a solution of sulfur in carbon disulfide.   Kapok oil gives a similar reaction. Processing (heating, refining, or hardening) reduces color intensity. Fats of animals fed cottonseed cake or meal may give a positive reaction.

Melting castor oil-containing oils or fats with dry potassium hydroxide, followed by precipitation of the salts from the aqueous solution of the soap and acidification, yields fatty acids from which characteristic sebacic acid crystallizes.

The difference in the melting point of the glycerides and fatty acids of lard is much greater than the difference in tallow or hardened fats.   The *Bomer value* (BV) is computed as

$$BV = A + 2(A - B) \qquad (36.6)$$

where A = melting point of glycerides

B = melting point of fatty acids

The BV is at least 73 in lard, and significantly lowered by a mixture of 10% or more of tallow and hardened fats.

Shaking a sesame oil-containing mixture with hydrochloric acid and an ethanol solution of furfural (or sucrose), gives a stable pink color that can be used for quantitative determination by the sensitive Baudouin reaction.   The Fittelson test for the detection of teaseed oil in olive oil is based on the formation of an unstable deep red color formed by adding acetic anhydride, chloroform, and sulfuric acid and followed by adding cold (5°C) ether.

The presence of adulterants in milk fat has been the subject of many investigations.   The adulteration can be established in two general ways.   One involves the identification of a foreign material in the fat, the other, the demonstration of an unusual concentration of specific regular milk fat components (Kurtz 1965).

The difference between the sterols of animal and vegetable fats affords a positive means of identifying adulteration by a vegetable fat. Sometimes, the identification is complicated by the presence of several phytosterols including those that have physical and chemical properties similar to those of cholesterol. The use of one of those methods was described earlier (Eisner *et al.* 1962).

Vitamin E exists in several forms. Brown (1952) found only α-tocopherol in cow's milk, whereas of the vegetable oils he examined also had β-, γ-, or σ-tocopherols.

The high proportion of the soluble, volatile fatty acids—particularly butyric acid—is the most prominent difference between that of milk fat and other animal fats. Tests such as the Reichert-Meissl value, Polenske number, and butyric acid determination can be used to establish gross adulteration of milk fat.

The Reichert-Meissl value is equivalent to the number of milliliters of a decinormal sodium hydroxide solution required to neutralize the volatile, soluble acids obtained under specified conditions (Fig. 36.7) from 5 gm of fat. Milk fat contains more of those acids than

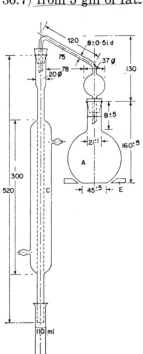

FIG. 36.7. DISTILLING ASSEMBLY FOR REICHERT-MEISSL AND POLENSKE DETERMINATION

any of the fats from which it might be desirable to distinguish it. The Canadian Standard for milk fat requires a Reicher-Meissl value of at least 24. The Polenske number equals the number of milliliters of decinormal solution required to neutralize the volatile, insoluble acids obtained from 5 gm of fat. The amount of such acids in milk fat is small, but is high in fats with a high caprylic acid content (i.e. coconut oil). The Canadian Standard requires that the Polenske number be not above 3.5, and not greater than 10% of the Reichert-Meissl value.

It is difficult to demonstrate on the basis of the Reichert-Meissl and Polenske determinations or direct butyric acid assays the adulteration of milk fat by small amounts of other fats, especially in view of some of the extreme values of those parameters in some unadulterated samples. Thus, Zehren and Jackson (1956) reported that 500 samples of butterfat from 42 locations from 25 States in the United States had Reichert-Meissl values of 24.2 to 33.6 (avg 29.0), and Polenske values of 1.1 to 3.0 (avg 1.9). The refractive index varied from 1.4531 to 1.4557 (avg 1.4540). Analyses by Keeney (1956) of the above samples showed a butyric acid content of 9.6 to 11.3 mole % (avg 10.9). The significance of the latter value is somewhat limited as high-concentrate, low-roughage diets may decrease significantly the proportion of butyric acid.

Automated Lipid Analyses

As in other areas, automated lipid analyses are based mainly on experience gained in clinical chemistry. Recent advances in auto-mated analyses of blood cholesterol, triglycerides, and phospholipids were reviewed by Levine (1967). Cholesterol can be determined by the method of Levine and Zak (1964) that involves heating the sample at 95°C with a solution containing ferric chloride, sulfuric acid, and acetic acid, and measuring the color at 520 nm.

For the determination of triglycerides, two methods are available. In the procedure of Lofland (1964), phospholipids are removed on an ion-exchange column and the eluate is saponified with alcoholic potassium hydroxide. After evaporation and addition of sulfuric acid, the glycerol formed is oxidized to formaldehyde with periodate. Excess periodate is destroyed with sodium arsenite, and the formaldehyde reacted at 95°C with chromotropic acid to give a colored product that is measured at 570 nm. The procedure of Kessler and Lederer (1965) is based on the reaction of Hantzsch involving condensation of an amine, beta diketone, and an aldehyde. A phospholipid-free extract is prepared by adding an ion-exchange resin to the isopropanol extract. In addition, Lloyds' reagent and a copper-lime mixture are added to remove chromogens and glucose interference. The extract in isopropanol is saponified with aqueous potassium hydroxide, and mixed with periodate and the acetylacetone reagent. The formaldehyde (from the oxidation of glycerol) is condensed with the acetylacetone reagent to produce 3,5-diacetyl-1,4-dihydrolutidine that is measured in a fluorometer. Up to 20 samples can be tested per hour.

The determination of phospholipids according to Zilversmit and Davis (1960) requires precipitation with 10% trichloroacetic acid and separation of the phospholipid as a complex with protein. The precipitate is digested with perchloric acid and the digest is assayed by a colorimetric molybdate method for phosphorus. Whitley and Alburn (1964) digested the precipitate in a continuous flow digest or with a mixture of perchloric and sulfuric acids in the presence of vanadium pentoxide as a catalyst. After neutralization, the digest is mixed with molybdate and hydrazine, and the color measured at 815 nm in a colorimeter equipped with silicon photocells.

Fat Stability and Rancidity

The term *rancidity* is used to describe development of objectionable flavors and odors. As a result of these changes, consumer acceptance of the food is lowered. Rancidity may be caused by either hydrolytic or oxidative changes in the fat. Hydrolytic rancidity involves chemical or enzymatic hydrolysis of fats into free fatty acids and glycerol. Oxidative rancidity involves the addition of atmospheric oxygen in the presence of enzymes or certain chemicals. Hydrolytic activity is important in dairy products and coconut items. Those foods contain glycerides of low-molecular weight fatty acids such as butyric, caproic, caprylic, or capric. Cereal lipids on the other hand, contain high molecular weight fatty acids which when hydrolyzed do not produce the same type of off-flavors and odors as are produced by the hydrolysis of low molecular weight glycerides.

Fat Acidity.—Deterioration of grain and milled products is accompanied by increased acidity. The acids formed include free fatty acids, acid phosphates, and amino acids; but at the early stages of deterioration, fat acidity increases at a much greater rate than either of the other two types or all types of acidity combined. Several workers have, therefore, suggested determining fat acidity as an index of grain soundness (Zeleny and Coleman 1938; Anderson and Alcock 1954; Baker *et al.* 1959; Pomeranz and Shellenberger 1966).

The organic acidity of fats and oils can be expressed in several ways:

(1)   Acid value—the number of milligrams of potassium hydroxide required to neutralize 1 gm of fat or oil.

(2)   The acidity of edible oils is sometimes expressed as milliliters N sodium hydroxide solution used to neutralize fatty acids in 100 gm fat.

(3)   The free fatty acid (FFA) content-percentage by weight of specified fatty acid (either oleic with a molecular weight of 282; or average, and specified molecular weight appropriate to the nature of the analyzed fat or oil).

Acid value can be converted to FFA (expressed as oleic acid) by the following formula

$$\text{Acid value} = 1.99 \times \% \text{ FFA} \qquad (36.7)$$

$$\% \text{ FFA} = 0.503 \times \text{acid value} \qquad (36.8)$$

For the determination of free fatty acids, 4 to 20 gm fat (depending on the expected acidity and normality of selected titrant) is dissolved in a neutralized solvent and titrated with a standardized aqueous solution of sodium hydroxide in the presence of phenolphthalein (or alkali-blue 6 B for dark-colored oils) as indicator.   The acidity of the oil is calculated from the amount of consumed titrant.

Originally, ethanol was (and in the AOCS and British Standard Methods is) used as solvent.   As many oils and fats are only slightly soluble in ethanol, the titration involves heating to 60–65°C, or even boiling.   At the high temperature required, mono- and diglycerides may be saponified and some fatty acids volatilized.   Using a mixture of solvents (generally ethanol-ether, 1:1) makes it possible to titrate at room temperature.   On the other hand, in large laboratories that use large amounts of solvents, it is simpler to recover a single solvent.   Also, the low solubility of the oil in ethanol is sometimes advantageous in the titration of dark-colored products.   Pyridine-denatured alcohol cannot be used as it has a strong buffering action.

The American Association of Cereal Chemists recommends three procedures for the determination of fat acidity in grain.   In the basic method, grain samples ground so that 90% or more pass a 40-mesh sieve are extracted with petroleum ether, within 1 hr of grinding to forestall enzymatic breakdown of lipids.   The extract is evaporated, redissolved in a benzene-ethanol mixture (1:1) containing 0.02%

phenolphthalein, and titrated with 0.0178 N potassium hydroxide to the endpoint, matching color of a standard prepared from dilute solutions of potassium dichromate and potassium permanganate. In the rapid method, preground material is extracted with benzene for 4 min in a special grinder-extractor; or the ground sample is shaken mechanically for 30 min, or intermittently by hand for 45 min. The extract is filtered, and an aliquot is titrated with standard potassium hydroxide. In the colorimetric method, based on the procedure developed by Baker (1961, 1964), the benzene extract (or petroleum ether extract after evaporation and dissolving in benzene) is mixed with a cupric acetate solution, and after filtration the color is determined at 640 nm. The fat acidity value (mg KOH required to neutralize FFA from 100 gm grain, dry matter basis) is calculated from a calibration curve of pure oleic acid passed through the colorimetric procedure.

Detection of Oxidation Products.—The peroxide value is the most commonly used assay of state of oxidation in fats and oils. Many methods have been devised for its measurement (Lea 1962). Barnard and Hargrave (1951) reviewed the value of the various methods and found the iodometric procedures were most commonly used. The original method of Lea (1931) has been improved by the exclusion of oxygen from reagents and reaction flask (Lea 1946).

The peroxide value is commonly determined by measuring the amount of iodine liberated from a saturated potassium iodide solution at room temperature, by fat or oil dissolved in a mixture of glacial acetic acid and chloroform (2:1). The liberated iodine is titrated with standard sodium thiosulfate, and the peroxide value is expressed in milliequivalents of peroxide-oxygen per kilogram fat. The *Lea value*, often mentioned in the literature, is expressed in millimoles of peroxide-oxygen per kilogram fat; and is numerically half the peroxide value. The method is applicable to oils, fats, and margarine. In the latter, the oil and water phase must be separated before analysis. The peroxide value is an indicator of the products of primary oxidation. It measures rancidity or degree of oxidation but not stability of a fat.

Stability Tests.—Several methods have been developed to determine the stability (or susceptibility) of an oil to the development of rancidity. The oven or Schaal Method (Pool 1931) was originally designed to evaluate the stability of shortening in baked products. The odor of a sample stored at 63°C is observed periodically until rancidity is detected organoleptically. In the accelerated test, 1 day of incubation at the elevated temperature, corresponds to 6 to 10 days of incubation at 21°C. In the oxygen absorption test, the sample is heated in the presence of oxygen until the rate of absorption of oxygen undergoes a definite increase. The method has the advantage of being a direct measure and does not depend on the rate of formation of decomposition products (Mehlenbacher 1958).

In the active oxygen method, AOM, the fat is heated, dried, and filtered air is blown through it until the peroxide value increases to some value that has been previously shown to indicate the onset of rancidity. The results are influenced by many factors including temperature, light, availability of oxygen, surface factors, the presence of natural and synthetic antioxidants, the nature of the fat, and the presence of impurities, especially trace metals. The peroxide values achieved by the AOM–tested samples, at which the fat is rancid by organoleptic tests, vary with the nature of the fat and range from about 20 meq per gm for lard to over 100 for vegetable oils (Dugan 1955). The AOM test is rapid; the results can be obtained within about $1/20$ of the time required in the oven test. The correlation between actual shelf-life is, however, better in the oven test than in the AOM test, because the temperature is closer to normal storage conditions, rancidity rather than peroxide values is measured, and extreme conditions of forced air circulation are avoided. Baumann (1959) pointed to difficulties in correlating peroxide values and chemical tests of fat stability with graders' judgments of off-flavor and off-odor in oil stored for 28 months.

Courtesy LaPine Scientific Co.

FIG. 36.8. AOM FAT STABILITY APPARATUS

Pohle et al. (1962, 1963) have compared several analytical methods for the prediction of relative stability of fats and oils, and proposed a rapid oxygen bomb method for evaluating the stability of fats and shortening. A comparison of data from several accelerated laboratory tests with organoleptic evaluation of samples stored at 85°F (Pohle et al. 1964) indicated that the different types of fat behaved differently. Consequently, the laboratory tests cannot be used as an index of shelf-life stability, except for a given type of formulation of fat for which the relationship between the laboratory test and shelf-life stability has been established.

Miscellaneous Tests.—The odors and flavors associated with typical oxidative rancidity are mostly due to carbonyl type compounds. The shorter chain aldehydes and ketones isolated from rancid fats are due to oxidative fission and are associated with advanced stages of oxidation. The carbonyl type compounds develop in low concentration early in the oxidative process. The Kreis test is a sensitive indicator of the early stages of the oxidative process. The substance responsible for this reaction is epihydrinaldehyde. The Kreis test often indicates changes that are not necessarily consistent with fat stability, as measured by other methods. Therefore, the test should be used as a supplementary index (and not instead of) other tests such as peroxide value. In addition to epihydrinaldehyde, several other compounds (i.e. malonic dialdehyde and acrolein treated with hydrogen peroxide) yield a colored product with phloroglucinol. However, as the presence of those compounds has not been demonstrated in oxidized fat, they do not appear in the currently accepted mechanism of oxidation (Patton et al. 1951); and they seem to result from secondary oxidation products formed from the decomposition of peroxides.

Numerous workers have attempted to make quantitative evaluations of rancidity through the use of carbonyl tests. The method of Lapin and Clark (1951) is based on the formation of a colored quinoidal ion of a 2,4-dinitrophenylhydrazone in a solution of a base. An extensive study and application of the method was made by Henick et al. (1954). Their method involved the formation of the 2,4-dinitrophenylhydrazones in benzene solutions with trichloroacetic acid as catalyst. The absorbance at 430 and 460 nm can be used for the simultaneous determination of saturated and allenic carbonyls.

A chemical test that deserves mention because of the great variety of uses to which it has been put, is the thiobarbituric acid test (TBA). Kohn and Liversedge (1944) have shown originally, that aerobic oxidation products of animal tissues gave a color reaction with thiobarbituric acid. The reaction was traced to the oxidation of unsaturated fatty acids, mainly linolenic (Bernheim et al. 1947; Wilbur et al. 1949). The pigment produced in the sensitive color reaction is a condensation product of two molecules of TBA and one of malonic dialdehyde. The TBA test was originally devised for the evaluation of dairy products (Patton and Kurz 1951). The test has been found useful in testing many vegetable and animal fats. The test is performed by treating the fat in a benzene or chloroform solution with the TBA reagent in an aqueous acetic acid solution. After shaking, the aqueous layer is separated and heated in a boiling water bath for 30 min to develop a pink chromogen with maximum absorbance at 532 nm. The intensity of the color is a measure of the degree of oxidation. During the early stages of oxidation, the amounts of TBA-reactive substances in oxidized unsaturated fatty acids are closely correlated with peroxide value, oxygen uptake, and diene conjugation (Dahle et al. 1962). Kwon and Olcott (1966) have shown that malonaldehyde is the principal TBA-reactive substance in oxidized methyl linolenate, fatty acid esters, and squalene. The kinds of TBA-reactive substances produced depend on the substrate and oxidation conditions. The compounds may undergo extensive modification at advanced stages of oxidation, and the test is useful as a measure of lipid oxidation only during the initial stages of oxidation. The test is most useful for detecting incipient oxidation of lipids rich in methylene interrupted, three or more, double bonds. The color formation is empirical, as the color yield from each fatty acid varies.

The test has been applied by many workers directly to a lipid-containing material without prior extraction of fat, the red pigment being subsequently extracted and measured. Such methods are, however, open to criticism as a variety of substances, other than the oxidation products of lipids, may give misleading color reactions with TBA. To overcome that difficulty, it may be advisable to determine the color reaction after purification of TBA-reactive compounds by steam distillations, and determination of the oxidation products in the distillate. The TBA reagent is unstable in the presence of acid, peroxide, and heat, conditions under which it is generally used. A procedure in which an aqueous extract of a food or emulsified fat is allowed to react with TBA at room temperature for 15 hr without acid (in place of 100°C for 10 to 50 min) has been recommended (Tarladgis et al. 1962; Lea 1962). Several improvements of the method were described by Yu and Sinnhuber (1967).

BIBLIOGRAPHY

AHRENS, E. H., and CRAIG, L. C. 1952. Separation of the higher fatty acids. J. Biol. Chem. *195*, 299–310.

ALEXANDER, D. E., SILVELA, L. S., COLINS, F. I., and RODGERS, R. C. 1967. Analysis of oil content of maize by wide-line NMR. J. Am. Oil Chemists' Soc. *44*, 555–558.

ALLEN, C. F., GOOD, P., DAVIS, H. F., CHISUM, P., and FOWLER, S. D. 1966. Methodology for separation of plant lipids and application to spinach leaf and chloroplast lamellae. J. Am. Oil Chemists' Soc. *43*, 223–231.

ALLEN, R. R. 1955. Determination of unsaturation. J. Am. Oil Chemists' Soc. *32*, 671–674.

ANDERSON, J. A., and ALCOCK, H. W. 1954. Storage of Cereal Grains and their Products. Am. Assoc. Cereal Chemists, St. Paul, Minn.

ANON. 1950. Deutsche Gesellschaft fur Fettwissenschaft. D. G. F. Einheits-methoden. Munster, Germany.

ANON. 1954. International Union of Pure and Applied Chemistry. Standard Methods for the Analysis of Oils and Fats. Paris, France.

ANON. 1960. Standard Methods for the Examination of Dairy Products, 11th Edition. Am. Public Health Assoc., New York.

ANON. 1962. Cereal Laboratory Methods, 7th Edition. Am. Assoc. Cereal Chemists, St. Paul, Minn.

ANON. 1964. Official and Tentative Methods of the American Oil Chemists Society, 2nd Edition, Amer. Oil Chemists Soc., Chicago.

ANON. 1965. Official Methods of Analysis of the Association of Official Anal. Chemists, 10th Edition. Assoc. Offic. Anal. Chemists, Washington, D.C.

ANON. 1967. IUPAC-IUB Commission on Biochemical Nomenclature (CBN). The nomenclature of lipids. European J. Biochem. *2*, 127–131.

BABICHEVA, O. I., and GORELIK, L. D. 1968. Determination of the fat content in canned meat and vegetable foods by a refractometric method. Konserv. Ovoshchesush. Prom. *23*, 35–37 (Chem. Abstr. *68*, 113376z).

BAKER, D. 1961. A colorimetric method for determining fat acidity in grain. Cereal Chem. *38*, 47–50.

BAKER, D. 1964. A colorimetric method for determining free fatty acids in vegetable oils. J. Am. Oil Chemists' Soc. *41*, 21–22.

BAKER, D., NEUSTADT, M. H., and ZELENY, L. 1959. Relationships between fat acidity values and types of damage in grain. Cereal Chem. *36*, 308–311.

BARNARD, D., and HARGRAVE, K. R. 1951. Analytical studies concerned with the reactions between organic peroxides and thio-esters. I. Analysis of organic peroxides. Anal. Chim. Acta *5*, 476–488.

BARRON, E. J., and HANAHAN, D. J. 1958. Observations on the silicic acid chromatography of the neutral lipides of rat livers, beef liver, and yeast. J. Biol. Chem. *231*, 493–503.

BAUMAN, L. F., CONWAY, T. F., and WATSON, S. A. 1963. Heritability of variations in oil content of individual oil kernels. Science *139*, 498–499.

BAUMANN, L. A. 1959. Evaluating refined cottonseed oils in storage. J. Am. Oil Chemists' Soc. *36*, 28–34.

BERNHEIM, F., BERNHEIM, M. L. C., and WILBUR, K. M. 1947. The reaction between thiobarbituric acid and the oxidation products of certain lipides. J. Biol. Chem. *174*, 257–264.

BIGGS, D. A. 1967. Milk analysis with the infrared milk analyzer. J. Dairy Sci. *50*, 799–803.

BLANK, M. L., SCHMIT, J. A., and PRIVETT, O. S. 1964. Quantitative analysis of lipids by thin-layer chromatography. J. Am. Oil Chemists' Soc. *41*, 371–376.

BLIGH, E. G., and DYER, W. J. 1959. A rapid method of total lipid extraction and purification. Can. J. Biochem. Physiol. *37*, 911–917.

BLOKSMA, A. H. 1966. Extraction of flour by mixtures of butanol-1 and water. Cereal Chem. *43*, 602–622.

BLOOR, 1925. Biochemistry of fats. Chem. Rev. *2*, 243–300.

BOEKENOOGEN, H. A. 1964. Analysis and Characterization of Oils, Fats, and Fat Products. Interscience Publishers, New York.

BOOTH, V. H. 1965. Mapping plant lipids by paper chromatography. Chromatog. Rev. *7*, 98–118.

BRAUN, W. Q. 1955. Dilatometric measurements. J. Am. Oil Chemists' Soc. *32*, 633–637.

BROWN, F. 1952. The estimation of vitamin E. Separation of tocopherol mixtures occurring in natural products by paper chromatography. Biochem. J. *51*, 237–239.

CAMPEN, W. A. C., and GEERLING, H. 1954. Fast and simple determination of the amount of crude fat and fatty acids in animal feed—a method of general application. Chem. Weekblad *50*, 385–393 (Chem. Abstr. *48*, 10952 i).

CANNON, J. H. 1964. Sterol acetate test for foreign fats in dairy products. J. Assoc. Offic. Agr. Chemists *47*, 577–580.

CASINOVI, C. G. 1963. A comprehensible bibliography of organic substances by countercurrent distribution. Chromatog. Rev. *5*, 161–207.

CHAPMAN, D. 1960. Infrared spectroscopic characterization of glycerides. J. Am. Oil Chemists' Soc. *37*, 73–77.

CHAPMAN, D. 1965. Infrared spectroscopy of lipids. J. Am. Oil Chemists' Soc. *42*, 353–371.

CHAPMAN, D., RICHARDS, R. E., and YORKE, R. W. 1960. A nuclear magnetic resonance study of the liquid/solid content of margarine fat. J. Am. Oil Chemists' Soc. *37*, 243–246.

COCKS, L. V., and VAN REDE, C. 1966. Laboratory Handbook for Oil and Fat Analysts. Academic Press, New York.

COLLINS, F. I., ALEXANDER, D. E., RODGERS, R. C., and SILVELLA, L. S. 1967. Analysis of oil content of soybeans by wide-line NMR. J. Am. Oil Chemists' Soc. *44*, 708–710.

CONWAY, T. F. 1960. Proc. Symposium on High-Oil Corn. Dept. Agronomy, Univ. Illinois. *Cited by* Alexander *et al.* 1967. Analysis of oil content of maize by wide-line NMR. J. Am. Oil Chemists' Soc. *44*, 555–558.

CONWAY, T. F., and JOHNSON, L. F. 1969. Nuclear magnetic resonance measurement of oil "unsaturation" in single viable corn kernels. Science *164*, 827–828.

CRAIG, L. C. 1944. Identification of small amounts of organic compounds by distribution studies. II. Separation by countercurrent distribution. J. Biol. Chem. *155*, 519–534.

CROPPER, F. R., and HEYWOOD, A. 1953. Analytical separation of the methyl esters of the C_{12}–C_{22} fatty acids by vapor-phase chromatography. Nature *172*, 1101–1102.

DAHLE, L. K., HILL, E. G., and HOLMAN, R. T. 1962. The thiobarbituric acid reaction and the autoxidation of polyunsaturated fatty acid methyl esters. Arch. Biochem. Biophys. *98*, 253–261.

DEUEL, H. J., JR. 1951. The Lipids, Their Chemistry and Biochemistry, Vol. 1. Interscience Publishers, New York.

DUGAN, L., Jr. 1955. Stability and rancidity. J. Am. Oil Chemists' Soc. *32*, 605–609.

DUTTON, H. J. 1954. Countercurrent fractionation of lipids. Progr. Chem. Fats Lipids *2*, 292–325.

DUTTON, H. J. 1955. The analysis of lipids by countercurrent distribution. J. Am. Oil. Chemists' Soc. *32*, 652–659.

DUTTON, H. J. 1961. Some applications of mass spectrometry to lipid research. J. Am. Oil Chemists' Soc. 38, 660–664.

DYACHENKO, P., and SAMSONOV, Y. 1964. Infrared spectra of dried milk and its components. Molochn. Prom. 25, No. 7, 11–13 (Chem. Abstr. 61, 15263 f).

EISNER, J., and FIRESTONE, D. 1963. Gas chromatography of unsaponifiable matter. II. Identification of vegetable oils by their sterols. J. Assoc. Offic. Agr. Chemists 46, 542–550.

EISNER, J., WONG, N. P., FIRESTONE, D., and BOND, J. 1962. Gas chromatography of unsaponifiable matter. I. Butter and margarine sterols. J. Assoc. Offic. Agr. Chemists 45, 337–342.

ENTENMANN, C. 1961. The preparation of tissue lipid extracts. J. Am. Oil Chemists' Soc. 38, 534–538.

ETTINGER, C. L., MALANOSKI, A., and KIRSCHENBAUM, H. 1965. Detection and estimation of animal fats in vegetable oils by gas chromatography. J. Assoc. Offic. Agr. Chemists 48, 1186–1191.

FEHR, W. F., COLLINS, F. I., and WEBER, C. R. 1968. Evaluation of methods for proteins and oil determination in soybean seed. Crop Sci. 8, 47–49.

FELDMAN, G. D. 1967. Human occular lipids: their analysis and distribution. Surv. Ophthalmol. 12, 207–243.

FERREN, W. P., and MORSE, R. E. 1963. Wide-line nuclear magnetic resonance determination of liquid-solid content of soybean oil at various degrees of hydrogenation. Food Technol. 17, No. 8, 112–114.

FITZGERALD, J. W., RINGS, G. R., and WINDER, W. C. 1961. Ultrasonic method for measurement of fluids-nonfat and milk fat in fluid milk. J. Dairy Sci. 44, 1165.

FOLCH, J., LEES, M., and STANLEY, G. H. S. 1957. A simple method for the isolation and purification of total lipides from animal tissues. J. Biol. Chem. 226, 497–509.

FONTELL, K., HOLMAN, R. T., and LAMBERTSEN, G. 1960. Some new methods for separation and analysis of fatty acids and other lipids. J. Lipid Res. 1, 391–404.

FREEMAN, C. P., and WEST, D. 1966. Complete separation of lipid classes on a single thin-layer plate. J. Lipid Res. 7, 324–327.

FREEMAN, N. K. 1957. Infrared spectroscopy of serum lipides. Ann. N. Y. Acad. Sci. 69, 131–144.

HANGAARD, G., and PETTINATI, J. D. 1959. Photometric milk fat determination. J. Dairy Sci. 42, 1255–1275.

HENICK, A. S., BENCA, M. F., and MITCHELL, J. H. 1954. Estimating carbonyl compounds in rancid fats and foods. J. Am. Oil Chemists' Soc. 31, 88–91.

HILDITCH, T. P., and WILLIAMS, P. N. 1964. The Chemical Constitution of Natural Fats, 4th Edition. Chapman and Hall, London.

HIRSCH, J., and AHRENS, E. H. JR. 1958. The separation of complex lipide mixtures by the use of silicic acid chromatography. J. Biol. Chem. 233, 311–320.

HOLMAN, R. T., LUNDBERG, W. O., and BURR, G. O. 1945. Spectrophotometric studies of the oxidation of fats. III. Ultraviolet absorption spectra of oxidized octadecatrienoic acids. J. Am. Chem. Soc. 67, 1390–1394.

HOLMAN, R. T., and RAHM, J. J. 1966. Analysis and characterization of polyunsaturated fatty acids. Progr. Chem. Fats Lipids 9, No. 1, 15–90.

HORNING, E. C. and VANDENHEUVEL, W. J. A. 1964. Gas chromatography in lipid investigations. J. Am. Oil Chemists' Soc. 41, 707–716.

HORNSTEIN, I., CROWE, P. F., and RUCK, J. B. 1967. Separation of muscle lipids into classes by nonchromatographic techniques. Anal. Chem. 39, 352–354.

HUNT, W. H., NEUSTADT, M. H., HART, J. R., and ZELENY, L. 1952. A rapid dielectric method for determining the oil content of soybeans. J. Am. Oil Chemists' Soc. 29, 258–261.

HUNT, W. H., NEUSTADT, M. H., SHURKUS, A. A., and ZELENY, L. 1951. A simple iodine-number refractometer for testing flaxseed and soybeans. J. Am. Oil Chemists' Soc. 28, 5–8.

JAMES, A. T., and MARTIN, A. J. P. 1952. Gas-liquid partition chromatography; the separation and microestimation of volatile fatty acids from formic acid to dodecanoic acid. Biochem. J. 50, 679–690.

JAMIESON, G. R., and REID, E. H. 1965. The analysis of oils and fats by gas chromatography. J. Chromatog. 17, 230–237.

JOHNSON, L. F., and SHOOLERY, J. N. 1962. Determination of unsaturation and average molecular weight of natural fats by nuclear magnetic resonance. Anal. Chem. 34, 1136–1139.

KATC, I., KEENEY, M., and BASSETTE, R. 1959. Colorimetric determination of fat in milk and the saponification number of a fat. J. Dairy Sci. 42, 903–906.

KAUFMAN, F. L. 1964. Infrared spectroscopy of fats and oils. J. Am. Oil Chemists' Soc. 41, No. 8, 4, 6, 21, 38, 42.

KAUFMANN, H. P. 1958. Analysis of Fat and Fat Products, Vols. 1 and 2. Springer-Verlag, Berlin. (German)

KAUFMANN, H. P., MANKEL, G., and LEHMANN, K. 1961. Gas chromatography of fatty compounds. I. General survey. Fette, Seifen, Anstrichmittel 63, 1109–1116. (Chem. Abstr. 56, 11729 h.)

KAUFMANN, H. P., SEHER, A., and MANKEL, G. 1962. Gas chromatography of fats. II. Quantitative applications. Fette, Seifen, Anstrichmittel 64, 501–509. (Chem. Abstr. 57, 7399 b.)

KEENEY, M. 1956. A survey of United States butterfat constants. II. Butyric acid. J. Assoc. Offic. Agr. Chemists 39, 212–225.

KESSLER, G., and LEDERER, H. 1965. Technicon Symp. Proc. N.Y. Cited by Levine, 1967.

KING, J. D. 1966. N. M. R. analysis of meat composition. Proc. Meat Ind. Res. Conf. Chicago, 149–157. (Chem. Abstr. 67, 115886 u.)

KOHN, H. I., and LIVERSEDGE, N. 1944. A new aerobic metabolite whose production by brain is inhibited by apomorphine, emetine, ergotamine, adrenaline, and menadione. J. Pharmacol. 82, 292–300.

KOHN, R. 1964. Application of gas chromatography in analyses of foods. Qual. Plant. Mat. Veg. 11, 150–167. (German)

KOHN, R. 1965. Application of infrared spectroscopy in food analysis. I. Infrared spectroscopy in the intermediate range of 3 to 15 μm. Z. Lebensm. Untersuch. Forsch. 129, 28–40. (German)

KOHN, R., and LAUFER-HEYDENREICH, S. 1966. Application of infrared spectroscopy in food analysis. Z. Lebensm. Untersuch. Forsch. 129, 92–97. (German)

KUKSIS, A. 1966. Quantitative lipid analysis by combined thin-layer and gas-liquid chromatographic systems. Chromatog. Rev. 8, 172–207.

KUKSIS, A., and McCARTHY, M. J. 1964. Triglyceride gas chromatography as a means of detecting butterfat adulteration. J. Am. Oil Chemists' Soc. 41, 17–21.

KUMMEROW, F. A. 1960. Fats in human nutrition. J. Am. Oil Chemists' Soc. 37, 503–509.

KURTZ, F. E. 1965. The lipids of milk-composition and properties. In Fundamentals of Dairy Chemistry, B. H. Webb and A. H. Johnson (Editors). Avi Publishing Co., Westport, Conn.

KWON, T. W., and OLCOTT, H. S. 1966. Thiobarbituric-acid-reactive substances from autoxidized or ultraviolet irradiated unsaturated fatty esters and squalene. J. Food Sci. 31, 552–557.

LAPIN, G. R., and CLARK, L. C.   1951.   Colorimetric method for determination of traces of carbonyl compounds.   Anal. Chem. 23, 541–543.

LEA, C. H.   1931.   Effect of light on the oxidation of fats.   Proc. Roy. Soc. (London), Ser. 108B, 175–179.

LEA, C. H.   1946.   The determination of the peroxide values of edible fats and oils: the iodometric method.   J. Soc. Chem. Ind. 65, 286–290.

LEA, C. H.   1956.   Biochemical Problems of Lipids.   Butterworths Scientific Publishing Co., London.

LEA, C. H.   1962.   The oxidative deterioration of food lipids.   In Symposium on Foods: Lipids and Their Oxidation, H. W. Schultz, E. A. Day, and R. O. Sinnhuber (Editors).   Avi Publishing Co., Westport, Conn.

LEE, C. F., AMBROSE, M. E., and SMITH, P., JR.   1966.   Determination of lipids in fish meal.   J. Assoc. Offic. Agr. Chemists 49, 946–949.

LEEMANS, F. A., and McCLOSKEY, J. M.   1967.   Combination gas chromatography-mass spectrometry.   J. Am. Oil Chemists' Soc. 44, 11–17.

LESTER, A. L.   1963.   Federation Proc. 22, 415.   Cited by Wells and Dittmer, 1966.   A microanalytical technique for the quantitative determination of twenty-four classes of brain lipids.   Biochem. 5, 3405–3408.

LEVINE, J. B.   1967.   Recent advances in automated lipid analysis.   J. Am. Oil Chemists' Soc. 44, 95–98.

LEVINE, J. B., and ZAK, B.   1964.   Automated determination of serum total cholesterol.   Clin. Chim. Acta 10, 381–384.

LEVOWITZ, D.   1967.   Determination of fat and total solids in dairy products.   In Laboratory Analysis of Milk and Milk Products.   U.S. Dept. Health, Educ., Welfare, Cincinnati.

LOFLAND, H. B.   1964.   A semiautomated procedure for the determination of triglycerides in serum.   Anal. Biochem. 9, 393–400.

MANGOLD, H. K.   1961.   Thin-layer chromatography of lipids.   J. Am. Oil Chemists' Soc. 38, 708–727.

MANGOLD, H. K.   1964.   Thin-layer chromatography of lipids.   J. Am. Oil Chemists' Soc. 41, 762–773.

MANGOLD, H. K., SCHMIDT, H. H. O., and STAHL, E.   1964.   Thin-layer chromatography (TLC).   Methods Biochem. Analy. 12, 393–451.

MARINETTI, G. V.   1962.   Chromatographic separation, identification, and analysis of phosphatides.   J. Lipid Res. 3, 1–20.

MARINETTI, G. V.   1964.   Chromatographic analysis of polar lipids on silicic acid impregnated paper.   In New Biochemical Separations, A. T. James, and L. J. Morris (Editors).   D. Van Nostrand Co., London.

MECHAM, D. K. and MOHAMMAD, A.   1955.   Extraction of lipids from wheat products.   Cereal Chem. 32, 405–415.

MEHLENBACHER, V. C.   1958.   Standard methods in the fat and oil industry.   Prog. Chem. Fats Lipids 5, 1–29.

MEHLENBACHER, V. C.   1960.   Analysis of Fats and Oils.   The Garrard Press, Champaign, Ill.

MITCHELL, D. J.   1967.   Collaborative studies of methods for butterfat in homogenized and chocolate milk.   J. Assoc. Offic. Anal. Chemists 50, 537–541.

MITCHELL, J. H., KRAYBILL, H. R., and ZSCHEILE, F. P.   1943.   Quantitative spectrum analysis of fats.   Ind. Eng. Chem. (Anal. Ed.) 15, 1–3.

MIWA, T. K., KWOLEK, W. F., and WOLFF, I. A.   1966.   Quantitative determination of unsaturation in oils by using an automatic-titrating hydrogenator.   Lipids 1, 152–157.

MOJONNIER, T., and TROY, H. C.   1925.   Technical Control of Dairy Products.   Mojonnier Bros. Co., Chicago.

MORRIS, L. J.   1964.   Specific separations by chromatography on impregnated adsorbents.   In New Biochemical Separations, A. T. James, and L. J. Morris (Editors).   D. Van Nostrand Co., London.

MORRIS, L. J. 1966. Separation of lipids by silver ion chromatography. J. Lipid Res. 7, 717–732.

MURPHY, M. F. and McGANN, T. C. A. 1967. Investigations on the use of the Miko-tester for routine estimation of fat content in milk. J. Dairy Res. 34, 65–72.

NADJ, L. J., and WEEDEN, D. G. 1966. Refractometric estimation of total fat in chocolate-type products. Anal. Chem. 38, 125–126.

NICHOLS, B. W. 1964. The separation of lipids by thin-layer chromatography. Lab. Practice 13, 299–305.

NUTTER, L. J., and PRIVETT, O. S. 1968. An improved method for the quantitative analysis of lipid classes via thin-layer chromatography employing charring and densitometry. J. Chromatog. 35, 519–525.

O'CONNOR, R. T. 1955. Ultraviolet absorption spectroscopy. J. Am. Oil Chemists' Soc. 32, 616–624.

O'CONNOR, R. T. 1961A. Near-infrared absorption spectroscopy—a new tool for lipid analysis. J. Am. Oil Chemists' Soc. 38, 641–648.

O'CONNOR, R. T. 1961B. Recent progress in the application of infrared absorption spectroscopy to lipid chemistry. J. Am. Oil Chemists' Soc. 38, 648–659.

PADLEY, F. B. 1964. Thin-layer chromatography of lipids. In Thin-layer Chromatography, G. B. Marini-Bettolo (Editor). Elsevier Publishing Co., Amsterdam, The Netherlands.

PATTON, S., KEENEY, M., and KURTZ, G. W. 1951. Compounds producing the Kreis color reaction with particular reference to oxidized milk fat. J. Am. Oil Chemists' Soc. 28, 391–393.

PATTON, S., and KURTZ, G. W. 1951. 2-Thiobarbituric acid as a reagent for detecting milk-fat oxidation. J. Dairy Sci. 34, 669–674.

PELICK, N., WILSON, T. L., MILLER, M. E., ANGELONI, F. M., and STEIN, J. M. 1965. Some practical aspects of thin-layer chromatography of lipids. J. Am. Oil Chemists' Soc. 42, 393–399.

PINTO, A. F., and ENAS, J. D. 1949. Rapid method of copra analysis and its application to the various oil seeds. J. Am. Oil Chemists' Soc. 26, 723–730.

PITT, G. A. J., and MORTON, R. A. 1957. Ultraviolet spectrophotometry of fatty acids. Progr. Chem. Fats Lipids 4, 227–278.

POHLE, W. D., GREGORY, R. L., and TAYLOR, J. R. 1962. A comparison of several analytical techniques for prediction of relative stability of fats and oils to oxidation. J. Am. Oil Chemists' Soc. 39, 226–229.

POHLE, W. D., GREGORY, R. L., and VAN GIESSEN, B. 1963. A rapid bomb method for evaluating the stability of fats and shortenings. J. Am. Oil Chemists' Soc. 40, 603–605.

POHLE, W. D. et al. 1964. A study of methods for evaluation of the stability of fats and shortenings. J. Am. Oil Chemists' Soc. 41, 795–798.

POMERANZ, Y. 1965. Thin layer chromatography in studies of cereal lipids. Qualitas Plant. Mater. Vegetabiles 12, 322–341.

POMERANZ, Y., and SHELLENBERGER, J. A. 1966. The significance of fatty acids in cereals. Am. Miller Processor 94, No. 6, 9–11.

POOL, P. O. 1931. Rancidity and stability in shortening products. Oil Fat Ind. 8, 331–336.

PRIVETT, O. S., and BLANK, M. L. 1964. Basic techniques and research applications of thin-layer chromatography. Offic. Dig. 36, 454–463.

PRIVETT, O. S., BLANK, M. L., CODDING, D. W., and NICKELL, E. C. 1965. Lipid analysis by quantitative thin-layer chromatography. J. Am. Oil Chemists' Soc. 42, 381–393.

RODEGHIER, A. A. 1955. Determination of impurities in fats and oils. J. Am. Oil Chemists' Soc. 32, 578–581.

ROUSER, G., GALLI, C., LIEBER, E., BLANK, M. L., and PRIVETT, O. S. 1964. Analytical fractionation of complex lipid mixtures: DEAE cellulose column chromatography combined with quantitative thin layer chromatography. J. Am. Oil Chemists' Soc. *41*, 836–840.

ROUSER, G., KRITCHEVSKY, G., GALLI, C., and HELLER, D. 1965. Determination of polar lipids: quantitative column and thin layer chromatography. J. Am. Oil Chemists' Soc. *42*, 215–227.

ROUSER, G., KRITCHEVSKY, G., HELLER, D., and LIEBER, E. 1963. Lipid composition of beef brain, beef liver, and the sea anemone: two approaches to quantitative fractionation of complex lipid mixtures. J. Am. Oil Chemists' Soc. *40*, 425–454.

RYHAGE, R., and STENHAGEN, E. 1960. Mass spectrometry in lipid research. J. Lipid Res. *1*, 361–390.

SAGER, O. S., SANDERS, G. P., NORMAN, G. H., and MIDDLETON, M. B. 1955. A detergent test for the milk fat content of dairy products. J. Assoc. Offic. Agr. Chemists *38*, 931–940.

SCHAIN, P. 1949. The use of detergents for quantitative fat determination. I. Determination of fat in milk. Science *110*, 121–122.

SCHMID, H. H. O., BAUMANN, W. J., CUBERO, J. M., and MANGOLD, H. K. 1966. Fractionation of lipids by successive adsorption and argentation chromatography on adjacent layers. Biochim. Biophys. Acta *125*, 189–196.

SCHOBER, B. 1967. Application of a refractometric method for lipid determination in fish and fish products. I. Determination of the fat content in herring. Fischereiforschung *5*, 121–124. (Chem. Abstr. *69*, 1825 j.)

SCHWARZ, H. P., DREISBACH, L., CHILDS, R., and MASTRANGELO, S. V. 1957. Infrared studies of tissue lipids. Ann. N. Y. Acad. Sci. *69*, 116–130.

SIAKOTOS, A. N., and ROUSER, G. 1965. Analytical separation of nonlipid water soluble substances and gangliosides from other lipids by dextran column chromatography. J. Am. Oil. Chemists' Soc. *42*, 913–919.

SKIPSKI, V. P., GOOD, J. J., BARCLAY, M., and REGGIO, R. B. 1968. Quantitative analysis of simple lipid classes by thin-layer chromatography. Biochim. Biophys. Acta *152*, 10–19.

SMITH, H. M. 1955. Melting point, solidification, and consistency. J. Am. Oil Chemists' Soc. *32*, 593–595.

STERN, I., and SHAPIRO, B. 1953. A rapid and simple method for the determination of esterified fatty acids and for total fatty acids in blood. J. Clin. Pathol. *6*, 158–160.

STILLMAN, R. C. 1955. Bleach and color methods. J. Am. Oil Chemists' Soc. *32*, 587–593.

TARLADGIS, B. G., PEARSON, A. M., and DUGAN, L. R. 1962. The chemistry of the 2-thiobarbituric acid test for the determination of oxidative rancidity in foods. I. Some important side reactions. J. Am. Oil. Chemists' Soc. *39*, 34–39.

TAYLOR, J. R., POHLE, W. D., and GREGORY, R. J. 1964. Measurement of solids in triglycerides by using nuclear resonance spectroscopy. J. Am. Oil Chemists' Soc. *41*, 177–180.

TRAPPE, W. 1940. Separation of biological fats from natural mixtures by means of adsorption columns. I. The eluotropic series of solvents. Biochem. Z. *305*, 150–161. (Chem. Abstr. *35*, 477[7].)

TREVELYAN, W. E. 1966. Determination of some lipid constituents of baker's yeast. J. Inst. Brewing *72*, 184–192.

TROENG, S. 1955. Oil determination of oilseed. Gravimetric routine method. J. Am. Oil Chemists' Soc. *32*, 124–126.

TSEN, C. C., LEVI, I., and HLYNKA, I. 1962. A rapid method for the extraction of lipids from wheat products. Cereal Chem. *39*, 195–203.

WELLS, M. A., and DITTMER, J. C. 1963. The use of Sephadex for the removal of nonlipid contaminants from lipid extracts. Biochem. 2, 1259–1263.

WELLS, M. A., and DITTMER, J. C. 1966. A microanalytical technique for the quantitative determination of twenty-four classes of brain lipids. Biochem. 5, 3405–3408.

WHEELER, D. H. 1954. Infrared absorption spectroscopy in fats and oils. Progr. Chem. Fats Lipids 2, 268–291.

WHITLEY, R. W., and ALBURN, H. E. 1964. Proc. Technicon Symp., Ardsley, N.Y. Paper 65. Cited by Levine, J. B. 1967. Recent advances in automated lipid analysis. J. Am. Oil Chemists' Soc. 44, 95–98.

WIJS, J. J. A. 1929. The Wijs method as the standard for iodine absorption. Analyst 54, 12–14.

WILBUR, K. M., BERNHEIM, F., and SHAPIRO, O. W. 1949. The thiobarbituric acid reagent as a test for the oxidation of unsaturated fatty acids by various reagents. Arch. Biochem. 24, 305–313.

WILLIAMS, K. A. 1950. Oils, Fats and Fatty Foods. Churchill Publishing Co., London, England.

WREN, J. J. 1961. Chromatography of lipids on silicic acid. Chromatog. Rev. 3, 111–133.

YU, T. C., and SINNHUBER, R. O. 1967. An improved 2-thiobarbituric acid (TBA) procedure for the measurement of autooxidation in fish oils. J. Am. Oil Chemists' Soc. 44, 256–258.

ZEHREN, V. L., and JACKSON, H. C. 1956. A survey of United States butterfat constants. I. Reichert-Meissl, Polenske and refractive index values. J. Assoc. Offic. Agr. Chemists 39, 194–212.

ZELENY, L., and COLEMAN, D. A. 1938. Acidity in cereals and cereal products, its determination and significance. Cereal Chem. 15, 580–595.

ZILVERSMIT, D. B., and DAVIS, A. K. 1960. Microdetermination of plasma phospholipides by trichloroacetic acid precipitation. J. Lab. Clin. Med. 35, 155–160.

ZIMMERMAN, D. C. 1962. The relationship between seed density and oil content in flax. J. Am. Oil Chemists' Soc. 39, 77–78.

CHAPTER 37

Nitrogenous Compounds

INTRODUCTION

The problem of providing adequate protein for an expanding world population is second only to the overall food problem. Table 37.1 (from Altschul 1962) summarizes the protein content of selected foodstuffs. Apart from their nutritional significance, proteins play a large part in the organoleptic properties in foods (Rhodes 1963; Schultz and Anglemier 1964). Proteins exert a controlling influence on the texture of foods from animal sources. Protein content of wheat and flour is considered one of the best single indices of bread-making quality. The protein test, although generally not included as a grading factor in grain standards, is accepted as a marketing factor (Pomeranz 1968).

TABLE 37.1

PROTEIN CONTENT (N × 6.25%) OF SELECTED FOODSTUFFS[1]

Animal Origin	Protein	Plant Origin	Protein
Milk		Rice, whole	7.5–9.0
Whole, dried	22–25	Rice, polished	5.2–7.6
Skimmed, dried	34–38	Wheat, flour	9.8–13.5
Beef		Corn, meal	7.0–9.4
Dried	81–90	Chick, pea	22–28
Roasted	72	Soybean	33–42
Egg		Peanut	25–28
Whole, dried	35	Walnut	15–21
Whole, dried,		Potato[2]	10–13
defatted	77	Tapioca[2]	1.3
Herring[2]	81	Alfalfa[2]	18–23
	69	Chlorella[2]	23–44
		Torula yeast[2]	38–55

[1] Unless stated otherwise, on as is basis.
[2] H_2O-free basis.

Proteins often occur in foods in physical or chemical combination with carbohydrates or lipids. The glycoproteins and lipoproteins affect the rheological properties of food solutions or have technical applications as edible emulsifiers. The aging of meat is associated with chemical changes in the proteins (Whitaker 1959). Pure native proteins have little flavor. During heating (boiling, baking, roasting) the amino acid side chains are degraded or interact with other

food components (i.e. lysine and reducing sugars), and give typical
flavors (Danehy and Pigman 1951). Excessive heating may, on the
other hand, reduce nutritive value (Rice and Beuk 1953). The role
of proteins in the processing and storage of various foods was reviewed by Feeney and Hill (1960).

The food analyst most commonly wishes to know the total protein
content of a food, even though that content is made up of a complex
mixture of proteins. At the present time, all methods of determining the total protein content of foods are empirical in nature. Isolation and direct weighing of the protein would provide an absolute
method. Such a method is sometimes used in biochemical investigation, but is completely impractical for food analysis.

The primary nutritional importance of protein is as a source of
amino acids. Twenty-four amino acids are generally thought to be
constituents of proteins. Some amino acids are essential to good
physical and mental health. Of the amino acids in food, eight are
known to be essential to man, that is, they must be supplied in the
diet to maintain growth and health. Table 37.2 summarizes the
contents of essential amino acids in proteins (Lichtfield and Sachsel
(1965). Proteins from some plant sources (i.e. cereal grains) are
deficient in certain amino acids (i.e. lysine). Deficient proteins must
be combined with those from other sources to provide an adequate
balance of the essential amino acids. Such a balance can be accomplished by a combination of wheat flour with dry skim milk or soy
flour. Detailed tables on the amino acid composition of foods were
presented by Block (1945), Orr and Watt (1957), and by the Food
and Agriculture Organization of the United Nations (Anon. 1968).
The methods of amino acid analysis of proteins were reviewed
among others by Block (1945, 1960), Dunn (1950), Martin and

TABLE 37.2
ESSENTIAL AMINO ACIDS IN PROTEINS
(Gm/100 Gm of Protein)

Amino Acid	FAO Reference	Skim Milk	Soya	Beef	Egg	Fish	Yeast
Lysine	4.2	8.6	6.8	8.3	6.3	6.6	6.8
Tryptophan	1.4	1.5	1.4	1.0	1.5	1.6	0.8
Phenylalanine	2.8	5.5	5.3	3.5	5.7	4.1	4.5
Methionine	2.2	3.2	1.7	2.8	3.2	3.0	2.6
Threonine	2.8	4.7	3.9	4.5	4.9	4.8	5.0
Leucine	4.8	11.0	8.0	7.2	9.0	10.5	8.3
Isoleucine	4.2	7.5	6.0	4.7	6.2	7.7	5.5
Valine	4.2	7.0	5.3	5.1	7.0	5.3	5.9

Synge (1945), and James and Morris (1964). The reviews give details of the preparation of samples for analysis, hydrolysis of proteins, and the separation and assay of amino acids in the hydrolysates. The latter include specific methods for certain amino acids (colorimetric, enzymatic, and microbiological) and methods for assaying all or most of the amino acids after preliminary separation (by paper, or thin-layer chromatography, column chromatography on ion exchange resins, and gas chromatography). The most powerful methods of quantitative amino acid analysis in protein hydrolysates are based on ion exchange chromatography (Spackman *et al.* 1958), and more recently by gas chromatography (Gehrke and Stalling 1967).

The food analyst is sometimes interested in knowing the content of a particular protein in a mixture. The classical investigations of Osborne (1907) on the differences in solubility in various solvents of wheat proteins still provide a useful separation and characterization tool. Changes in protein solubility are useful in determining the length and severity of heat treatment in processing milk, soy products, or animal products. The milk proteins of chocolate can be determined from the amount of proteins precipitated from an oxalate solution with tannic acid, provided the chocolate has not been subjected in processing to elevated temperatures (Motz 1968). Finally, some of the more modern tools of fractionation such as adsorption, partition, exclusion, and exchange chromatography, and particularly electrophoresis (paper, starch gel, and polyacrylamide) can help ascertain the purity of an isolated protein and identify the source of a protein. The usefulness of such techniques was demonstrated by the identification of fish species by starch gel electrophoresis of protein extracts (Thompson 1962), or paper chromatographic characterization of milk proteins of various animal species (Hilpert and Enkelmann 1963). Stark (1962) described a method to classify plant nitrogenous compounds according to their reactivity to ion exchange resins and various eluants. The procedure is useful in determining nitrogen distribution, removing undesirable substances prior to other chromatographic procedures, or in isolating specific compounds. For a detailed study, the protein generally must be extracted from natural sources by maceration (with various homogenizers, mills, or mechanical pestles), by disruption of cells (by alternate freezing and thawing, ultrasonic vibration, or fine grinding by colloidal milling), digestion of enzymes, or by solvent extraction (dilute salts, aqueous or anhydrous organic solvents, or surface active agents). The methods of isolation and separation of proteins were described by Keller and Block (1960), James and Morris (1964), and Pfleiderer (1967).

The significance of nonprotein, organic, nitrogenous compounds in foods has been appreciated only in recent years. They include amino acids, amines, amides, quarternary nitrogen compounds, purines, and pyrimidines. They contribute to the nutritional value, flavor, color (especially in baked or roasted products), and other important food attributes. They provide a source of nutrients and growth factors that are important in malting, brewing, and panary fermentation. On the other hand, excessively high levels of free amino acids may result from proteolytic degradation of proteins in cereal grains stored at elevated moisture levels and temperatures. Similarly, degradation products of animal proteins are indices of incipient deterioration, and are used to ascertain storability and soundness of foods. Inorganic nitrogenous compounds (ammonia, nitrate, and nitrite) are determined to establish the sanitary status of foods, follow aging and processing of cheese, or ascertain the absence of excessive amounts of undesirable pickling components in processed meats. They are determined in many investigations of nitrogen metabolism in animals and plants.

TOTAL PROTEIN CONTENT

The most common procedure for a protein assay depends on determining a specific element or group in the protein, and calculating the protein content by using an experimentally established factor. Methods based on the analysis for constituents of proteins include those for determining carbon or nitrogen, certain amino acids, or the peptide linkage. In some proteins, certain constituents (iron in hemoglobin, iodine in thyroglobulin) can serve as a basis for protein assay. In all of the above methods, it is assumed that the constituent determined is present entirely in the protein fraction. Thus, any nonprotein carbon-containing matter must be removed if the protein content is to be determined from the carbon content; and if the Kjeldahl method is used, protein-nitrogen only should be measured. The common practice of estimating the protein content of foods from the total nitrogen assay is, therefore, not always correct. The presence of nonprotein nitrogen compounds is as, a rule, generally small compared to the protein content of most sound foods.

Elementary Analysis

Carbon analysis has several advantages for determining the protein content of foods. The digestion can be accomplished easier than for a nitrogen determination, and a high percentage of carbon minimizes experimental error and provides a relatively constant conversion factor. The difficulty of a complete and quantitative

separation of protein from nonprotein carbon-containing components
is practically unsurmountable.

A nitrogen determination for a protein assay is the most commonly
used procedure. It is generally assumed that a mixture of pure
proteins will contain 16% nitrogen. Thus the protein content of a
sample is obtained by multiplying the determined nitrogen by the
factor 6.25 = (100/16). For the approximate analysis of foods containing an unknown distribution of proteins of unknown composition,
this is a practical and widely accepted procedure. Much confusion
has resulted in reporting protein content: protein levels above
100% in pure or highly concentrated protein fractions, and differences among laboratories orginating from the use of various conversion factors. Much of the confusion could be eliminated by reporting the nitrogen rather than the calculated protein content. The food
industry and trade will probably be reluctant to make such a
change and will continue to report the calculated protein content.

Yet the inherent limitations of the procedure must be realized; the
results are affected by nonprotein nitrogen, the nitrogen content of a
particular protein mixture is seldom known precisely, and the methods of a nitrogen determination are wrought with some difficulties.
For practical purposes, unless shown to the contrary, the effects of
nonprotein nitrogen can be assumed of little consequence. The general factor of 6.25 is used for most foods. For wheat, milk, and
gelatin the factors of 5.70, 6.38, and 5.55, respectively, are used
(Jones 1931). The availability of better methods for protein isolation and characterization (including amino acid composition) provides the basis for a continuous reexamination of the conversion
factors (Tkachuk 1966). Although extremes in nitrogen content
range from 4.2% in beta lipoproteins to 30% in protamine, most
foods contain about 16% nitrogen. Difficulties in the nitrogen
determination are outlined in the following discussion of the two
main methods used in food analysis, the Kjeldahl method, and the
Dumas procedure.

The Kjeldahl Method.—The Danish investigator, Kjeldahl,
worked out in 1883 a method for determining organic nitrogen in
his studies on protein changes in grain used in the brewing industry.
Since the first publication of Kjeldahl, the method has undergone
many changes (Bradstreet 1940, 1965). Basically, the sample is
heated in sulfuric acid and digested till the carbon and hydrogen are
oxidized, and the protein nitrogen is reduced and transformed into
ammonium sulfate. Then concentrated sodium hydroxide is added
and the digest heated to drive off the liberated ammonia into a

known volume of a standard acid solution. The unreacted acid is determined and the results are transformed, by calculation, into a percentage of protein in the original sample.

Kjeldahl originally used potassium permanganate for the oxidation, but this was discontinued as the results were unsatisfactory. In 1885, Wilforth found that a digestion with sulfuric acid was accelerated by adding some catalysts. Gunning in 1889 suggested adding potassium sulfate to raise the boiling point of the digestion mixture to shorten the reaction. The test is, therefore, generally known as the Kjeldahl-Wilforth-Gunning method.

Various factors are known to influence the completeness and speed of the conversion of protein-nitrogen into ammonia by the sulfuric acid digestion. Thus, in some proteins it is more difficult to convert the organic nitrogen to ammonia. Histidine and tryptophan-rich proteins generally require long or harsh digestion conditions. Excessive ratios of potassium or sodium sulfate (added to raise the boiling point) to acid may result in heat decomposition and the loss of ammonia. Generally, digestion temperatures of 370° to 410°C are best.

Nearly all of the likely elements of the periodic table have been tried for their effect on Kjeldahl digestion. Mercury, copper, and selenium have been widely employed. Mercury is superior to copper, though an additional step is required—precipitation of mercury with sodium thiosulfate, to decompose the mercury-ammonia complex formed during digestion. The most controversial catalyst is selenium. It has a more rapid effect than mercury and unlike mercury it requires no further treatment before distillation. Nevertheless, loss of nitrogen can occur if too much selenium is used or the digestion temperature is not carefully controlled; the conditions are more critical than with copper or mercury.

In commercial practice where large numbers of samples are run daily, many time-saving devices are used (Neill 1962). In wheat or flour analysis, digesting a 1-g sample, using a known amount of standardized 0.1253 N sulfuric acid, and titrating (by an inverse-reading buret) with 0.1253 N sodium hydroxide, makes it possible to report percent protein directly from the buret reading. The use of automatic pipets for dispensing the receiver acid solutions is an advantage in testing large numbers of samples. If the catalyst digestion mixture contains mercury, a mercury precipitant is incorporated with the sodium hydroxide solution at the time of its preparation. An antibumper or pumice stone is blended with the catalyst powder mixture for a one-shot addition. Heating levels are adjusted so that digestion of 1 gm with 25 ml concentrated sulfuric acid and a catalyst mixture (potassium sulfate, mercuric oxide, and copper sulfate) is completed within 35 to 45 min.

The boric acid modification is accurate and has the advantage that only one standard solution (of titrating acid) is required. Neither the amount (about 50 ml) nor the concentration (about 4%) of boric acid in the receiving bottle have to be precise. If small samples are available (10 to 30 mg), a microKjeldahl modification employing the boric acid procedure and steam distillation of the liberated ammonia is used. Figures 37.1 and 37.2 show a battery for macroKjeldahl digestion and distillation, and units for digestion and distillation on a micro scale (Parnas 1938).

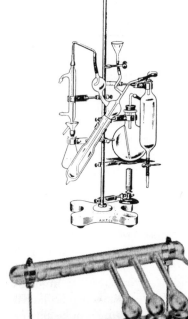

FIG. 37.1.   MICRO KJELDAHL
DIGESTION (BOTTOM),
AND DISTILLATION APPARATUS (TOP)

Courtesy A. H. Thomas (top), and La Pine Scientific Co. (bottom)

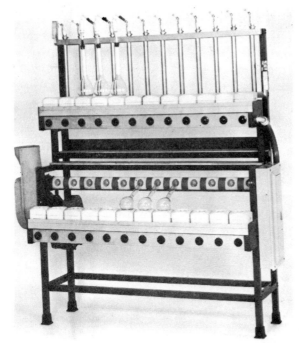

Courtesy Labconco Co.

FIG. 37.2. KJELDAHL DIGESTION AND
DISTILLATION APPARATUS

Several methods are available to determine the ammonium sulfate in the digest. The digest may be alkalized and the liberated ammonia absorbed in acid and measured titrimetrically or colorimetrically. The colorimetric method is based on the procedure of Van Slyke and Hiller (1933). It consists in reacting a solution containing ammonium ions with alkaline phenol and hypochlorite. On heating the solution an intense blue color is produced, which is closely related to that of indophenol (Mann 1963; Varley 1966). Thymol hypobromite can also be used to determine nitrogen in the form of ammonia (Glebko *et al.* 1967). For the microdetermination of ammonia, the Conway microdiffusion procedure can be used. It depends on the transfer, by diffusion, of the ammonia from an alkalized solution to a standard acid, followed by titration (Conway 1957).

Because of the high sensitivity and simplicity, Nessler's colorimetric method directly applied to Kjeldahl digests can be applied to determining nitrogen in foods (Hettrick and Whitney 1949; Williams 1964).   However, the conditions required for the optional color reaction and stability are rigorous.   The colored complex does not form a molecular solution, some cations form a precipitate with the alkali, and minute amounts of some heavy metals produce a considerable inhibition.   The addition of potassium cyanide to the Nessler reagent avoids some of the difficulties (Minari and Zilversmit 1963).

With the development of a continuous digestion module (Ferrari 1960), it is possible to determine nitrogen in biological fluids or suspensions within several min.   The digestion is followed by a colorimetric determination in a neutralized digest to which alkaline phenol and sodium hypochlorite are added.   The use of an automated Kjeldahl analyzer for the determination of nitrogen in biological material was described by Siriwardene *et al.* (1966).   The procedure is capable of handling 20 samples an hour, reduces the labor involved to a minimum, and still maintains a high degree of accuracy and reproducibility.

The Dumas Method.—In the classical Dumas procedure (1831), nitrogen was freed by pyrolysis, and the freed elemental nitrogen determined volumetrically.   In recent years, major improvements in both the pyrolysis and nitrogen determination have ensured precise and accurate analyses of nitrogen in organic materials by the method (Sternglanz and Kollig 1962).   Improved catalysts and rapid gas chromatographic methods for the nitrogen determination make it possible to make an assay on a microscale in 2 min.

Neutron Activation.—Recently, a method was developed for analyzing the nitrogen content of foods by fast neutron activation analysis (Wood 1965).   Despite the large initial cost of installation, it shows promise for the precise and rapid (about 5 min) assay of protein in various foods.

The Biuret Method

The biuret method was proposed first by Riegler in 1914.   It is based on the observation that substances containing two or more peptide bonds form a purple complex with copper salts in alkaline solutions.   The biuret procedure is simple, rapid, and inexpensive. It involves a reaction with the peptide linkage and, therefore, furnishes an accurate estimate of protein.   The Kjeldahl procedure measures total nitrogen and does not distinguish between protein and nonprotein nitrogen (Miller and Johnson 1954).

The biuret method seems less subject to criticism as the base of a colorimetric method for protein analysis than nearly any other such method (Kirk 1947).   There are practically no substances other than protein that are normally present in biological materials and that give much interference.   Although the color development with various proteins is not identical, deviations are encountered less frequently than with other colorimetric methods.   The results may be, however, affected by the presence of lipids and interfering opalescence from components of biological origin.

On the other hand, more material is required for the biuret than for most colorimetric assay methods of protein.   The biuret procedure is not an absolute method and the color must be standardized against known protein or against another method, i.e. the Kjeldahl analysis of nitrogen.

Many modifications of the originally published biuret method have been proposed for routine determinations of small amounts of protein in physiological fluids, plant and animal tissues, and chromatographically separated protein fractions.   The use of the method for the determination of protein in meats was described by Torten and Whitaker (1964), and in oilseeds by Pomeranz (1965). Sober *et al.* (1965) pointed out that adding sodium tetraborate diminished the reducing action of lactose in milk, and the biuret reaction could be applied without the separation of lactose.

Use of the biuret method in testing cereals is of particular interest. Originally, Pinckney (1949) described a procedure in which aliquots of a cleared alkaline protein extract were combined with measured amounts of the alkaline copper reagent.   The color was measured after a given reaction period.   In the modified method (Pinckney 1961) the stabilized reagent containing potassium hydroxide, sodium potassium tartarate, and copper sulfate is applied directly to the weighed sample for the simultaneous extraction and reaction. Jennings (1961) determined protein in white cereal grains by a modified biuret method that involved the simultaneous extraction of protein and color development in an alkaline copper tartarate solution. In dark-colored cereal grains, the extract prepared by the biuret extraction procedure was treated with the Folin-Ciocalteau phenol reagent.   The results agreed well with the Kjeldahl nitrogen determination.   In the second procedure, the extracted brown components in dark grains did not affect the precision of the determination as in the first method.

The Phenol-Reagent Method

One of the most widely used methods for the determination of protein in solutions is based on the interaction of proteins with the phenol reagent and copper under alkaline conditions. The color reaction involves a copper-catalyzed oxidation of aromatic amino acids and other groups by a heteropolyphosphate (phosphotungstic-phosphomolybdic) reagent. The basis of the method has been established by Wu (1922), and a number of modifications appeared before the important modifications of Folin and Ciocalteau (1927) and of Lowry et al. (1951). Chow and Goldstein (1960) have recently shown that many of the functional groups found in proteins (in addition to the well-known reactivity of tyrosine and tryptophan) are responsible for the final blue color concentration. The procedure (generally known as the Lowry method) is highly valued because of its sensitivity.

The method is 10 to 20 times as sensitive as the ultraviolet absorbance and up to 100 times the biuret method. The method is relatively specific, since few substances encountered in biological materials cause serious interference (Solecka et al. 1968). The results are affected little by the turbidity of the original protein solutions. Yet, the empirical nature of the method must be recognized. The color intensity varies with the amino acid composition of the protein and with the analytical conditions. The color is not strictly proportional to the protein concentration. The Lowry method is more time-consuming than direct absorbance measurement at 280 nm, is destructive, and requires multiple operations on each sample and incubation between the addition of reagents. The procedure is sensitive to high concentrations of sucrose, such as are used in sucrose gradient ultracentrifugation. Attempts to overcome the sucrose inhibition were only partly successful (Schuel and Schuel 1967).

Direct Spectrophotometric Methods

Most proteins exhibit a distinct ultraviolet absorption maximum at 280 nm, due primarily to the presence of tyrosine, tryptophan, and phenylalanine. Since the content of these amino acids in proteins from some sources differs within a reasonably narrow range, the absorption peak at 280 nm has been used by Warburg and Christian (1941) as a rapid and fairly sensitive test of protein concentration.

The measurement of ultraviolet absorption is a rapid, easily accomplished, and generally nondestructive method for the assay of protein contents of biological fluids and protein solutions from fractionation procedures. The determination can be made under conditions where other methods fail (in the presence of ammonium ions and certain salts). The results must be, however, interpreted with care. Since protein (and especially fractions) differ considerably in their amino acid composition, absorption at 280 nm may vary by a factor of 5 or more for equal concentrations of protein (Sober et al. 1965). Nucleic acid has a strong ultraviolet absorption band at 280 nm, but it absorbs much stronger at 260 than at 280 nm. With proteins, the reverse is true. This fact is used to eliminate, by calculation, the interference of nucleic acids in the determination of protein. Yet, considerable error may be present in such calculations, since different proteins and nucleic acids do not always have the same absorption.

Nakai et al. (1964) described a procedure for the protein determination in milk based on absorbance measurement at 280 nm. The method can be used in testing regular, defatted, pasteurized, and sterilized milk and cream; but not for whey or casein solutions. The preparation prior to an absorbance measurement is somewhat lengthy, and partly eliminates the advantage of the absorbance procedure.

The peptide bond has a strong absorption band at about 180 nm. Unfortunately, the technical difficulties of measurements at such a low wavelength are almost insurmountable. A compromise is often made by measuring at 210 or 220 nm; the measurement still being up to 20 times more sensitive than at 280 nm. At the low wavelengths, many of the commonly used buffers, particularly those containing carboxyl groups, absorb strongly and the selection of buffers is restricted.

Fluorescence measurements can detect and determine some proteins on a nondestructive basis and at very low levels. As the ultraviolet fluorescence is primarily due to tryptophan residues, with lesser contributions from tyrosine and alanine, the response from different proteins varies over a considerable range. Fox et al. (1963) evaluated the effects of pH, temperature, fluorescing impurities, and protein aggregation on determining protein in milk by fluorescence. By use of a urea solution containing citrate and phosphate, protein aggregation was eliminated and reproducibility was improved. The direct fluorimetric method was found precise and rapid. However, it requires a high dilution, good temperature control, and fails in the presence of certain impurities.

A photometric method to determine wheat protein content was described (Zeleny 1941; Zeleny et al. 1942). The sample is extracted with ethanol; the extract is treated with base and buffer and absorbance is measured at 530 nm. Correlation coefficients between the percentage of Kjeldahl protein and the absorbance were -0.969 for wheat and -0.987 for flour protein.

Nephelometric or Turbidimetric Methods

Turbidity produced when a protein is mixed with low concentrations of any of the common precipitants can be used as an index of protein concentration. The commonly used precipitants are trichloroacetic acid, potassium ferricyanide, and sulfosalicylic acid (Layne 1957).

For proteins in a solution, the methods are rapid and convenient. For solid foods, extraction of the protein into a solution both lengthens and complicates an analysis. The methods yield different values with different proteins, and do not permit differentiation between protein and other components precipitated under the employed experimental conditions. The methods are generally highly dependent on reproduction of conditions and calibration against other empirical methods.

A light-scattering technique for estimating soluble protein was described by Tappau (1966). The automatic determination is performed by treating sample solutions with acetic acid, potassium ferricyanide, and in some cases sodium tungstate and measuring the light scattered by the precipitate. A turbidimetric method based on precipitating with sulfosalicylic acid of the alkaline extract of wheat or wheat flour was described by Feinstein and Hart (1959).

Dye-Binding Methods

In recent years dye-binding methods are finding increasing use in determining the protein content of foods. Specific group reagents for proteins were reviewed by Olcott and Fraenkel-Conrat (1947) and by Rosenberg and Klotz (1960).

Fraenkel-Conrat and Cooper (1944) were the first to report that proteins bind quantitatively, under specified conditions with certain organic dyes. The dye binding can be used to determine total acidic and basic groups of proteins. Those observations lead to numerous investigations on the applicability of the dye-binding capacity method for protein determination in foods.

Earlier investigations (Udy 1954, 1956) estimated protein content from the amount of the disulfonic anionic dye, orange G, bound at pH 2.2. This dye binds specifically under acidic conditions to free amino groups, lysine, the imidazole group of histidine, and the guanidyl group of arginine. It has been subsequently shown that the estimation of protein by dye binding can be improved by using the dye acid orange 12, that is structurally identical to orange G, with the exception that acid orange 12 has only one sulfonic acid group. The analyzed food is treated with an excess dye. The dye and protein in the food react quantitatively to form an insoluble complex that can be separated either by centrifugation or filtration. From the concentration of unbound dye (measured colorimetrically), the binding capacity can be calculated. A quantitative relationship between the amount of bound dye and the protein content of the sample permits the construction, for each food, of a conversion table from which the percentage protein is read. A commercial instrumental setup is available. The dye is mixed with the sample in a special reactor or on a laboratory shaker and is transferred to a squeeze-type polyethylene bottle fitted with a fiberglass filter disc in the dropper cap. Light transmittance through the filtered dye solution is determined in a special colorimeter as the filtrate is transferred dropwise into a flow-through cuvette. The dye-binding procedure is rapid and eliminates the problems of skillful manipulation and corrosive reagents of the Kjeldahl procedure. The binding of acid orange 12 can be used to determine the protein contents of sound and normal samples of cereal grains, oil seeds, legumes, and animal and dairy products. Additional dyes that have been recommended, mainly for protein determination in meat and milk products, include cochineal red A, buffalo black, and amido black 10B.

Amido black 10B was shown (Tarassuk et al. 1967) to give a greater change in optical density (per unit of milk protein) than orange G. The binding of amido black was affected by the presence of potassium dichromate and formaldehyde but not by mercuric chloride. The dye binding capacity of milk protein was not affected by homogenizing, condensing, or heating to 90° for 15 min. Extensive proteolysis increases dye binding, whereas heating to browning reduces it. The dye-binding test is considered suitable for normal milk samples, but not for atypical milk such as colostrum, mastitic, and very late lactation milks.

Stoichiometric binding and precipitation of bromosulfalein by proteins was originally suggested as a basis for protein determination by Greif (1950). The method is widely used in the determination of protein, on a macro- and microscale, in biological fluids, and in histochemical and cytochemical investigations. Use of the dye in determining protein content of food has been limited.

Miscellaneous

Amino acids that may be readily determined offer a simple approach to protein analysis. Of those, tyrosine and tryptophan have received the most attention. Less effort has been directed to analyzing for protein content by determining the sulfur-containing amino acids cystine and cysteine.

The Folin reaction (as modified by Lowry) was described earlier in this chapter. After heating a protein solution with concentrated nitric acid a white precipitate that slowly turns yellow (the xantoproteic reaction) is formed. The yellow color results from the nitration of aromatic amino acids. On neutralization, the color turns orange. The xantoproteic acid reaction is simpler, but less sensitive and more difficult to standardize than the Lowry method.

The Millon reaction is specific for tyrosine. The reagent is prepared by dissolving mercury in excess nitric acid and diluting the solution. Reaction with proteins gives, upon boiling, a brick-red color.

Proteins exert relatively large effects on certain physical properties of their solutions. Determination of these effects is used as a basis of estimating the protein content. The basic assumption in all of these tests is that the effect of nonprotein solutes is negligible. To the extent that the assumption is valid, determination of physical properties can be used to determine rapidly, accurately, and reproducibly protein content.

A refractive index determination provides a good general physical method, as the refractive index is nearly the same for most proteins. Determination of bovine plasma protein from a refractive index was described by Weeth and Speth (1968). The advantages of refractometric assays of protein in milk were outlined by Andrievskaya (1966). Certain amino acids, polypeptides, and proteins dissolved in a cobalt-containing buffer of suitable pH, produce a catalytic reaction at the dropping mercury electrode. The reaction has been named after its discoverer, Brdicka. Application of the reaction in biochemical analysis was reviewed by Muller (1963). The use of infrared spectroscopy in the analysis of proteins in foods was discussed by Kohn (1965); features of an infrared milk analyzer were described by Biggs (1967). Other physical parameters that have been proposed for protein determination include specific gravity, viscosity, surface tension, absorption of radiation, conductivity, and polarization.

Formol-binding methods have gained little popularity due to the low precision (Kirk 1947). The methods measure the increase in acidic groups that can be titrated with base, when amino acids, proteins, and polypeptides are reacted with neutral formaldehyde. Recent studies (Drux and Bauer 1964; Hill and Stone 1964) indicate that formol titration can be used in determining the protein content of natural and processed milk (including ice cream).

Christian (1966) described a direct titration method of the native proteins in solution. The titrant is coulometrically generated hypobromite and the end point is detected amperometrically.

Many European cereal chemists run gluten determinations as an estimate of protein content. The gluten test is based on the fact that when wheat flour is made into a dough, the starch can be washed away and a cohesive mass of gluten is obtained. Pelshenke and Bolling (1962) studied the relation between gluten contents and Kjeldahl protein. The formula

$$\% \text{ wet gluten} = \frac{\% \text{ protein} - a}{b} \tag{37.1}$$

where $a = 7.34$ and $b = 0.2271$ was obtained for a large number of samples studied. Dry gluten was about $1/_3$ of wet gluten, and the factor b used for correlating dry gluten with protein was 0.63. The gluten test can be used as a measure of both protein quantity and quality (from the breadmaking standpoint). The test is used little in the United States for several reasons. The determination is subject to a large analytical error; attempts to standardize the test using salt solutions and a mechanical gluten washer have reduced the error only partly. Gluten contains, depending on assay conditions, variable amounts of protein (generally 80% on a dry basis), and the proteins contain only the water-insoluble components. Gluten can be washed easily from flour but not from wheat; consequently, it is of limited value in plant breeding or wheat purchasing. The gluten test is not suited for large-scale routine determinations.

Several indirect methods of protein determination in milk were reviewed by Cavagnol (1961). Milk protein heated with sodium hydroxide releases a consistent amount of ammonia; the method is rapid and linearly related to the Kjeldahl-nitrogen determination. By measuring the volume of phosphotungstic acid-precipitated serum proteins, it is possible to measure the serum protein in milk. The method correlates well with the Kjeldahl procedure provided the acid precipitable casein and denatured serum protein are removed first.

NONPROTEIN NITROGENOUS SUBSTANCES

The least empirical procedures for separating proteins from non-protein nitrogenous compounds are dialysis and ultrafiltration with suitable membranes. Heat coagulation is likely to free most juices or extracts from protein, though some proteins (casein, gelatin) are not heat-coagulable. Heat coagulation and some protein precipitants have the advantages of rapidly inactivating proteolytic enzymes and reducing artifacts. The use of protein precipitants as picric acid,

sulfosalicylic acid, trichloroacetic acid, etc. is empirical.   The use of
aqueous ethanol for obtaining protein-free extracts cannot be applied
to gliadin-containing materials.   In most separations, possible retention on the precipitated protein of small nonprotein molecules by
adsorption, ion exchange, etc. should be considered (Synge 1955).

In a critical study of the methods for the determination of nonprotein nitrogen, Bell (1963) used the following methods to separate
protein from nonprotein nitrogenous substances in skim milk, serum,
a water extract of flour, and a water extract of bran: dialysis, heat,
tungstic acid, copper hydroxide, ferric oxide hydrasol, lead acetate,
trichloroacetic acid, phosphotungstic acid, metaphosphoric acid,
tannic acid, sulfosalicylic acid, ethanol, mercuric chloride, a mixture
of chloroform and octyl alcohol (8:1), and a mixture of phenol,
acetic acid, and water (1:1:1).   Nonprotein fractions obtained by
the various protein separation methods, differed widely.   Dialysis
or related techniques appeared to achieve separations most closely
related to nonprotein nitrogen, as theoretically defined.   The binding of amino acids to proteins was small.

Varner *et al.* (1953) described a method to determine ammonium,
amide, nitrate, and nitrite nitrogen in protein-free plant extracts.
The extracts are buffered at pH 10 with borate, placed in a modified
semimicro Kjeldahl distillation unit, and the ammonium nitrogen is
removed by vacuum distillation at 40°C.   Concentrated alkali is
added to the distillation flask and the amide nitrogen is removed by
steam distillation at 100°C.   After adding ferrous sulfate as a reducing agent, the nitrite nitrogen is removed as ammonia.   The nitrate
nitrogen is reduced to ammonia by ferrous sulfate after addition of
silver sulfate as catalyst.   The procedure requires about 20 min.   To
overcome the interference of glucose, an ion-exchange resin mixture
may be used and the total procedure requires about 2 hr.   If the
nitrate concentration is very low, a colorimetric determination is
recommended.   Modifications to overcome some of the limitations of
the above procedure (amino nitrogen is not estimated and nitrate
reduction is incomplete) were suggested by Barker and Volk (1964).
Christianson *et al.* (1965A, B) described chromatographic and specific spectrographic methods to identify and determine nonprotein
nitrogenous substances in corn grain and corn steep liquor.

Several methods are available to determine the aggregate of free
amino acids in protein-free extracts.   Titrimetric methods respond
to the carboxyl groups of acids and other compounds and to phenolic
groups of glycosides.   The formol titration is fairly specific for
amino groups, but does not distinguish between amino acids and

other amines.   Gasometric methods in which the nitrogen liberated after adding nitrous acid is measured manometrically are somewhat more specific in this respect, but give negative results with proline and other secondary amines.   There is also interference by the amide group of glutamine, and glutathione gives anomalous results.

Methods based on the complex formation with copper, such as that of Pope and Stevens (1939), are useful for semiquantitative determination as the colored complexes are fairly specific for free amino acids.   The color reaction of amino acids with ninhydrin is used widely though it has two disadvantages: the color and its intensity vary for different amino acids, and the procedure is not specific for amino acids.   The high specificity of the ninhydrin-carbon dioxide method (Van Slyke et al. 1941) makes it attractive for the determination of total free amino acids.

Some free amino acids can be determined by specific colorimetric tests.   Separation of free amino acids by two-dimensional paper chromatography, elution from the paper and determination by spectrophotometry or direct absorbance measurement on the paper, are useful for semiquantitative estimations.   Enzymatic methods employing specific decarboxylases can be used for some amino acids. Microbiological assays give sometimes poorer results with free amino acids than with protein hydrolysates, as the extracts of the former may contain growth inhibitors or stimulants and peptides that affect the results (Synge 1955).   Best quantitative results can be obtained by fractionation on ion-exchange columns.

Several methods are available to detect and determine ammonia in protein-free extracts.   They are based on the basic properties of the volatile ammonia, on its oxidation to elementary nitrogen, on conversion of the ammonium ion with formaldehyde to hexamethylenetetramine, and on the Nessler reaction.   In most cases, it is essential to eliminate the interference from nitrogen-containing organic compounds that might give secondary reactions leading to the formation of ammonia.

Nitrite can be detected by the sulfanilic acid-α-naphthylamine reaction of Grau and Mirna (1957).   High concentrations of nitrite can be determined by oxidative titration with permanganate or cerium sulfate.   Bromometric and especially iodometric titrations are somewhat more complicated than other oxidative methods, but more specific (Niedermaier 1967).   Titration with hydrazine sulfate; the capacity of ammonium chloride, urea, or thioura to react with nitrites and give nitrogen that can be determined gasometrically; and the reaction between nitrite and alcohols (methanol or ethanol) in an acidic medium to give a volatile ester—are some of the other available quantitative recations.   For small amounts, colorimetric reactions are recommended.   Rivanol (2-ethoxy-6,9-diaminoacridinium chloride) gives with nitrite in the presence of dilute hydrochloric acid a stable color with a maximum absorbance at 515 nm; the relation is linear for 0 to 60 γ nitrite per 50 ml solution (Svach and Zyka 1955).   Benzidine in acetic acid gives in the presence of nitrite, depending on the concentration, a red to orange-red color. The reaction is specific, and the color intensity is proportional to the nitrite concentration (Barakat and Sadek 1964).   Sawicki *et al.* (1963) reviewed numerous spectrophotometric methods for the determination of nitrite, and described new and more sensitive procedures that could be potentially used in the analysis of foods.

In many instances, the detection of nitrates is affected by the presence of nitrites.   If no interference is anticipated, reaction with diphenylamine, or brucine, or the detection after conversion of nitrate to nitrite—are recommended.   For a quantitative assay, a gasometric determination after the reduction to nitrogen with an acidified ferrous chloride solution, precipitation with the organic nitron reagent, color reaction with brucine (Baker 1967), an indirect polarographic method (Davidkova and Davidek 1960), determination by the nitrate electrode method (Paul and Carlson 1968) or by infrared spectroscopy (Magee *et al.* 1964), and an enzymatic method (Lowe and Hamilton 1967)—can be used.

A differential colorimetric method for nitrite and nitrate was described by Lambert and Zitomer (1960), Davidek *et al.* (1962), and Kamm *et al.* (1965).   A method that has been used most extensively in meat products was originally described by Grau and Mirna (1957). A borate extract is cleared with a zinc sulfate solution.   Nitrite is determined directly, and nitrate after reduction to nitrite with cadmium in a slightly ammoniacal solution.   For the color formation, the nitrite is reacted with sulfanilic acid in acetic acid and naphthylamine in acetic acid, mixed shortly before the assay.   The method was simplified and improved by Follett and Ratcliff (1963).   In the modified procedure, the protein-free aqueous solution is acidified and passed through a cadmium column to reduce nitrate to nitrite, and the nitrite is determined colorimetrically before and after reduction. Landmann (1966) reported that the modified procedure gave satisfactory results.   As naphthylamine is carcinogenic, Adnaanse and Robbers (1969) suggested the use of 1-naphthylamine-7-sulfonic acid as a substitute.

BIBLIOGRAPHY

ADNAANSE, A., and ROBBERS, J. E. 1969. Determination of nitrite and nitrate in some horticultural and meat products and in samples of oil. J. Sci. Food Agr. 20, 321–325.

ALTSCHUL, A. M. 1962. Seed proteins and world food problems. Econ. Botany 16, 2–13.

ANDRIEVSKAYA, L. 1966. Refractometric analyzer for milk. Proc. Intern. Dairy Congr. Proc. 17th, Munich 2, 187–189. (Chem. Abstr. 66, 9618, 1967.)

ANON. 1968. Amino Acid Content of Food and Biological Data on Proteins. Food Agr. Organ. U.N., Rome, Italy.

BAKER, A. S. 1967. Colorimetric determination of nitrate in soil and plant extracts with brucin. Agr. Food Chem. 15, 802–806.

BARAKAT, M. Z., and SADEK, I. 1964. Determining nitrite in sausage. Food Technol. 18, 242–244.

BARKER, A. V., and VOLK, R. J. 1964. Determination of ammonium, amide, amino, and nitrate nitrogen in plant extracts by a modified Kjeldahl method. Anal. Chem. 36, 439–441.

BELL, P. M. 1963. A critical study of method for the determination of nonprotein nitrogen. Anal. Biochem. 5, 443–451.

BIGGS, D. A. 1967. Milk analysis with the infrared milk analyzer. J. Dairy Sci. 50, 799–803.

BLOCK, R. J. 1945. The amino acid composition of food proteins. Advan. Protein Chem. 2, 119–154.

BLOCK, R. J. 1960. Amino acid analysis of protein hydrolysates. In A Laboratory Manual of Analytical Methods of Protein Chemistry, Vol. I, The Separation and Isolation of Proteins, P. Alexander, and R. J. Block (Editors). Pergamon Press, London.

BRADSTREET, R. B. 1940. A review of the Kjeldahl determination of organic nitrogen. Chem. Rev. 27, 331–350.

BRADSTREET, R. B. 1965. The Kjeldahl Method for Organic Nitrogen. Academic Press, New York.

CAVAGNOL, A. 1961. Food. Anal. Chem. 33, 50R–60R.

CHOW, S., and GOLDSTEIN, A. 1960. Chromogenic groups in the Lowry protein determination. Biochem. J. 75, 109–115.

CHRISTIAN, G. D. 1966. Coulometric titration of proteins with electrogenerated hypobromite. Anal. Biochem. 14, 183–190.

CHRISTIANSON, D. D., CAVINS, J. F., and WALL, J. S. 1965A. Steep liquor constituents. Identification and determination of nonprotein nitrogenous substances in corn steep liquor. Agr. Food. Chem. 13, 277–280.

CHRISTIANSON, D. D., WALL, J. S., and CAVINS, J. F. 1965B. Nutrient distribution in grain. Location of nonprotein nitrogenous substances in corn grain. Agr. Food Chem. 13, 272–276.

CONWAY, E. J. 1957. Microdiffusion Analysis and Volumetric Error. Crosby, Lockwood, and Son, London, England.

DANEHY, J. P., and PIGMAN, W. W. 1951. Reactions between sugars and nitrogenous compounds and their relationship to certain food problems. Advan. Food Res. 3, 241–290.

DAVIDEK, J., KLEIN, S., and ZACKOVA, A. 1962. Colorimetric determination of nitrate and nitrite in biological materials. Z. Lebensmittel Unters. Forsch. 119, 342–346. (German)

DAVIDKOVA, E., and DAVIDEK, J. 1960. Indirect polarographic determination of nitrate in biological material. Z. Lebensmittel Unters. Forsch. 111, 477–483. (German)

DRUX, A., and BAUER, H. J. 1964. Contribution to determination of protein

content in milk by formol titration.   Nahrung *8*, 99–103.   (German)
DUMAS, J. B. A.   1831.   Letter of M. Dumas to M. Gay Lussac on a procedure
of organic analysis.   Annales de Chemie (2), *47*, 198–213.   (French)
DUNN, M. S.   1950.   Determination of amino acids.   Advan. Chem. Ser. *3*,
13–28.
FEENEY, R. E., and HILL, R. M.   1960.   Protein chemistry and food research.
Advan. Protein Chem. *10*, 23–73.
FEINSTEIN, L., and HART, J. R.   1959.   A simple method for determining the
protein content of wheat and flour samples.   Cereal Chem. *36*, 191–193.
FERRARI, A.   1960.   Nitrogen determination by a continuous digestion and
analysis system.   Ann. N. Y. Acad. Sci. *87*, 792–800.
FOLIN, O., and CIOCALTEAU, V.   1927.   Tyrosine and tryptophan determina-
tion of proteins.   J. Biol. Chem. *73*, 627–650.
FOLLETT, M. J., and RATCLIFF, P. W.   1963.   Determination of nitrite and
nitrate in meat products.   J. Sci. Food. Agr. *14*, 138–144.
FOX, K. K., HOLSINGER, V. H., and PALLANSCH, M. J.   1963.   Fluorimetry as
a method of determining protein content of milk.   J. Dairy Sci. *46*, 302–310.
FRAENKEL-CONRAT, H., and COOPER, M.   1944.   The use of dyes for the determination of acid and basic groups in proteins.   J. Biol. Chem. *154*, 239–246.
GEHRKE, C. W., and STALLING, D. L.   1967.   Quantitative analysis of the
twenty natural protein acids by gas-liquid chromatography.   Separ. Sci. *2*,
101–138.
GLEBKO, L. I., ULKINA, J. I., and VASKOVSKY, V. E.   1967.   Spectrophoto-
metrical method for determination of nitrogen in biological preparations based
on thymol-hypobromite reaction.   Anal. Biochem. *20*, 16–23.
GRAU, R., and MIRNA, A.   1957.   Determination of nitrite, nitrate, and sodium
chloride in meat products and gravies.   Z. Anal. Chem. *158*, 182–189.
GREIF, R. L.   1950.   Use of bromosulfalein for the measurement of proteolytic
activity.   Proc. Soc. Exptl. Biol. Med. *75*, 813–815.
HETTRICK, J. H., and WHITNEY, R. M.   1949.   Determination of nitrogen in
milk by direct nesslerization of the digested sample.   J. Dairy Sci. *32*, 111–
112.
HILL, R. L., and STONE, W. K.   1964.   Procedure for determination of protein
in ice milk and ice cream by formol titration.   J. Dairy Sci. *47*, 1014–1015.
HILPERT, H., and ENKELMANN, D.   1963.   Paper chromatographic characterization of whey proteins of various animals.   Milchwiss. *18*, 26–29.
JAMES, A. T., and MORRIS, L. J.   1964.   New Biochemical Separations.   D.
van Nostrand Publishing Co., London.
JENNINGS, A. C.   1961.   Determination of the nitrogen content of cereal grain
by colorimetric methods.   Cereal Chem. *38*, 467–478.
JONES, D. B.   1931.   Factors for converting percentages of nitrogen in foods
and feeds into percentages of proteins.   U.S. Dept. Agr. Circ. *183*, 1–21.
KAMM, L., McKEOWN, G. G., and SMITH, D. M.   1965.   Colorimetric method
for the determination of the nitrate and nitrite content of baby foods.   J.
Assoc. Offic. Agr. Chemists *48*, 892–897.
KELLER, S., and BLOCK, R. J.   1960.   Separation of proteins.   *In* A Laboratory
Manual of Analytical Methods of Protein Chemistry, Vol. I, The Separation
and Isolation of Proteins, P. Alexander, and R. J. Block (Editors).   Pergamon
Press, London.
KIRK, P. L.   1947.   The chemical determination of proteins.   Advan. Protein
Chem. *3*, 139–167.
KJELDAHL, C.   1883.   New method for determination of nitrogen in organic
materials.   Z. Analyt. Chem. *22*, 366–382.   (German)
KOHN, R.   1965.   Application of infrared spectroscopy in food analysis. I.
Infrared spectroscopy in the intermediate range of 3 to 15 μm.   Z. Lebens-
mittel Unters. Forsch. *129*, 28–40.   (German)

LAMBERT, J. L., and ZITOMER, F. 1960. Differential colorimetric determination of nitrite and nitrate ions. Anal. Chem. *32*, 1684–1686.

LANDMANN, W. A. 1966. Determination of nitrite and nitrate in meat products. J. Assoc. Offic. Agr. Chemists *49*, 875–877.

LAYNE, E. 1957. Spectrophotometric and turbidimetric methods for measuring proteins. *In* Methods of Enzymology, Vol. 3, S. P. Colowick, and N. O. Kaplan (Editors). Academic Press, New York.

LICHTFIELD, J. H., and SACHSEL, G. F. 1965. Technology and protein malnutrition. Cereal Sci. Today *10*, 458, 460–462, 464, 472.

LOWE, R. H., and HAMILTON, J. L. 1967. Rapid method for determination of nitrate in plant and soil extracts. Agr. Food Chem. *15*, 359–361.

LOWRY, O. H. ROSEBROUGH, N. J., FARR, A. L., and RANDALL, R. J. 1951. Protein measurement with the Folin phenol reagent. J. Biol. Chem. *193*, 265–275.

MAGEE, R. J., SHAHINE, S. A. E. F., and WILSON, C. L. 1964. The determination of nitrate by infrared spectroscopy. Microchim. Acta, 1019–1022.

MANN, L. T. 1963. Spectrophotometric determination of nitrogen in total micro-Kjeldahl digests. Application of phenol-hypochlorite reaction to microgram amounts of ammonia in total digest of biological material. Anal. Chem. *35*, 2179–2183.

MARTIN, A. J. P., and SYNGE, R. L. M. 1945. Analytical chemistry of the proteins. Advan. Protein Chem. *2*, 1–83.

MILLER, B. S., and JOHNSON, J. A. 1954. A review of methods for determining the quality of wheat and flour for breadmaking. Kansas Agr. Expt. Sta. Tech. Bull. *76*.

MINARI, O., and ZILVERSMIT, D. B. 1963. Use of KCN for stabilization of color in direct Nesslerization of Kjeldahl digests. Anal. Biochem. *6*, 320–327.

MOTZ, R. J. 1968. A critical evaluation of the AOAC method for the determination of milk protein in milk chocolate when applied to crumb chocolate. Analyst *93*, 116–117.

MULLER, O. H. 1963. Polarographic analysis of proteins, amino acids, and other compounds by means of the Brdicka reaction. Methods Biochem. Analy. *11*. 329–403.

NAKAI, S., WILSON, H. K., and HERREID, E. O. 1964. Spectrophotometric determination of protein in milk. J. Dairy Sci. *47*, 356–358.

NEIDERMAIER, T. 1967. Nitrogenous compounds, ammonia, nitric acid and nitrous acid. *In* Handbuch der Lebensmittelchemie, Vol. II, Part 2, Analytik der Lebensmittel, Nachweis und Bestimmung von Lebensmittel—Inhaltstoffen, J. Schormuller (Editor). Springer-Verlag, Berlin. (German)

NEILL, C. D. 1962. The Kjeldahl protein test. Cereal Sci. Today *7*, 6–8, 10, 12.

OLCOTT, H. S., and FRAENKEL-CONRAT, H. 1947. Specific group reagents for proteins. Chem. Rev. *41*, 151–197.

ORR, M. L., and WATT, B. K. 1957. Amino acid content of foods. Home Econ. Res. Rept. *4*, U.S. Dept. Agr., Washington, D.C.

OSBORNE, T. B. 1907. The proteins of the wheat kernel. Carnegie Institute of Washington, Washington, D.C.

PARNAS, J. K. 1938. The Kjeldahl nitrogen determination by a modification of Parnas and Wagner. Z. Anal. Chem. *114*, 261–275. (German)

PAUL, J. L., and CARLSON, R. A. 1968. Nitrate determination in plant extracts by the nitrate electrode. Agr. Food Chem. *16*, 766–768.

PELSHENKE, P. F., and BOLLING, H. 1962. Relation between protein and wet gluten. Getreide und Mehl *12*, 29–33. (German)

PFLEIDERER, G. 1967. Nitrogenous compounds, proteins, amino acids, and amines. *In* Handbuch der Lebensmittelchemie, Vol. II, Part 2, Analytik der Lebensmittel, Nachweis und Bestimmung von Lebensmittel-Inhaltstoffen, J. Schormuler (Editor). Springer-Verlag, Berlin. (German)

PINCKNEY, A. J. 1949. Wheat protein and the biuret reaction. Cereal Chem. 26, 423–439.

PINCKNEY, A. J. 1961. The biuret test as applied to the estimation of wheat protein. Cereal Chem. 38, 501–506.

POMERANZ, Y. 1965. Evaluation of factors affecting the determination of nitrogen in soya products by the biuret and orange G binding methods. J. Food Sci. 30, 307–311.

POMERANZ, Y. 1968. Relation between chemical composition and bread-making potentialities of wheat flour. Advan. Food Res. 16, 335–455.

POPE, C. G., and STEVENS, M. F. 1939. Determination of amino nitrogen using a copper method. Biochem. J. 33, 1070–1077.

RHODES, D. N. 1963. Protein biochemistry. In Recent Advances in Food Science, Vol. 3, J. M. Leitch, and D. N. Rhodes (Editors). Butterworths, London.

RICE, E. E., and BEUK, J. F. 1953. The effects of heat upon the nutritive value of protein. Advan. Food Res. 4, 233–279.

RIEGLER, E. 1914. A colorimetric method for determination of albumin. Z. Anal. Chem. 53, 242–245.

ROSENBERG, R. M., and KLOTZ, I. M. 1960. Dye-binding methods. In A Laboratory Manual of Analytical Methods of Protein Chemistry, Vol. I, The Separation and Isolation, P. Alexander, and R. J. Block (Editors). Pergamon Press, London.

SAWICKI, E., STANLEY, T. W., PFAFF, J., and JOHNSON, H. 1963. Sensitive new methods for autocatalytic spectrophotometric determination of nitrite through free radical chromogens. Anal. Chem. 35, 2183–2191.

SCHOBER, R., NICLAUS, W., and CHRIST, W. 1964. Possibility of using the biuret reaction for determining protein in milk. Milchwissenschaft 19, 75–78. (Anal. Abstr. 12, 1987, 1965.)

SCHUEL, H., and SCHUEL, R. 1967. Automated determination of protein in the presence of sucrose. Anal. Biochem. 20, 86–93.

SCHULTZ, H. W., and ANGLEMIER, A. F. 1964. Symposium on Foods: Proteins and Their Reactions. Avi Publishing Co., Westport, Conn.

SIRIWARDENE, J. A., EVANS, R. A., THOMAS, A. J., and AXFORD, R. F. E. 1966. Use of an automated Kjeldahl analyzer for determination of nitrogen in biological material. Proc. Technicon Symp. Automation in Analytical Chemistry.

SOBER, H. A., HARTLEY, R. W., CARROLL, W. R., and PETERSON, E. A. 1965. Fractionation of proteins. In The Proteins, Composition, Structure and Function, H. Neurath (Editor). Academic Press, New York.

SOLECKA, M., ROSS, J. A., and MILLIKAN, D. F. 1968. Evidence of substances interfering with the Lowry test for protein in plant leaf tissue. Phytochem. 7, 1293–1295.

SPACKMAN, D. H., STEIN, W. H., and MOORE, S. 1958. Automatic recording apparatus for use in the chromatography of amino acids. Anal. Chem. 30, 1190–1206.

STARK, J. B. 1962. Use of ion-exchange resins to classify plant nitrogenous compounds in beet molasses. Anal. Biochem. 4, 103–109.

STERNGLANZ, P. D., and KOLLIG, H. 1962. Evaluation of an automatic nitrogen analyzer for tractable and refractory compounds. Anal. Chem. 34, 544–547.

SVACH, M., and ZYKA, J. 1955. Photometric determination of nitrite with rivanol. Z. Anal. Chem. 148, 1–2.

SYNGE, R. L. M. 1955. Peptides (bound amino acids) and free amino acids. In Modern Methods of Plant Analysis, Vol. I, K. Paech, and M. V. Tracey (Editors). Springer-Verlag, Berlin.

TAPPAU, D. V. 1966. A light scattering technique for measuring protein concentration. Anal. Biochem. *14*, 171–182.

TARASSUK, N. P., ABE, N., and MOATS, W. A. 1967. The dye binding of milk proteins. U.S. Dept. Agr. Tech. Publ.

THOMPSON, R. 1962. Identification of fish species by starch gel zone electrophoresis of protein extracts. J. Assoc. Offic. Agr. Chem. *45*, 275–276.

TKACHUK, R. 1966. Note on the nitrogen to protein conversion factor for wheat flour. Cereal Chem. *43*, 223–225.

TORTEN, J., and WHITAKER, J. R. 1964. Evaluation of the biuret and dye-binding methods for protein determination in meats. J. Food Sci. *29*, 168–174.

UDY, D. C. 1954. Dye binding capacities of wheat flour protein fractions. Cereal Chem. *31*, 389.

UDY, D. C. 1956. A rapid method for estimating total protein in milk. Nature *178*, 314–315.

VAN SLYKE, D. D., and HILLER, A. 1933. Determination of ammonia in blood. J. Biol. Chem. *102*, 499–504.

VAN SLYKE, D. D., DILLON, R. T., MACFADYEN, D. A., and HAMILTON, P. 1941. Gasometric determination of carboxyl groups in free amino acids. J. Biol. Chem. *141*, 627–669.

VARLEY, J. A. 1966. Automatic methods for the determination of nitrogen, phosphorus and potassium in plant material. Analyst *91*, 119–126.

VARNER, J. E., BULEN, W. A., VANECKO, S., and BURRELL, R. C. 1953. Determination of ammonium, amide, nitrite, and nitrate nitrogen in plant extracts. Anal. Chem. *25*, 1528–1529.

WARBURG, O., and CHRISTIAN, W. 1941. Isolation and crystallization of the fermentation enzyme enolase. Biochem. Z. *310*, 384–421.

WEETH, H. J., and SPETH, C. F. 1968. Estimation of bovine plasma protein from refractive index. J. Animal Sci. *27*, 146–149.

WHITAKER, J. R. 1959. Chemical changes associated with aging of meat with emphasis on the proteins. Advan. Protein Chem. *9*, 1–60.

WILLIAMS, P. C. 1964. Determination of total nitrogen in feeding stuffs. Analyst *89*, 276–281.

WOOD, D. E. 1965. Fast neutron activation analysis for nitrogen in grain products. AEC Accession No. 7007, Rept. *KN-65-186*.

WU, H. 1922. A new colorimetric method for the determination of plasma proteins. J. Biol. Chem. *51*, 33–39.

ZELENY, L. 1941. A simple photometric method for determining the protein content of wheat flour. Cereal Chem. *18*, 86–92.

ZELENY, L., NEUSTADT, M. H., and DIXON, H. B. 1942. Further developments on the photometric determination of wheat protein. Cereal Chem. *19*, 1–11.

CHAPTER 38

Objective versus Sensory Evaluation of Foods

The overall quality of food can be divided into three main categories: quantitative, hidden, and sensory (Kramer 1966). Some quantitative aspects of food quality are primarily of interest to the processor, i.e. yield of product obtained from a raw material; others are of interest both to the consumer and manufacturer, i.e. the ratio of more expensive to less expensive foods or components in a processed food. In some cases the ratio can be evaluated roughly by sensory methods. Hidden quality attributes include nutritional value of a food or the presence of toxic compounds that, generally, cannot be determined by sensory evaluation. They include, for instance, the vitamin C content of juices, or the presence of trace amounts of pesticides from spraying fruits and vegetables. Sensory attributes of quality guide the consumer in his selection of foods. Such attributes are measured by the processor to determine consumer preference in order to manufacture an acceptable product at maximum production economy. Sensory attributes are measured also in determining the conformity of a food with established government or trade standards and food grades.

The selection, acceptance, and digestibility of a food are largely determined by its sensory properties. Evaluation of sensory properties, is, however, affected by personal preference that is influenced by factors ranging from the caprices of fashion to the prevalence of dentures; social, cultural, and religious patterns; psychological factors; variations in climate and in the general physical status of the individual; availability; and nutritional education (Amerine et al. 1965).

To minimize the effects of such factors, different procedures for sensory evaluation have been devised and the results are evaluated by statistical methods. Large consumer groups are generally used to determine consumer reaction. Highly trained experts are employed for evaluating small differences in high-quality foods. Laboratory tests may be conducted to study the human perception of food attributes; to correlate sensory attributes with chemical and physical measurements; to evaluate raw material selection; to study processing effects and the means of maintaining uniform quality; to establish shelf-life stability, and to reduce costs. The aim of such laboratory tests may be to establish differences between samples, determine directional differences, or to determine quality-preference differences. Reasons for selecting a specific test; the experimental design; the mechanics of selection and the training of judges; and details of conducting, recording, and interpreting the results are outside the scope of this book. They can be found in several books including those published recently by Kramer and Twigg (1966), Amerine *et al.* (1965), and in the three ASTM publications (Anon. 1968A, B and C).

According to Kramer (1966) sensory attributes include: (1) appearance (color, size, shape, absence of defects, and the consistency of liquid and semisolid products;) (2) kinesthetics (texture, consistency, and viscosity), and (3) flavor (taste and odor).

Color is an important appearance factor. In agriculture, color development or color changes are used in assessing the maturity of fruits and vegetables. The color of a food often affects our perception of and evaluation by other senses. Discoloration or the fading of color is often accompanied or identified by consumers as being associated with undesirable changes in texture and flavor. In addition to color, the appearance characteristics of gloss, sheen, transparency, and turbidity are often important. Many objective methods are available to measure most of the appearance characteristics. For a color determination, spectrophotometric or color standards are used. In some instances, however, objective measurements cannot evaluate correctly the composite visual appearance, and sensory tests are still used widely.

Objective methods for the determination of kinesthetic attributes were discussed in detail previously.

The available evidence points to the existence of four basic taste modalities: sweet, sour, salt, and bitter, perceived by receptors on the tongue. The sweet taste is produced by a variety of non-ionized aliphatic hydroxy compounds, the most important of which are sugars. Saccharin, the best known synthetic sweetening agent, is 200 to 700 times as sweet as sucrose. Not all acids are sour: i.e. some amino acids are actually sweet. The threshold perception for weak organic acids is at a pH of about 3.7 to 3.9; for strong organic acids at about 3.4 to 3.5. Sugar may enhance or depress the perception of sourness. A pure salty taste is typified in foods by sodium chloride. Generally, other salts give a mixed taste. The typical bitter stimuli are given by alkaloids such as caffeine or quinine; also some amides are bitter. Tannins (i.e. in tea and alcoholic beverages) contribute to both bitterness and astringency.

Some compounds are unique. Thus, creatine—a constituent of muscle—is tasteless to some individuals and bitter to others. Similarly, sodium benzoate—a food preservative—is variously sweet, sour, salty, bitter, or tasteless. Monosodium glutamate's flavor-enhancing properties are utilized widely by the food industry. Pure glutamates are odorless, but have a pleasant, mild flavor with a sweet-salty taste, and impart some tactile sensation. A specific, synergistic flavor-enhancing action exists between monosodium glutamate and some of the 5'-ribonucleotides.

Taste responses to many organic compounds are highly specific; thus anomers of some sweet-tasting sugars are bitter. Taste sensations and thresholds are affected by many factors, including food temperature, overall food composition, concentrations of individual components, and age and individual variations among tasters.

The use of polarimetry in establishing the purity of sugars or the identity of optically active ingredients, that effect the taste of foods, is well-established. Determination of the refractive index is used widely to determine viscosity, sweetness, or the total solids of processed fruit and vegetable products. The refractive index has been used as an aid to detect the watering of milk or gross adulteration of butter with other fats. To determine the addition of small amounts of salt to juices or purées, actual assays of sodium chloride may be required.

Objective measurements are useful in detecting the presence or development of bitter components, the chemical identity and assay of which have been established.

In many instances, several objective determinations are required; in others, computing the ratios between several parameters, i.e. the sugar-titratable acidity ratio, are required to establish a satisfactory correlation with sensory evaluation. The ratio depends, however, on the type, variety, and maturity of the raw fruit, and on the composition and concentration of the syrup.

The sensation of odor—olfaction—has its receptors, in man, restricted to a small portion of the olfactory mucosa. As many as several thousand different odors can be distinguished. Classification of the odors has been the subject of many controversial investigations. Thus far, no satisfactory and widely-accepted system has been described. Odor in foods is most commonly found in organic compounds containing sulfur, nitrogen, and certain halogens. A functional group that imparts odor to an odorless compound is called an *osmophore*. Strong osmophores include inorganic (lead, arsenic, sulfur, chlorine, bromine) or organic (esters, carbonyls, amines, imines, lactones) entities. Double bonds, hydroxyl groups, and ring structures affect odor. Compounds with high molecular weight (above 300) and of low volatility are generally odorless. In addition, small differences in structure can significantly affect odor.

Gas-liquid chromatography (GLC) has been most useful in detecting small amounts of volatile components. Its usefulness has been increased by combining it with mass spectrometry or infrared spectroscopy for the identification of column effluents. Despite these contributions, GLC has certain important limitations. Some components that contribute to odor have a limited volatility (i.e. the boiling point of vanillin is 285°C); others are highly volatile at room temperature (i.e. hydrogen sulfide). Whereas the response of the chromatograph is linear, there is good evidence that the relationship between stimulus magnitude and response magnitude of our nose is a power function (Wick 1965).

It has been, thus far, difficult to recombine the eluates from the column or to prepare a synthetic mixture of individual components that duplicates the flavor of a natural product. Hopefully, with better separation technics (more sensitive and less degradative) and with a better understanding of the interaction between the individual components, preparation of such synthetic mixtures will be possible.

But even if GLC separated and identified clearly the components of flavor, we still have to use sensory methods to determine which of the components are indispensable to the production of desirable flavors. Such determinations must be made in the parent food, as its major (nonvolatile) components and texture affect significantly our evaluation of flavor.

Admittedly, evaluation of flavor by objective tests is a most challenging and difficult problem facing the food analyst (Anon. 1968C). According to Farber and Lerke (1958), several requirements should be met in the objective evaluation of freshness of fishery products. These include a high correlation with sensory evaluation, usefulness over a wide range of quality scores—from incipient spoilage to advanced deterioration—applicability to all kinds of fish products—including factory-processed, significant instrumental or analytical response to small quality differences, and relative speed and simplicity for use in routine testing.

How well can such requirements be met? According to Kramer and Twigg (1966) in a quality control system, where objective methods are relied upon, the use of a taste-testing panel is an admission of failure, except in a case where an objective method is being tested for conformance with human evaluation.

As mentioned earlier, acceptance of a food by consumers is generally affected by its various attributes. Consequently, several objective tests may be required for overall evaluation. A list of such tests is given in Table 38.1 (Kramer 1965). The main difficulty is, that once all the individual variables are determined objectively,

TABLE 38.1

CLASSIFICATION OF QUALITY FACTORS AND THEIR MEASUREMENT

	Fruits	Vegetables	Objective Methods of Measurement
Appearance Factors			
Size	Diameter, drained weight	Sieve size, drained weight	Scales, screens, micrometers
Shape	Height/width ratio	Straightness	Dimension ratios, displacement, angles
Wholeness	Cracked pieces	Cracked pieces	Counts, percent whole, photographs, models
Pattern defects	Blemishes, bruises, spots, extraneous matter	Blemishes, spots, bruises, extraneous matter	Photographs, drawings, models
Finish	Finish, gloss		Goniophotometers, gloss meters
Color	Color	Color	Color cards, dictionaries, reflectance and transmittance meters
Consistency	Consistency	Consistency	Consistometers, viscometers, flow meters, spread meters
Kinesthetic Factors			
	Texture, firmness, grit, character	Texture, mealiness, succulence, fiber, maturity	Tenderometers, texture meters; compressing, penetrating, cutting instruments; tests for solids, moisture, grit, fiber
Odor and Flavor Factors			
	Flavor, aroma, ripeness	Flavor, sweetness	Hydrometers, refractometers, pH meters, determination of sugar, sodium chloride, acid, sugar/acid ratio, enzymes, volatile substances, amines, chromatography

they must be integrated into an overall regression equation that correlates all the individual characteristics of quality to overall consumer preferences. Introducing the individual objective determination into the overall regression equation is wrought with many difficulties. They include a nonlinear relation between sensory and chemical or objective determinations, shifting standards in sensory appraisal, establishment of meaningful limits, difference in detection and significance of threshold and indifference limits in objective and subjective tests, and varying significance of specific objective factors in various foods (Pilgrim 1957). In practice, there is undoubtedly an interaction among the various components. The magnitude and direction of that interaction has been studied little. From the standpoint of chemical analysis, the difficulty lies in assigning a definite value to each of the components and their interaction products. That value is variable.

Although on one hand the limitation of objective tests must be recognized, the large cost, the large amount of work involved, and the uncertainty in evaluating foods consistently and meaningfully by sensory tests are well-established. In addition, objective assays are applicable at higher concentrations than subjective assessments without the danger of fatigue effects. Consequently, in practice, whenever a physical or chemical test can be run to obtain an objective measure of food quality, the objective test is preferred. As the final criterion of quality is human evaluation, the value of objective measurements must be evaluated by their correlation with sensory measurements. This correlation will generally be higher, the better our understanding of the nature of the physical and chemical parameters involved, and the better the available tools and methods for their reproducible and precise measurement. According to Kramer and Twigg (1966), a correlation of 0.90, or better, is desirable though in some cases useful information can be obtained if correlations are as low as 0.80. For predicting responses in production, very high correlations are necessary.

BIBLIOGRAPHY

AMERINE, M. A., PANGBORN, R. M., and ROESSLER, E. B. 1965. Principles of Sensory Evaluation of Food. Academic Press, New York.

ANON. 1968A. Basic Principles of Sensory Evaluation, Spec. Tech. Publ. 433. American Society for Testing and Materials, Philadelphia.

ANON. 1968B. Manual on Sensory Testing Methods, Spec. Tech. Publ. 434. American Society for Testing and Materials, Philadelphia.

ANON. 1968C. Correlation of Subjective-Objective Methods in the Study of Odors and Tastes, Spec. Tech. Publ. 440. American Society for Testing and Materials, Philadelphia.

FARBER, L., and LERKE, P. A.   1958.   A review of the value of volatile reducing
  substances for the chemical assessment of the freshness of fish and fish products.
  Food Technol. *12*, 677–680.
KRAMER, A.   1965.   Food quality.   Am. Assoc. Advan. Sci. Publ. *77*, G. W.
  Irving, and S. R. Hoover (Editors).   Am. Assoc. Advan. Sci., Washington,
  D.C.
KRAMER, A.   1966.   Parameters of quality.   Food Technol. *20*, 1147–1148.
KRAMER, A. and TWIGG, B. A.   1966.   Fundamentals of Quality Control for
  the Food Industry, 2nd Edition.   Avi Publishing Co., Westport, Conn.
KRAMER, A., and TWIGG, B. A.   1970.   Quality Control for the Food Industry,
  3rd Edition, Vol. 1.   Avi Publishing Co., Westport, Conn.
PILGRIM, F. J.   1957.   The components of food acceptance and their measure-
  ment.   Am. J. Clin. Nutr. *5*, 171–175.
WICK, E. L.   1965.   Chemical and sensory aspects of the identification of odor
  constituents in foods.   Food Technol. *19*, 827–833.

Index

Other AVI Books

AGRICULTURAL AND FOOD CHEMISTRY: PAST, PRESENT,
   FUTURE   *Teranishi*
BASIC FOOD CHEMISTRY
   Lee
CITRUS SCIENCE AND TECHNOLOGY
   Vols. 1 and 2
   Nagy, Shaw and Veldhuis
ENCYCLOPEDIA OF FOOD SCIENCE
   Peterson and Johnson
EVALUATION OF PROTEINS FOR HUMANS
   Bodwell
FOOD ANALYSIS LABORATORY EXPERIMENTS
   Meloan and Pomeranz
FOOD CHEMISTRY
   Aurand and Woods
FOOD CHEMISTRY
   Meyer
FOOD COLLOIDS
   Graham
FOOD COLORIMETRY: THEORY AND APPLICATIONS
   Francis and Clydesdale
FOOD PROTEINS
   Whitaker and Tannenbaum
IMMUNOLOGICAL ASPECTS OF FOODS
   Catsimpoolas
INTRODUCTORY FOOD CHEMISTRY
   Garard
LABORATORY MANUAL IN FOOD CHEMISTRY
   Woods and Aurand
PRINCIPLES OF FOOD CHEMISTRY
   deMan
RHEOLOGY AND TEXTURE IN FOOD QUALITY
   deMan, Voisey, Rasper and Stanley
SUGAR CHEMISTRY
   Shallenberger and Birch
THE TECHNOLOGY OF FOOD PRESERVATION
   4th Edition   *Desrosier and Desrosier*

710